MATLAB®-Based Electromagnetics

Branislav M. Notaroš

Department of Electrical and Computer Engineering
Colorado State University

PEARSON

Boston Columbus Indianapolis New York San Francisco Upper Saddle River
Amsterdam Cape Town Dubai London Madrid Milan Munich Paris Montreal Toronto
Delhi Mexico City Sao Paulo Sydney Hong Kong Seoul Singapore Taipei Tokyo

Vice President and Editorial Director, ECS: *Marcia J. Horton*
Executive Editor: *Andrew Gilfillan*
Editorial Assistant: *William Opaluch*
Executive Marketing Manager: *Tim Galligan*
Marketing Assistant: *Jon Bryant*
Permissions Project Manager: *Karen Sanatar*
Senior Managing Editor: *Scott Disanno*
Production Project Manager / Editorial Production Manager: *Greg Dulles*
Cover Designer: *Suzanne Behnke*

ISBN-10: 0-13-285794-4
ISBN-13: 978-0-13-285794-9

Printed in the United States of America
3 17

Library of Congress Cataloging-in-Publication Data on File

Pearson Education Ltd., London
Pearson Education Singapore, Pte. Ltd
Pearson Education Canada, Inc.
Pearson Education–Japan
Pearson Education Australia PTY, Limited
Pearson Education North Asia, Ltd., Hong Kong
Pearson Educación de Mexico, S.A. de C.V.
Pearson Education Malaysia, Pte. Ltd.
Pearson Education Upper Saddle River, New Jersey

To Olivera, Jelena, and Milica

CONTENTS

PREFACE

Electromagnetic theory is a fundamental underpinning of technical education, but, at the same time, one of the most difficult subjects for students to master. In order to help address this difficulty and contribute to overcoming it, here is a textbook on electromagnetic fields and waves completely based on MATLAB®, and so entitled, simply, *MATLAB®-Based Electromagnetics.* This text provides engineering and physics students and other users with an operational knowledge and firm grasp of electromagnetic fundamentals aimed toward practical engineering applications, by teaching them "hands on" electromagnetics through a unique and comprehensive collection of MATLAB computer exercises and projects. Essentially, the book unifies two themes: it presents and explains electromagnetics using MATLAB on one side, and develops and discusses MATLAB for electromagnetics on the other.

The book is designed primarily (but by no means exclusively) for junior-level undergraduate students in electrical and computer engineering, physics, and similar departments, for either two-semester course sequences or one-semester courses, and/or equivalent quarter arrangements. However, it can also be used earlier and later in the curriculum. It covers all important theoretical concepts, methodological procedures, and solution tools in electromagnetic fields and waves for undergraduates – organized in 12 chapters on electrostatic fields; steady electric currents; magnetostatic fields; time-varying electromagnetic fields; uniform plane electromagnetic waves; transmission lines; waveguides and cavity resonators; and antennas and wireless communication systems. It largely follows the organization of *Electromagnetics* by Branislav M. Notaroš, published in 2010 by *Pearson Education.*

On the other hand, the book allows for flexibility in coverage of the material, including the transmission-lines-early and transmission-lines-first approaches. Namely, Chapter 10 (Circuit Analysis of Transmission Lines) discusses only pure circuit-theory concepts, so that it can be taken at any time, along with Sections 6.7 and 6.8, which introduce phasors and complex representatives of time-harmonic voltages and currents.

MATLAB®-Based Electromagnetics is a self-contained textbook that can be used either as a supplement to any available electromagnetics text (e.g., [1]–[17] in the Bibliography) or as an independent resource. In other words, it is designed either to complement another (currently used or to be adopted) textbook and a variety of teaching styles, as a comprehensive companion adding a very significant MATLAB component to the course, or to serve as a principal resource for a MATLAB-based electromagnetic fields and waves course. In either way, MATLAB exercises are designed to strongly enforce and enhance both theoretical concepts and problem-solving techniques in electromagnetics.

In addition, respective parts of almost all chapters of the book can be effectively incorporated in higher-level courses on antennas, microwave theory and engineering, wave propagation and guidance, advanced electromagnetic theory, computational electromagnetics, electrical machines, signal integrity, etc. (for instance, see [18]–[34]). Furthermore, the book may be used by students outside of any particular course arrangements and by practicing engineers and scientists as well – to review and solidify the knowledge of fundamentals of electromagnetic fields and waves or certain aspects of electromagnetic theory and applications, now in a MATLAB context. Some may use it, on the other hand, to build and enhance the understanding and command of MATLAB syntax, functionality, and programming, in the framework of electromagnetics. Finally, because of its project-based philosophy and format, the book may be useful for distance learning, online courses, and other forms of nontraditional course delivery.

Overall distinguishing features of MATLAB®-Based Electromagnetics:

- The book provides two interwoven themes: presentation and study of *electromagnetics using MATLAB* and development and discussion of *MATLAB for electromagnetics*
- Can be used to either *complement* available electromagnetics text, or as an *independent* resource
- Contains *389 MATLAB computer exercises and projects*, covering and reinforcing practically all important theoretical concepts, methodologies, and problem-solving techniques in electromagnetic fields and waves (see Appendix 3 for a full *List of MATLAB Exercises*)
- Designed primarily *for undergraduate electromagnetics*, but can also be used *in follow-up courses* on antennas, propagation, microwaves, advanced electromagnetic theory, computational electromagnetics, electrical machines, signal integrity, etc.
- Provides a *theoretical overview* at the start of *each section* within each chapter of the book
- Maintains a favorable balance of MATLAB exercises between *static (one third) and dynamic (two thirds) topics*
- Offers *MATLAB exercises at all levels of difficulty*, from a few lines of MATLAB code, to those requiring a great deal of initiative and exploration
- Contains *125 TUTORIALS with detailed solutions* merged with *listings of MATLAB codes (m files)*; a demo tutorial for every class of MATLAB problems and projects is provided
- Gives *98 HINTS with guidance on the solution*, equations, and programming, often with portions of the code and/or resulting graphs and movie snapshots for validation
- Features *48 3-D and 2-D movies developed and played in MATLAB*, which are extremely valuable for interactive visualizations of fields and waves
- Displays *133 figures generated in MATLAB* with plots of geometries of structures, vector fields, guided and unbounded waves, wave polarization curves, Smith charts, transient signals, antenna patterns, etc.
- Presents *16 graphical user interfaces (GUIs) built in MATLAB* to calculate and display parameters and characteristics of various electromagnetic structures, materials, and systems, selected from a pop-up menu

- Offers *130 MATLAB exercises recommended to be done also "by hand"* – i.e., not using MATLAB, thus serving as traditional written problems

MATLAB® (by *MathWorks, Inc.*) is chosen not only for its very high quality and versatility, but because it represents a generally accepted standard in science and engineering education worldwide (e.g., [35]–[44]). Assignments of computer exercises along with traditional "by hand" problems help students develop a stronger intuition and a deeper understanding of electromagnetics. Moreover, this approach actively challenges and involves the student, providing additional benefit as compared to a passive computer demonstration. This book provides abundant opportunities for instructors to assign in-class and homework projects, and for students to engage in independent learning. MATLAB exercises are also ideal for interactive in-class explorations and discussions (*active teaching and learning*), and for team work and peer instruction (*collaborative teaching/learning*).

On the other side, studying and practicing these diverse exercises allows students to gain comprehensive and operational knowledge and skills in concepts and techniques of MATLAB use and programming. These skills can then be used effectively and implemented in other areas of study, including other courses in the curriculum.

To make this text self-contained and easy to use, each section within each chapter begins with a brief theoretical overview, which is followed by MATLAB exercises strongly coupled to the theory. Some theoretical details, along with abundant applications, are introduced through exercises themselves.

Each book section contains a large number of tutorial exercises with detailed solutions merged with listings of MATLAB codes (m files). TUTORIALS explain every step, with ample discussions of approaches, programming strategies, MATLAB formalities, and alternatives. They are written so that even a reader with no prior experience with MATLAB can fully understand them. Most importantly, all new concepts, approaches, and techniques in MATLAB programming as applied to electromagnetic fields and waves are covered with TUTORIALS. With a total of 125 TUTORIALS, for each class and type of MATLAB problems and projects there is always a demo exercise or set of exercises with complete detailed tutorials and code listings to provide students with guidance for completing similar exercises on their own. This is especially notable for exercises of intermediate and advanced levels of difficulty.

In addition to exercises with TUTORIALS, there are 98 exercises with HINTS, which provide guidance on the solution, equations, and programming, sometimes with most critical portions of MATLAB codes for the problem. In many cases, they present the resulting graphs and movie snapshots, so that readers can see what exactly they are expected to do and can verify and validate their codes. The exercises include 48 movies developed and played in MATLAB, 133 figures generated in MATLAB, and 16 GUIs built in MATLAB.

However, even the exercises with TUTORIALS can be assigned for homework or classwork, as their completion requires putting together a MATLAB code from the provided portions of the code listing (note that in some tutorials, some standard parts of the code, like data input, are omitted); actual running of the code; and generation and presentation of results. It is recommended that these exercises be a part of every homework assignment.

MATLAB exercises require a great deal of analytical preparation, manipulation with equations, and conceptual, strategic, and calculative analysis of problem situations. Thus, they are a great complement or often an adequate substitute for traditional written problems. Moreover, 130 exercises are designated as two-purpose problems: as computer-based MATLAB problems and as conventional "by hand" computational problems (these exercises are marked by H in the book). Of course, the reader can compare and cross-validate the results from MATLAB and those obtained "by hand." In addition, these problems (and their variations) can be given as traditional problems for homework and on exams. Note, however, that many other (unmarked) exercises can entirely, or in large part, be solved without MATLAB. In general, solving the problems and studying the topics both analytically and using MATLAB is most beneficial.

MATLAB codes described (and listed) in TUTORIALS or proposed in other exercises provide prolonged benefits of learning. By running codes; generating results, figures, and diagrams; playing movies and animations; and solving a large variety of problems in MATLAB, in class, with peers in study groups, or individually, students gain a deep understanding of electromagnetics. Most importantly, this book capitalizes on a win-win combination of pedagogical benefits of MATLAB-based electromagnetics education and students' computer-related skills and interests. As opposed to other available electromagnetics texts that feature some use of MATLAB, this text is by far the most complete and ambitious attempt to use MATLAB across all elements of electromagnetics education.

Some specific technical and pedagogical features of the content of MATLAB®-Based Electromagnetics and its MATLAB exercises, projects, codes, and solutions:

- *Field computation and visualization in MATLAB:*
    - ◇ MATLAB codes for computing and plotting electric and magnetic forces and fields (vectors) due to arbitrary 3-D arrays of stationary and moving charges; movie of electron travel in a magnetic field
    - ◇ Calculation and visualization of boundary conditions for oblique and horizontal boundary planes between arbitrary media, without and with surface charges/currents on the plane
    - ◇ Calculations and movies of electromagnetic induction due to moving and rotating loops in various magnetic fields
    - ◇ Graphical representation of complex numbers and movies of voltage and current phasor rotation in the complex plane
    - ◇ Symbolic computation of E and H fields and transmitted power for arbitrary TE and TM modes in a rectangular metallic waveguide, and of fields and stored energy in a rectangular cavity resonator
- *Computation and visualization of uniform plane waves in MATLAB:*
    - ◇ 2-D and 3-D movies visualizing attenuated and unattenuated traveling and standing uniform plane electromagnetic waves in different media

  - ◇ 2-D and 3-D movies and plots of circularly polarized (CP) and elliptically polarized (EP) waves; analysis and movie visualization of changes of wave polarization and handedness due to travel through anisotropic crystals
  - ◇ 3-D and 2-D movies of incident, reflected, and transmitted (refracted) plane waves for both normal and oblique incidences on both perfectly-conducting and dielectric boundaries, transient processes and steady states
  - ◇ Computation and visualization in MATLAB of angular dispersion of a beam of white light into its constituent colors in the visible spectrum using a glass prism

- *Symbolic and numerical programming in MATLAB:*
  - ◇ Symbolic differentiation and integration in all coordinates, symbolic Maxwell's equations, volumetric power/energy computations, conversion from complex to time domain, radiation integrals, etc.
  - ◇ Numerical differentiation and integration, various types of finite differences and integration rules, vector integrals, Maxwell's equations, optimizations, numerical solutions to nonlinear equations, etc.

- *Field and circuit analysis of transmission lines in MATLAB:*
  - ◇ GUI for primary and secondary circuit parameters of multiple transmission lines
  - ◇ MATLAB analysis and design (synthesis) of microstrip and strip lines with fringing
  - ◇ Numerical solutions and complete designs in MATLAB of impedance-matching transmission-line circuits with shunt and series short- and open-circuited stubs, including finding the stub location

- *Transmission-line analysis and design using the Smith chart in MATLAB:*
  - ◇ Construction of the Smith chart in MATLAB, adding dots of data on the chart, movies of Smith chart calculations on transmission lines, movies finding load impedances using the Smith chart
  - ◇ Searching for a desired impedance along a line in a numerical fashion and complete design in a Smith chart movie of impedance-matching transmission-line circuits with series stubs

- *MATLAB calculation of transients on transmission lines with arbitrary terminations:*
  - ◇ General MATLAB code for calculation of transients on transmission lines; plotting transient snapshots and waveforms; transient responses for arbitrary step/pulse excitations and matching conditions
  - ◇ Numerical simulation in MATLAB of a bounce diagram: bounce-diagram matrix; extracting signal waveforms/snapshots from the diagram; complete MATLAB transient analysis using bounce diagrams
  - ◇ Complete transient analysis in MATLAB of transmission lines with reactive loads and pulse excitation, with the use of an ordinary differential equation (ODE) solver; generator voltage computation

- *MATLAB analysis and visualization of antennas, wireless systems, and antenna arrays:*
  - ◇ Computation of the near field, radiation integral, far field, radiation and ohmic resistances, radiation efficiency, directivity, and gain of wire antennas
  - ◇ Functions in MATLAB for generating 3-D polar pattern plots of arbitrary radiation functions and for cutting a 3-D pattern in three characteristic planes to obtain and plot 2-D polar radiation patterns
  - ◇ Playing a movie to visualize the dependence of the radiation pattern on the electrical length of wire antennas
  - ◇ 3-D visualization of a wireless system with arbitrarily positioned and oriented wire dipole antennas; complete MATLAB analysis of systems with nonaligned antennas, including CP and EP transmitting antennas
  - ◇ Computation of the array factor of arbitrary linear arrays of point sources, generation of 3-D radiation pattern plots and 2-D pattern cuts in characteristic planes; complete MATLAB analysis of linear arrays
  - ◇ Implementation and visualization of the pattern multiplication theorem for antenna arrays in xy-, xz-, and yz-planes; complete MATLAB analysis of uniform and nonuniform arrays of arbitrary antennas (not just point sources)
- *MATLAB solutions to nonlinear problems:*
  - ◇ Graphical and numerical solutions for a simple nonlinear electric circuit
  - ◇ Complete numerical solutions in MATLAB for simple and more complex nonlinear magnetic circuits, movies of magnetization–demagnetization processes
- *Computational electromagnetic techniques in MATLAB:*
  - ◇ MATLAB codes based on the method of moments (MoM) for 3-D numerical analysis of charged metallic bodies (plates, boxes, and a parallel-plate capacitor); preprocessing and postprocessing
  - ◇ MATLAB codes for 2-D finite-difference (FD) numerical solution of Laplace's equation, based on both iterative and direct solutions of FD equations; potential, field, and charge computations

Pearson eText of the book, with full-color figures, is available on the Companion Website (CW):

www.pearsonhighered.com/notaros

I thank the reviewers of the manuscript for their extremely detailed, useful, and competent comments, which helped me to improve the quality of the book, including: Prof. Jacob Adams, North Carolina State University; Prof. Charles F. Bunting, Oklahoma State University; Prof. Gregory David Durgin, Georgia Institute of Technology; Prof. Ahmed Helmy, University of Florida;

Dr. Ming-Shih Huang, Penn State University; Prof. Leo C. Kempel, Michigan State University; Prof. Sudarshan R. Nelatury, Penn State University; Prof. L. Wilson Pearson, Clemson University; Prof. Ahmad Safaai-Jazi, Virginia Tech; Prof. Ergun Simsek, George Washington University; Prof. Jing Wang, University of South Florida; and Prof. Y. Ethan Wang, University of California, Los Angeles.

I am grateful to my Pearson Prentice Hall editor Andrew Gilfillan, who has been very helpful and supportive, and whose input was essential at many stages in the development of the book.

I would like to acknowledge and express special thanks and sincere gratitude to my Ph.D. students Ana Manić, Nada Šekeljić, and Sanja Manić for their truly outstanding work and invaluable help with writing MATLAB computer exercises, tutorials, and codes.

Please send comments, suggestions, questions, and/or corrections to notaros@colostate.edu.

Branislav M. Notaroš

Fort Collins, Colorado

m Files on Instructor Resources

On Instructor Resources (IR), the book provides MATLAB codes (m files) for *all* MATLAB exercises, separated into 12 folders (chapter folders). Code listings in TUTORIALS, where they appear merged with the narratives of the solutions to exercises, are obtained directly from the corresponding m files provided on IR. All figures appearing in MATLAB exercises, including the snapshots of MATLAB movies, GUIs, etc., as well as all other numerical and textual results given in the exercises, are generated using the m files from these folders.

There are about 560 m files – for 389 MATLAB exercises; some exercises have multiple m files, as their solutions consist of more than one function or the main program and a function, etc. Files with functions (written in MATLAB) have descriptive names, e.g., *bounceDiagram.m.* Files with main MATLAB programs are named according to the numeration/labeling of exercises: for instance, main program for *MATLAB Exercise 12.32* is in file *ME12_32.m.* So, the "recipe" for finding the m file for a given MATLAB exercise is simple: (1) if the exercise features a main MATLAB program, find the m file with the name matching the exercise label (number) in the respective chapter folder; (2) in the case of a function exercise, look for the function name in the statement of the exercise, and find the m file with that name. For convenience, however, the names of respective m files are explicitly specified in all MATLAB exercises in the book. *List of MATLAB Exercises* in Appendix 3 contains the full list of m files as well.

Files for GUIs built in MATLAB are stored in separate folders (subfolders) named after the corresponding MATLAB exercises, within the respective chapter folders. For example, files for the GUI from *MATLAB Exercise 2.13* are in subfolder *ME2_13(GUI)* in *Chapter_2* folder. A GUI folder always includes the GUI control function file (e.g., *capCalc1.m*) and the GUI layout file with the same name and fig extension (*capCalc1.fig*). It may also include one or more other functions (m files), for specific calculations or other tasks. Some GUI folders contain a number of png files (e.g., *coaxcable.png*), with pictures (drawings) of structures (previously created by another computer program for drawing), which are imported in the GUI control function.

The provided codes are run in MATLAB in a standard fashion: by selecting, opening, and running the appropriate m file in the appropriate chapter (or GUI) folder.

The material on IR, including all 12 chapter folders with m files, is meant *for instructors only.* In addition to serving as a *Solutions Manual* for the book, it provides an invaluable resource for lectures, recitations, and class demonstrations. Namely, any and all MATLAB codes for the current class topic, including codes for movies; calculators; GUIs; numerical solvers; data processors and tabulators; diagram and figure plotters; conceptual demonstrators; etc., can readily be run and discussed, in support of a theoretical presentation, or as a problem session.

Any of the provided m files on Instructor Resources and any of the included MATLAB codes or any part of a code may be used only for educational purposes associated with the book, MATLAB®-Based Electromagnetics.

1 ELECTROSTATIC FIELD IN FREE SPACE

Introduction:

Electrostatics is the branch of electromagnetics that deals with phenomena associated with static electricity, which are essentially the consequence of a simple experimental fact – that charges exert forces on one another. These forces are called electric forces, and the special state in space due to one charge in which the other charge is situated and which causes the force on it is called the electric field. Any charge distribution in space with any time variation is a source of the electric field. The electric field due to time-invariant charges at rest (charges that do not change in time and do not move) is called the static electric field or electrostatic field. This is the simplest form of the general electromagnetic field, and its physics and mathematics represent the foundation of the entire electromagnetic theory. On the other hand, a clear understanding of electrostatics is essential for many practical applications that involve static electric fields, charges, and forces in electrical and electronic devices and systems.

1.1 Coulomb's Law

Coulomb's law states that the electric force $\mathbf{F}_{\mathrm{e}12}$ on a point charge Q_2 due to a point charge Q_1 in a vacuum or air (free space) is given by[1] (Fig.1.1)

$$\mathbf{F}_{\mathrm{e}12} = \frac{1}{4\pi\varepsilon_0}\frac{Q_1 Q_2}{R^2}\hat{\mathbf{R}}_{12} \quad \text{(Coulomb's law)} , \tag{1.1}$$

where $\mathbf{R}_{12}$ denotes the position vector of Q_2 relative to Q_1, $R = |\mathbf{R}_{12}|$ is the distance between the two charges, $\hat{\mathbf{R}}_{12} = \mathbf{R}_{12}/R$ is the unit vector[2] of the vector $\mathbf{R}_{12}$ ($|\hat{\mathbf{R}}_{12}| = 1$), and ε_0 is the permittivity of free space,

$$\varepsilon_0 = 8.8542 \text{ pF/m} \quad \text{(permittivity of free space)} . \tag{1.2}$$

By point charges we mean charged bodies of arbitrary shapes whose dimensions are much smaller than the distance between them. The SI (International System of Units) unit for charge is the coulomb (abbreviated C). This is a very large unit of charge. The charge of an electron turns out to be

$$Q_{\text{electron}} = -1.6022 \times 10^{-19} \text{ C} \quad \text{(charge of electron)} . \tag{1.3}$$

The unit for force ($\mathbf{F}$) is the newton (N).

---

[1] In typewritten work, vectors are commonly represented by boldface symbols, e.g., $\mathbf{F}$, whereas in handwritten work, they are denoted by placing a right-handed arrow over the symbol, so as $\vec{F}$.

[2] All unit vectors in this text will be represented using the "hat" notation, so the unit vector in the x-direction (in the rectangular coordinate system), for example, is given as $\hat{\mathbf{x}}$ (note that some of the alternative widely used notations for unit vectors would represent this vector as $\mathbf{a}_x$, $\mathbf{i}_x$, and $\mathbf{u}_x$, respectively).

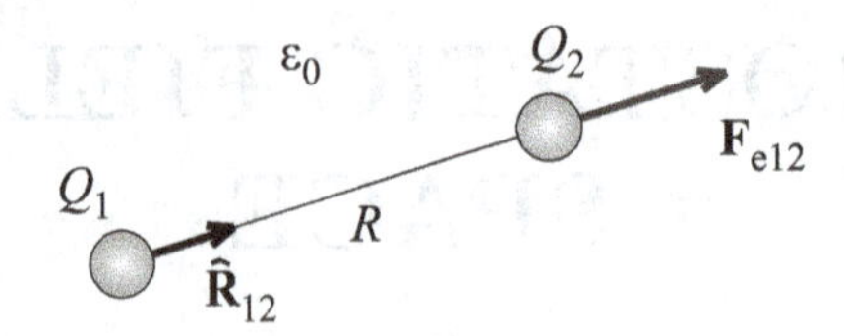

Figure 1.1 Notation in Coulomb's law, given by Eq.(1.1).

If we have more than two point charges, we can use the principle of superposition to determine the resultant force on a particular charge – by adding up vectorially the partial forces exerted on it by each of the remaining charges individually.

MATLAB EXERCISE 1.1 **Vector magnitude.** Using MATLAB, write a function `vectorMag()` that calculates the magnitude of a given vector. The input to the function is either a one-dimensional (1-D) vector, $\mathbf{a} = a_x\,\hat{\mathbf{x}}$, or a 2-D one, $\mathbf{a} = a_x\,\hat{\mathbf{x}} + a_y\,\hat{\mathbf{y}}$, or a 3-D vector,

$$\mathbf{a} = a_x\,\hat{\mathbf{x}} + a_y\,\hat{\mathbf{y}} + a_z\,\hat{\mathbf{z}} \quad \text{(Cartesian vector components)} \tag{1.4}$$

– in a Cartesian coordinate system (Fig.1.2), and it is specified as either a row or column array, named `vector`, with the elements of the array representing respective Cartesian components of the vector **a**. *(vectorMag.m on IR)*[3]

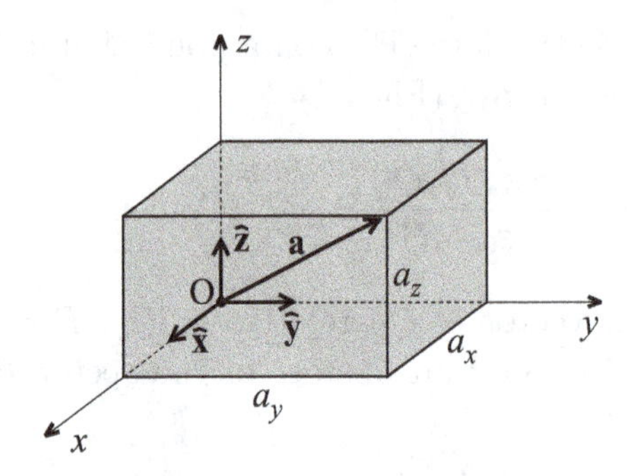

Figure 1.2 Decomposition of a vector (**a**) onto components (a_x, a_y, and a_z) in a Cartesian coordinate system, Eq.(1.4); for MATLAB Exercise 1.1.

TUTORIAL:
First, in an `if-else` branching, we check if the input `vector` is of proper dimensions, namely, if it is a column/row array. In a case of improper dimensions, the function displays an error message in the MATLAB Command Window. Otherwise, it calculates the vector magnitude, `mag`, using the relationship (see Fig.1.2)

$$a = |\mathbf{a}| = \sqrt{a_x^2 + a_y^2 + a_z^2}\,, \tag{1.5}$$

and the code can be written as follows:

```
function mag = vectorMag(vector)
```

[3] *IR* = Instructor Resources (for the book).

```
[m,n]= size(vector);
if (m ~= 1) && (n ~= 1)
mag = 0;
disp('Error - vector of improper dimensions');
else
mag = sqrt(sum(vector.^2));
end;
```

Note that, in MATLAB, `a ~= b` means $a \neq b$, and that && represents a logical AND operation. Note also that the use of . following a matrix symbol (quantity), e.g., `vector.^2`, implies that the operation is applied to each element of the matrix.

For additional help (if needed) on any topic in MATLAB programming, see [44] (or [39–43]) in the Bibliography.

MATLAB EXERCISE 1.2 **2-D vector plot.** Write a function `vecPlot2D()` in MATLAB that, for a given 2-D vector as input, specified by coordinates of its starting and ending points in the xy-plane, plots the vector. *(vecPlot2D.m on IR)*

HINT:

The code is the same as for the next MATLAB exercise, just in two instead of three dimensions.

MATLAB EXERCISE 1.3 **3-D vector plot.** Use MATLAB to write a function `vecPlot3D()` that – for an input 3-D vector given by coordinates of its starting and ending points – plots the vector in a 3-D Cartesian coordinate system. Other input parameters are: information whether or not scaling is to be performed, the color of the vector, and, since the function will later be used to plot the electric force on a charge, whether there should be a dot representing the charge "target" at the tail of the force vector. *(vecPlot3D.m on IR)*

TUTORIAL:

To plot vectors, we use MATLAB function `quiver()`, namely, its 3-D version `quiver3()`. Note that this function is used in MATLAB Exercise 1.13 to plot vector field lines in space (its main application is for drawing multiple vectors in a space mesh); however, here it is implemented for a single vector. The input to `quiver3` contains Cartesian coordinates of the starting (tail) point of the vector, and the three vector components. In addition, it is possible to include as input a scaling factor for the components, with a zero value of the factor meaning no scaling. The purpose of this factor is to enable fitting of the vector plots within the geometry of the problem.

So, from Cartesian coordinates or components of the position vector ($\mathbf{r} = x\,\hat{\mathbf{x}} + y\,\hat{\mathbf{y}} + z\,\hat{\mathbf{z}}$) of the starting and ending points of the vector, `Vi` and `Vf`, respectively, we first calculate the Cartesian components of the vector, `u`, `v`, and `w`, and save the coordinates of the starting point as `(x,y,z)`,

```
function vecPlot3D(Vi,Vf,scaling,color,origin)
u = Vf(1) - Vi(1);
v = Vf(2) - Vi(2);
w = Vf(3) - Vi(3);
```

```
x = Vi(1);
y = Vi(2);
z = Vi(3);
```

Then, if so specified at the input, we draw a large dot at the tail of the arrow (e.g., a charge "target") employing MATLAB function `plot3()` with a proper input. MATLAB command `hold on` is used to add the plot on the already existing figure without erasing it.

```
if (origin ~= 0)
plot3(x,y,z,'o','MarkerSize',10,'MarkerFaceColor',color);
hold on;
end;
```

Finally, `quiver3` is used to add the vector, and the figure properties are specified as white background and equal axes, which makes the same lengths along different axes appear equal,

```
quiver3(x,y,z,u,v,w,scaling,'LineWidth',2,'Color',color);
whitebg('white');
axis equal;
grid on;
hold off;
```

MATLAB EXERCISE 1.4 **Electric force due to multiple charges.** Write a program in MATLAB that uses functions from the previous three MATLAB exercises and calculates and plots the electric force on a point charge due to N other point charges in free space. The input to the program consists of N, coordinates of charge points, and charges Q_1, Q_2, ..., Q_N, as well as coordinates and charge of the point charge on which the force is evaluated. *(ME1_4.m on IR)*

TUTORIAL:

At the beginning – of **all main programs** in MATLAB – it is important to **CLEAR previously assigned memory** (`clear all`) and **CLOSE all figures** (`close all`), as well as to define constants that will be used in the program,

```
clear all;
close all;
EPS0 = 8.8542*10^(-12);
```

where `EPS0` is the permittivity of free space, given in Eq.(1.2).

Then we use MATLAB command `a = input(text)`, as it allows the user to input data after reading an additional explanation in `text`. `N` represents the number of "source" charges, while `x,y,z,Q` are the arrays, of length `N`, with given coordinates and charges. Since `N` can be arbitrary, the input of array data is given within a `for` loop, in which every element in the array is entered separately. In addition, `xp,yp,zp` and `Qp` correspond to the "target" point, where the force is calculated. Input of all data is implemented as follows:

```
N = input('Enter the number of charges (positive integer):  ');
for i = 1:N
```

```
x(i) = input(['Enter the X coordinate in cm for charge ', int2str(i) ,':  ']);
y(i) = input(['Enter the Y coordinate in cm for charge ', int2str(i),':  ']);
z(i) = input(['Enter the Z coordinate in cm for charge ', int2str(i),':  ']);
Q(i) = input(['Enter the charge in nC for charge ', int2str(i),':  ']);
end
disp('Properties of the point at which the force is calculated');
xp = input('Enter the X coordinate in cm :  ');
yp = input('Enter the Y coordinate in cm :  ');
zp = input('Enter the Z coordinate in cm :  ');
Qp = input('Enter the "target" charge in nC: ');
```

Note that it is usually helpful to specify bounds on entry values in the explanations in **text** of the **input(text)** command, so that the user is reminded to not enter physically impossible values (e.g., the number of charges N must be a positive integer). Even better would be to check in the program whether the entered value is physically meaningful (allowed), i.e., whether it is within the allowed bounds, and to repeat the request for input if it is not. We scale all input data to the basic SI units (i.e., cm to m, and nC to C),

```
x = x * 10^(-2);
y = y * 10^(-2);
z = z * 10^(-2);
Q = Q * 10^(-9);
xp = xp * 10^(-2);
yp = yp * 10^(-2);
zp = zp * 10^(-2);
Qp = Qp * 10^(-9);
```

Next, we calculate the distance between each "source" charge and the "target" charge, with which we fill an array **r**, and the unit vector ($\hat{\mathbf{R}}_{12}$) along the line connecting charges (Fig.1.1) – as required by the electric force expression in Eq.(1.1), where **uVec** is a $3 \times N$ matrix whose N columns represent N vectors $\hat{\mathbf{R}}_{12}$ along these distances. In particular, the three elements of every column, **ux,uy,uz**, equal the three Cartesian components [as in Eq.(1.4)] of the respective unit vector. Recall that, in MATLAB, the use of . following a matrix symbol (quantity) implies that the operation is applied to each element of the matrix, and not to the whole matrix as a matrix operation; for example, while a^ 2 gives the square of the matrix a (product of matrices a and a), a.^ 2 results in a matrix whose elements equal the squares of the respective elements of the matrix a.

```
r = sqrt((xp - x).^2 + (yp - y).^2 + (zp - z).^2);
ux = (xp - x)./r;
uy = (yp - y)./r;
uz = (zp - z)./r;
uVec = [ux; uy; uz];
```

Once these quantities are computed, it is a simple matter to implement Eq.(1.1) in a matrix form and find the force due to each charge,

```
F = (ones(3,1)*(Qp*Q./(4*pi*EPS0*r.^2))).*uVec;
```

where F represents a $3 \times N$ matrix with elements of each column being the Cartesian components of one of the force vectors.

In order to obtain the components of the total force vector, all forces are summed vectorially; namely, `Ftot` is the column vector obtained by adding up all columns of the matrix `F`, and the magnitude of the total force vector is evaluated invoking function `vectorMag` (from MATLAB Exercise 1.1),

```
Ftot = sum(F,2);
Fmag = vectorMag(Ftot);
Fuv = (Ftot/Fmag)';
```

The results are displayed as:

```
fprintf(['Magnitude of the resultant force at point P is %f mN.\n'],Fmag*1000);
disp('Unit vector of the resultant force :');
disp(Fuv);
```

Finally, we plot individual forces on the "target" charge – due to `N` "source" charges – in blue, and the total force in red,

```
figure(1);
plot3(0,0,0,'k');
hold on;
for i=1:N
plot3(x(i),y(i),z(i),'o','MarkerSize',10,'MarkerFaceColor','b');
line ([xp , x(i)],[yp,y(i)], [zp,z(i)],'LineStyle',':');
hold on;
vecPlot3D([xp yp zp],[xp yp zp] + F(:,i)',1/Fmag/100,'b',0);
hold on;
end;
vecPlot3D([xp yp zp],[xp yp zp] + Ftot',1/Fmag/100,'r',1);
text(1.1*xp,1.1*yp,1.1*zp,'Total electric force','Color','r');
hold off;
axis equal;
xlabel('x');
ylabel('y');
zlabel('z');
title('Force from ',int2str(N),'-point charges');
```

where MATLAB functions `plot3` and `line` are used to plot the charge points and to draw a line with specified starting and ending points, respectively, while `vecPlot3D` is the function created in MATLAB Exercise 1.3.

MATLAB EXERCISE 1.5 **Four equal point charges at tetrahedron vertices.** Figure 1.3(a) shows four point charges $Q = 1$ nC positioned in free space at four vertices of a regular (equilateral) tetrahedron with the side length $a = 1$ cm. Using MATLAB, find the resultant electric

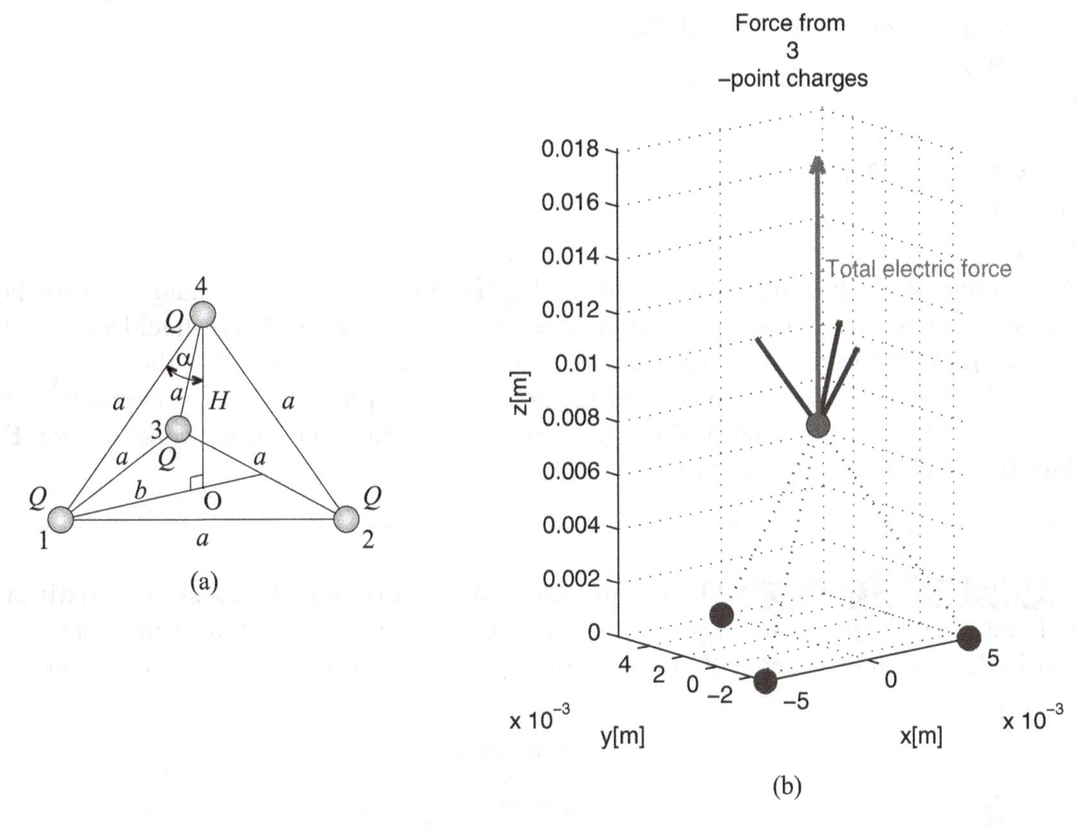

Figure 1.3 (a) Four point charges at tetrahedron vertices and (b) 3-D visualization of a MATLAB solution for individual forces on charge 4 due to charges 1, 2, and 3, respectively, and the resultant force – obtained using the program from MATLAB Exercise 1.4; for MATLAB Exercise 1.5. *(color figure on CW)*[4]

force on one of the charges – for example, on the charge at the top of the tetrahedron. *(ME1_5.m on IR)* **H**[5]

HINT:
Apply the program written in the previous MATLAB exercise, with `N` = 3, charges 1, 2, and 3 in Fig.1.3(a) as the "source" charges, and charge 4 as the "target" charge. In particular, place the origin of the Cartesian coordinate system at the center of the base of the tetrahedron [point O in Fig.1.3(a)]. To determine the coordinates of the charges, note that the distance b between charge 1 and point O and, from it, the height H of the tetrahedron are [the orthocenter of an equilateral triangle partitions its heights (h) into segments $2h/3$ and $h/3$ long, where $h = \sqrt{3}a/2$]

$$b = \frac{2}{3}\left(\frac{\sqrt{3}}{2}a\right) = \frac{\sqrt{3}}{3}a \quad \longrightarrow \quad H = \sqrt{a^2 - b^2} = \sqrt{\frac{2}{3}}\,a \,. \tag{1.6}$$

Based on these relationships, implement the following lines of MATLAB code to input data:

```
x = [0 -0.5 0.5];
```

[4] CW = Companion Website (of the book).
[5] **H** = recommended to be done also "by hand," i.e., not using MATLAB.

```
y = [sqrt(3)/3 -sqrt(3)/6 -sqrt(3)/6];
z = [0 0 0];
xp = 0;
yp = 0;
zp = sqrt(2)/sqrt(3);
Q = [1 1 1];
Qp = 1;
```

As a result of computation, the plot of individual forces on the top charge due to the three charges at the tetrahedron base and the total force ($\mathbf{F}_{e4} = \mathbf{F}_{e14} + \mathbf{F}_{e24} + \mathbf{F}_{e34}$) should look like the one in Fig.1.3(b). By means of the rotation button in the MATLAB figure, rotate the 3-D visualization of the solution to better see the forces and the geometrical representation of the problem. Finally, invoking Eq.(1.1) and the superposition principle, obtain the analytical expression for $\mathbf{F}_{e4}$ and validate the MATLAB result against that expression.

MATLAB EXERCISE 1.6 **Three unequal point charges in Cartesian coordinate system.** Point charges $Q_1 = 1\ \mu\text{C}$, $Q_2 = -2\ \mu\text{C}$, and $Q_3 = 2\ \mu\text{C}$ are situated in free space at points defined by Cartesian coordinates $(1\text{ m}, 0, 0)$, $(0, 1\text{ m}, 0)$, and $(0, 0, 1\text{ m})$, respectively (see Fig.1.4).

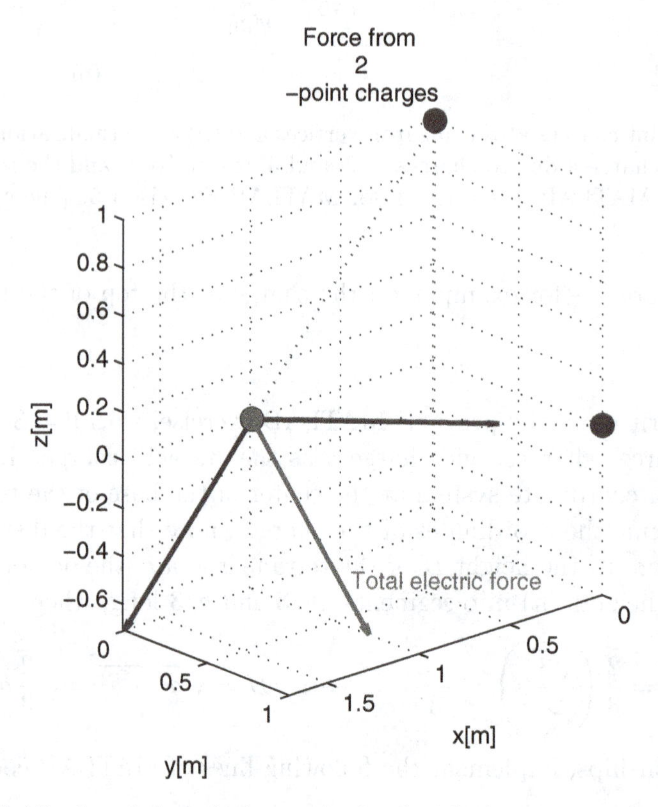

Figure 1.4 3-D MATLAB plot of the solution – obtained with the program from MATLAB Exercise 1.4 – for the vector summation of Coulomb forces in the Cartesian coordinate system; for MATLAB Exercise 1.6. *(color figure on CW)*

In MATLAB, compute the resultant electric force on charge Q_1, and graphically represent partial and total force vectors $\mathbf{F}_{e21}$, $\mathbf{F}_{e31}$, and $\mathbf{F}_{e1} = \mathbf{F}_{e21} + \mathbf{F}_{e31}$. *(ME1_6.m on IR)* **H**

HINT:

Use the program developed in MATLAB Exercise 1.4, with the following to input data:

```
x = [0 0]*100;
y = [1 0]*100;
z = [0 1]*100;
xp = 100;
yp = 0;
zp = 0;
Q = [-2 2]*1000;
Qp = 1000;
```

The scaling factor in function `vecPlot3D` should be changed from `1/Fmag/100` (see MATLAB Exercise 1.4) to `1/Fmag` to be able to properly plot and see forces within the current coordinate mesh.

The resulting graph is shown in Fig.1.4.

1.2 Electric Field Intensity Vector Due to Given Charge Distributions

To quantitatively describe the electric field, we introduce a vector quantity called the electric field intensity vector, **E**. By definition, it is equal to the electric force $\mathbf{F}_e$ on a small probe (test) point charge Q_p placed in the electric field, divided by Q_p, that is,

$$\mathbf{E} = \frac{\mathbf{F}_e}{Q_p} \quad (Q_p \to 0) \quad \text{(definition of the electric field intensity vector; unit: V/m)} . \tag{1.7}$$

The unit for the electric field intensity is volt per meter (V/m). From the definition in Eq.(1.7) and Coulomb's law, Eq.(1.1), we obtain the expression for the electric field intensity vector of a point charge Q at a distance R from the charge (Fig.1.5)

$$\mathbf{E} = \frac{1}{4\pi\varepsilon_0} \frac{Q}{R^2} \hat{\mathbf{R}} \quad \text{(electric field due to a point charge)} , \tag{1.8}$$

where $\hat{\mathbf{R}}$ is the unit vector along R directed from the center of the charge (source point) toward the point at which the field is (to be) determined (field or observation point).

By superposition, the electric field intensity vector produced by N point charges $(Q_1, Q_2, \ldots, Q_N)$ at a point that is at distances $R_1, R_2, \ldots, R_N$, respectively, from the charges can be obtained as

$$\mathbf{E} = \mathbf{E}_1 + \mathbf{E}_2 + \cdots + \mathbf{E}_N = \frac{1}{4\pi\varepsilon_0} \sum_{i=1}^{N} \frac{Q_i}{R_i^2} \hat{\mathbf{R}}_i , \tag{1.9}$$

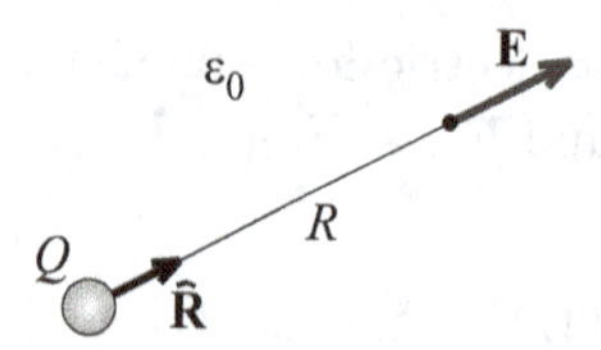

Figure 1.5 Electric field intensity vector due to a point charge in free space.

where $\hat{\mathbf{R}}_i$, $i = 1, 2, \ldots, N$, are the corresponding unit vectors.

In the general case, charge can be distributed throughout a volume, on a surface, or along a line. Each of these three characteristic continuous charge distributions is described by a suitable charge density function. The volume charge density (in a volume v) is defined as [Fig.1.6(a)]

$$\rho = \frac{\mathrm{d}Q}{\mathrm{d}v} \quad \text{(volume charge density; unit: C/m}^3\text{)}\,, \tag{1.10}$$

the surface charge density (on a surface S) is given by [Fig.1.6(b)]

$$\rho_\mathrm{s} = \frac{\mathrm{d}Q}{\mathrm{d}S} \quad \text{(surface charge density; unit: C/m}^2\text{)}\,, \tag{1.11}$$

and the line charge density (along a line l) is [Fig.1.6(c)]

$$Q' = \frac{\mathrm{d}Q}{\mathrm{d}l} \quad \text{(line charge density; unit: C/m)}\,. \tag{1.12}$$

Note that the symbol ρ_v is sometimes used instead of ρ, σ instead of ρ_s, and ρ_l instead of Q'. In addition, by Q' ($Q' = \text{const}$) we also represent the so-called charge per unit length (p.u.l.) of a long uniformly charged structure (e.g., thin or thick cylinder), defined as the charge on one meter (unit of length) of the structure divided by 1 m,

$$Q' = Q_\text{p.u.l.} = \frac{Q_{\text{along }l}}{l} = \frac{Q_{\text{for 1 m length}}}{1\,\text{m}} \quad \text{(charge per unit length, in C/m)}\,, \tag{1.13}$$

and hence Q' numerically equals the charge on each meter of the structure.

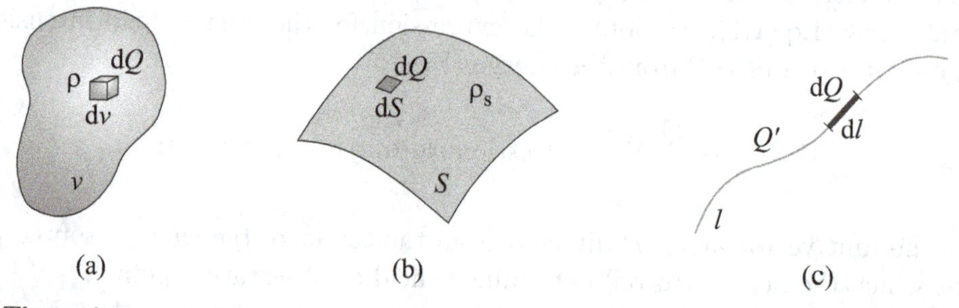

Figure 1.6 Three characteristic continuous charge distributions and charge elements: (a) volume charge, (b) surface charge, and (c) line charge.

By virtue of the superposition principle, the electric field intensity vector due to each of the charge distributions ρ, ρ_s, and Q' can be regarded as the vector summation of the field intensities contributed by the numerous equivalent point charges making up the charge distribution. Thus, by replacing Q in Eq.(1.8) with charge element $\mathrm{d}Q = Q'\,\mathrm{d}l$ in Fig.1.6(c) and integrating, we get

$$\mathbf{E} = \frac{1}{4\pi\varepsilon_0}\int_l \frac{Q'\,\mathrm{d}l}{R^2}\hat{\mathbf{R}} \quad \text{(electric field due to line charge)}\ , \tag{1.14}$$

and similar expressions for the electric field due to surface and volume charges.

MATLAB EXERCISE 1.7 **Electric field due to multiple charges.** In MATLAB, calculate and plot the electric field intensity vector due to N point charges placed at arbitrary locations in a Cartesian coordinate system, in free space, based on Eq.(1.9) and the program developed in MATLAB Exercise 1.4. *(ME1_7.m on IR)*

MATLAB EXERCISE 1.8 **Three charges at rectangle vertices.** Three small charged bodies of equal charges $Q = -1$ nC are placed at three vertices of a rectangle with sides $a = 4$ cm and $b = 2$ cm. Using the program from the previous MATLAB exercise, find the direction and magnitude of the electric field vector at the remaining vertex of the rectangle. *(ME1_8.m on IR)* **H**

HINT:
Implement `1/Emag/10` as the scaling factor in function `vecPlot3D` (see MATLAB Exercise 1.4). For the viewpoint, use MATLAB function `view` with a zero azimuthal angle and 90° elevation angle, so `view(0,90)`.

MATLAB EXERCISE 1.9 **Charged ring and an equivalent point charge.** Consider a line charge of uniform charge density Q' distributed around the circumference of a ring of radius a in air. Due to symmetry, the electric field intensity vector along the axis of the ring normal to its plane, at the point P in Fig.1.7(a), has a z-component only, which, applying Eq.(1.14), comes out to be

$$\begin{aligned} E = E_z &= \oint_C \mathrm{d}E_z = \oint_C \mathrm{d}E\cos\alpha = \oint_C \frac{Q'\,\mathrm{d}l}{4\pi\varepsilon_0 R^2}\frac{z}{R} = \frac{Q'z}{4\pi\varepsilon_0 R^3}\oint_C \mathrm{d}l = \frac{Q'az}{2\varepsilon_0 R^3} \\ &= \frac{Qz}{4\pi\varepsilon_0\left(z^2+a^2\right)^{3/2}} \quad (Q = Q'2\pi a)\ , \end{aligned} \tag{1.15}$$

where Q is the total charge of the ring. Plot in MATLAB the dependence of E on the coordinate z [in Fig.1.7(a)], for $z > 0$. In addition, show that far away along the z-axis, the charged ring in Fig.1.7(a) is equivalent to a point charge with the same amount of charge located at the coordinate origin. Finally, plot, along the z-axis, the electric field due to this equivalent point charge and compare the two graphs. *(ME1_9.m on IR)*

HINT:
For $|z| \gg a$, $z^2 + a^2 \approx z^2$, with which Eq.(1.15) becomes

$$E \approx \frac{Q}{4\pi\varepsilon_0 z^2} \quad (|z| \gg a)\ . \tag{1.16}$$

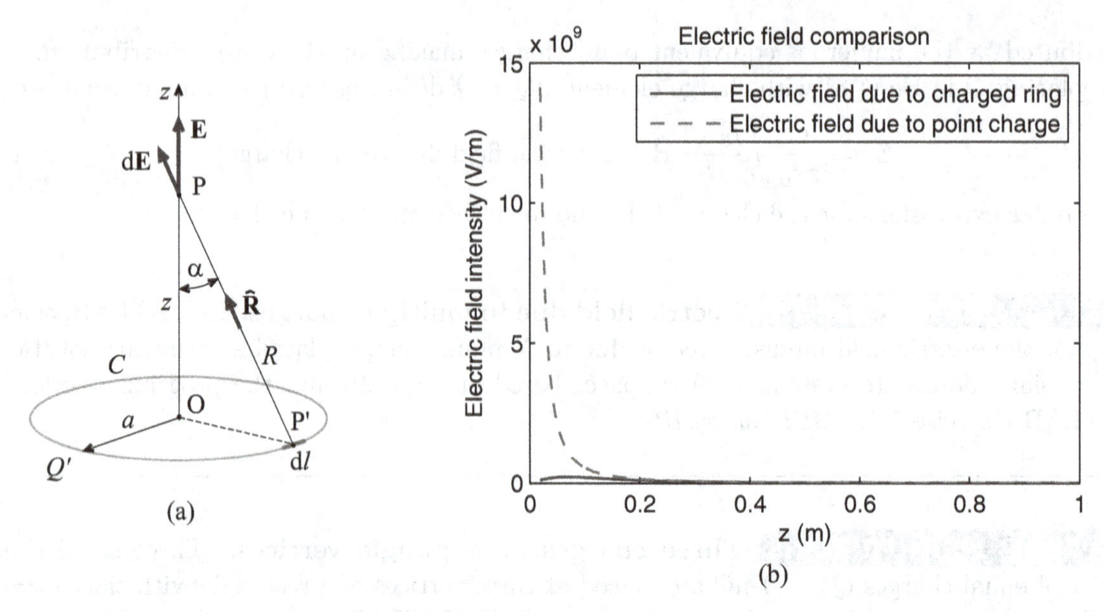

Figure 1.7 (a) Evaluation of the electric field along the axis of a charged ring normal to its plane, Eq.(1.15), and (b) MATLAB comparison of electric field intensities along the z-axis due to the ring of charge and the equivalent (for $|z| \gg a$) point charge, respectively, for $Q' = 1$ mC/m and $a = 10$ cm; for MATLAB Exercise 1.9. *(color figure on CW)*

So, indeed, far away from the ring, its charge is equivalent to a point charge Q located at its center. In other words, when the distance of the field point from the ring is much larger than the ring dimensions, the ring can be considered as a point charge and the actual shape of the ring (or any other charged body) does not matter. This in fact is the definition of a point charge or a small charged body.

The resulting graphical comparison of the expressions in Eqs.(1.15) and (1.16) using MATLAB is shown in Fig.1.7(b).

MATLAB EXERCISE 1.10 **Symbolic integration.** MATLAB supports operations with symbolic variables, that is, all kinds of calculations using symbols in place of real-valued (numerical) variables. Symbolic integration is implemented in MATLAB through function `int()`. Write your own function named `integral()` that invokes `int` and has the following input data [`integral(f,t,r,a,b)`] in order to compute the integral $\int_a^b ft\,\mathrm{d}r$: `f` and `t` (their product represents the function to be integrated), `r` (independent variable of integration), and `a` and `b` (integration limits). *(integral.m on IR)*

MATLAB EXERCISE 1.11 **Charged disk – symbolic and analytical solutions.** Consider a very thin charged disk (i.e., a circular sheet of charge), of radius a and a uniform surface charge density ρ_s, in free space, and the electric field it generates at a point P along the z-axis in Fig.1.8(a). By subdividing the disk into elemental rings of width $\mathrm{d}r$, as shown in Fig.1.8(a),

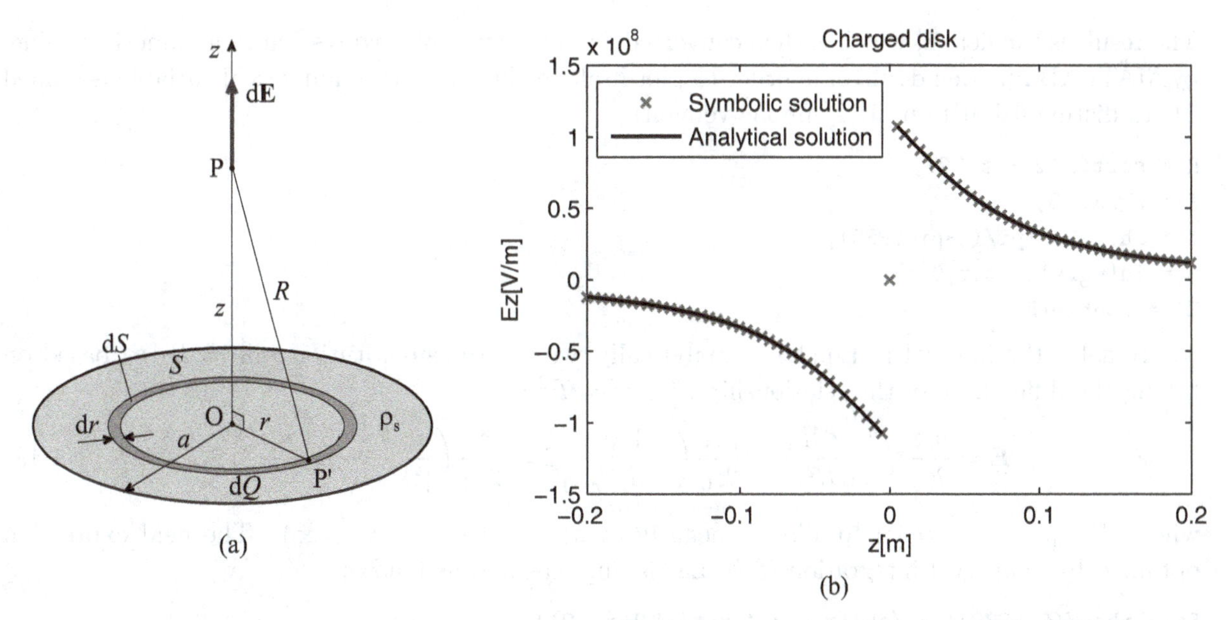

Figure 1.8 (a) Evaluation of the electric field due to a charged disk and (b) field intensity along the axis (z-axis) of the disk obtained by symbolic integration in Eq.(1.17) using function `integral` (from the previous MATLAB exercise) and by analytical integration in Eq.(1.18), respectively; for MATLAB Exercise 1.11. *(color figure on CW)*

applying Eq.(1.15) for the field at the point P due to a ring of radius r $(0 \leq r \leq a)$ and charge $\mathrm{d}Q = \rho_\mathrm{s}\,\mathrm{d}S$, with $\mathrm{d}S = 2\pi r\,\mathrm{d}r$ being the surface area of the ring (calculated as the area of a thin strip of length equal to the ring circumference, $2\pi r$, and width $\mathrm{d}r$), and superposition, the total electric field vector is given by

$$\mathbf{E} = \int_S \mathrm{d}\mathbf{E} = \int_S \frac{\mathrm{d}Q z}{4\pi\varepsilon_0 R^3}\,\hat{\mathbf{z}} = \frac{\rho_\mathrm{s} z}{2\varepsilon_0}\,\hat{\mathbf{z}} \int_{r=0}^{a} \frac{r\,\mathrm{d}r}{R^3}\,, \qquad R = \sqrt{r^2+z^2}\,. \tag{1.17}$$

Using this expression and function `integral` (written in the previous MATLAB exercise), compute $\mathbf{E}$ by symbolic integration. Also, solve the integral analytically and plot both solutions for $\rho_\mathrm{s} = 2\ \mathrm{mC/m^2}$, $a = 10$ cm, and $-2a \leq z \leq 2a$. *(ME1_11.m on IR)*

TUTORIAL:

First, we declare `r` [the radius of the elemental ring in Fig.1.8(a)] and `z` (the range of z-coordinates along the disk axis at which the field is computed) as symbolic variables – in MATLAB, defining symbolic variables is done by command `syms`, so we type `syms r z` – and specify the values of constants ρ_s, a, and ε_0,

```
syms r z
rhos = 0.002;
a = 0.1;
EPS0 = 8.8542 * 10^(-12);
z = -2*a:0.05*a:2*a;
```

Then, we define the variable `R` in Fig.1.8(a) and functions `f` and `t` constituting the field due to the elemental ring of charge, in Eq.(1.17), to be integrated symbolically – invoking function `integral`.

The result is the field E, which is then converted from the symbolic expression to a numerical value, by MATLAB function `double`, in order to plot it along the z-axis (the numerical variable is named E1, to distinguish it from its symbolic version).

```
R = sqrt(r^2 + z.^2);
f = z./R.^3;
t = rhos*2*r*pi/(4*pi*EPS0);
E = integral(f,t,r,0,a);
E1 = double(E);
```

To solve the integral in Eq.(1.17) analytically, we use the substitution $r\,\mathrm{d}r = R\,\mathrm{d}R$, based on taking the differential of the relationship $r^2 + z^2 = R^2$,

$$\mathbf{E} = \frac{\rho_s z}{2\varepsilon_0}\int_{r=0}^{a}\frac{\mathrm{d}R}{R^2}\,\hat{\mathbf{z}} = \frac{\rho_s z}{2\varepsilon_0}\left(-\frac{1}{R}\right)\Bigg|_{r=0}^{a}\hat{\mathbf{z}} = \frac{\rho_s}{2\varepsilon_0}\left(\frac{z}{|z|} - \frac{z}{\sqrt{a^2+z^2}}\right)\hat{\mathbf{z}}\,, \tag{1.18}$$

where $R|_{r=0} = \sqrt{z^2} = |z|$ to allow a negative z as well $(-\infty < z < \infty)$. The field expression obtained by analytical integration (field Ea) is implemented as follows:

```
Ea = rhos/(2*EPS0)*(z./abs(z)- z./sqrt(a^2+z.^2));
Ea1 = double(Ea);
plot (z,E1,'rx',z,Ea1,'k','LineWidth',1.4);
xlabel('z[m]');
ylabel('Ez[V/m]');
legend('Symbolic solution','Analytical solution');
title('Charged disk');
```

The two solutions and their mutual agreement are shown in Fig.1.8(b).

MATLAB EXERCISE 1.12 **Charged hemisphere – numerical integration.** The purpose of this MATLAB exercise is to introduce numerical integration as the third (often the only available) way to solve integrals, besides symbolic MATLAB integration and analytical solutions. Consider the uniformly charged hemispherical surface in Fig.1.9 and use MATLAB and numerical integration to compute the electric field intensity vector at an arbitrary point (for any z) along the z-axis. The simplest numerical integration formula is

$$\int_a^b f(x)\,\mathrm{d}x \approx \sum_{i=1}^{N} f(x_i)\Delta x\,, \quad \Delta x = \frac{b-a}{N}\,, \tag{1.19}$$

where N denotes the number of integration segments (increments) and x_i $(i = 1, 2, \ldots, N)$ are coordinates of centers of segments. *(ME1_12.m on IR)* **H**

HINT:

We subdivide the hemisphere into thin rings, as depicted in Fig.1.9. The radius of a ring whose position on the hemisphere is defined by an angle θ $(0 \le \theta \le \pi/2)$ is $a_r = a\sin\theta$ and its charge is given by

$$\mathrm{d}Q = \rho_s\,\mathrm{d}S = \rho_s\,\underbrace{2\pi a\sin\theta}_{C_r}\,\underbrace{a\,\mathrm{d}\theta}_{\mathrm{d}l_r}\,, \tag{1.20}$$

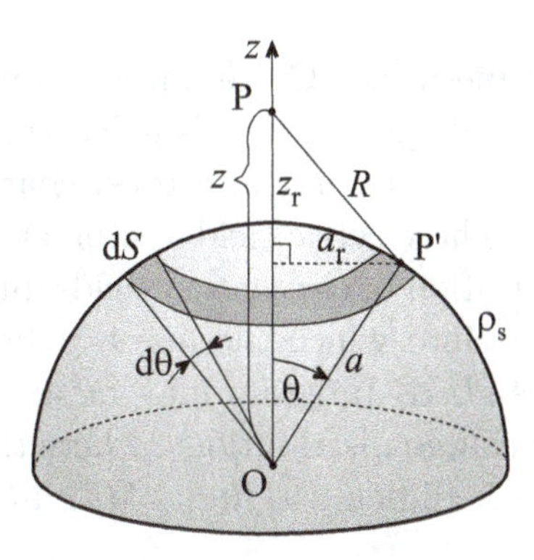

Figure 1.9 Evaluation of the electric field at an arbitrary point (P) along the axis of a hemispherical surface charge, carrying out numerical integration, in θ, in MATLAB; for MATLAB Exercise 1.12.

where C_r and dl_r denote the ring circumference and width, respectively. From Fig.1.9, the local z-coordinate of the field point P with respect to the ring center is $z_r = z - a\cos\theta$ and the distance of the point P from the source point P′ on the ring equals, by means of the cosine formula,

$$R = \sqrt{z^2 + a^2 - 2az\cos\theta}\ . \tag{1.21}$$

So, based on Eq.(1.15) and the superposition principle, the electric field intensity $d\mathbf{E}$ at the point P due to the charge dQ and the total field $\mathbf{E}$ can be expressed as

$$d\mathbf{E} = \frac{dQ\,(z - a\cos\theta)}{4\pi\varepsilon_0 R^3}\,\hat{\mathbf{z}} \quad\longrightarrow\quad \mathbf{E} = \frac{\rho_s a^2}{2\varepsilon_0}\int_{\theta=0}^{\pi/2} \frac{(z - a\cos\theta)\sin\theta\, d\theta}{(z^2 + a^2 - 2az\cos\theta)^{3/2}}\,\hat{\mathbf{z}}\ . \tag{1.22}$$

To solve the integral in θ in Eq.(1.22) by numerical integration, according to Eq.(1.19), the main `for` loop is as follows:

```
for theta = 0:dtheta:(pi/2)
ds = 2*pi*a^2*sin(theta)*dtheta;
dE = (ros/(4*pi*EPS0)*ds*(z - a*cos(theta)))/(sqrt(z^2 + a^2 - 2*a*z*cos(theta)))^3;
E = E + dE;
end
```

Compare the result, for different positions of the field point along the hemisphere axis (different values of the coordinate z in Fig.1.9) and different values of a and ρ_s, to the field expression obtained by analytical integration [with the substitution given by Eq.(1.21)], which reads

$$\mathbf{E} = \frac{\rho_s a^2}{2\varepsilon_0 z^2}\left(\frac{a}{\sqrt{z^2 + a^2}} + \frac{z - a}{|z - a|}\right)\hat{\mathbf{z}}\ . \tag{1.23}$$

For different values of the integration increment `dtheta` in the above `for` loop, evaluate the relative error of numerical integration, relative to the exact (analytical) solution.

MATLAB EXERCISE 1.13 Vector numerical integration and field visualization using `quiver`. MATLAB function `quiver` is used for visualization of field vectors in space. Input data are coordinates of nodes in a mesh in a Cartesian coordinate system and intensities of field components at the nodes. Implement **quiver** to visualize the electric field distribution due to a uniform straight line charge of finite length a and total charge Q placed along the x-axis and centered at the coordinate origin. The electric field vector at each node of the mesh should be computed by vector numerical integration of elementary fields due to equivalent point charges along the line representing short segments into which the line is subdivided. With such an integration (superposition) procedure, this MATLAB program may be applicable, with minor modifications, to many similar and more complex charge distributions, where the analytical expression for electric field components is not available or is difficult to find. *(ME1_13.m on IR)*

TUTORIAL:

The line charge density is $Q' = Q/a$ (quantity **rho** in our MATLAB code). We subdivide the line into elementary segments of length **dl** and charge **dQ = rho*dl**, with **xline** and **yline** being the x- and y-coordinates of segment centers where the equivalent point charges **dQ** are assumed to reside.

```
rho = Q/a ;
dl = 0.001*a;
xline = -a/2:dl:a/2;
yline = zeros (1,length (xline));
```

Next, we define a 2-D mesh of field points **[x,y]** in the xy-plane, employing MATLAB function **meshgrid**, with the mesh step adopted independently from the length of the integration segment, **dl**,

```
v = -0.825*a:0.15*a:0.975*a;
u = -0.525*a:0.15*a:0.675*a;
[x,y] = meshgrid(v,u);
[M,N] = size(x);
```

(MATLAB function **size** checks dimensions of a matrix). Note that the mesh is generated such that no field points coincide with the line charge.

In the part of the MATLAB code that follows, **r** is an **M**× **N**× **length(xline)**×2 matrix with each element corresponding to the electric field vector at one of the **M**× **N** field points due to one of the **length(xline)** integration segments of the charge line, i.e., one of the equivalent point charges along the line. Matrices **rabs** and **runit** contain magnitudes and unit vectors, respectively, for the computation of fields due to point charges, based on Eq.(1.8), to fill the matrix **E**. Then, x- and y-components of the total electric field vector at each node in the mesh are obtained as a sum of all elemental field components [e.g., **Etotx(i,j) = sum(E(i,j,:,1))** for the x-component] – vector numerical integration.

```
for i = 1:M
for j = 1:N
for t = 1:length(xline)
r(i,j,t,:) = [x(i,j)-xline(t) y(i,j)-yline(t)];
rabs(i,j,t) = vectorMag(r(i,j,t,:));
runit(i,j,t,:) = r(i,j,t,:)/rabs(i,j,t);
E(i,j,t,:) = rho*dl*runit(i,j,t,:)/(4*pi*EPS0*rabs(i,j,t)^2);
```

```
end;
Etotx(i,j) = sum(E(i,j,:,1));
Etoty(i,j) = sum(E(i,j,:,2));
end;
end;
```

Finally, we use quiver to visualize the resultant field distribution,

```
figure(1);
line([-a/2 a/2],[0 0]);
hold on;
quiver(v,u,Etotx,Etoty);
hold off;
title('Field distribution');
xlabel('x[m]');
ylabel('y[m]');
```

as shown in Fig.1.10.

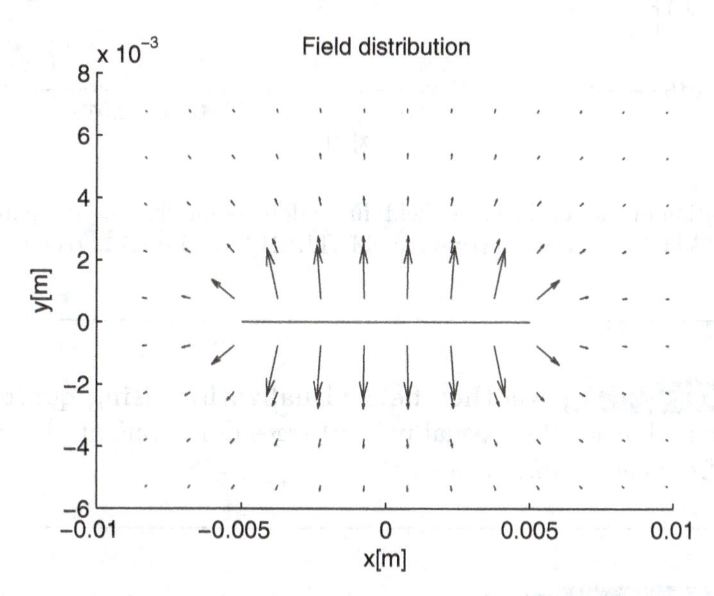

Figure 1.10 Visualization of the electric field distribution due to a uniform straight line charge of finite length ($a = 1$ cm and $Q = 1$ nC) using MATLAB function quiver, where the electric field vector at mesh nodes is previously computed, in MATLAB, as a sum of elementary fields due to equivalent point charges modeling the line; for MATLAB Exercise 1.13. *(color figure on CW)*

MATLAB EXERCISE 1.14 Visualization of the electric field due to four point charges. With the use of MATLAB function quiver, visualize the total electric field vector due to four equal point charges Q placed at vertices of a square of edge length a in free space. *(ME1_14.m on IR)*

HINT:
See the tutorial to the previous MATLAB exercise. The result is displayed in Fig.1.11.

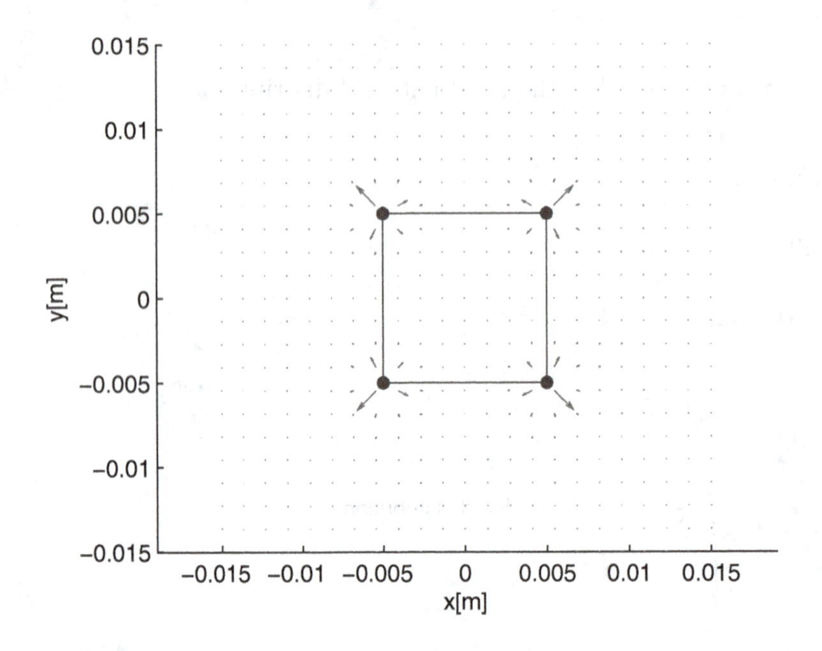

Figure 1.11 Visualization of the total electric field due to four point charges at square vertices ($Q = 1$ nC and $a = 1$ cm) using MATLAB function `quiver`; for MATLAB Exercise 1.14. *(color figure on CW)*

MATLAB EXERCISE 1.15 **Another field visualization using `quiver`.** Repeat the previous MATLAB exercise but for three equal point charges Q residing at vertices of an equilateral triangle of side a in free space. *(ME1_15.m on IR)*

MATLAB EXERCISE 1.16 **Fields due to line charges of finite and infinite lengths.** Write a MATLAB program that compares the electric field intensity E due to a straight line charge of finite length L to that due to an infinite line charge of the same uniform density Q' (charge per unit length) in free space, based on the following field expression, given for the situation in Fig.1.12(a) and obtained applying Eq.(1.14):

$$\mathbf{E} = \frac{Q'}{4\pi\varepsilon_0 D}\left[(\sin\theta_2 - \sin\theta_1)\,\hat{\mathbf{x}} + (\cos\theta_2 - \cos\theta_1)\,\hat{\mathbf{z}}\right] \quad \text{(line charge of finite length)}, \tag{1.24}$$

where the geometry of the problem is defined by the perpendicular distance from the line charge to the field point, D, and angles θ_1 and θ_2 (for the particular position of the point P shown in Fig.1.12(a), $\theta_1 < 0$ and $\theta_2 > 0$). Note that a numerical solution for this case is carried out in MATLAB Exercise 1.13. The field should be computed along the symmetry line of the finite line

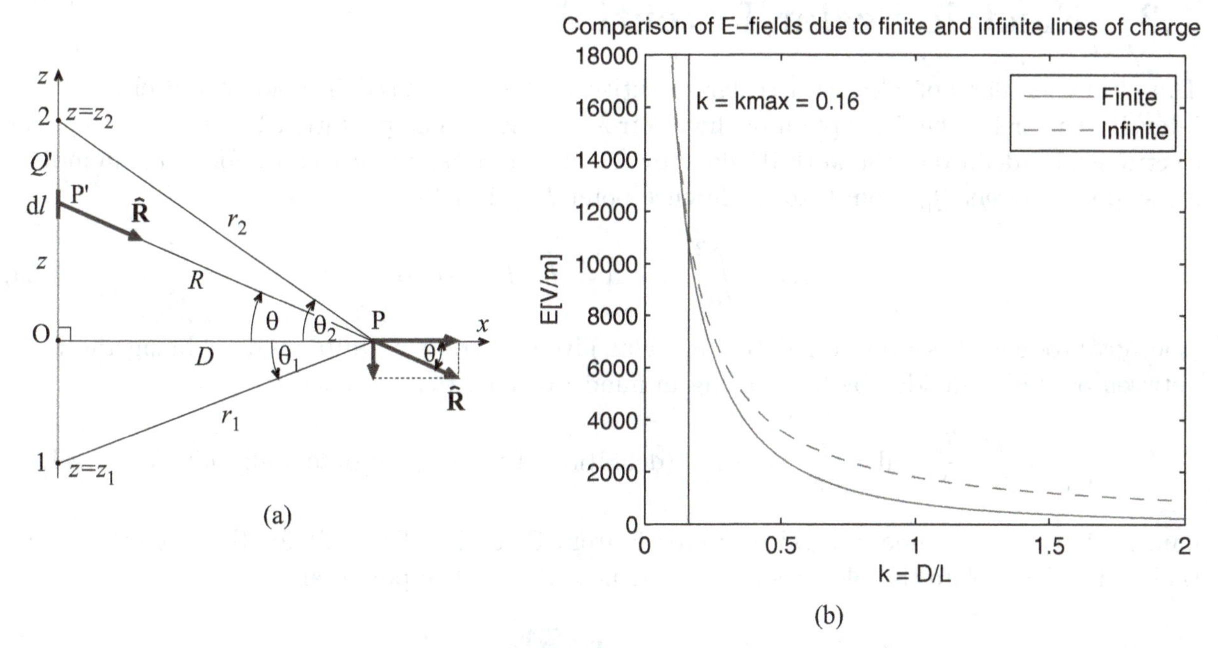

Figure 1.12 (a) Analytical evaluation of the electric field vector due to a straight line charge of finite length and (b) comparison of field intensities, calculated using MATLAB and Eqs.(1.24) and (1.25), due to line charges of finite ($L = 1$ cm) and infinite lengths and determination of the distance $D_{\rm max}$ ($k_{\rm max} = D_{\rm max}/L$) from the charges beyond which the relative error when approximating the finite charge by an infinite one is above 5%; for MATLAB Exercise 1.16. *(color figure on CW)*

charge that is perpendicular to it, so that the vector $\mathbf{E}$ is radial with respect to the charge. The input data are L and Q'. As an output, the program plots the dependence of E on the distance D from each of the line charges, and finds the distance $D_{\rm max}$ beyond which the relative error when approximating the finite charge by an infinite one is above 5% or, equivalently, below which the line charge of length L appears, in terms of the electric field it produces, as infinite (within the 5% error margin). *(ME1_16.m on IR)*

HINT:

The expression for the electric field vector due to an infinite line charge of uniform density Q' is a special case of that in Eq.(1.24), with $\theta_1 \to -\pi/2$ and $\theta_2 \to \pi/2$ (the line extends to $z \to -\infty$ and $z \to \infty$, respectively), so it reads

$$\mathbf{E} = \frac{Q'}{2\pi\varepsilon_0 D}\,\hat{\mathbf{x}} \quad \text{(infinite line charge)}\,. \tag{1.25}$$

Upon implementing Eqs.(1.24) and (1.25) in MATLAB, Fig.1.12(b) shows the respective field distributions and comparison for $L = 1$ cm and $Q' = 1$ nC/m, where it turns out that $D_{\rm max} = 0.16L = 1.6$ mm.

1.3 Electric Scalar Potential

The electric scalar potential is a scalar quantity that can be used instead of the electric field intensity vector for the description of the electrostatic field. The potential, V, at a point P in an electric field is defined as the work W_e done by the field, that is, by the electric force, $\mathbf{F}_e$, in moving a test point charge, Q_p, from P to a reference point $\mathcal{R}$ (Fig.1.13),

$$W_e = \int_P^{\mathcal{R}} \mathbf{F}_e \cdot d\mathbf{l} = \int_P^{\mathcal{R}} F_e \, dl \cos\alpha \tag{1.26}$$

(the dot product of vectors **a** and **b** is a scalar given by $\mathbf{a} \cdot \mathbf{b} = |\mathbf{a}||\mathbf{b}|\cos\alpha$, α being the angle between **a** and **b**), divided by Q_p. Having in mind Eq.(1.7), this becomes

$$V = \frac{W_e}{Q_p} = \int_P^{\mathcal{R}} \frac{\mathbf{F}_e}{Q_p} \cdot d\mathbf{l} = \int_P^{\mathcal{R}} \mathbf{E} \cdot d\mathbf{l} \quad \text{(definition of the electric potential; unit: V)}\,, \tag{1.27}$$

namely, V equals the line integral of vector **E** from P to $\mathcal{R}$.[6] The unit for the potential is volt (abbreviated V). Note that Φ is also used to denote the electric potential.

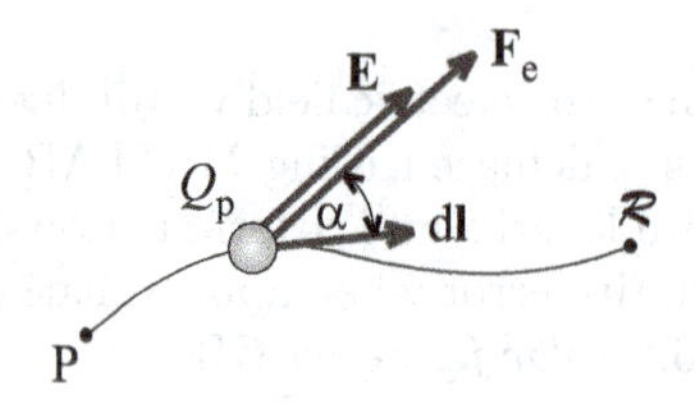

Figure 1.13 Displacement of a test charge in an electrostatic field.

By the principle of conservation of energy, the net work done by the electrostatic field in moving Q_p from a point A to some point B and then moving it back to A along a different path is zero (because after the round trip, the system is the same as at the beginning). This means that the line integral of the electric field intensity vector along an arbitrary closed path (contour) is zero,

$$\oint_C \mathbf{E} \cdot d\mathbf{l} = 0 \quad \text{(conservative nature of the electrostatic field)}\,, \tag{1.28}$$

which constitutes Maxwell's first equation for the electrostatic field.

From Fig.1.14 and Eqs.(1.27) and (1.8), the electric scalar potential at a distance R from a point charge Q in free space with respect to the reference point at infinity is

$$V = \int_P^{\mathcal{R}} \mathbf{E} \cdot d\mathbf{l} = \int_{x=R}^{\infty} E \, dx = \int_R^{\infty} \frac{Q}{4\pi\varepsilon_0 x^2} \, dx \quad \longrightarrow \quad V = \frac{Q}{4\pi\varepsilon_0 R}$$

$$\text{(electric potential due to a point charge)}\,. \tag{1.29}$$

[6] The line integral of a vector function (field) **a** along a line (curve) l, from a point A to a point B, is defined as $\int_l \mathbf{a} \cdot d\mathbf{l} = \int_A^B \mathbf{a} \cdot d\mathbf{l}$, where $d\mathbf{l}$ is the differential length vector tangential to the curve (as in Fig.1.13) oriented from A toward B. If the line is closed (for example, a circle or a square), we call it contour (and usually mark it C), and the corresponding line integral, $\oint_C \mathbf{a} \cdot d\mathbf{l}$, is termed the circulation of **a** along C. The reference direction of $d\mathbf{l}$ coincides with the orientation of the contour.

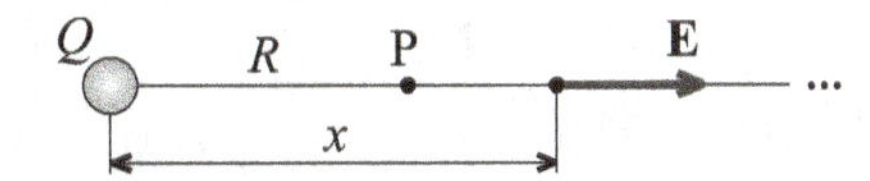

Figure 1.14 Evaluation of the electric potential due to a point charge in free space.

Starting with Eq.(1.29) and applying the superposition principle, the expression for the resultant electric potential, similar to the field expression in Eq.(1.9), is obtained for the system of N point charges, which reads

$$V = \frac{1}{4\pi\varepsilon_0} \sum_{i=1}^{N} \frac{Q_i}{R_i} , \tag{1.30}$$

as well as for the three characteristic continuous charge distributions in Fig.1.6. For instance, the potential expression corresponding to that in Eq.(1.14) for the field vector is given by

$$V = \frac{1}{4\pi\varepsilon_0} \int_l \frac{Q' \,\mathrm{d}l}{R} \quad \text{(potential due to line charge)} . \tag{1.31}$$

MATLAB EXERCISE 1.17 **Dot product of two vectors.** Write a function `dotProduct()` in MATLAB that computes the dot product of two vectors in the Cartesian coordinate system,

$$\mathbf{a} \cdot \mathbf{b} = (a_x \,\hat{\mathbf{x}} + a_y \,\hat{\mathbf{y}} + a_z \,\hat{\mathbf{z}}) \cdot (b_x \,\hat{\mathbf{x}} + b_y \,\hat{\mathbf{y}} + b_z \,\hat{\mathbf{z}}) = a_x b_x + a_y b_y + a_z b_z . \tag{1.32}$$

Test the function for vectors $\mathbf{a} = \hat{\mathbf{x}} + 2\,\hat{\mathbf{y}} + 3\,\hat{\mathbf{z}}$ and $\mathbf{b} = \hat{\mathbf{x}} + \hat{\mathbf{y}} + \hat{\mathbf{z}}$ ($\mathbf{a} \cdot \mathbf{b} = 6$). *(dotProduct.m on IR)*

MATLAB EXERCISE 1.18 **Numerical integration of a line integral.** Write a function `LineIntegral()` in MATLAB for numerical integration of the line integral of a vector **a** along a given path between points A and B in space. The integral is (approximately) calculated as

$$\int_{\mathrm{A}}^{\mathrm{B}} \mathbf{a} \cdot \mathrm{d}\mathbf{l} \approx \sum_{i=1}^{N} \mathbf{a}_i \cdot \mathrm{d}\mathbf{l}_i , \tag{1.33}$$

where $\mathbf{a}_i$ and $\mathrm{d}\mathbf{l}_i$ are given (approximate) values of **a** and d**l** at N points along the path (the input to the program are $\mathbf{a}_i$ and $\mathrm{d}\mathbf{l}_i$, $i = 1, 2, \ldots, N$). *(LineIntegral.m on IR)*

MATLAB EXERCISE 1.19 **Work in the field of a point charge.** A point charge $Q_1 =$ 10 nC is positioned at the center of a square contour $a = 10$ cm on a side, as shown in Fig.1.15. In MATLAB, find the work done by electric forces in carrying a charge $Q_2 = -1$ nC from the point M_1 to the point M_2 marked in the figure – using function `LineIntegral` from the previous MATLAB exercise. *(ME1_19.m on IR)* **H**

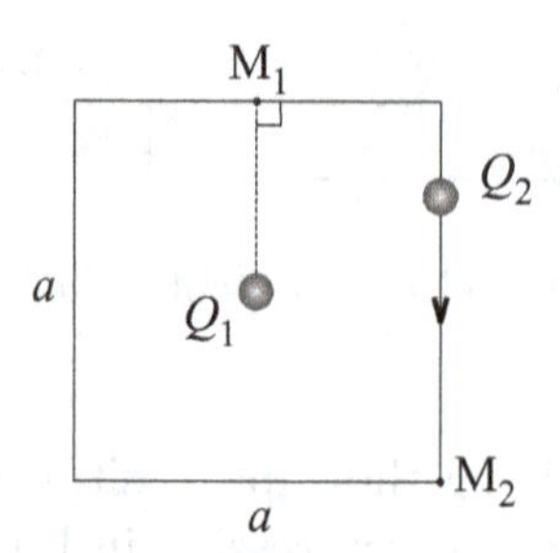

Figure 1.15 Movement of a charge Q_2 in the field of a charge Q_1 positioned at the center of a square contour; for MATLAB Exercise 1.19.

TUTORIAL:
By function `LineIntegral`, we numerically solve the line integral in the expression for the work in Eq.(1.26), that is, the line integral

$$W_e = \int_{M_1}^{M_2} \mathbf{F}_e \cdot \mathrm{d}\mathbf{l} \, . \tag{1.34}$$

Let us adopt the integration path consisting of a horizontal straight line segment from the point M_1 to the right in Fig.1.15 and a vertical one down to the point M_2. Let `da` denote the elementary segment (step) for integration along the lines, and let `x0,y0` be the coordinates of the charge Q_1 in Fig.1.15, while `x1,y1` and `x2,y2` represent arrays of x- and y-coordinates defining the integration points along the horizontal and vertical line segments, respectively.

```
da = 0.000001*a;
x0 = 0;
y0 = 0;
x1 = 0:da:a/2;
y1 = a/2*ones(1,length(x1));
y2 = a/2:-da:-a/2;
x2 = a/2*ones(1,length(y2));
x = [x1 x2];
y = [y1 y2];
```

In the lines of MATLAB code that follow, `r` is a matrix of distances from the stationary charge Q_1 to each of the integration points, that is, to the moving charge Q_2 (Fig.1.15), matrix `F` contains the x- and y-components of the electric force [Eq.(1.1)] on Q_2 (due to Q_1) at integration points [vector $\mathbf{F}_e$ in Eq.(1.34)], and `dl` represents components of the vector $\mathrm{d}\mathbf{l}$ in Eq.(1.34) at integration points, computed as a difference between the corresponding adjacent points. Finally, we invoke function `LineIntegral` to compute the total work, `W`.

```
r = sqrt((x-x0).^2+(y-y0).^2);
F = [(x-x0)./r.^3 ; (y-y0)./r.^3]*(Q1*Q2/(4*pi*EPS0));
N = length(x);
dx = x-[-da x(1:N-1)];
dy = y-[a/2-da y(1:N-1)];
dl = [dx ; dy];
```

```
W = LineIntegral(F,dl);
```

The result for `W` is −526.5 nJ.

MATLAB EXERCISE 1.20 **Numerical proof that E-field is conservative – movie.** By numerical integration in MATLAB, prove that the circulation (line integral) of the electric field intensity vector due to a point charge Q in free space along a square contour with the charge at its center is zero. *(ME1_20.m on IR)*

TUTORIAL:

This tutorial provides a movie created in MATLAB that visualizes the process of numerical integration (solution of the line integral) of the electric field vector (due to the point charge) along the square contour. The result confirms that Eq.(1.28) is satisfied – in a numerically exact fashion (the result may be not exactly zero but a very small number, due to numerical errors).

Let the square lie in the xy-plane, with its edges parallel to the coordinate axes, and let the charge Q be at the coordinate origin. In our MATLAB code, `N` is introduced as the number of elementary segments (steps) for integration along the square, whereas `x,y` contain x- and y-coordinates of integration points (centers of elementary segments). The array `r` is filled with distances between the charge and integration points, while `ux` and `uy` consist of x- and y-components of the corresponding unit vectors (along distances `r`), as required by the field expression in Eq.(1.8).

```
N = 300;
n = N/4;
delta = a/n;
temp = -a/2+delta/2:delta:a/2-delta/2;
x = [-a/2*ones(1,n),temp,a/2*ones(1,n),-temp];
y = [temp,a/2*ones(1,n),-temp,-a/2*ones(1,n)];
r = sqrt(x.^2+y.^2);
ux = x./r;
uy = y./r;
```

Similarly, `dl` represents components `dx,dy` of the vector $\mathbf{dl}$ in Eq.(1.28) at integration points,

```
dx = [zeros(1,n),delta*ones(1,n),zeros(1,n),-delta*ones(1,n)];
dy = [delta*ones(1,n),zeros(1,n),-delta*ones(1,n),zeros(1,n)];
dl = [dx;dy];
```

Next, we initialize the integration (`Int = 0`), and draw the structure (Fig.1.16): the point charge (magenta star), integration contour, and electric field vector at a point of the contour (by means of function `vecPlot2D`, from MATLAB Exercise 1.2),

```
Int = 0;
plot(x,y,'g');
hold on;
plot(0,0,'m*')
hold on;
for i = 1:N
```

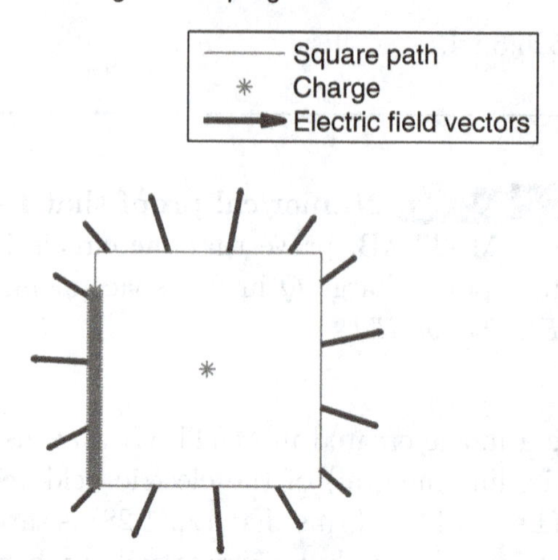

Figure 1.16 Snapshot of the MATLAB movie visualizing the process of numerical integration (line integral) of the electric field vector due to a point charge along a square contour with the charge at its center [Eq.(1.28)]; for MATLAB Exercise 1.20. *(color figure on CW)*

```
E(:,i) = Q/r(i)^2/(4*pi*EPS0)*[ux(i),uy(i)];
if mod(i,20)==0
vecPlot2D([x(i),y(i)],[x(i)+E(1,i),y(i)+E(2,i)],abs(1/E(1,1))/10,'b',0);
hold on;
end;
end;
legend('Square path' ,'Charge', 'Electric field vectors');
```

(note the use of MATLAB function `legend`). Finally, we calculate the line integral of **E** along C (using function `dotProduct`, from MATLAB Exercise 1.17). Within the same `for` loop, we plot a mark (red star) on the contour of the current position in the integration process. At the end of each cycle in the loop, MATLAB function `getframe` takes the current figure and merges it into a movie with previously taken figures. For reference, we compute as well the potential (due to the charge Q) at a vertex of the square contour, so that it is obvious that – in comparison to this value – the obtained result of integration along C is negligibly small – zero, i.e., that Eq.(1.28) holds true.

```
title('Integration in progress...');
for i = 1:N
Int = Int + dotProduct(E(:,i),dl(:,i));
plot(x(i),y(i),'r*');
hold on;
axis equal;
axis off;
if i==N
title(['Result of numerical line integration of E along square contour is 0!']);
```

```
end;
xlim([-1.5*a,1.5*a]);
ylim([-1.5*a,1.5*a]);
M(i)= getframe();
end;
hold off;
Vvertex = Q/(4*pi*EPS0*a/2*sqrt(2));
fprintf(['\n Potential at square vertex due to charge Q is %e V while circulation '...
'of E is %e V'],Vvertex,Int);
fprintf(['\n Being aware of numerical errors, we conclude that the obtained result of'...
' integration is negligibly small -- zero']);
```

Note that the three dots (...) are used, in MATLAB, to break a code line (here, the purpose is simply to represent the code line that would otherwise be exceedingly long for the book format). A snapshot (frame) of the MATLAB movie is shown in Fig.1.16.

MATLAB EXERCISE 1.21 **Circulation of E-vector along a contour of complex shape.** Figure 1.17 shows a contour consisting of two semicircular parts, of radii a and b ($a < b$), and two linear parts, each of length $b - a$, situated in free space. A point charge Q is placed at the contour center (point O). As in the previous MATLAB exercise, find numerically the line integral of the electric field intensity vector (due to Q) along the contour. *(ME1_21.m on IR)* **H**

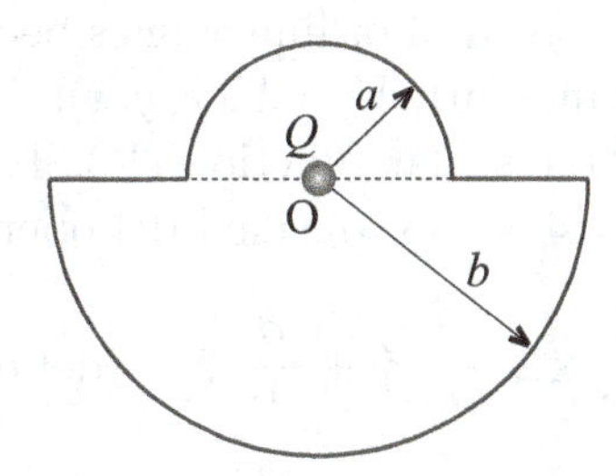

Figure 1.17 Contour with two semicircular and two linear parts, and a point charge (Q) at its center; for MATLAB Exercise 1.21.

MATLAB EXERCISE 1.22 **Electric potential due to multiple charges.** Write a program in MATLAB that calculates the electric scalar potential at an arbitrary point in space (air) due to N point charges placed arbitrarily in the Cartesian coordinate system. *(ME1_22.m on IR)*

HINT:
Use Eq.(1.30) and modify (simplify – given that the potential is a scalar) the program developed in MATLAB Exercise 1.4.

MATLAB EXERCISE 1.23 **Electric potential due to a charged ring.** From Eq.(1.31) and Fig.1.7(a), the electric potential at an arbitrary point along the axis of a charged ring normal to its plane is given by

$$V = \frac{1}{4\pi\varepsilon_0}\oint_C \frac{Q'\,dl}{R} = \frac{Q'}{4\pi\varepsilon_0 R}\oint_C dl = \frac{Q'a}{2\varepsilon_0\sqrt{z^2+a^2}} \quad (-\infty < z < \infty)\,. \tag{1.35}$$

Using this expression, repeat MATLAB Exercise 1.9 but for the potential due to the ring, including a comparison with the potential due to the equivalent (for $|z| \gg a$) point charge. *(ME1_23.m on IR)*

1.4 Differential Relationship Between the Field and Potential in Electrostatics, Gradient

Equation (1.27) represents an integral relationship between the electric field intensity vector and the potential in electrostatics, which enables us to determine V if we know $\mathbf{E}$. An equivalent, differential, relationship between these two quantities, that can be used for evaluating $\mathbf{E}$ from V, is given by

$$\mathbf{E} = E_x\,\hat{\mathbf{x}} + E_y\,\hat{\mathbf{y}} + E_z\,\hat{\mathbf{z}} = -\left(\frac{\partial V}{\partial x}\,\hat{\mathbf{x}} + \frac{\partial V}{\partial y}\,\hat{\mathbf{y}} + \frac{\partial V}{\partial z}\,\hat{\mathbf{z}}\right), \tag{1.36}$$

where we have partial derivatives instead of ordinary ones because the potential is a function of all three coordinates (multivariable function), $V = V(x, y, z)$. The expression in the parentheses in Eq.(1.36) is called the gradient of the scalar function (V). It is sometimes written as $\operatorname{grad} V$, but much more frequently we write it using the so-called del operator or nabla operator, defined as

$$\nabla = \frac{\partial}{\partial x}\,\hat{\mathbf{x}} + \frac{\partial}{\partial y}\,\hat{\mathbf{y}} + \frac{\partial}{\partial z}\,\hat{\mathbf{z}} \quad \text{(del operator)}\,. \tag{1.37}$$

So, we have

$$\mathbf{E} = -\operatorname{grad} V = -\nabla V \quad (\mathbf{E}\text{ from }V\text{ in electrostatics})\,, \tag{1.38}$$

where, in the Cartesian coordinate system (Fig.1.2),

$$\operatorname{grad} V = \nabla V = \frac{\partial V}{\partial x}\,\hat{\mathbf{x}} + \frac{\partial V}{\partial y}\,\hat{\mathbf{y}} + \frac{\partial V}{\partial z}\,\hat{\mathbf{z}} \quad \text{(gradient in Cartesian coordinates)}\,. \tag{1.39}$$

The other two best-known and most commonly used coordinate systems are the cylindrical and the spherical. An arbitrary point (M) in the cylindrical coordinate system is represented as (r, ϕ, z), as shown in Fig.1.18(a) (note that ρ, in place of r, is used to denote the cylindrical radial coordinate in some texts). The coordinate unit vectors, $\hat{\mathbf{r}}$, $\hat{\boldsymbol{\phi}}$, and $\hat{\mathbf{z}}$, are all mutually perpendicular, the vector $\mathbf{E}$ can be expressed as $\mathbf{E} = E_r\hat{\mathbf{r}} + E_\phi\hat{\boldsymbol{\phi}} + E_z\,\hat{\mathbf{z}}$, and the gradient of $V = V(r, \phi, z)$ is given by

$$\operatorname{grad} V = \nabla V = \frac{\partial V}{\partial r}\,\hat{\mathbf{r}} + \frac{1}{r}\frac{\partial V}{\partial \phi}\,\hat{\boldsymbol{\phi}} + \frac{\partial V}{\partial z}\,\hat{\mathbf{z}} \quad \text{(gradient in cylindrical coordinates)}\,. \tag{1.40}$$

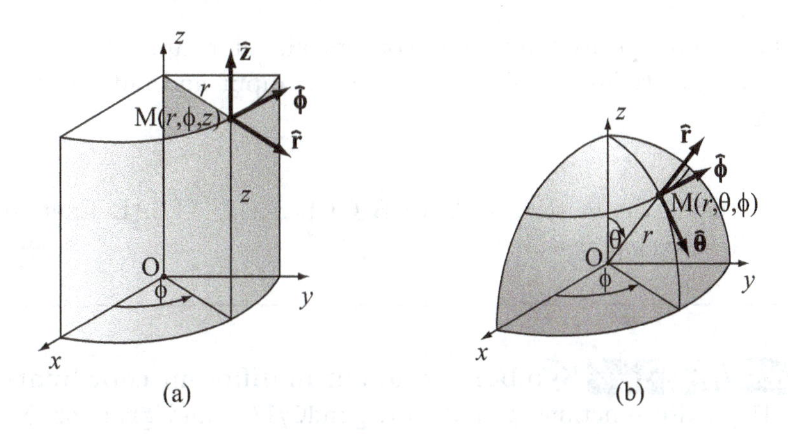

Figure 1.18 (a) Point $\mathrm{M}(r, \phi, z)$ and coordinate unit vectors in the cylindrical coordinate system. (b) Point $\mathrm{M}(r, \theta, \phi)$ and coordinate unit vectors in the spherical coordinate system.

In the spherical coordinate system, a point M is defined by (r, θ, ϕ), as illustrated in Fig.1.18(b) (R instead of r is used is some texts). The component representation of $\mathbf{E}$ reads $\mathbf{E} = E_r\hat{\mathbf{r}} + E_\theta\hat{\boldsymbol{\theta}} + E_\phi\hat{\boldsymbol{\phi}}$, where $\hat{\mathbf{r}}$, $\hat{\boldsymbol{\theta}}$, and $\hat{\boldsymbol{\phi}}$ are mutually perpendicular coordinate unit vectors, and the gradient of $V = V(r, \theta, \phi)$ is obtained as

$$\operatorname{grad} V = \nabla V = \frac{\partial V}{\partial r}\,\hat{\mathbf{r}} + \frac{1}{r}\frac{\partial V}{\partial \theta}\,\hat{\boldsymbol{\theta}} + \frac{1}{r\sin\theta}\frac{\partial V}{\partial \phi}\,\hat{\boldsymbol{\phi}} \quad \text{(in spherical coordinates)}\,. \tag{1.41}$$

For a given scalar field f (not necessarily electrostatic potential), the magnitude of ∇f at a point in space equals the maximum space rate of change in the function f per unit distance $[|\nabla f| = (\mathrm{d}f/\mathrm{d}l)_{\max}]$ and ∇f points in the direction of the maximum space rate of change in f. So, the gradient of f is a vector that provides us with both the direction in which f changes most rapidly and the magnitude of the maximum space rate of change.

MATLAB EXERCISE 1.24 **Cartesian to cylindrical and spherical coordinate conversions.** (a) Write functions `car2Cyl()` and `car2Sph()` in MATLAB that convert the coordinates of a point given in the Cartesian coordinate system to the coordinates in the associated cylindrical and spherical coordinate systems [Fig.1.18(a) and (b)], respectively. Assume that all lengths are expressed in meters, while the angles ϕ and θ should be expressed in radians. Check cases when some of the input data are zero. (b) Repeat (a) but for the conversions in the opposite direction, namely, write functions `cyl2Car()` and `sph2Car()` to convert cylindrical and spherical coordinates, respectively, to Cartesian coordinates of a point. *(car2Cyl.m, car2Sph.m, cyl2Car.m, and sph2Car.m on IR)*

MATLAB EXERCISE 1.25 **GUI for different coordinate conversions.** Create a graphical user interface (GUI) in MATLAB that combines the functions for different coordinate conversions written in the previous MATLAB exercise, as well as the corresponding functions to convert cylindrical to spherical coordinates of a point and vice versa [functions `cyl2Sph()` and `sph2Cyl()`].

The user should be able to specify the type of conversion, enter input data, and run the function, upon which the program displays results. All angles at input and output are in degrees. *[folder ME1_25(GUI) on IR]*

HINT:
Build an interface – using the existing MATLAB GUI. See MATLAB Exercise 2.1 (in the next chapter) – for the GUI development.

MATLAB EXERCISE 1.26 **Symbolic gradient in different coordinate systems.** Based on Eqs.(1.39)–(1.41), write functions `gradCar()`, `gradCyl()`, and `gradSph()` in MATLAB that take as input the symbolic expression for a function $f(x, y, z)$ and return its gradient (∇f) in the Cartesian, cylindrical, and spherical coordinate systems, respectively, as a symbolic vector function. *(gradCar.m, gradCyl.m, and gradSph.m on IR)*

HINT:
See MATLAB Exercise 1.33.

MATLAB EXERCISE 1.27 **Field from potential, in three coordinate systems.** With the use of functions from the previous MATLAB exercise and Eq.(1.38), write a MATLAB program that calculates the electric field intensity vector (**E**) from the symbolic expression for the potential (V) as input. Once the user selects one of the three coordinate systems, the program should offer the proper input for a symbolic expression describing the potential (in that coordinate system). *(ME1_27.m on IR)*

MATLAB EXERCISE 1.28 **Direction of the steepest ascent.** The terrain elevation in a region is given by a function $h(x, y) = 100x \ln y$ [m] (x, y in km), where x and y are coordinates in the horizontal plane and 1 km $\leq x, y \leq$ 10 km. Write a code in MATLAB [use function `gradCar` (from MATLAB Exercise 1.26)] to answer the following questions: What is the direction of the steepest ascent at (3 km, 3 km)? How steep, in degrees, is that ascent? *(ME1_28.m on IR)* **H**

1.5 Electric Dipole

An electric dipole is a very important, fundamental electrostatic system consisting of two point charges Q of opposite polarities separated by a distance d. Introducing a spherical coordinate system whose origin is at the dipole center as shown in Fig.1.19 and using Eq.(1.29) and superposition, the electrostatic scalar potential due to the dipole at a point P, at a large distance compared to d, comes out to be

$$V = \frac{p \cos\theta}{4\pi\varepsilon_0 r^2} \quad (\text{for } r \gg d) \,, \quad \mathbf{p} = Q\mathbf{d} \quad (p = Qd) \quad (\text{electric dipole potential}) \,, \qquad (1.42)$$

where $\mathbf{p}$ is the dipole moment, the unit for which is C · m. Applying the formula for the gradient in spherical coordinates, Eq.(1.41), to the expression for V in Eq.(1.42) yields the associated electric field expression for the dipole:

$$\mathbf{E} = -\nabla V = \frac{p}{4\pi\varepsilon_0 r^3}\left(2\cos\theta\,\hat{\mathbf{r}} + \sin\theta\,\hat{\boldsymbol{\theta}}\right) \quad \text{(electric dipole field)} . \tag{1.43}$$

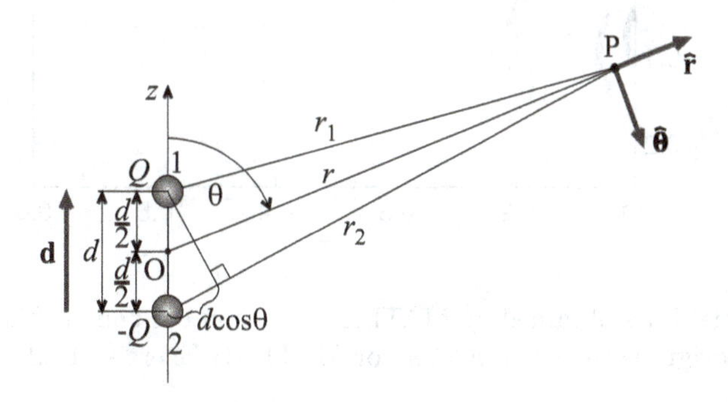

Figure 1.19 Electric dipole.

MATLAB EXERCISE 1.29 **Equipotential lines for a small electric dipole.** Write a MATLAB program that displays the distribution of the electric potential due to an electric dipole (Fig.1.19) with a moment $\mathbf{p} = p\,\hat{\mathbf{z}}$ located at the origin of a spherical coordinate system. As output, the program provides two plots in the plane defined by $y = 0$ in the associated Cartesian coordinate system: one representing the potential by means of MATLAB function `pcolor` (that uses color to visualize the third dimension) and the other showing equipotential lines (cuts in a specific plane of equipotential surfaces, namely, surfaces having the same potential, $V =$ const, at all points) with the help of MATLAB function `contour`. *(ME1_29.m on IR)*

HINT:
The electric dipole potential is given by Eq.(1.42), where, for $y = 0$, $\cos\theta = z/r$ and $r = \sqrt{x^2 + z^2}$. Figure 1.20 shows the resulting equipotential lines.

MATLAB EXERCISE 1.30 **Visualizing the electric dipole field.** For the electric dipole from the previous MATLAB exercise, write a program in MATLAB that displays its electric field intensity [magnitude of $\mathbf{E}$ in Eq.(1.43)] in the plane $y = 0$ using MATLAB function `pcolor`. *(ME1_30.m on IR)*

MATLAB EXERCISE 1.31 **Symbolic expression for the electric dipole field.** For the electric dipole in Fig.1.19, obtain the expression for the electric field vector in Eq.(1.43) from the expression for the electric potential in Eq.(1.42) – using MATLAB. In particular, apply the

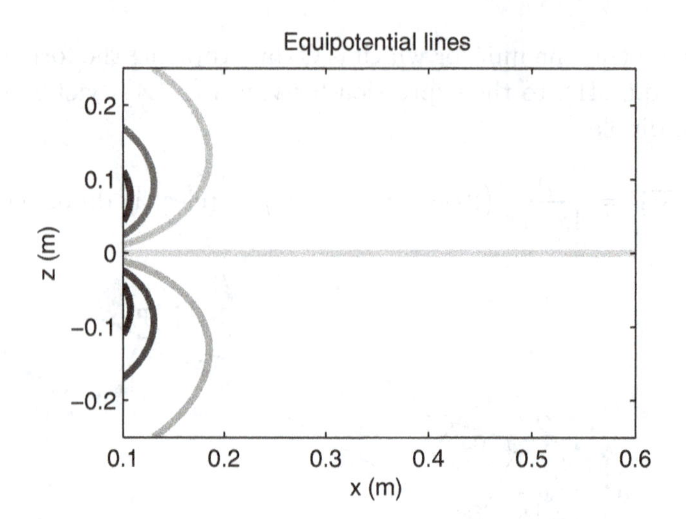

Figure 1.20 Equipotential lines obtained by MATLAB function `contour` in the plane $y = 0$ for an electric dipole at the coordinate origin with $\mathbf{p} = 1$ pCm $\hat{\mathbf{z}}$; for MATLAB Exercise 1.29. *(color figure on CW)*

symbolic function for the gradient in spherical coordinates, function **gradSph** (from MATLAB Exercise 1.26), to the expression for V. *(ME1_31.m on IR)* **H**

1.6 Gauss' Law in Integral Form

Gauss' law represents one of the fundamental laws of electromagnetism. It involves the flux (surface integral) of the vector **E** through a closed mathematical surface,[7] and can equivalently be formulated in a differential form, which is based on a differential operation called divergence on **E** (to be introduced in the next section). This important equation, in either form, provides an easy means of calculating the electric field due to highly symmetrical charge distributions, including problems with spherical, cylindrical, and planar symmetry, respectively.

Gauss' law (in integral form) states that the outward flux of the electric field intensity vector through any closed surface in free space is equal to the total charge enclosed by that surface, Q_S, divided by ε_0, namely,

$$\oint_S \mathbf{E} \cdot \mathrm{d}\mathbf{S} = \frac{Q_S}{\varepsilon_0} \quad \text{(Gauss' law)} . \tag{1.44}$$

The most general case of continuous charge distributions is the volume charge distribution, Fig.1.21(a), in terms of which Gauss' law can be written as

$$\oint_S \mathbf{E} \cdot \mathrm{d}\mathbf{S} = \frac{1}{\varepsilon_0} \int_v \rho \, \mathrm{d}v \quad \text{(Gauss' law for volume charge)} , \tag{1.45}$$

with v denoting the volume enclosed by the surface S and ρ the volume charge density. This

[7]The flux of a vector function **a** through an open or closed surface S is defined as $\int_S \mathbf{a} \cdot \mathrm{d}\mathbf{S}$, where $\mathrm{d}\mathbf{S}$ is the vector element of the surface perpendicular to it, and directed in accordance to the orientation of the surface.

particular form of Gauss' law is usually referred to as Maxwell's third equation for the electrostatic field in free space.

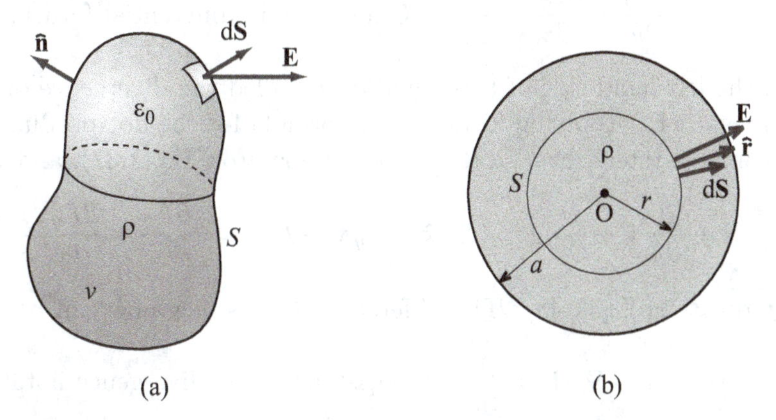

Figure 1.21 (a) Arbitrary closed surface containing a volume charge distribution in free space. (b) Application of Gauss' law to a problem with spherical symmetry.

To illustrate the application of Gauss' law, let us consider a sphere of radius a filled with a uniform volume charge of density ρ in free space, and determine the electric field distribution both inside and outside the sphere. Due to spherical symmetry, the electric field vector at an arbitrary point in space is of the form $\mathbf{E} = E(r)\,\hat{\mathbf{r}}$, as shown in Fig.1.21(b). The Gaussian surface S is a spherical surface of radius r centered at the origin, so that Gauss' law, Eq.(1.45), results in

$$\oint_S \mathbf{E}\cdot \mathrm{d}\mathbf{S} = \oint_S E(r)\,\mathrm{d}S = E(r)\oint_S \mathrm{d}S = E(r)\underbrace{4\pi r^2}_{S} = \frac{1}{\varepsilon_0}\left\{\begin{array}{ll} \rho 4\pi r^3/3 & \text{for } r < a \\ \rho 4\pi a^3/3 & \text{for } r \geq a \end{array}\right., \tag{1.46}$$

where, since $\rho = \text{const}$ (uniform charge distribution), the charge enclosed by S is computed as ρ times the corresponding volume, which is that of either the entire domain inside S, for $r < a$, or the charged sphere of radius a (there is no charge outside this sphere), for $r \geq a$. Hence, the field distribution everywhere is given by

$$E(r) = \left\{\begin{array}{ll} \rho r/(3\varepsilon_0) & \text{for } r < a \\ \rho a^3/(3\varepsilon_0 r^2) & \text{for } r \geq a \end{array}\right. . \tag{1.47}$$

MATLAB EXERCISE 1.32 **Sphere with a uniform volume charge.** In MATLAB, plot the dependence on the radial distance from the sphere center (r) of the charge density and electric field intensity, respectively, inside and outside a sphere of radius $a = 1$ m with a uniform volume charge density $\rho = 1\ \mathrm{nC/m^3}$ in free space, based on Eq.(1.47). *(ME1_32.m on IR)*

1.7 Differential Form of Gauss' Law, Divergence

Gauss' law in Eq.(1.45) represents an integral relationship between the electric field intensity vector, **E**, and the volume charge density, ρ. An equivalent, differential, relationship between **E** and ρ, that is, the differential form of Gauss' law, is given by

$$\frac{\partial E_x}{\partial x} + \frac{\partial E_y}{\partial y} + \frac{\partial E_z}{\partial z} = \frac{\rho}{\varepsilon_0} \quad \text{(Gauss' law in differential form)} . \tag{1.48}$$

The expression on the left-hand side of this equation is called the divergence of a vector function (**E**), and is written as div **E**. Applying formally the formula for the dot product of two vectors in the Cartesian coordinate system, Eq.(1.32), to the del operator, Eq.(1.37), and vector **E**, we get

$$\nabla \cdot \mathbf{E} = \left(\frac{\partial}{\partial x}\,\hat{\mathbf{x}} + \frac{\partial}{\partial y}\,\hat{\mathbf{y}} + \frac{\partial}{\partial z}\,\hat{\mathbf{z}} \right) \cdot (E_x\,\hat{\mathbf{x}} + E_y\,\hat{\mathbf{y}} + E_z\,\hat{\mathbf{z}}) = \frac{\partial E_x}{\partial x} + \frac{\partial E_y}{\partial y} + \frac{\partial E_z}{\partial z} , \tag{1.49}$$

and this is exactly div **E**, in Eq.(1.48). The differential Gauss' law now can be written in a short form

$$\operatorname{div} \mathbf{E} = \frac{\rho}{\varepsilon_0} \quad \text{or} \quad \nabla \cdot \mathbf{E} = \frac{\rho}{\varepsilon_0} \quad \text{(Gauss' law using divergence notation)} , \tag{1.50}$$

where, in the Cartesian coordinate system,

$$\operatorname{div} \mathbf{E} = \nabla \cdot \mathbf{E} = \frac{\partial E_x}{\partial x} + \frac{\partial E_y}{\partial y} + \frac{\partial E_z}{\partial z} \quad \text{(divergence in Cartesian coordinates)} . \tag{1.51}$$

The corresponding formula for the cylindrical coordinate system [Fig.1.18(a)] is

$$\nabla \cdot \mathbf{E} = \frac{1}{r}\frac{\partial}{\partial r}(rE_r) + \frac{1}{r}\frac{\partial E_\phi}{\partial \phi} + \frac{\partial E_z}{\partial z} \quad \text{(divergence in cylindrical coordinates)} , \tag{1.52}$$

and that for spherical coordinates [Fig.1.18(b)]

$$\nabla \cdot \mathbf{E} = \frac{1}{r^2}\frac{\partial}{\partial r}\left(r^2 E_r\right) + \frac{1}{r\sin\theta}\frac{\partial}{\partial \theta}(\sin\theta E_\theta) + \frac{1}{r\sin\theta}\frac{\partial E_\phi}{\partial \phi} \quad \text{(spherical coordinates)} . \tag{1.53}$$

MATLAB EXERCISE 1.33 **Symbolic divergence in Cartesian coordinates.** Using the symbolic programming option in MATLAB, write a function `divCar()` that takes as input symbolic expressions for `fx`, `fy`, and `fz` representing the x-, y-, and z-components, respectively, of a vector function in the Cartesian coordinate system and returns the expression for the divergence of the function. *(divCar.m on IR)*

TUTORIAL:
In MATLAB, differentiation is carried out invoking function `diff`. We define symbolic variables `x,y,z`, by `syms x,y,z`, so that the function `diff(fx,x)` yields – in symbolic form – the derivative with respect to `x` of a given function `fx` (f_x). Hence, based on Eq.(1.51), the symbolic divergence is obtained as follows:

```
function F = divCar(fx,fy,fz)
syms x y z
F = diff(fx,x) + diff(fy,y) + diff(fz,z);
```

MATLAB EXERCISE 1.34 **Charge from field, in three coordinate systems.** Employing the function written in the previous MATLAB exercise, as well as the corresponding functions `divCyl()` and `divSph()` that, respectively, give the symbolic divergence in cylindrical and spherical coordinates [based on Eqs.(1.52) and (1.53)], and Eq.(1.50), write a MATLAB program that finds the charge density (ρ) from the symbolic expression for the electric field intensity vector ($\mathbf{E}$) as input (the medium is air). Once the user chooses one of the three coordinate systems, the program should offer the proper input for symbolic expressions describing the components of $\mathbf{E}$ (in that coordinate system). *(ME1_34.m on IR)*

MATLAB EXERCISE 1.35 **Differential Gauss' law – spherical symmetry.** Using the program developed in the previous MATLAB exercise, show that the electric field with a radial spherical component only given by Eq.(1.47) is produced by a uniformly charged sphere of radius a and charge density ρ in free space. *(ME1_35.m on IR)* **H**

1.8 Method of Moments for Numerical Analysis of Charged Metallic Bodies

Consider a charged metallic body of an arbitrary shape situated in free space. Let the electric potential of the body with respect to the reference point at infinity be V_0. Our goal is to determine the charge distribution of the body. The potential at an arbitrary point on the body surface, S, can be expressed in terms of the surface charge density, ρ_s, over the entire S [as in Eq.(1.31) – for the line charge]. On the other hand, this potential equals V_0 (interior and surface of a conductor are equipotential, $V = \text{const}$), and hence

$$\frac{1}{4\pi\varepsilon_0}\int_S \frac{\rho_s \, \mathrm{d}S}{R} = V_0 \quad \text{(surface integral equation for charge distribution)}\,, \tag{1.54}$$

at an arbitrary point on S. This is an integral equation with the function ρ_s over S as unknown quantity, to be determined.

Equation (1.54) cannot be solved analytically – in a closed form, but only numerically, with the aid of a computer. The method of moments (MoM) is a common numerical technique used in solving integral equations such as Eq.(1.54) in electromagnetics and in other disciplines of science and engineering. MoM can be implemented in numerous ways, but the simplest MoM solution in this case implies subdivision of the surface S into small patches ΔS_i, $i = 1, 2, \ldots, N$, with a constant approximation of the unknown function ρ_s on each patch. That is, we assume that each patch is uniformly charged,

$$\rho_s \approx \rho_{si} \quad (\text{on } \Delta S_i)\,, \quad i = 1, 2, \ldots, N \quad \text{(piece-wise constant approximation)}\,. \tag{1.55}$$

With this approximation, we reduce Eq.(1.54) to its approximate form:

$$\sum_{i=1}^{N} \rho_{si} \int_{\Delta S_i} \frac{\mathrm{d}S}{4\pi\varepsilon_0 R} = V_0\,, \tag{1.56}$$

in which the unknown quantities are N charge-distribution coefficients, ρ_{s1}, ρ_{s2}, ..., ρ_{sN}. Shown in Fig.1.22 is an example of the application of this method to a metallic cube, where square patches are used [for the particular subdivision shown in the figure, $N = 6 \times (5 \times 5) = 150$, which is a rather coarse model].

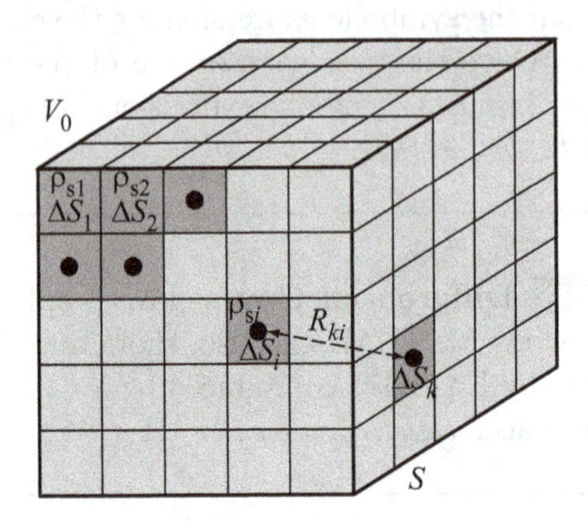

Figure 1.22 Method of moments (MoM) for analysis of charged metallic bodies: approximation of the surface charge distribution on a cube by means of N small square patches with constant charge densities.

By stipulating that Eq.(1.56) be satisfied at centers of every small patch, individually, we obtain[8]

$$\begin{aligned} A_{11}\rho_{s1} + A_{12}\rho_{s2} + \cdots + A_{1N}\rho_{sN} &= V_0 \quad (\text{at the center of } \Delta S_1) , \\ A_{21}\rho_{s1} + A_{22}\rho_{s2} + \cdots + A_{2N}\rho_{sN} &= V_0 \quad (\text{at the center of } \Delta S_2) , \\ &\vdots \\ A_{N1}\rho_{s1} + A_{N2}\rho_{s2} + \cdots + A_{NN}\rho_{sN} &= V_0 \quad (\text{at the center of } \Delta S_N) . \end{aligned} \tag{1.57}$$

This is a system of N linear algebraic equations in N unknowns, ρ_{s1}, ρ_{s2}, ..., ρ_{sN}. In matrix form,

$$[A][\rho_s] = [B] \quad (\text{MoM matrix equation}) , \tag{1.58}$$

where $[\rho_s]$ is a column matrix whose elements are the unknown charge-distribution coefficients, whereas elements of the column matrix $[B]$ are known and all equal V_0. Elements of matrix $[A]$, which is a square matrix, are given by

$$A_{ki} = \int_{\Delta S_i} \frac{dS}{4\pi\varepsilon_0 R} \quad (\text{at the center of } \Delta S_k) , \quad k, i = 1, 2, \ldots, N \quad (\text{point-matching}) , \tag{1.59}$$

and they can be computed irrespective of the particular charge distribution. Physically, A_{ki} is the potential at the center of patch ΔS_k due to patch ΔS_i that is uniformly charged with unit (1 C/m^2) surface charge density. In the case of commonly used square or triangular flat patches, this potential can be evaluated analytically (exactly), while it is evaluated numerically (approximately) if some

[8]The variant of the method of moments in which the left-hand side and the right-hand side of an integral equation [in our case, Eq.(1.54)] are "matched" to be equal at specific points of the definition domain of the equation (surface S in our case) is called the point-matching method. The idea of point-matching is similar to the concept of taking moments in mechanics, and hence the generic name method of moments.

other surface elements are used (for example, curvilinear quadrilateral or triangular patches). Once matrix $[A]$ is filled, i.e., all its elements computed, we can use matrix inversion,

$$[\rho_\mathrm{s}] = [A]^{-1}[B] \,, \tag{1.60}$$

or any other standard method for solving systems of linear algebraic equations (e.g., Gaussian elimination method), to obtain the numerical results for the charge-distribution coefficients, which constitute an approximate surface charge distribution of the body, and a numerical solution to the integral equation, Eq.(1.54). The larger number of subdivisions, N, the more accurate (but more computationally demanding in terms of computer resources) solution.

The crudest approximation in computing the elements of matrix $[A]$ is given by

$$A_{ki} = \begin{cases} \Delta S_i/(4\pi\varepsilon_0 R_{ki}) & \text{for } k \neq i \\ \sqrt{\Delta S_i}/(2\sqrt{\pi}\varepsilon_0) & \text{for } k = i \end{cases} . \tag{1.61}$$

Here, all nondiagonal elements of $[A]$ ($k \neq i$) are evaluated by approximating the charged patch ΔS_i by an equivalent point charge, $\Delta Q_i = 1 \ (\mathrm{C/m^2}) \times \Delta S_i$, placed at the patch center, and using the expression for the electric potential due to a point charge in free space, Eq.(1.29), with R_{ki} being the distance between centers of patches ΔS_k and ΔS_i (Fig.1.22). In evaluating diagonal elements (self terms) of $[A]$ ($k = i$), the potential due to a (square or triangular) patch ΔS_i at the center of that same patch is evaluated by approximating the patch by the equivalent circular patch of the same surface area and radius $\sqrt{\Delta S_i/\pi}$, and employing the expression for the potential at the center of a circular uniformly charged patch[9] with charge density $\rho_\mathrm{s} = 1 \ \mathrm{C/m^2}$.

Starting from the result for $[\rho_\mathrm{s}]$, we can now obtain any other quantity of interest (potential and field at any point in space, etc.). For instance, the total charge of the body can be found using the approximate version of the corresponding integral expression for the surface charge distribution in Fig.1.6(b):

$$Q = \int_S \rho_\mathrm{s} \, \mathrm{d}S \quad \longrightarrow \quad Q \approx \sum_{i=1}^{N} \rho_{\mathrm{s}i} \Delta S_i \,. \tag{1.62}$$

MATLAB EXERCISE 1.36 **Main MoM matrix, for arbitrary charged body.** Write a function `matrixA()` that computes the matrix $[A]$ in Eq.(1.58) for the method-of-moments analysis of an arbitrary charged metallic body – based on Eq.(1.61). The function is independent of the geometry of the body; rather, it takes as input the previously calculated arrays of surfaces of small patches ΔS_i, $i = 1, 2, \ldots, N$, and coordinates of their centers (see Fig.1.22, for example). However, the program distinguishes between the following three cases: (i) charged strips (centers along a straight line), (ii) charged plates (centers in one plane), and (iii) charged 3-D objects (centers in 3-D space). This function can then be used in MoM codes for different geometries, as long as the approximation in Eq.(1.61) is used. *(matrixA.m on IR)*

TUTORIAL:

In the case that input vectors (arrays) are not of the same size (N), the function writes an error message and returns zero as output. Otherwise, we check the number of input arguments (to

[9] Using the same subdivision of a charged disk (circular patch) into elemental rings as in Fig.1.8(a) and integrating (analytically) the electric potential due to these rings, which, in turn, is given by Eq.(1.35) for $z = 0$, we obtain the following expression for the potential at the disk center: $V = \rho_\mathrm{s} a/(2\varepsilon_0)$.

distinguish between the strip, plate, and 3-D cases); this is done by the local variable **nargin** in MATLAB – it shows the number of input arguments of the function within which it is called. We then calculate the distances between points (centers of patches) by the appropriate expression for the given case. Finally, elements of the $N \times N$ matrix $[A]$ are evaluated by means of Eq.(1.61), irrespective of the case.

```
function A = matrixA(EPS,dS,x,y,z)
N = length(x);
if (length(dS)== N)
for i = 1:N
for j = 1:N
if nargin == 3
r = sqrt((x(j)-x(i))^2);
end;
if nargin == 4
r = sqrt((x(j)-x(i))^2+(y(j)-y(i))^2);
end;
if nargin == 5
r = sqrt((x(j)-x(i))^2+(y(j)-y(i))^2+(z(j)-z(i))^2);
end;
if (i==j)
A(i,j) = sqrt(dS(j))/(2*sqrt(pi)*EPS);
else
A(i,j) = dS(j)/(4*pi*EPS*r);
end;
end;
end;
else
A = 0;
disp ('Incorrect input data in function matrixA');
end;
```

MATLAB EXERCISE 1.37 Preprocessing of geometrical data for the MoM matrix. Write a function `localCoordinates()` in MATLAB that – for the input `n,m,a,b` – subdivides a rectangular surface of sides a and b into $n \times m$ patches and returns arrays of local coordinates of centers of patches, p and q, as well as of surface areas of patches, ΔS. The patches should be numerated from $(1, 1)$ in one corner of the rectangle to (n, m) in the diagonally opposite corner. This function can then be used to prepare the geometrical data (so-called preprocessing) for computing the elements of the associated MoM matrix $[A]$. *(localCoordinates.m on IR)*

MATLAB EXERCISE 1.38 Total charge, based on the MoM analysis. Write a function `totalCharge()` in MATLAB that takes as input an array of surfaces of small patches ΔS_i and an array of associated surface charge densities ρ_{si}, $i = 1, 2, \ldots, N$, obtained by the MoM analysis, and finds, in postprocessing, the total charge Q of the body using Eq.(1.62). More precisely, Q is obtained by matrix multiplication of a row matrix of surfaces of patches, `S`, and a column matrix of charge-distribution coefficients, `rhos`. *(totalCharge.m on IR)*

MATLAB EXERCISE 1.39 **MoM-based MATLAB program for a charged square plate.** Using the method of moments and functions developed in previous three MATLAB exercises, write a computer program in MATLAB to determine the charge distribution on a very thin charged square plate of edge length a at a potential V_0, in free space. Subdivide the plate into N square patches, and assume that $a = 1$ m and $V_0 = 1$ V. (a) Tabulate and plot the results for the surface charge density (ρ_s) of the patches, taking $N = 100$ (ten partitions in each dimension). (b) Compute the total charge of the plate, taking (i) $N = 16$, (ii) $N = 36$, (iii) $N = 64$, and (iv) $N = 100$, respectively. *(ME1_39.m on IR)*

TUTORIAL:

In the first step, we fix the values of input parameters `N,a,V0` within the code, instead of the option with the user entering them during the run of the simulation (of course, these numbers can be changed in the code, for different simulations), and evaluate the mesh increment, `dl`,

```
N = 100;
a = 1;
V0 = 1;
EPS0 = 8.8542*10^(-12);
n = sqrt(N);
dl = a/n;
```

where `n` is the number of subdivisions of the plate in each dimension ($N = n^2$). Next, we call function `localCoordinates` (from MATLAB Exercise 1.37) to calculate coordinates of centers and surface areas of patches on the plate, `[x,y,S]`, which are then used as input to function `matrixA` (developed in MATLAB Exercise 1.36) to fill the MoM matrix `A` ($[A]$). After the column matrix `B` ($[B]$) in Eq.(1.58) is filled as well (note that `X'` means the transpose of `X` in MATLAB), the column matrix of charge-distribution coefficients, `rhos` ($[\rho_s]$), is found by matrix inversion, Eq.(1.60), which is performed in MATLAB as `A\B` [alternatively, it can be done as `rhos = inv(A)*B`].

```
[x,y,S] = localCoordinates(n,n,a,a);
A = matrixA(EPS0,S,x,y);
B = V0*ones(1,N)';
rhos = A\B;
```

In order to realistically visualize the obtained charge distribution over the plate, we repack the `1` $\times$ `N` matrix `rhos` into an `n` $\times$ `n` matrix `rhos2D` (for the numeration of patches, see MATLAB Exercise 1.37),

```
for k = 1:n
rhos2D(k,:) = rhos((k-1)*n+1:k*n);
end;
```

The charge distribution is displayed by means of MATLAB function **surf**, which gives a 3-D graph with both the vertical dimension and color representing the value of the function (charge density), as shown in Fig.1.23.

```
figure(1);
[x2D,y2D] = meshgrid(0:1/(n-1):1);
surf(x2D,y2D,rhos2D*10^12);
colormap('cool');
shading interp;
title('Surface charge distribution of the plate');
xlabel('x [m]');
ylabel('y [m]');
zlabel('\rhos [^{pC}/{m^2}]');
zlim([0,150]);
```

It can alternatively be displayed using MATLAB function **imagesc**, which is a 2-D view with the value of the function represented by color only,

```
figure(2);
imagesc(rhos2D);
colormap('cool');
axis equal;
shading interp;
title('Surface charge distribution of the plate');
xlabel('x');
ylabel('y');
axis off;
```

Finally, the total charge of the plate, `Qtot`, is found by function `totalCharge` (from the previous MATLAB exercise),

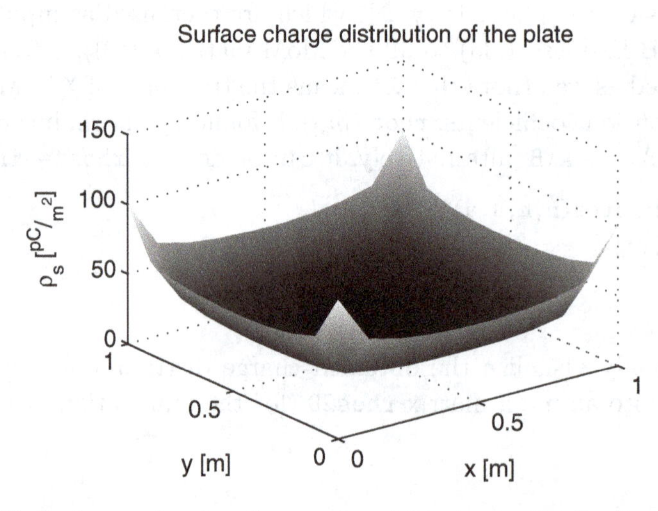

Figure 1.23 MATLAB display – using function **surf** – of the charge distribution of a square plate obtained – in MATLAB – by the method of moments; for MATLAB Exercise 1.39. *(color figure on CW)*

```
Qtot = totalCharge(S,rhos);
fprintf(['\n Total charge of the plate is %2.3d pC/m^2'],Qtot*10^12);
```

MATLAB EXERCISE 1.40 **MoM program for a rectangular charged plate.** Repeat the previous MATLAB Exercise but for a rectangular charged plate of side lengths a and b ($a \neq b$), with the respective numbers of subdivisions along sides amounting to n and m. *(ME1_40.m on IR)*

MATLAB EXERCISE 1.41 **MoM-based MATLAB program for a charged cube.** Write a computer program in MATLAB for the method-of-moments analysis of a charged metallic cube, Fig.1.22, with edge length $a = 1$ m, and compute the total charge of the cube for $V_0 = 1$ V and ten, or as many as possible (given available computational resources), subdivisions per cube edge ($N = 600$ if ten subdivisions per edge are adopted). *(ME1_41.m on IR)*

HINT:
See the tutorial to MATLAB Exercise 1.39. The following lines of MATLAB code perform meshing of the surface of the cube. We mesh each side of the cube in the same way, with `Nside` standing for the number of patches on a side (`Nside` = n^2). The local coordinates of centers of patches are calculated only once, and then properly used to compute the global `x,y,z` coordinates for all sides (on each side, one of the global coordinates is constant), which are merged (for all sides) together.

```
[p,q,dS] = localCoordinates(n,n,a,a);
temp0 = zeros(1,Nside);
temp1 = a*ones(1,Nside);
x = [p,p,p,p,temp0,temp1];
y = [q,q,temp0,temp1,p,p];
z = [temp0,temp1,q,q,q,q];
S = [dS,dS,dS,dS,dS,dS];
```

Once the cube surface is meshed, most lines of the program written in MATLAB Exercise 1.39 can be directly applied. Note that, upon the MoM matrix equation is solved, parts of the column matrix `rhos` corresponding to individual cube sides should be extracted from it and repacked into a square matrix in order to represent them graphically (as in Fig.1.23). Note also that the obtained charge distributions on all the sides should come out to be the same, due to the symmetry of the problem.

MATLAB EXERCISE 1.42 **MoM program for a charged parallelepiped.** Repeat the previous MATLAB exercise but for a charged parallelepiped (rectangular box). Parallelepiped edge lengths are a, b, and c, and the respective numbers of subdivisions along edges are n, m, and k. *(ME1_42.m on IR)*

MATLAB EXERCISE 1.43 **Field computation in postprocessing of the MoM solution.** Write a function `fieldE()` in MATLAB that evaluates, in postprocessing, the electric field intensity vector at an arbitrary point in space due to a charged body (e.g., the cube in Fig.1.22), whose charge distribution, approximately described by Eq.(1.55), is determined by the MoM analysis. The input to the program contains `EPS0,x0,y0,z0,rhos,S,x,y,z`, where `EPS0` is the permittivity of air (or another dielectric surrounding the body) and `(x0,y0,z0)` are Cartesian coordinates of the field point, while `rhos,S,x,y,z` are arrays containing the surface charge densities and surface areas of MoM patches, and coordinates of their centers. *(fieldE.m on IR)*

HINT:

We approximate the field integral for a surface charge distribution [Fig.1.6(b)], analogous to the one in Eq.(1.14), in the same way the potential integral in Eq.(1.54) is reduced to its approximate form in Eq.(1.56), based on the charge density approximation in Eq.(1.55), as follows:

$$\mathbf{E} = \frac{1}{4\pi\varepsilon_0}\int_S \frac{\rho_s\,\mathrm{d}S}{R^2}\,\hat{\mathbf{R}} \approx \frac{1}{4\pi\varepsilon_0}\sum_{i=1}^{N}\rho_{si}\int_{\Delta S_i}\frac{\mathrm{d}S}{R^2}\,\hat{\mathbf{R}}\,, \tag{1.63}$$

and then approximate the integrand $\hat{\mathbf{R}}/R^2$ by its value at the center of the patch ΔS_i ($i = 1, 2, \ldots, N$), which results in

$$\mathbf{E} \approx \frac{1}{4\pi\varepsilon_0}\sum_{i=1}^{N}\frac{\rho_{si}}{R_i^2}\,\hat{\mathbf{R}}_i\int_{\Delta S_i}\mathrm{d}S = \frac{1}{4\pi\varepsilon_0}\sum_{i=1}^{N}\frac{\rho_{si}\Delta S_i}{R_i^2}\,\hat{\mathbf{R}}_i\,,\quad \hat{\mathbf{R}}_i = \frac{\mathbf{R}_i}{R_i}\,. \tag{1.64}$$

MATLAB code based on this equation is similar to that written in MATLAB Exercise 1.36.

MATLAB EXERCISE 1.44 **Field computation in plate and cube problems.** Using function `fieldE` (written in the previous MATLAB exercise), as well as programs from MATLAB Exercises 1.39 and 1.41, compute (i) the electric field along the axis of the plate from MATLAB Exercise 1.39 perpendicular to its plane at points that are $a/2$, $2a$, and $100a$, respectively, distant from the plate surface (for $N = 100$) and (ii) the field inside the cube from MATLAB Exercise 1.41, at a quarter of its space diagonal (body diagonal) and at its center. *(ME1_44.m on IR)*

---

2 ELECTROSTATIC FIELD IN DIELECTRICS

Introduction:

Dielectrics or insulators are nonconducting materials, having very little free charges inside them (theoretically, perfect dielectrics have no free charges). However, another type of charges, called bound or polarization charges, exist in a polarized dielectric, as atoms and molecules in the dielectric behave like microscopic electric dipoles (see Fig.1.19). In electrostatic systems containing both conductors and dielectrics, the equivalent electric-field sources are both free and bound charges, considered to reside in free space. By introducing the concept of dielectric permittivity, we are left, in turn, to deal with free charges in the system only, while the contribution of bound charges to the field is properly added through the permittivity. In continuation, we perform the electrostatic analysis of capacitors and transmission lines, composed of both conductors and dielectrics, to determine their capacitance and breakdown characteristics, as a culmination of our study of the theory and applications of the electrostatic field.

2.1 Characterization of Dielectric Materials

In the most general electrostatic system, in place of Eqs.(1.44), (1.45), and (1.50) we have the corresponding forms of the generalized Gauss' law:

$$\oint_S \mathbf{D} \cdot d\mathbf{S} = Q_S \,, \quad \oint_S \mathbf{D} \cdot d\mathbf{S} = \int_v \rho \, dv \,, \quad \nabla \cdot \mathbf{D} = \rho \quad \text{(Generalized Gauss' law)}\,, \tag{2.1}$$

where $\mathbf{D}$ denotes the electric flux density vector (also known as the electric displacement vector or electric induction vector), the unit of which is $\mathrm{C/m^2}$, Q_S is the total free charge enclosed by an arbitrary closed surface S, and ρ is the free charge density. For linear dielectrics,

$$\mathbf{D} = \varepsilon \mathbf{E} \,, \quad \varepsilon = \varepsilon_r \varepsilon_0 \quad \text{(permittivity of a linear dielectric)}\,, \tag{2.2}$$

with ε being the permittivity and ε_r the relative permittivity of the medium ($\varepsilon_r \geq 1$). The unit for ε is farad per meter (F/m), while ε_r is dimensionless. Table 2.1 shows values of the relative permittivity of a number of selected materials, for electrostatic or low-frequency time-varying (time-harmonic) applied electric fields,[1] at room temperature (20°C). For nonlinear dielectrics, the relation between $\mathbf{D}$ and $\mathbf{E}$, $\mathbf{D} = \mathbf{D}(\mathbf{E})$, is nonlinear.

[1] At higher frequencies, when viewed over very wide frequency ranges, the permittivity generally (for most materials) is not a constant, but depends on the operating frequency of electromagnetic waves propagating through the material.

Table 2.1 Relative permittivity of selected materials*

Material	ε_r	Material	ε_r
Vacuum	1	Quartz	5
Freon	1	Diamond	5–6
Air	1.0005	Wet soil	5–15
Styrofoam	1.03	Mica (ruby)	5.4
Polyurethane foam	1.1	Steatite	5.8
Paper	1.3–3	Sodium chloride (NaCl)	5.9
Wood	2–5	Porcelain	6
Dry soil	2–6	Neoprene	6.6
Paraffin	2.1	Silicon nitride (Si_3N_4)	7.2
Teflon	2.1	Marble	8
Vaseline	2.16	Alumina (Al_2O_3)	8.8
Polyethylene	2.25	Animal and human muscle	10
Oil	2.3	Silicon (Si)	11.9
Rubber	2.4–3	Gallium arsenide	13
Polystyrene	2.56	Germanium	16
PVC	2.7	Ammonia (liquid)	22
Amber	2.7	Alcohol (ethyl)	25
Plexiglass	3.4	Tantalum pentoxide	25
Nylon	3.6–4.5	Glycerin	50
Fused silica (SiO_2)	3.8	Ice	75
Sulfur	4	Water	81
Glass	4–10	Rutile (TiO_2)	89–173
Bakelite	4.74	Barium titanate ($BaTiO_3$)	1,200

* For static or low-frequency applied electric fields, at room temperature.

Another concept in characterization of materials is homogeneity. A material is said to be homogeneous when its properties do not change from point to point in the region being considered. In a linear homogeneous dielectric, ε is a constant independent of spatial coordinates. Otherwise, the material is inhomogeneous [e.g., $\varepsilon = \varepsilon(x, y, z)$ in the region].

Next, we introduce the concept of isotropy in classifying dielectric materials. In a linear isotropic dielectric, ε is a scalar quantity, and hence **D** and **E** are always collinear and in the same direction, regardless of the orientation of **E**. In an anisotropic medium, however, individual components of **D** depend differently on different components of **E**, so that Eq.(2.2) becomes a matrix equation,

$$\begin{bmatrix} D_x \\ D_y \\ D_z \end{bmatrix} = \begin{bmatrix} \varepsilon_{xx} & \varepsilon_{xy} & \varepsilon_{xz} \\ \varepsilon_{yx} & \varepsilon_{yy} & \varepsilon_{yz} \\ \varepsilon_{zx} & \varepsilon_{zy} & \varepsilon_{zz} \end{bmatrix} \begin{bmatrix} E_x \\ E_y \\ E_z \end{bmatrix}, \tag{2.3}$$

and $[\varepsilon]$ is the permittivity tensor.

Finally, we note that the electric field intensity, E, in a dielectric cannot be increased indefinitely: if a certain value is exceeded, the dielectric becomes conducting; it temporarily or permanently loses its insulating property, and is said to break down. The breaking field value, i.e., the maximum electric field intensity that an individual dielectric material can withstand without breakdown is termed the dielectric strength of the material and is denoted by E_{cr} (critical field intensity). For air, $E_{cr0} = 3$ MV/m. The values of E_{cr} for some selected dielectric materials are presented in Table 2.2.

Table 2.2 Dielectric strength of selected materials*

Material	E_{cr} (MV/m)	Material	E_{cr} (MV/m)
Air (atmospheric pressure)	3	Bakelite	25
Barium titanate ($BaTiO_3$)	7.5	Glass (plate)	30
Freon	~ 8	Paraffin	~ 30
Germanium	~ 10	Silicon (Si)	~ 30
Wood (douglas fir)	~ 10	Alumina	~ 35
Porcelain	11	Gallium arsenide	~ 40
Oil (mineral)	15	Polyethylene	47
Paper (impregnated)	15	Mica	200
Polystyrene	20	Fused quartz (SiO_2)	~ 1000
Teflon	20	Silicon nitride (Si_3N_4)	~ 1000
Rubber (hard)	25	Vacuum	∞

* At room temperature.

MATLAB EXERCISE 2.1 **GUI – pop-up menu for the permittivity table of materials.** Create a graphical user interface (GUI) in MATLAB to show values of the relative permittivity of selected materials given in Table 2.1 as a pop-up menu. *[folder ME2_1(GUI) on IR]*[2]

TUTORIAL:

To start, type **guide** in the Command Window, in MATLAB, to get into the GUI development environment, select Blank GUI, and click OK. To build the GUI, choose one Pop-up Menu and two Static Text components. In the Pop-up Menu, list (type) in alphabetical order all (or some) materials given in Table 2.1 [left click, choose Property Inspector, and type names of all (or some) materials in the String section, leaving the first row blank]. In addition, leave one Static Text component blank, and use the other as a title, typing "Relative permittivity of selected materials." Upon the GUI is saved, for instance, as `RelPermittivity` (`GUIname` in general), MATLAB will create two files, `RelPermittivity.fig` and `RelPermittivity.m` (`GUIname.fig` and `GUIname.m`), containing the layout (fig file), shown in Fig.2.1, and all functions that build and control the GUI (m file).

Next, we fill in the values of relative permittivities from Table 2.1. In the m file (`RelPermittivity.m`), in its section `function RelPermittivity_OpeningFcn (hObject, eventdata, handles, varargin)`, the ε_r values should be entered, right after the `varargin` line, as follows:

```
handles.Air = '1.005';
handles.Alcohol = '25';
handles.Alumina = '8.8';
handles.Amber = '2.7';
handles.Ammonia = '22';
handles.Animal = '10';
```

[2] *IR* = Instructor Resources (for the book).

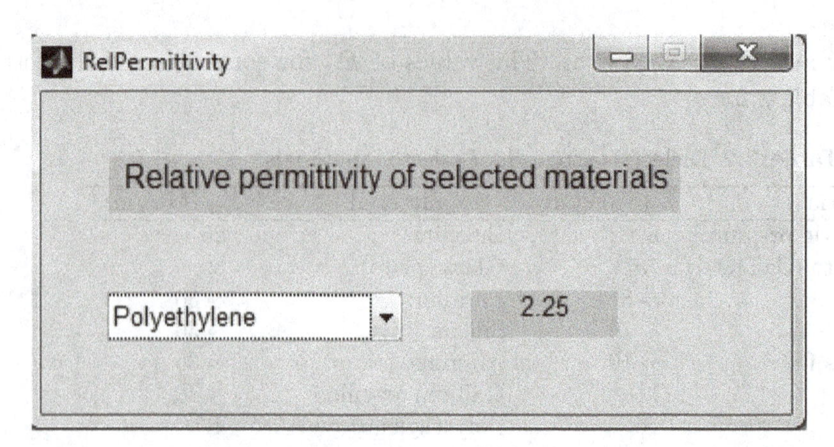

Figure 2.1 Layout of a graphical user interface (GUI) created in MATLAB to show ε_r values of different materials from Table 2.1 as a pop-up menu; for MATLAB Exercise 2.1. *(color figure on CW)*[3]

```
handles.Bakelite = '4.74';
handles.Barium = '1,200';
handles.Diamond = '5-6';
handles.DrySoil = '2-6';
handles.Freon = '1';
handles.FusedSilica = '3.8';
...
```

Then, to enable the proper pop-up menu functionality, the `case-switch` command is used, and the following lines should be included in the section `function popupmenu1_Callback(hObject, eventdata, handles)` of the m file:

```
switch get(handles.popupmenu1,'Value')
case 1
handles.currentdata = handles.blank;
set(handles.text2,'String',handles.currentdata);
case 2
handles.currentdata = handles.Air;
set(handles.text2,'String',handles.currentdata);
case 3
handles.currentdata = handles.Alcohol;
set(handles.text2,'String',handles.currentdata);
case 4
handles.currentdata = handles.Alumina;
set(handles.text2,'String',handles.currentdata);
...
end
```

Finally, after the GUI code has been completed, we perform some interventions in it in order to set-up the GUI figure to appear properly on any computer screen, and not only on the one

[3] *CW* = Companion Website (of the book).

used in creating the code, so that the GUI is transferable to any newly used computer and screen. Basically, we change the code to find the size of the newly used screen to be able to center the GUI figure in the screen and we keep the size of the GUI figure constant, independently of the screen used, by setting it in inches rather than in pixels (the size in pixels would vary across different screens). So, if it is not already open, reenter the GUI development environment using command **guide RelPermittivity** (**guide GUIname** in general) in the Command Window, go to Tools, and GUI Options, and change the Resize Behavior option to Proportional. This will allow resizing of the GUI frame with the objects inside the GUI figure resizing proportionally. To find the size of the GUI figure in inches, type the following lines of code in the Command Window:

```
h = figure(RelPermittivity);
set(h,'units','inches');
get(h,'position')
```

The first line assigns a handle to the GUI figure [handle is an identifier (number) that MATLAB assigns to the object, which is then used with **set** and **get** functions]. Of course, in general, **RelPermittivity** should be replaced by **GUIname**. The second line sets the units for the figure information at output to inches. The third line finds the size of the figure in inches; namely, the **get** command with the **position** option returns a four-element vector with the third and fourth elements being, respectively, the width and height of the figure. Note that a semicolon is not included in the command **get(h,'position')**, so that the return of the command is printed in the Command Window to the screen, to be read. Once the GUI figure size is determined, the following code lines are inserted in the m file (**RelPermittivity.m**) at the end of the **RelPermittivity_OpeningFcn** function (**GUIname_OpeningFcn** function in general) of the GUI code [right after the **guidata (hObject, handles)** line of the code]:

```
set(0,'units','inches');
screenSize = get(0,'screensize');
set(hObject,'units','inches','position',[screenSize(3)/2-(width/2),screenSize(4)/2...
-(height/2),width,height]);
```

(note again that the three dots, ..., are used to break the code line), where **width** and **height** should be replaced by the concrete values (numbers) obtained by the **get(h,'position')** command above (for this particular GUI figure as it was developed, these numbers came out to be **width** = 3.9375 and **height** = 1.5729); however, these numbers can as well be changed by the user. The first line sets the units to inches for the screen information to be obtained. The second line gets the size of the screen; namely, MATLAB function **get(0,'screensize')** returns a four-element vector whose third and fourth elements are width and height, respectively, of the screen, in inches. The third line employs MATLAB function **set()** to set the position and size of the GUI figure, with **hObject** standing for the handle to the figure. With units put to inches and using the **position** option, the position of the lower left corner of the figure is set by the first two elements of the four-element vector in the **set** function, so by **screenSize(3)/2-(width/2)** and **screenSize(4)/2-(height/2)**, while the width and height of the figure are set by the last two elements of the vector, **width** and **height**. With these added lines of code, the GUI figure placement and size will remain the same for every screen used.

MATLAB EXERCISE 2.2 **Permittivity tensor of an anisotropic medium.** Based on Eq.(2.3), compute in MATLAB the electric flux density vector, **D** (`D`), in an anisotropic dielectric if the electric field intensity vector and the relative-permittivity tensor are given by `E = [1 1 1]` V/m and `epsr = [2.51 0 0; 0 2.99 0; 0 0 4.11]`, respectively. Using MATLAB function `quiver3`, plot vectors **E** and **D** (see MATLAB Exercise 1.3). *(ME2_2.m on IR)* **H**[4]

MATLAB EXERCISE 2.3 **GUI for the dielectric-strength table of materials.** Repeat MATLAB Exercise 2.1 but for the values of the dielectric strength (E_{cr}) of selected materials given in Table 2.2. *[folder ME2_3(GUI) on IR]*

2.2 Dielectric–Dielectric Boundary Conditions

Let us consider a dielectric–dielectric boundary surface, shown in Fig.2.2. Let $\mathbf{E}_1$ and $\mathbf{D}_1$ be, respectively, the electric field intensity vector and electric flux density vector close to the boundary in medium 1, whereas $\mathbf{E}_2$ and $\mathbf{D}_2$ stand for the same quantities in medium 2. Equations (1.28) and (2.1) result in the following boundary conditions for tangential components of **E** and normal components of **D** on the boundary:

$$E_{1t} = E_{2t} \,, \quad D_{1n} - D_{2n} = \rho_s \quad \text{(dielectric–dielectric boundary conditions)} \,, \tag{2.4}$$

where the normal components are defined with respect to the unit normal $\hat{\mathbf{n}}$ directed from region 2 to region 1 (Fig.2.2), and ρ_s is the free surface charge density that may exist on the surface. In the absence of charge,

$$D_{1n} = D_{2n} \quad \longrightarrow \quad \varepsilon_1 E_{1n} = \varepsilon_2 E_{2n} \quad (\rho_s = 0) \,. \tag{2.5}$$

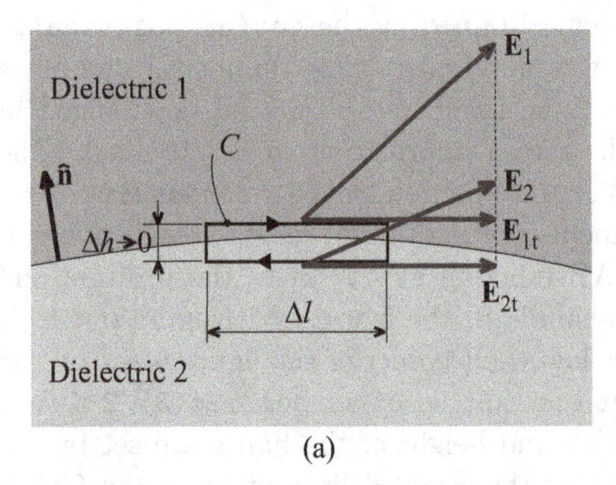

(a)

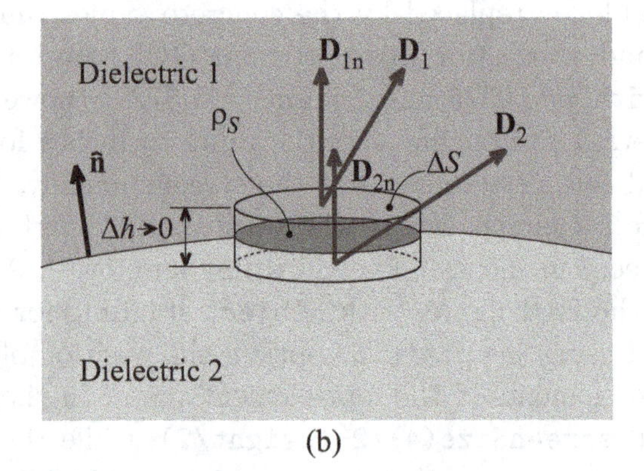

(b)

Figure 2.2 Dielectric–dielectric boundary surface: boundary conditions for (a) tangential components of **E** and (b) normal components of **D**.

---

[4]**H** = recommended to be done also "by hand," i.e., not using MATLAB.

MATLAB EXERCISE 2.4 **Dielectric–dielectric boundary conditions, oblique plane.** Write a program in MATLAB that uses dielectric–dielectric boundary conditions in Eqs.(2.4) and (2.5) to find the electric field intensity vector in medium 2 near the boundary ($\mathbf{E}_2$) for a given electric field intensity vector in medium 1 near the boundary ($\mathbf{E}_1$), if no free charge exists on the boundary surface ($\rho_s = 0$). The program should be written for an arbitrarily positioned (oblique) boundary plane between dielectric media 1 and 2, which does not necessarily coincide with one of the coordinate planes of a Cartesian coordinate system; however, the plane contains the coordinate origin. Assume that relative permittivities ε_{r1} and ε_{r2} of the media are also known, as well as the Cartesian components of the normal on the boundary, directed from region 2 to region 1. *(ME2_4.m on IR)*

TUTORIAL:

The simplest way to input the given vectors is to enter each of their components separately,

```
NORMALx = input('Enter x-component of the normal:  ');
NORMALy = input('Enter y-component of the normal:  ');
NORMALz = input('Enter z-component of the normal:  ');
Ex1 = input('Enter x-component of E-field in medium 1, in V/m:  ');
Ey1 = input('Enter y-component of E-field in medium 1, in V/m:  ');
Ez1 = input('Enter z-component of E-field in medium 1, in V/m:  ');
EPSR1 = input('Enter the relative permittivity of medium 1:  ');
EPSR2 = input('Enter the relative permittivity of medium 2:  ');
```

We check, in an `if-else` format, if the normal is zero, in which case a message is displayed, e.g., with a `disp` command:

```
if (NORMALx ~= 0 || NORMALy ~= 0 || NORMALz ~= 0)
...computation...
else
disp('Error - normal cannot be zero');
end
```

(recall that, in MATLAB, `a ~= b` means $a \neq b$). Otherwise, we compute the normal unit vector ($\hat{\mathbf{n}}$) on the boundary and the angle between $\mathbf{E}_1$ and $\hat{\mathbf{n}}$ (using the dot product). In a similar fashion, we could have checked in the program whether $\varepsilon_{r1} \geq 1$ and $\varepsilon_{r2} \geq 1$ at the entry. In addition, we decompose $\mathbf{E}_1$ onto its normal and tangential components.

```
NORMALmag = sqrt(NORMALx^2 + NORMALy^2 + NORMALz^2);
NORMALx = NORMALx/NORMALmag;
NORMALy = NORMALy/NORMALmag;
NORMALz = NORMALz/NORMALmag;
NORMAL = [NORMALx, NORMALy, NORMALz];
Emag = sqrt(Ex1^2 + Ey1^2 + Ez1^2);
E1 = [Ex1,Ey1,Ez1];
alphaAngle = acos((dot(NORMAL,E1'))/Emag);
E1normal = Emag*cos(alphaAngle).*NORMAL;
E1tangential = E1 - E1normal;
```

(note that `E1normal` and `E1tangential` are vectors, like `E1`).

From boundary conditions in Eqs.(2.4) and (2.5), we find $\mathbf{E}_2$:

```
E2normal = E1normal.*EPSR1/EPSR2;
E2tangential = E1tangential;
E2 = E2normal + E2tangential;
disp('E-field in medium 2, in V/m, is:');
fprintf('(%.3f)*ux',E2(1));
fprintf(' + (%.3f)*uy',E2(2));
fprintf(' + (%.3f)*uz\n',E2(3));
```

In the last part of this tutorial, we plot (although not required by the text of the exercise) the results for vectors $\mathbf{E}_1$, $\mathbf{E}_2$, and $\hat{\mathbf{n}}$, along with the boundary plane. We first determine the limits of the mesh for plotting based on the maximum value among all vector components,

```
A = [E1(1),E1(2),E1(3),E2(1),E2(2),E2(3),NORMALx,NORMALy,NORMALz];
B = max(abs(A)) + 0.1;
```

The general equation of the boundary plane reads

$$n_x(x - x_0) + n_y(y - y_0) + n_z(z - z_0) = 0 , \tag{2.6}$$

where n_x, n_y, and n_z are the components of $\hat{\mathbf{n}}$ and $(x_0, y_0, z_0) = (0, 0, 0)$ (the plane contains the coordinate origin). Based on this equation, we plot the plane by MATLAB function `surf(x,y,z)` (also used in MATLAB Exercise 1.39), upon performing the respective `if-else` checks of possible zero values for some components of $\hat{\mathbf{n}}$.

```
figure(1);
if NORMALz ~= 0
[x,y] = meshgrid(-B:2*B:B,-B:2*B:B);
Bz = -1/NORMALz.*(NORMALx.*x + NORMALy.*y);
h = surf(x,y,Bz); alpha(0.4); axis equal; hold on;
elseif NORMALy ~= 0
[x,z] = meshgrid(-B:2*B:B,-B:2*B:B);
By = -1/NORMALy.*(NORMALx.*x + NORMALz.*z);
h = surf (x,By,z);alpha(0.4);axis equal; hold on;
elseif NORMALx ~= 0
[y,z] = meshgrid(-B:2*B:B,-B:2*B:B);
Bx = -1/NORMALx.*(NORMALy.*y + NORMALz.*z);
h = surf (Bx,y,z);alpha(0.4);axis equal; hold on;
end
```

The vectors ($\mathbf{E}_1$, $\mathbf{E}_2$, and $\hat{\mathbf{n}}$) are plotted by MATLAB function `quiver3` (MATLAB Exercise 1.3),

```
plot3(0,0,0,'ko','MarkerFaceColor','k'); hold on;
quiver3(0,0,0,NORMALx, NORMALy, NORMALz,0,'r','LineWidth',2);
text (NORMALx/2, NORMALy/2, NORMALz/2,'n');
quiver3(0,0,0,E1(1),E1(2),E1(3),0,'b', 'LineWidth',2);
text (E1(1)/2,E1(2)/2,E1(3)/2,'E1');
quiver3(0,0,0,E2(1),E2(2),E2(3),0,'g','LineWidth',2);
```

```
text (E2(1)/2,E2(2)/2,E2(3)/2,'E2');
xlabel('x [m]'); ylabel('y [m]'); zlabel('z [m]');
```

Figure 2.3 shows $\mathbf{E}_1$, $\mathbf{E}_2$, $\hat{\mathbf{n}}$, and the boundary plane for $\varepsilon_{r1} = 2.1$, $\varepsilon_{r2} = 4$, `E1 = [1 2 1]` V/m, and `n = [1 0 1]`.

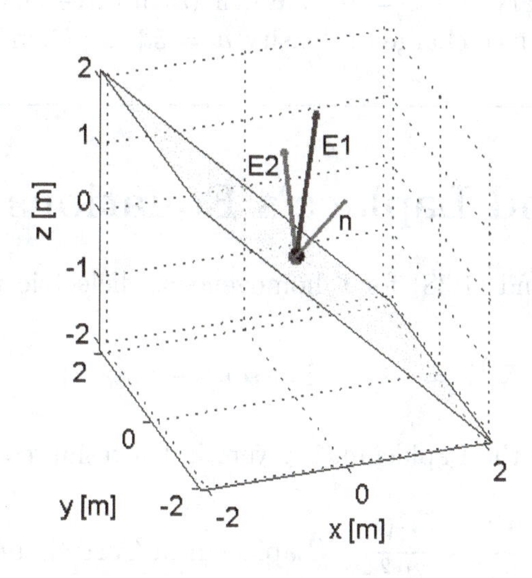

Figure 2.3 MATLAB computation of dielectric–dielectric boundary conditions for an arbitrarily positioned (oblique) boundary plane between media 1 and 2 ($\rho_s = 0$ on the boundary); for MATLAB Exercise 2.4. *(color figure on CW)*

MATLAB EXERCISE 2.5 **Oblique boundary plane with nonzero surface charge.** Repeat the previous MATLAB exercise but for an oblique boundary plane with nonzero surface charge of density ρ_s on it. *(ME2_5.m on IR)*

MATLAB EXERCISE 2.6 **Horizontal charge-free boundary plane.** Repeat MATLAB Exercise 2.4 but for a charge-free boundary plane coinciding with the xy-plane ($z = 0$) of the Cartesian coordinate system (in Fig.2.3). In specific, write a separate MATLAB program specialized for this particular boundary plane (and no other plane). Compare the results with those obtained by the general program for an oblique plane run with `n = [0 0 1]`. *(ME2_6.m on IR)*

MATLAB EXERCISE 2.7 **Horizontal boundary plane with surface charge.** Repeat the previous MATLAB exercise but for a horizontal boundary plane with free surface charge ($\rho_s \neq 0$). *(ME2_7.m on IR)*

MATLAB EXERCISE 2.8 **MATLAB computations of boundary conditions.** Assume that the plane $z = 0$ separates medium 1 ($z > 0$) and medium 2 ($z < 0$), with relative permittivities $\varepsilon_{r1} = 4$ and $\varepsilon_{r2} = 2$, respectively. The electric field intensity vector in medium 1 near the boundary (for $z = 0^+$) is $\mathbf{E}_1 = (4\,\hat{\mathbf{x}} - 2\,\hat{\mathbf{y}} + 5\,\hat{\mathbf{z}})$ V/m. In MATLAB, find the electric field intensity vector in medium 2 near the boundary (for $z = 0^-$), $\mathbf{E}_2$, if (a) no free charge exists on the boundary ($\rho_s = 0$) and (b) there is a surface charge of density $\rho_s = 53.12$ pC/m^2 on the boundary. **H**

2.3 Poisson's and Laplace's Equations

Combining Eqs.(2.1), (2.2), and (1.38) for a homogeneous dielectric region of permittivity ε, we obtain Poisson's equation:

$$\nabla^2 V = -\frac{\rho}{\varepsilon} \quad \text{(Poisson's equation)}\,, \tag{2.7}$$

where the ∇^2 operator, called the Laplacian, is given, in Cartesian coordinates, by

$$\nabla^2 V = \frac{\partial^2 V}{\partial x^2} + \frac{\partial^2 V}{\partial y^2} + \frac{\partial^2 V}{\partial z^2} \quad \text{(Laplacian in Cartesian coordinates)}\,. \tag{2.8}$$

In a charge-free region ($\rho = 0$),

$$\nabla^2 V = 0 \quad \text{(Laplace's equation)}\,, \tag{2.9}$$

which is known as Laplace's equation. In cylindrical coordinates, the Laplacian of V comes out to be

$$\nabla^2 V = \frac{1}{r}\frac{\partial}{\partial r}\left(r\frac{\partial V}{\partial r}\right) + \frac{1}{r^2}\frac{\partial^2 V}{\partial \phi^2} + \frac{\partial^2 V}{\partial z^2} \quad \text{(Laplacian in cylindrical coordinates)}\,. \tag{2.10}$$

In a spherical coordinate system,

$$\nabla^2 V = \frac{1}{r^2}\frac{\partial}{\partial r}\left(r^2\frac{\partial V}{\partial r}\right) + \frac{1}{r^2 \sin\theta}\frac{\partial}{\partial \theta}\left(\sin\theta\frac{\partial V}{\partial \theta}\right) + \frac{1}{r^2 \sin^2\theta}\frac{\partial^2 V}{\partial \phi^2} \quad \text{(spherical coordinates)}\,. \tag{2.11}$$

MATLAB EXERCISE 2.9 **Symbolic Laplacian in different coordinate systems.** By symbolic programming in MATLAB, write functions `LaplaceCar()`, `LaplaceCyl()`, and `LaplaceSph()` that find Laplacian in the Cartesian, cylindrical, and spherical coordinate systems, respectively, based on Eqs.(2.8), (2.10), and (2.11) (see MATLAB Exercise 1.33). Test the codes with $f_{\text{Cartesian}} = 3x^2y^3z$, $f_{\text{cylindrical}} = r^2\cos\phi\, z^3$, and $f_{\text{spherical}} = r^2\sin\theta\cos\phi$. *(LaplaceCar.m, LaplaceCyl.m, LaplaceSph.m, and ME2_9.m on IR)* **H**

2.4 Finite-Difference Method for Numerical Solution of Laplace's Equation

In many practical cases, Poisson's or Laplace's equation cannot be solved analytically, but only numerically. The most popular and perhaps the simplest numerical method for solving these equations (and other types of differential equations generally) is the finite-difference (FD) method. It consists of replacing the derivatives in the differential equation by their finite-difference approximations and solving the resulting algebraic equations. To illustrate this, consider an air-filled coaxial cable with conductors of square cross section, shown in Fig.2.4(a). Let the cross-sectional dimensions of the cable be a and b ($a < b$). The cable is charged by time-invariant charges, and the potentials of the conductors, V_a and V_b, are known. Our goal is to determine the distribution of the potential V in the space between the cable conductors.

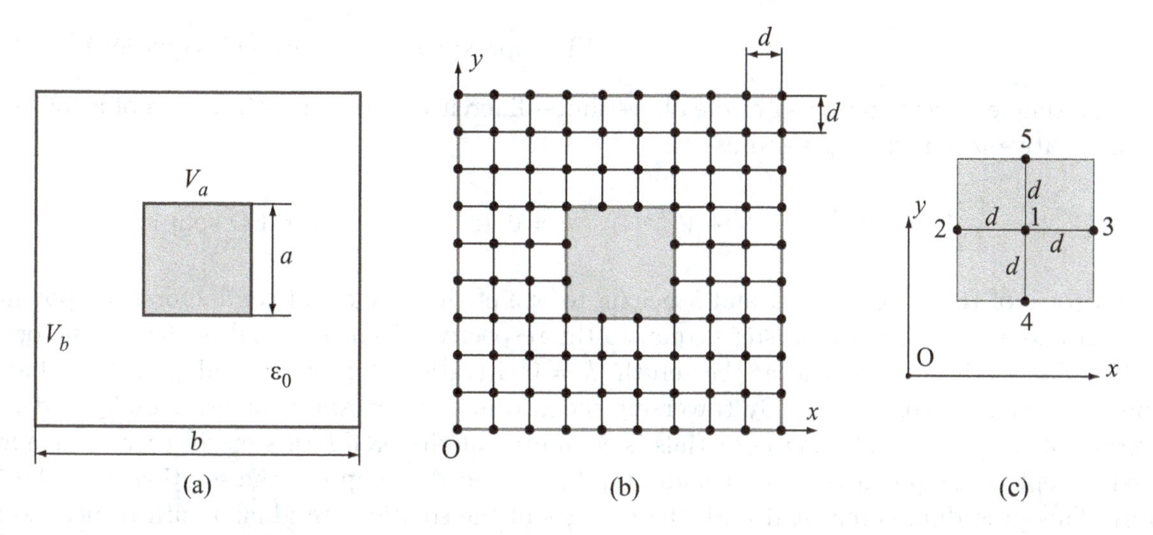

Figure 2.4 Finite-difference analysis of a coaxial cable of square cross section: (a) structure geometry, (b) nodes with discrete potential values as unknowns, and (c) detail of the grid for approximating Laplace's equation in terms of finite differences.

This is a two-dimensional electrostatic problem for which, according to Fig.2.4(b), Laplace's equation, Eqs.(2.9) and (2.8), is given by

$$\Delta V = \frac{\partial^2 V}{\partial x^2} + \frac{\partial^2 V}{\partial y^2} = 0\,. \tag{2.12}$$

We discretize the region between the conductors by introducing a grid with cells of sides d [Fig.2.4(b)], and employ the FD method in order to approximately compute the potentials at the grid points (nodes). Obviously, the accuracy of the computation depends on the grid resolution, i.e., the smaller spacing of the grid, d, the more accurate (but computationally slower) solution.

Figure 2.4(c) shows a detail of the grid in Fig.2.4(b). At the node 1, the backward-difference approximation for the first partial derivative of V with respect to the x-coordinate turns out to be

$$\left.\frac{\partial V}{\partial x}\right|_1 \approx \frac{V_1 - V_2}{d} \quad \text{(finite-difference approximation of a derivative)}\,, \tag{2.13}$$

which combined with the forward-difference approximation for the second partial derivative with respect to x yields (and analogously for the y direction):

$$\left.\frac{\partial^2 V}{\partial x^2}\right|_1 = \frac{\partial}{\partial x}\left(\frac{\partial V}{\partial x}\right)\bigg|_1 \approx \frac{1}{d}\left(\left.\frac{\partial V}{\partial x}\right|_3 - \left.\frac{\partial V}{\partial x}\right|_1\right) \approx \frac{1}{d}\left(\frac{V_3 - V_1}{d} - \frac{V_1 - V_2}{d}\right) = \frac{V_2 + V_3 - 2V_1}{d^2} ,$$

$$\left.\frac{\partial^2 V}{\partial y^2}\right|_1 \approx \frac{V_4 + V_5 - 2V_1}{d^2} , \tag{2.14}$$

so that the FD approximation for Laplace's differential equation at the point 1 in Fig.2.4(c) is

$$\Delta V|_1 \approx \frac{V_2 + V_3 + V_4 + V_5 - 4V_1}{d^2} = 0 \longrightarrow V_1 = \frac{1}{4}(V_2 + V_3 + V_4 + V_5)$$

$$\text{(FD approximation of Laplace's equation)} , \tag{2.15}$$

The simplest technique to solve the above finite-difference equation with the aid of a computer is an iterative technique expressed as

$$V_1^{(k+1)} = \frac{1}{4}\left[V_2^{(k)} + V_3^{(k)} + V_4^{(k)} + V_5^{(k)}\right] , \quad k = 0, 1, \ldots \quad \text{(iterative FD solution)} . \tag{2.16}$$

When some of the nodes 2, 3, 4, and 5 belong to one of the surfaces of conductors, the potential at such nodes is in all iteration steps equal to the respective given potential of the conductor (V_a or V_b). For the initial solution, at the zeroth ($k = 0$) iteration step, we can adopt $V^{(0)} = 0$ at all nodes between the conductors. By traversing the grid in a systematic manner, node by node, the average of the four neighboring potentials is computed at the $(k + 1)$th step for each node and is used to replace the potential at that node, Eq.(2.16), and thus improve the solution from the kth step. This procedure is repeated until the changes of the solution (residuals) with respect to the previous iteration at all nodes are small enough, i.e., until a final set of values for the unknown potentials consistent with the criterion

$$\left|V_1^{(k+1)} - V_1^{(k)}\right| < \delta_V \tag{2.17}$$

is obtained, where δ_V stands for the specified tolerance of the potential.

Once the approximate solution for the potential distribution is known, numerical results for electric field intensity vector, $\mathbf{E}$, at the grid nodes can be obtained by approximating the gradient operator involved in Eqs.(1.38) and (1.39) in terms of finite differences. For example, $\mathbf{E}$ at the node 1 in Fig.2.4(c) is computed approximately as

$$\mathbf{E}_1 = -\nabla V|_1 = -\left.\frac{\partial V}{\partial x}\right|_1 \hat{\mathbf{x}} - \left.\frac{\partial V}{\partial y}\right|_1 \hat{\mathbf{y}} \approx \frac{V_2 - V_3}{2d}\,\hat{\mathbf{x}} + \frac{V_4 - V_5}{2d}\,\hat{\mathbf{y}} \tag{2.18}$$

(central-difference approximation). Additionally, the surface charge density, ρ_s, on the conducting surfaces can be found by means of the boundary condition for the normal components of the vector $\mathbf{D}$ in Eqs.(2.4), which, since $\mathbf{E} = 0$ in the cable conductors, gives $\rho_s = D_n = \varepsilon_0 E_n$, where E_n is the normal component of the electric field vector near the surface. For example, assuming that the node 4 in Fig.2.4(c) belongs to the surface of the inner conductor in Fig.2.4(a), ρ_s at that point can be approximately evaluated as

$$\rho_{s4} = \varepsilon_0 E_{4n} = \varepsilon_0\, \hat{\mathbf{y}} \cdot \mathbf{E}_4 = \varepsilon_0\, E_{y4} = -\varepsilon_0 \left.\frac{\partial V}{\partial y}\right|_4 \approx -\varepsilon_0 \frac{V_1 - V_4}{d} = \varepsilon_0 \frac{V_a - V_1}{d} . \tag{2.19}$$

Finally, the total charge per unit length of each of the conductors, Eq.(1.13), can be found by numerically integrating ρ_s, that is, summing ρ_{si}, along the individual conductor contours [see Eq.(1.62)]. Thus, the per-unit-length charge of the inner conductor is given by

$$Q'_a = \oint_{C_a} \rho_s \, dl \approx \sum_{i=1}^{N_a} \rho_{si} d , \tag{2.20}$$

where N_a denotes the total number of nodes along the contour C_a of the conductor, and similarly for the outer conductor.

MATLAB EXERCISE 2.10 **FD-based MATLAB code – iterative solution.** Write a computer program in MATLAB for the iterative finite-difference analysis of a coaxial cable of square cross section, Fig.2.4, based on Eq.(2.16). Assume that $a = 1$ cm, $b = 3$ cm, $V_a = 1$ V, and $V_b = -1$ V. Plot the results for the distribution of the potential and the electric field intensity in the space between the conductors, and the surface charge density on the surfaces of conductors, taking the grid spacing to be $d = a/10$ and the tolerance of the potential $\delta_V = 10^{-8}$ V. Compute the total charge per unit length of the inner and the outer conductor, taking $d = a/N$ and $N = 2, 3, 5, 7, 9, 10, 12,$ and 25, respectively. *(ME2_10.m on IR)*

TUTORIAL:

See Figs.2.5–2.7. The first part of the code is data input, for the structure geometry and "excitation," tolerance of the potential, maximum number of iterations (`maxIter`), and array of values for the number of subdivisions of the inner conductor edge (`N`),

```
a = 0.01;
b = 0.03;
Va = 1;
Vb = -1;
deltaV = 10^(-8);
EPS0 = 8.8542*10^(-12);
maxIter = 500000;
N = [2 3 5 7 9 10 12 25];
```

We perform several complete cycles of FD analysis (`for` loop with `m` as variable), for different grid spacings, $d = a/N$, where $N = 2, 3, \ldots, 25$, respectively. Potentials of nodes belonging to surfaces of conductors equal the respective given potential of the conductor (V_a or V_b) in all iteration steps. The initial values, at the zeroth iteration step, of potentials at all nodes between the conductors are set to $(V_a + V_b)/2$, which is zero in our case.

```
for m = 1:length(N)
d = a/N(m);
```

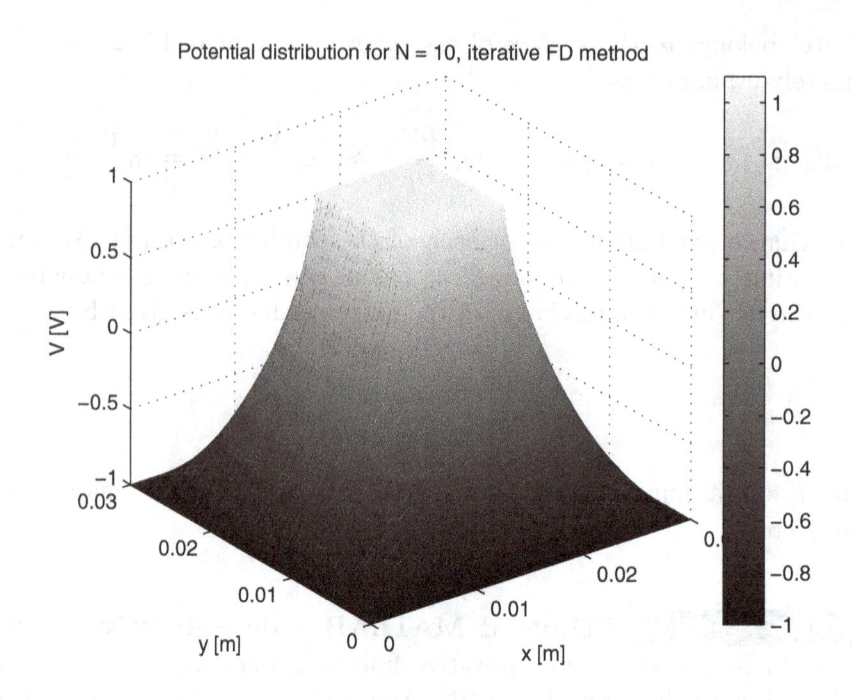

Figure 2.5 Electric potential in the space between the conductors of the coaxial cable of square cross section in Fig.2.4 – results by the iterative finite-difference technique implemented in MATLAB; for MATLAB Exercise 2.10. *(color figure on CW)*

```
N1 = N(m)+1;
N2 = b/a*N(m)+1;
V = ones(N2,N2)*(Va+Vb)/2;
V(1,:)  = Vb; V(:,1) = Vb; V(:,N2) = Vb; V(N2,:)  = Vb;
V((N2-N1)/2+1:(N2+N1)/2,(N2-N1)/2+1:(N2+N1)/2) = Va;
```

The core of the code is the iteration process based on Eq.(2.16), with the stopping criterion in Eq.(2.17) and that limiting the total number of iterations implemented in a `while` loop,

```
iterationCounter = 0;
maxError = 2*deltaV;
while (maxError > deltaV) && (iterationCounter < maxIter)
Vprev = V;
for i = 2:N2-1
for j = 2:N2-1
if V(i,j) ~= Va
V(i,j) = (Vprev(i-1,j)+Vprev(i,j-1)+Vprev(i+1,j)+Vprev(i,j+1))/4;
end;
end;
end;
difference = max(abs(V-Vprev));
maxError = max(difference);
```

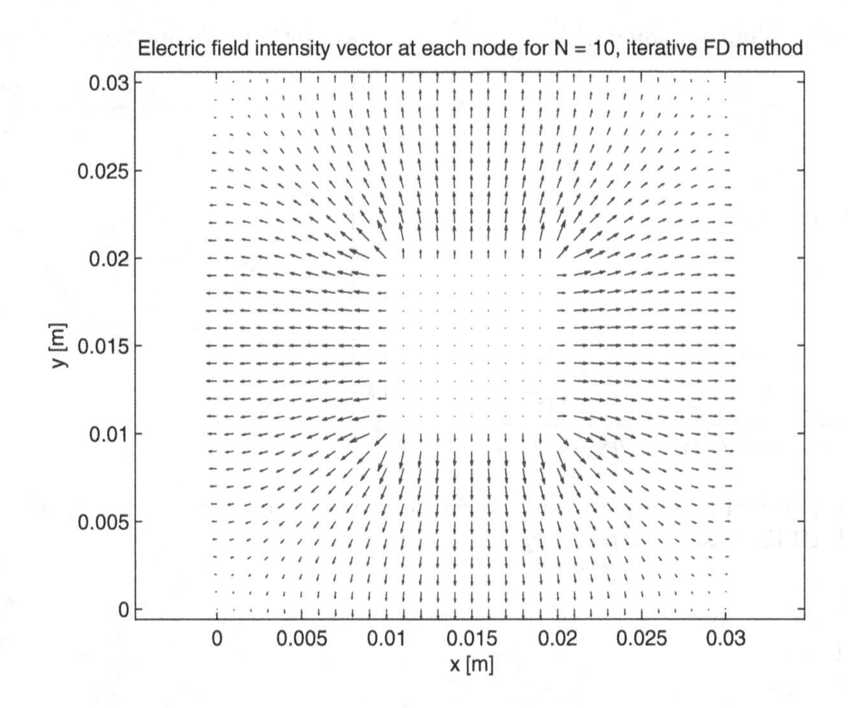

Figure 2.6 Electric field intensity vector corresponding to the potential in Fig.2.5; for MATLAB Exercise 2.10. *(color figure on CW)*

```
iterationCounter = iterationCounter+1;
end;
```

The electric field intensity vector in the structure is calculated, from the potential, using Eq.(1.38) and MATLAB function **gradient**. For the improved accuracy of computing the surface charge densities on the inner and outer conductors of the cable, over that with implementing Eq.(2.19), ρ_s at point i on the conductor surface is computed as the average of ρ_s values obtained from the electric field using the boundary condition $\rho_s = \varepsilon_0 E_n$ at that point and at point $i+1$ (next point on the surface). The electric field intensities in the boundary condition are, in turn, evaluated applying the second-order forward-difference approximation for the derivative of the potential, given by

$$\frac{\partial V}{\partial y} \approx \frac{-V(y+2d)+4V(y+d)-3V(y)}{2d} . \tag{2.21}$$

Total charges per unit length of the conductors are obtained from Eq.(2.20).

```
[x,y]= meshgrid(0:d:b);
[Ex,Ey] = gradient(-V,d,d);
sigmaOut = zeros(1,N2-1);
sigmaIn = zeros(1,N1-1);
for i = 1:N2-1
sigmaOut(i) = EPS0/2/d*(3/2*V(1,i)-2*V(2,i)+1/2*V(3,i)+3/2*V(1,i+1)-2*V(2,i+1)...
+1/2*V(3,i+1));
end;
```

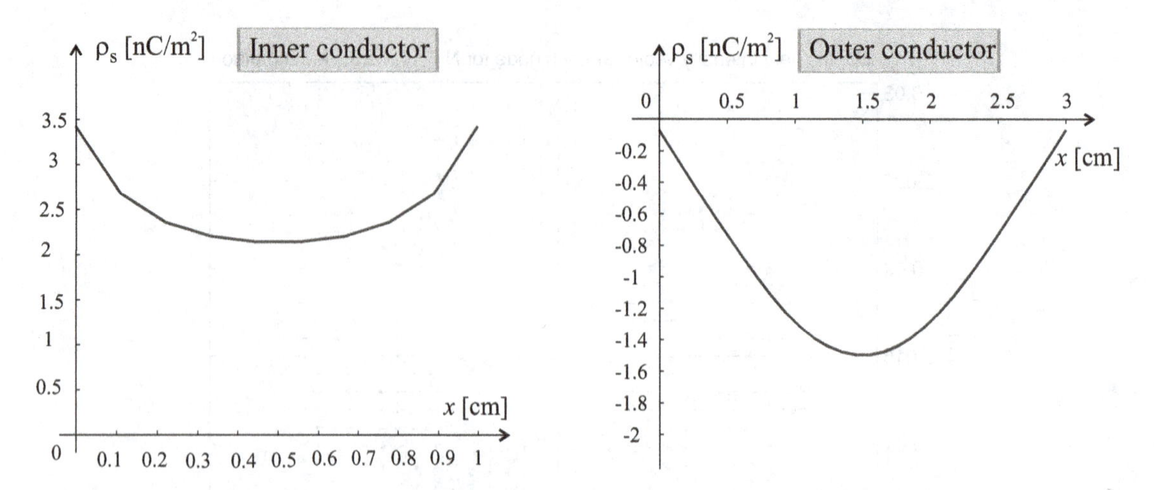

Figure 2.7 Computed surface charge density on the surfaces of conductors of the square coaxial cable in Fig.2.4; for MATLAB Exercise 2.10. *(color figure on CW)*

```
k = (N2-N1)/2+1;
for i = k:(N2+N1)/2-1
sigmaIn(i-k+1)=EPS0/2/d*(3/2*V(k,i)-2*V(k-1,i)+1/2*V(k-2,i)+3/2*V(k,i+1)-2*V(k-1,i+1)...
+1/2*V(k-2,i+1));
end;
Qouter(m) = 4*d*sum(sigmaOut);
Qinner(m) = 4*d*sum(sigmaIn);
```

Simulation results for the distribution of the potential and the electric field intensity in the space between the conductors and for the charge distribution of the conductors of the cable are plotted using MATLAB functions `surf`, `quiver`, and `plot`, respectively,

```
figure(4*m - 3);
quiver (x,y,Ex,Ey); xlabel('x [m]'); ylabel('y [m]');
title(['Electric field intensity vector at each node for N = ',num2str(N(m)),...
', iterative FD method']);axis equal;
figure(4*m - 2);
surf(x,y,V); shading interp; colorbar;
xlabel('x [m]'); ylabel('y [m]');
title(['Potential distribution for N = ', num2str(N(m)),', iterative FD method']);
figure(4*m-1);
dinner = a/(length(sigmaIn)-1);
innerCond = 0:dinner:a;
plot(innerCond, sigmaIn);
xlabel('x [m]'); ylabel('\rhoin [C/m^2]');
figure(4*m);
douter = b/(length(sigmaOut)-1);
outerCond = 0:douter:b;
plot(outerCond,sigmaOut);
```

```
xlabel('x [m]'); ylabel('\rhoout [C/m^2]');
clear Vstart V ;
end;
```

and the plots, for $d = a/10$ and $\delta_V = 10^{-8}$ V, are shown in Figs.2.5–2.7. The computed total per-unit-length charges of conductors, taking $d = a/N$ and $N = 2, 3, 5, 7, 9, 10, 12,$ and 25, respectively, are tabulated in Table 2.3.

Table 2.3 MATLAB FD results for conductor p.u.l. charges (Fig.2.4); for MATLAB Exercise 2.10.

N	$Q'_{\rm inner}$[C/m]	$Q'_{\rm outer}$[C/m]
2	0.8854×10^{-10}	-1.0625×10^{-10}
3	0.9391×10^{-10}	-1.0983×10^{-10}
5	0.9847×10^{-10}	-1.1084×10^{-10}
7	1.0066×10^{-10}	-1.1083×10^{-10}
9	1.0201×10^{-10}	-1.1074×10^{-10}
10	1.0252×10^{-10}	-1.1069×10^{-10}
12	1.0332×10^{-10}	-1.1060×10^{-10}
25	1.0579×10^{-10}	-1.1032×10^{-10}

MATLAB EXERCISE 2.11 **Computation of matrices for a direct FD method.** As an alternative to the iterative technique based on Eq.(2.16), the finite-difference analysis of a square coaxial cable, in Fig.2.4(a), can be carried out by directly solving the system of linear algebraic equations with the potentials at interior grid nodes in Fig.2.4(b) as unknowns [applying Eq.(2.15) to each interior grid node, we get a set of simultaneous equations the number of which equals the number of unknown potentials]. The system of equations, in which known potentials at nodes on the surface of conductors appear on the right-hand side of equations, is solved by the Gaussian elimination method (or by matrix inversion). In this MATLAB exercise, write a function `mACfd()` that establishes the system of equations, i.e., that computes matrices $[A]$ and $[C]$ of the matrix equation, for the direct FD analysis of the cable. *(mACfd.m on IR)*

TUTORIAL:

The input data to the function are the starting matrix of potentials (`Vstart`) with potentials at nodes that belong to surfaces of cable conductors equaling the respective given potential of the conductor (V_a or V_b) and (unknown) potentials at interior nodes set to zero, and the total number of nodes per one side of the outer conductor (`N2`); these data are supplied by the main MATLAB code for the direct FD analysis. All elements at the main diagonal of the matrix $[A]$ (`A`) equal unity, while non-diagonal elements of the matrix are either $-1/4$ or zero, according to Eq.(2.15). The matrix $[C]$ (`C`), which appears on the right-hand side of the matrix equation in the main MATLAB code, is filled with known potentials at nodes on the surfaces of conductors, from `Vstart`.

```
function[A,C] = mACfd(Vstart,N2)
C = zeros(N2^2,1);
```

```
for i = 1:N2
for j = 1:N2
k = (i-1)*N2+j;
A(k,k) = 1;
if (Vstart(i,j) == 0)
A(k,k-N2) = -1/4;
A(k,k+N2) = -1/4;
A(k,k+1) = -1/4;
A(k,k-1) = -1/4;
else C(k) = Vstart(i,j);
end;
end;
end;
```

MATLAB EXERCISE 2.12 **FD-based MATLAB code – direct solution.** Write the main MATLAB code for the direct finite-difference analysis of a square coaxial cable, Fig.2.4, based on Eq.(2.15) – see the previous MATLAB exercise. Compute and plot the same quantities as in MATLAB Exercise 2.10, and compare the results obtained by the two programs. *(ME2_12.m on IR)*

TUTORIAL:

In the following sequence of input and preparatory lines of the code, `N1` and `N2` are the numbers of nodes along an edge of the inner and the other conductors, respectively, of the cable in Fig.2.4(a).

```
a = 0.01;
b = 0.03;
Va = 1;
Vb = -1;
EPS0=8.8542*10^(-12);
N = [2 3 5 7 9 10 12 25];
for m = 1:length(N)
d = a/N(m);
N1 = N(m)+1;
N2 = b/a*N(m)+1;
```

Next, we fill the starting matrix of potentials `Vstart`, which is explained in the previous MATLAB exercise,

```
Vstart = zeros(N2,N2);
Vstart(1,:)  = Vb;
Vstart(:,1) = Vb;
Vstart(N2,:)  = Vb;
Vstart(:,N2) = Vb;
lim1=(N2-N1)/2+1;
lim2=(N2+N1)/2;
```

```
Vstart(lim1:lim2,lim1:lim2) = Va;
```

Using function `mACfd.m`, created in the previous MATLAB exercise, we then compute matrices $[A]$ (`A`) and $[C]$ (`C`), solve the resulting direct FD matrix equation (system of linear algebraic equations) by matrix inversion – by MATLAB function `inv`, and repack the solution, the column matrix `V`, into a square matrix, as follows:

```
[A,C] = mACfd(Vstart,N2);
V = inv(A)*C;
for i = 1:N2
V2D(i,:)  = V((i-1)*N2+1:i*N2);
end;
```

The electric field in the dielectric (air) and charge density and net charges of conductors are evaluated in the same way as in MATLAB Exercise 2.10.

Plots of the computed potential, field, and charge distributions, using the direct FD technique, look identical to those in Figs.2.5–2.7, and the total p.u.l. charges of conductors are given in Table 2.4, so an excellent agreement with the results by the iterative FD technique (MATLAB Exercise 2.10) is observed.

Table 2.4 Conductor p.u.l. charges (in Fig.2.4) obtained by the direct FD code; for MATLAB Exercise 2.12.

N	Q'_{inner}[C/m]	Q'_{outer}[C/m]
2	0.8854×10^{-10}	-1.0625×10^{-10}
3	0.9391×10^{-10}	-1.0983×10^{-10}
5	0.9847×10^{-10}	-1.1084×10^{-10}
7	1.0066×10^{-10}	-1.1083×10^{-10}
9	1.0201×10^{-10}	-1.1074×10^{-10}
10	1.0252×10^{-10}	-1.1069×10^{-10}
12	1.0332×10^{-10}	-1.1060×10^{-10}
25	1.0580×10^{-10}	-1.1032×10^{-10}

2.5 Evaluation of Capacitances of Capacitors and Transmission Lines

Figure 2.8 shows a capacitor – consisting of two metallic bodies (electrodes) embedded in a dielectric, and charged with equal charges of opposite polarities, Q and $-Q$. In linear capacitors (filled with linear dielectrics), Q is linearly proportional to the capacitor voltage, which, in turn, is evaluated as (Fig.2.8)

$$V = \int_1^2 \mathbf{E} \cdot \mathrm{d}\mathbf{l} \,. \tag{2.22}$$

Based on this proportionality, the capacitance of the capacitor is defined as

$$C = \frac{Q}{V} \quad \text{(capacitance of a capacitor)} \ . \tag{2.23}$$

It is always positive ($C > 0$), and the unit is the farad (F). For two-conductor transmission lines (two-body systems with very long conductors of uniform cross section), we define the capacitance per unit length of the line,

$$C' = C_{\text{p.u.l.}} = \frac{C}{l} = \frac{Q'}{V} \quad \text{(capacitance p.u.l. of a transmission line)} \ , \tag{2.24}$$

where C, l, and Q' are the total capacitance, length, and charge per unit length of the structure [see Eq.(1.13)]. The unit for C' is F/m.

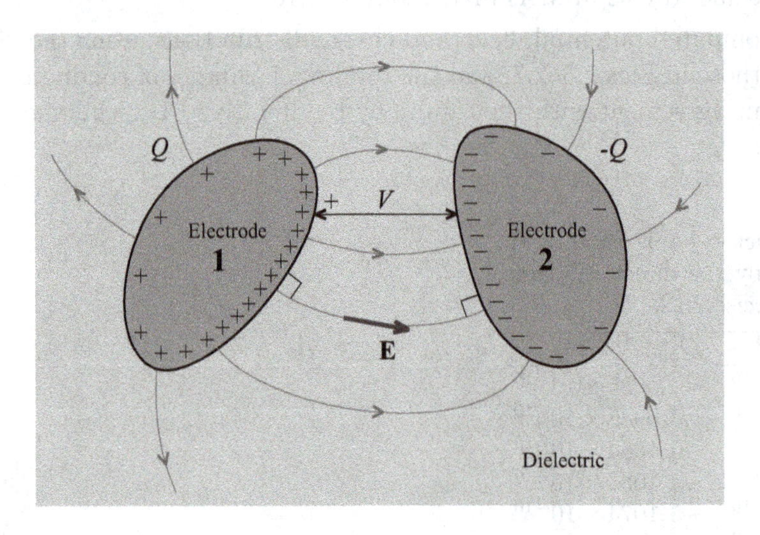

Figure 2.8 Capacitor.

Shown in Fig.2.9 are some of the most frequently used capacitors and transmission lines. For a spherical capacitor, Fig.2.9(a), applying the generalized Gauss' law in integral form, Eqs.(2.1), to a spherical surface S of radius r ($a < r < b$) positioned concentrically with the capacitor electrodes, similarly to the computation in Eq.(1.46), and using Eqs.(2.22) and (2.23), we obtain [note that $\varepsilon = \varepsilon_r \varepsilon_0$, Eq.(2.2)]:

$$E(r) = \frac{Q}{4\pi\varepsilon r^2} \quad (a < r < b) \quad \longrightarrow \quad V = \int_a^b E(r)\,\mathrm{d}r = \frac{Q}{4\pi\varepsilon}\left(\frac{1}{a} - \frac{1}{b}\right) \quad \longrightarrow \quad C = \frac{Q}{V} = \frac{4\pi\varepsilon ab}{b-a}$$

$$\text{(capacitance of a spherical capacitor)} \ . \tag{2.25}$$

In an analogous fashion, the capacitance per unit length [Eq.(2.24)] of a coaxial cable, Fig.2.9(b), is found to be

$$C' = \frac{Q'}{V} = \frac{2\pi\varepsilon}{\ln(b/a)} \quad \text{(capacitance p.u.l. of a coaxial cable)} \ , \tag{2.26}$$

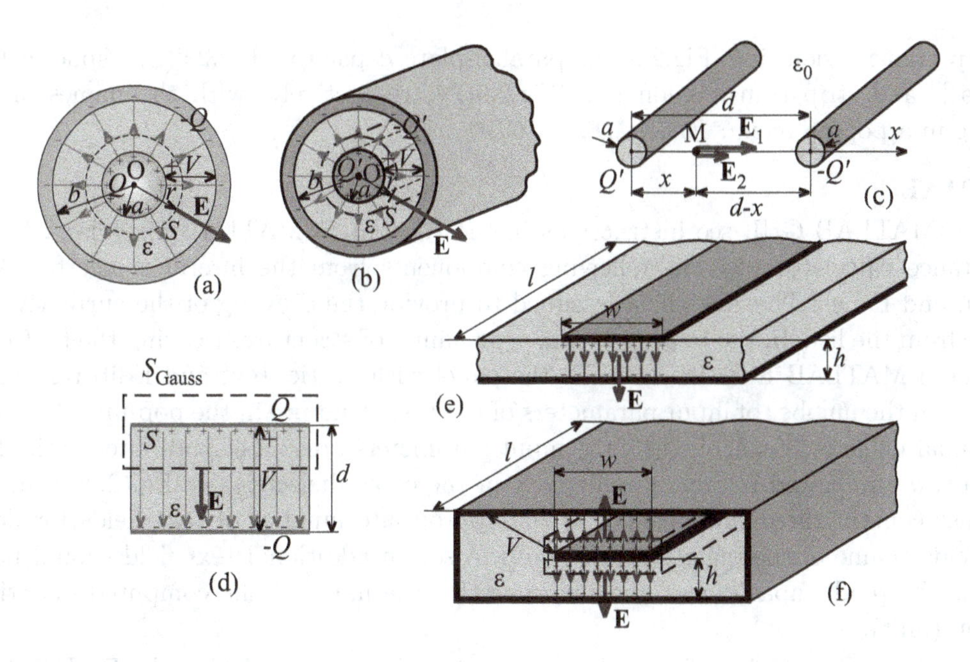

Figure 2.9 Examples of capacitors and transmission lines: (a) spherical capacitor, (b) coaxial cable, (c) thin two-wire transmission line, (d) parallel-plate capacitor, (e) microstrip transmission line, and (f) strip transmission line; in cases (d)–(f), fringing effects neglected.

while that of a thin symmetrical two-wire transmission line (with $d \gg a$), Fig.2.9(c), equals

$$C' = \frac{\pi\varepsilon}{\ln(d/a)} \quad \text{(capacitance p.u.l. of a thin two-wire line)} \,. \tag{2.27}$$

To analyze a parallel-plate capacitor with the fringing effects neglected (we assume that the electric field in the dielectric is uniform and that there is no field outside the dielectric), we apply the generalized Gauss' law to the rectangular Gaussian surface shown in Fig.2.9(d), which results in

$$C = \varepsilon\frac{S}{d} \quad \text{(capacitance of a parallel-plate capacitor, fringing neglected)} \,. \tag{2.28}$$

By a similar token, the capacitance p.u.l. of a microstrip transmission line, Fig.2.9(e), comes out to be

$$C' = \varepsilon\,\frac{w}{h} \quad \text{(capacitance p.u.l. of a microstrip line, fringing neglected)} \,, \tag{2.29}$$

and that of a strip transmission line, Fig.2.9(f), amounts to

$$C' = \frac{2\varepsilon w}{h} \quad \text{(capacitance p.u.l. of a strip line, fringing neglected)} \,. \tag{2.30}$$

MATLAB EXERCISE 2.13 **Capacitance calculator and GUI for multiple structures.** Create a capacitance calculator in the form of a graphical user interface (GUI) in MATLAB to calculate and show the capacitance or per-unit-length capacitance of a coaxial cable [Fig.2.9(b)], microstrip transmission line [Fig.2.9(e)], parallel-plate capacitor [Fig.2.9(d)], spherical capacitor [Fig.2.9(a)], and strip transmission line [Fig.2.9(f)], respectively, with the names of structures appearing in a pop-up menu. *[folder ME2_13(GUI) on IR]*

TUTORIAL:

To start a MATLAB GUI, see instructions in the tutorial to MATLAB Exercise 2.1. To create a capacitance calculator, use the following components from the interface palette: Axes, Pop-up Menu, and Panel. The Axes field is aimed to provide the drawing of the currently considered structure from the list. In the Pop-up Menu, type names of structures (leaving the first row blank), as is done in MATLAB Exercise 2.1. Fill the panel with Static Text and Edit Text components depending on the number of input parameters of the first structure in the pop-up list. For example, if the coaxial cable is first in the list, the input parameters are: inner and outer radii of the cable conductors, a and b, and relative permittivity of the cable dielectric, ε_{r} [Eq.(2.26)], and thus the panel must contain three edit fields, and the appropriate number of text fields for descriptions of edit fields (name of the quantity, unit, etc.) Also, an additional text field should be provided for the result (p.u.l. capacitance) and two more for the name of the computed quantity and its dimension (unit).

Capacitance computation of the selected structure is done by clicking the Push Button, which is also included in the Panel section. Set the panel to invisible regime in the Property Inspector and save the GUI as capCalc1.fig and capCalc1.m (see MATLAB Exercise 2.1).

Next, we import drawings of structures (previously created by an available computer program for drawing, and saved as *.png files) – see Fig.2.10. In the m file (`capCalc1.m`), in its section `function capCalc1_OpeningFcn(hObject, eventdata, handles, varargin)`, right after the `varargin` line, the drawings are imported using MATLAB command `imread` as follows (note that the structures are organized in the Pop-up Menu with their names in alphabetical order):

```
handles.coaxcable = imread('coaxcable.png');
handles.microstrip = imread('microstrip.png');
handles.ppcap = imread('ppcap.png');
handles.sphcap = imread('sphcap.png');
handles.stripline = imread('stripline.png');
```

In the same section, we set the Panel and Axes fields to invisible regime by typing:

```
set(handles.uipanel1,'Visible','off');
set(handles.axes2,'Visible','off');
```

Setting-up the position and size of the GUI figure to appear properly on any computer screen is carried out in the same fashion as in MATLAB Exercise 2.1, where, for this GUI figure, the `get(h,'position')` command gives `width` = 6.5 and `height` = 4.3229, to be used in filling the elements of the four-element vector in the `set` function, in file `capCalc1.m`.

To activate and control the Pop-up Menu, we use the `case-switch` command in the section `popupmenu1_Callback(hObject, eventdata, handles)` of the m file. Also in this section, we introduce a global variable `i` that indicates the current choice [for instance, `i = 1` if the current structure under consideration is a coaxial cable, `i = 2` for the microstrip line, and so on]. Command `set(handles.uipanel1,'Visible','off')` ensures that when the user chooses blank in

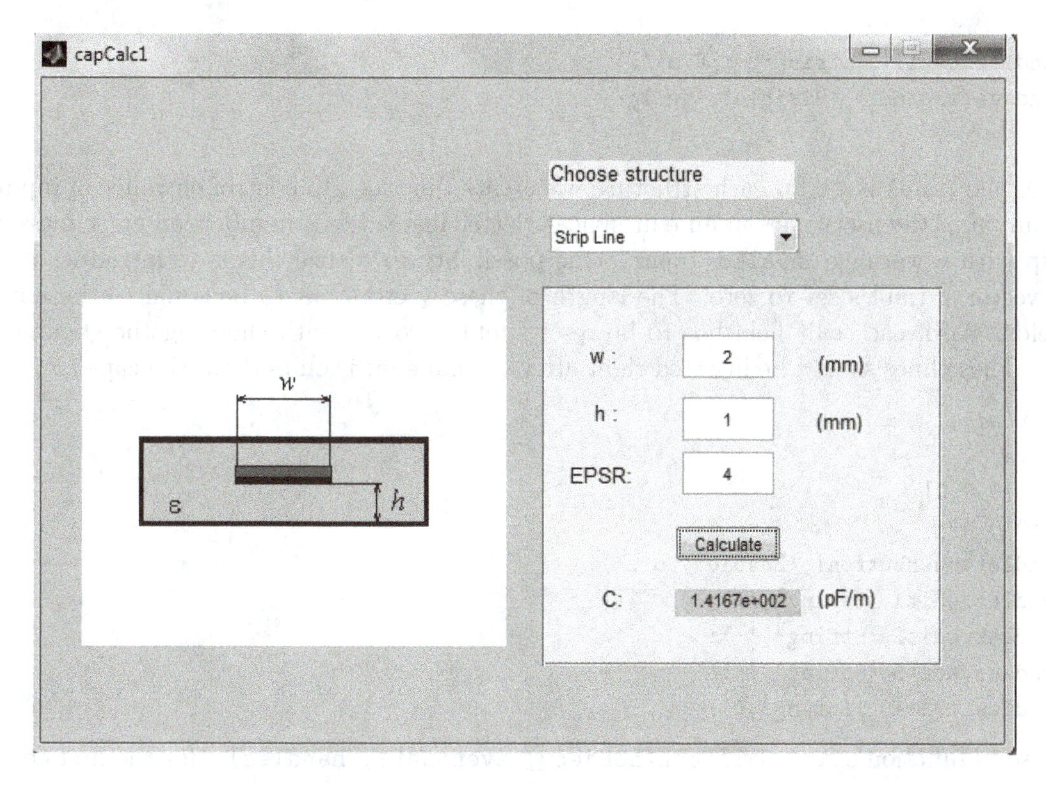

Figure 2.10 MATLAB capacitance calculator and graphical user interface for multiple structures: GUI in the case a strip line is selected in the pop-up menu; for MATLAB Exercise 2.13. *(color figure on CW)*

the pop-up list panel, the associated text and edit fields become invisible, while commands `cla reset; axis off` reset and remove the axes field.

```
switch get(handles.popupmenu1,'Value')
case 1
set(handles.uipanel1,'Visible','off');
cla reset; axis off;
```

The following `case` corresponds to the choice of a coaxial cable. When the user marks any structure from the list, the Panel must appear in the GUI, and this is done by `set(handles.uipanel1,'Visible','on')`. In addition, we have to rename all labels – text fields by the `set` command in accordance to the chosen structure. For example, `set(handles.text1,'String','a :')` means that field `text1` (for any component, its tag is defined in `Property Inspector`) is a string defined as `a`, and that the user is expected to enter the inner radius of the cable in the associated edit field.

```
case 2
handles.currentdata = handles.coaxcable;
imshow(handles.currentdata);
set(handles.text1,'String','a :');
set(handles.text4,'String','b :');
set(handles.text6,'String','(mm)');
```

```
set(handles.text11,'String','(pF/m)');
set(handles.uipanel1,'Visible','on');
i = 1;
```

After the Panel is set for each structure, we create functions to control entrance of input data. For example, if the user types in an edit space a letter instead of a number, an error message box pops up with a warning Invalid input. One possibility to realize this is to introduce two more global vector variables set to zero. The length of these vectors has to be equal to the number of edit fields. Also, each edit field has to be reset simultaneously with choosing the structure from the list. These lines should be inserted right after the case-switch part (in the capCalc1.m file):

```
global ready;
global var;
ready = [0 0 0];
var = [0 0 0];
set(handles.pushbutton1,'Enable','off');
set(handles.edit1,'String','');
set(handles.edit2,'String','');
set(handles.edit3,'String','');
set(handles.text10,'String','');
```

and these in function edit1_Callback(hObject, eventdata, handles) – for the first edit field:

```
handles.edit1 = str2double(get(hObject,'String'));
global ready;
global var;
if (isnan(handles.edit1));
msgbox('Invalid input','Error');
ready(1)= 0;
else
ready(1)= 1;
var(1) = handles.edit1;
end; if (ready == [1 1 1])
set(handles.pushbutton1,'Enable','on');
else set(handles.pushbutton1,'Enable','off');
end;
```

Note that the check whether the typed input is a number is performed using MATLAB function isnan(handles.edit1). Note also that until and unless all edit fields are correctly filled, the Push Button is not active. Callback functions for other edit fields are done analogously.

Finally, we implement the respective capacitance expressions, from Eqs.(2.26), (2.29), (2.28), (2.25), and (2.30), in function pushbutton1_Callback (hObject, eventdata, handles), and set the text field to display the output results:

```
global var;
global i;
EPS0 = 8.8542*10^(-12);
mm2m = 10^(-3);
```

```
mmsq2msq = 10^(-6);
if i == 1;
a = var(1)*mm2m;
b = var(2)*mm2m;
EPSR = var(3);
EPS = EPS0*EPSR;
C = capacitanceCoaxCable(EPS,a,b);
else if i == 2;
w = var(1)*mm2m;
h = var(2)*mm2m;
EPSR = var(3);
EPS = EPS0*EPSR;
C = capacitanceMicrostrip(EPS,w,h);
else if i == 3;
S = var(1)*mmsq2msq;
d = var(2)*mm2m;
EPSR = var(3);
EPS = EPS0*EPSR;
C = capacitancePPCapacitor(EPS,S,d);
else if i == 4;
a = var(1)*mm2m;
b = var(2)*mm2m;
EPSR = var(3);
EPS = EPS0*EPSR;
C = capacitanceSphCapacitor(EPS,a,b);
else
w = var(1)*mm2m;
h = var(2)*mm2m;
EPSR = var(3);
EPS = EPS0*EPSR;
C = capacitanceStripline(EPS,w,h);
end;
end;
end;
end;
C = C*10^12;
set(handles.text10,'String',num2str(C,'%.4e'));
```

The capacitances are computed via respective functions **capacitanceCoaxCable**, **capacitanceMicrostrip**, **capacitancePPCapacitor**, **capacitanceSphCapacitor**, and **capacitanceStripline**, separately written for the convenience of their later use independently from this GUI, with the first one finding the capacitance p.u.l. of a coaxial cable as

```
function C = capacitanceCoaxCable(EPS,a,b)
C = 2*pi*EPS/(log(b/a));
```

and so on. Note that MATLAB uses "log" in place of "ln" to denote the natural logarithm.

Figure 2.10 shows the GUI if, for instance, the strip line is selected in the pop-up menu.

MATLAB EXERCISE 2.14 **RG-55/U coaxial cable.** An RG-55/U coaxial cable has conductor radii $a = 0.5$ mm and $b = 3$ mm. The dielectric is polyethylene ($\varepsilon_\mathrm{r} = 2.25$). Determine the capacitance per unit length of the cable using the capacitance calculator from the previous MATLAB exercise. **H**

MATLAB EXERCISE 2.15 **Parallel-plate capacitor model of a thundercloud.** A typical thundercloud can be approximately represented, as far as its electrical properties are concerned, as a parallel-plate capacitor with horizontal plates of area $S = 15$ km^2 and vertical separation $d = 1$ km. Neglecting the fringing effects, find the capacitance of this capacitor by the capacitance calculator from MATLAB Exercise 2.13. **H**

MATLAB EXERCISE 2.16 **Capacitance of a metallic cube, using MoM MATLAB code.** Find the capacitance of the metallic cube numerically analyzed by the method of moments in MATLAB Exercise 1.41, and compare the result with capacitances of the following metallic spheres, respectively: (a) the sphere inscribed in the cube, (b) the sphere overscribed about the cube, (c) the sphere whose radius is the arithmetic mean of the radii of spheres in (a) and (b), (d) the sphere having the same surface as the cube, and (e) the sphere with the same volume as the cube.

TUTORIAL:

The capacitance of the metallic cube analyzed by the MoM MATLAB code amounts to $C = Q/V_0 = 73.27$ pF. The capacitance of a spherical conductor of radius r in free space is given by Eq.(2.25) with $b \to \infty$, $a = r$, and $\varepsilon = \varepsilon_0$, so it equals $C = 4\pi\varepsilon_0 r$. Hence, the capacitances of the metallic sphere inscribed in the cube (sphere radius $r_\mathrm{a} = a/2$), the sphere overscribed about the cube ($r_\mathrm{b} = a\sqrt{3}/2$), the one with $r_\mathrm{c} = (r_\mathrm{a} + r_\mathrm{b})/2$, the sphere having the same surface as the cube [$r_\mathrm{d} = a\sqrt{3/(2\pi)}$], and that with the same volume as the cube $\{r_\mathrm{e} = a[3/(4\pi)]^{1/3}\}$ come out to be $C_\mathrm{a} = 55.63$ pF, $C_\mathrm{b} = 96.36$ pF, $C_\mathrm{c} = 76$ pF, $C_\mathrm{d} = 76.88$ pF, and $C_\mathrm{e} = 69.02$ pF, respectively. We see that the capacitance of the sphere whose radius equals the arithmetic mean of the radii of spheres inscribed in and overscribed about the cube ($C_\mathrm{c} = 76$ pF) is quite close in value to the capacitance of the cube.

MATLAB EXERCISE 2.17 **Capacitance computation using FD MATLAB codes.** Compute the capacitance per unit length of the coaxial cable of square cross section numerically analyzed by a finite-difference technique in MATLAB Exercises 2.10 and 2.12, respectively.

TUTORIAL:

The capacitance per unit length of the coaxial cable of square cross section in Fig.2.4 analyzed by the iterative finite-difference MATLAB code written in MATLAB Exercise 2.10 is obtained by adding the following line at the end of the code:

```
C = Qouter/(Vb - Va);
```

and the result is $C'_{\text{iterative}} = 51.26006$ pF/m, while the direct FD MATLAB code from MATLAB Exercise 2.12 gives $C'_{\text{direct}} = 51.26008$ pF/m, so practically the same result, with the tolerance of the potential of $\delta_V = 10^{-8}$ V in the iterative solution and the grid spacing of $d = a/10$ in both solutions. As a reference, the per-unit-length capacitance of a (standard) coaxial cable (of circular cross section) having the same ratio of conductor radii ($b/a = 3$) and dielectric (air) as the square cable is $C'_{\text{standard coax}} = 50.61$ pF/m [from Eq.(2.26)].

MATLAB EXERCISE 2.18 **Main MoM matrix for a parallel-plate capacitor.** Consider the parallel-plate capacitor shown in Fig.2.11, and write a function `matrixACap()` in MATLAB that computes the matrix [A] in Eq.(1.58) for the method-of-moments analysis of the structure, assuming that the upper and lower plates are at potentials V and $-V$, respectively. *(matrixACap.m on IR)*

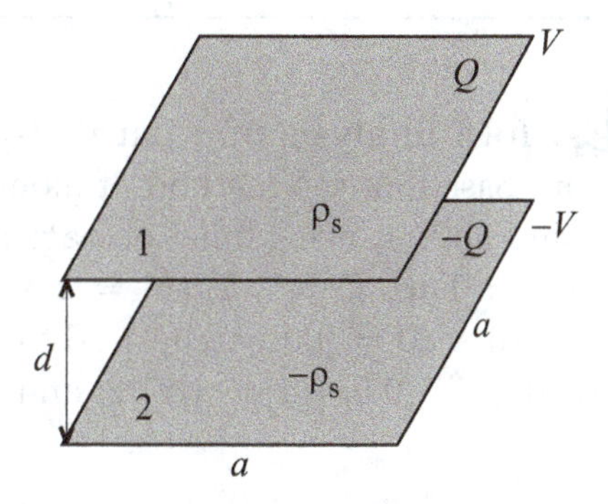

Figure 2.11 Air-filled parallel-plate capacitor with square plates; for MATLAB Exercise 2.18.

TUTORIAL:

We subdivide each of the plates in Fig.2.11 into N square patches, and assume constant charge densities on individual patches (as in Fig.1.22). Since the charge densities of pairs of corresponding patches that are right above/below each other on the two plates are equal in magnitude and opposite in polarity, the unknowns in the MoM procedure are charge densities ρ_{s1}, ρ_{s2}, ..., ρ_{sN} on the upper plate only, and matching points, at which the potentials are computed, are centers of the same patches. However, the potentials are due to pairs of patches on both plates, with charge densities ρ_{si} and $-\rho_{si}$ ($i = 1, 2, \ldots, N$). So, the matrix [A] is evaluated for the upper plate only, but with contributions of charges of patches on the lower plate taken into account as well. Accordingly, we modify function `matrixA` from MATLAB Exercise 1.36, as follows, with `r1` and `r2` standing, respectively, for the distance between two points (patch centers) on the upper plate and the distance between the first point and the "image" of the second point on the lower plate (d is the separation between plates).

```
function A = matrixACap(EPS,dS,x,y,d)
N = length(x);
if (length(dS) == N)
for i = 1:N
for j = 1:N
r1 = sqrt((x(j)-x(i))^2 + (y(j)-y(i))^2);
r2 = sqrt((x(j)-x(i))^2 + (y(j)-y(i))^2 + d^2);
if (i==j)
A(i,j) = sqrt(dS(j))/(2*sqrt(pi)*EPS) - dS(j)/(4*pi*EPS*r2);
else
A(i,j) = dS(j)/(4*pi*EPS*r1) - dS(j)/(4*pi*EPS*r2);
end;
end;
end;
else
A = 0;
disp ('Incorrect input data in function matrixACap');
end;
```

MATLAB EXERCISE 2.19 **MoM analysis of a parallel-plate capacitor in MATLAB.** Write the main MATLAB program based on the method of moments to evaluate the capacitance (C) of the parallel-plate capacitor in Fig.2.11, using function `matrixACap`, developed in the previous MATLAB exercise. Assume that $a = 1$ m, $V = 1$ V ($V_1 = 1$ V and $V_2 = -1$ V), and $N = 100$ (each plate is subdivided into $N = 10 \times 10 = 100$ patches). By means of this MoM program, find C for the following d/a ratios: (i) 0.1, (ii) 0.5, (iii) 1, (iv) 2, and (v) 10. *(ME2_19.m on IR)*

TUTORIAL:

After defining input parameters like in MATLAB Exercise 1.39, we call function `localCoordinates`, from MATLAB Exercise 1.37, to calculate coordinates of centers and surface areas of patches on the upper plate of the parallel-plate capacitor in Fig.2.11, which are then used as input to function `matrixACap`, to fill the main MoM matrix $[A]$ (A),

```
[x,y,S] = localCoordinates(n,n,a,a);
A = matrixACap(EPS0,S,x,y,d);
```

The rest of the code is practically the same as in MATLAB Exercise 1.39. The computed surface charge distribution over the upper plate is shown in Fig.2.12.

We calculate the total charge of the plate by function `totalCharge` from MATLAB Exercise 1.38, and the capacitance of the capacitor as

```
Qtot = totalCharge(S,rhos);
C = Qtot/(2*V);
```

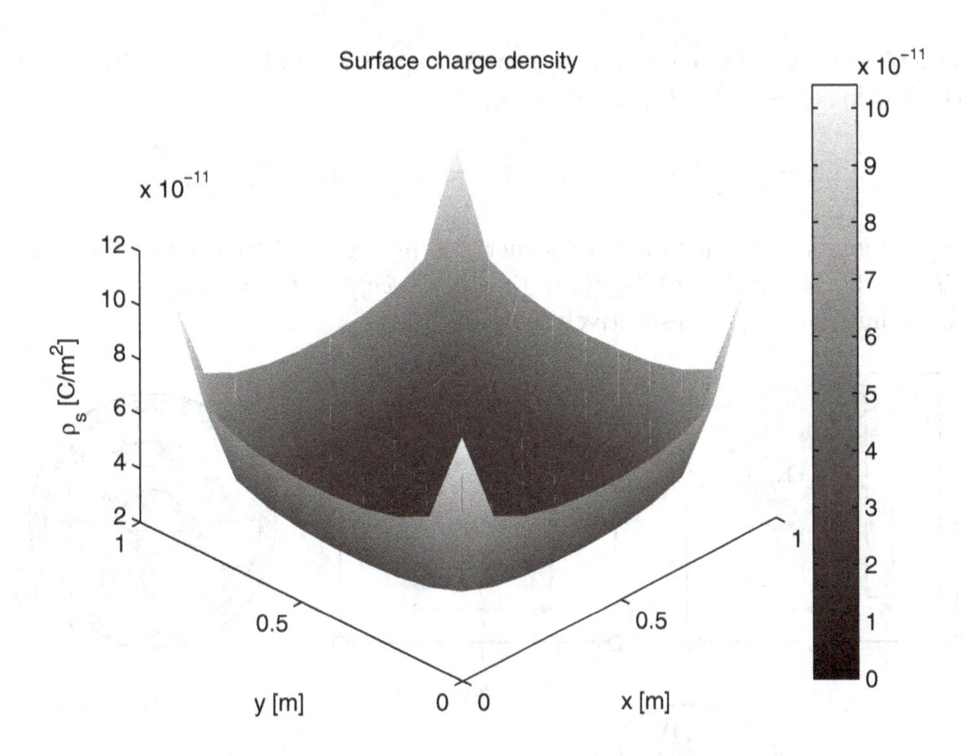

Figure 2.12 MATLAB display of the charge distribution of the upper plate of the parallel-plate capacitor in Fig.2.11 computed by a method-of-moments MATLAB code; for MATLAB Exercise 2.19. *(color figure on CW)*

For d/a ratios of 0.1, 0.5, 1, 2, and 10, C turns out to be 117 pF, 38.3 pF, 28.7 pF, 24 pF, and 20.6 pF, respectively. Note that the corresponding C values obtained from Eq.(2.28), which neglects the fringing effects, are 88.5 pF, 17.7 pF, 8.85 pF, 4.43 pF, and 0.885 pF.

2.6 Capacitors with Inhomogeneous Dielectrics

Often we deal with capacitors and transmission lines containing inhomogeneous dielectrics; Fig.2.13 shows some characteristic examples. As an illustration of the capacitance evaluation for such electrostatic systems, the field analysis of the parallel-plate capacitor with two dielectric layers in Fig.2.13(a), based on applying the generalized Gauss' law [Eqs.(2.1)] to a rectangular closed surface enclosing the plate charged with Q, with the right-hand side positioned in either one of the dielectrics, gives

$$D_1 = D_2 = D = \frac{Q}{S} \quad \longrightarrow \quad E_1 = \frac{D}{\varepsilon_1} = \frac{Q}{\varepsilon_1 S} \quad \text{and} \quad E_2 = \frac{D}{\varepsilon_2} = \frac{Q}{\varepsilon_2 S}$$

$$\longrightarrow \quad V = E_1 d_1 + E_2 d_2 = \frac{Q}{S}\left(\frac{d_1}{\varepsilon_1} + \frac{d_2}{\varepsilon_2}\right) \quad \longrightarrow \quad C_a = \frac{Q}{V} = \frac{\varepsilon_1 \varepsilon_2 S}{\varepsilon_2 d_1 + \varepsilon_1 d_2}\,. \tag{2.31}$$

Alternatively, C_a can be obtained as the equivalent capacitance of two capacitors (with homogeneous dielectrics) in series, using Eq.(2.28) twice,

$$C_a = \frac{C_1 C_2}{C_1 + C_2}, \quad \text{where} \quad C_1 = \varepsilon_1 \frac{S}{d_1} \quad \text{and} \quad C_2 = \varepsilon_2 \frac{S}{d_2}. \tag{2.32}$$

In a similar fashion, the capacitances of structures in Figs.2.13(b)–(e) are found to be $C_b = (\varepsilon_1 S_1 + \varepsilon_2 S_2)/d$, $C_c = 4\pi[(b-a)/(\varepsilon_1 ab) + (c-b)/(\varepsilon_2 bc)]^{-1}$, $C_d = 2\pi(\varepsilon + \varepsilon_0)ab/(b-a)$, and $C'_e = \pi[\ln 2/\varepsilon + \ln(d/2a)/\varepsilon_0]^{-1}$, respectively.

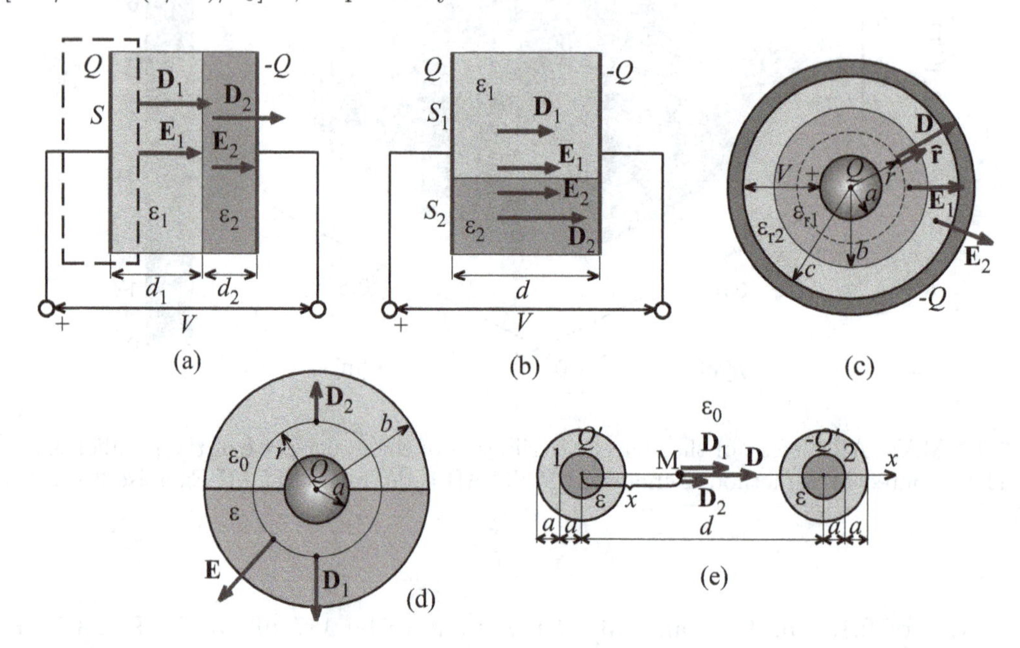

Figure 2.13 Examples of capacitors and transmission lines with inhomogeneous dielectrics: (a) parallel-plate capacitor with two dielectric layers (fringing neglected), (b) parallel-plate capacitor with two dielectric sectors (fringing neglected), (c) spherical capacitor with two concentric dielectric layers, (d) spherical capacitor half filled with a liquid dielectric, and (e) two-wire line with dielectrically coated conductors ($d \gg a$).

MATLAB EXERCISE 2.20 **GUI for capacitors with inhomogeneous dielectrics.** Repeat MATLAB Exercise 2.13 but for the five types of capacitors with inhomogeneous dielectrics in Fig.2.13. *[folder ME2_20(GUI) on IR]*

2.7 Dielectric Breakdown in Electrostatic Systems

We shall now analyze capacitors and transmission lines in high-voltage applications, i.e., in situations where the electric field in the dielectric is so strong that there is a danger of dielectric breakdown in the structure. Dielectric breakdown occurs when the largest field intensity in the dielectric reaches the critical value for that particular material – dielectric strength of the material,

$E_{\rm cr}$ (see Table 2.2). In structures with nonuniform electric field distributions, the principal task is to identify the most vulnerable spot for dielectric breakdown, and to relate the corresponding largest electric field intensity to the breakdown voltage of the capacitor or line; this is the highest possible voltage that can be applied to the structure (before it breaks down). The task is more complex for structures containing multiple dielectric regions.

As an illustration, consider a spherical capacitor with two concentric dielectric layers, Fig.2.13(c). From the generalized Gauss' law [Eqs.(2.1)], the electric flux density in both layers is given by $D(r) = Q/(4\pi r^2)$, so that the electric field intensities in the layers are

$$E_1(r) = \frac{Q}{4\pi\varepsilon_{\rm r1}\varepsilon_0 r^2} \quad (a < r < b) \quad \text{and} \quad E_2(r) = \frac{Q}{4\pi\varepsilon_{\rm r2}\varepsilon_0 r^2} \quad (b < r < c)\ . \tag{2.33}$$

We do not know in advance which dielectric layer would break down first after a voltage of critical value is applied across the capacitor electrodes. Since $E_1(a^+)$ is the largest field intensity in the inner layer, this layer would break down if

$$E_1(a^+) = E_{\rm cr1} \quad \text{(breakdown in the inner dielectric layer)}\ , \tag{2.34}$$

while an eventual breakdown in the outer layer would occur if

$$E_2(b^+) = E_{\rm cr2} \quad \text{(breakdown in the outer dielectric layer)}\ . \tag{2.35}$$

Having in mind Eqs.(2.33), the capacitor charges in cases in Eqs.(2.34) and (2.35), respectively, amount to

$$Q_{\rm cr}^{(1)} = 4\pi\varepsilon_{\rm r1}\varepsilon_0 a^2 E_{\rm cr1} \quad \text{and} \quad Q_{\rm cr}^{(2)} = 4\pi\varepsilon_{\rm r2}\varepsilon_0 b^2 E_{\rm cr2}\ . \tag{2.36}$$

As Q becomes larger and larger, the breakdown occurs when Q reaches the smaller of the two charges in Eqs.(2.36). The critical value of the capacitor charge is thus

$$Q_{\rm cr} = \min\{Q_{\rm cr}^{(1)}, Q_{\rm cr}^{(2)}\}\ . \tag{2.37}$$

For instance, for the numerical data for the capacitor in Fig.2.13(c) given by $a = 3$ cm, $b = 5$ cm, $c = 8$ cm, $\varepsilon_{\rm r1} = 2.5$, $\varepsilon_{\rm r2} = 5$, $E_{\rm cr1} = 50$ MV/m, and $E_{\rm cr2} = 30$ MV/m, $Q_{\rm cr}^{(1)} < Q_{\rm cr}^{(2)}$ [$Q_{\rm cr}^{(1)} = 0.3Q_{\rm cr}^{(2)}$], meaning that the breakdown occurs in the inner dielectric layer. Hence, $Q_{\rm cr} = Q_{\rm cr}^{(1)} = 12.52\ \mu$C, and the breakdown voltage of the capacitor comes out to be $V_{\rm cr} = Q_{\rm cr}/C = 769$ kV, where $C = 4\pi\varepsilon_0[(b-a)/(\varepsilon_{\rm r1}ab) + (c-b)/(\varepsilon_{\rm r2}bc)]^{-1} = 16.28$ pF is the capacitance of the capacitor (see MATLAB Exercise 2.20).

MATLAB EXERCISE 2.21 **Breakdown in a spherical capacitor with a multilayer dielectric.** Consider a spherical capacitor with N concentric dielectric layers. The inner radius, relative permittivity, and the dielectric strength of the ith layer are a_i, $\varepsilon_{{\rm r}i}$, and $E_{{\rm cr}i}$, respectively ($i = 1, 2, \ldots, N$). The inner radius of the outer conductor of the capacitor is b. Write a MATLAB program to find which dielectric layer would break down first after a voltage of critical value is applied across the capacitor electrodes and to compute the breakdown voltage of the capacitor. Test the program for $N = 3$, $a_1 = 2.5$ cm, $a_2 = 5$ cm, $a_3 = 7$ cm, and $b = 9$ cm, if the dielectrics constituting layers 1, 2, and 3 are polystyrene, quartz, and silicon, respectively (use GUI's from MATLAB Exercises 2.1 and 2.3 to get the values of material parameters). *(ME2_21.m on IR)* **H**

TUTORIAL:
The data input is realized as in MATLAB Exercise 1.4. We implement a generalization of Eqs.(2.36) and (2.37) to the case of an arbitrary number (N) of layers, and, instead of only two charges $Q_{cr}^{(1)}$ and $Q_{cr}^{(2)}$, we now have an array of such charges. For finding the minimum charge in the array, we use MATLAB function `min`, which also returns as output the position (index) of the minimum value in the array.

```
Q = 4*pi*EPS0.*EPSR.*Ecr.*a.^2;
[Qmin,index] = min(Q);
```

The corresponding breakdown voltage is computed as $V_{cr} = Q_{cr}/C_1 + Q_{cr}/C_2 + \ldots + Q_{cr}/C_N$, with C_i being the capacitance of the spherical capacitor with a homogeneous dielectric – that of the ith layer ($i = 1, 2, \ldots, N$), given by Eq.(2.25):

```
Vcr = 0;
if N > 1
for i=1:N-1
Vcr = Vcr + Qmin/4/pi/EPS0/EPSR(i)*(a(i+1)-a(i))/a(i+1)/a(i);
end
end
Vcr = Vcr + Qmin/4/pi/EPS0/EPSR(N)*(b-a(N))/b/a(N);
```

MATLAB EXERCISE 2.22 **Breakdown in a coaxial cable with a multilayer dielectric.** Repeat the previous MATLAB exercise but for a coaxial cable with N coaxial dielectric layers. *(ME2_22.m on IR)* **H**

MATLAB EXERCISE 2.23 **Parallel-plate capacitor with multiple layers.** Repeat MATLAB Exercise 2.21 but for a parallel-plate capacitor with N dielectric layers placed like in Fig.2.13(a). The thicknesses of layers are d_i ($i = 1, 2, \ldots, N$) and fringing effects can be neglected. *(ME2_23.m on IR)*

MATLAB EXERCISE 2.24 **Parallel-plate capacitor with multiple sectors.** Repeat the previous MATLAB exercise but for a parallel-plate capacitor with N dielectric sectors placed like in Fig.2.13(b). *(ME2_24.m on IR)*

---

3 STEADY ELECTRIC CURRENTS

Introduction:

So far, we have dealt with electrostatic fields, associated with time-invariant charges at rest. We now consider the charges in an organized macroscopic motion, which constitute an electric current. Our focus in this chapter is on the steady flow of free charges in conducting materials (for conducting properties of materials, see Table 3.1), i.e., on steady (time-invariant) electric currents, whose macroscopic characteristics (like the amount of current through a wire conductor) do not vary with time. Steady currents are also called direct currents, abbreviated dc.

3.1 Continuity Equation, Conductivity, and Ohm's Law in Local Form

The current intensity, I, is defined as a rate of movement of charge passing through a surface (e.g., cross section of a cylindrical conductor),

$$I = \frac{\mathrm{d}Q}{\mathrm{d}t} \quad \text{(current intensity or, simply, current; unit: A)} , \tag{3.1}$$

i.e., I equals the total amount of charge that flows through the surface during an elementary time $\mathrm{d}t$, divided by $\mathrm{d}t$. The unit for current intensity, which is usually referred to as, simply, current, is ampere or amp (A), equal to C/s. The current density vector, $\mathbf{J}$, is a vector that is directed along the current lines and whose magnitude, with reference to Fig.3.1(a), is given by

$$J = \frac{\mathrm{d}I}{\mathrm{d}S} \quad \text{(current density; unit: A/m}^2\text{)} , \tag{3.2}$$

where $\mathrm{d}I$ is the current flowing through an elementary surface of area $\mathrm{d}S$.

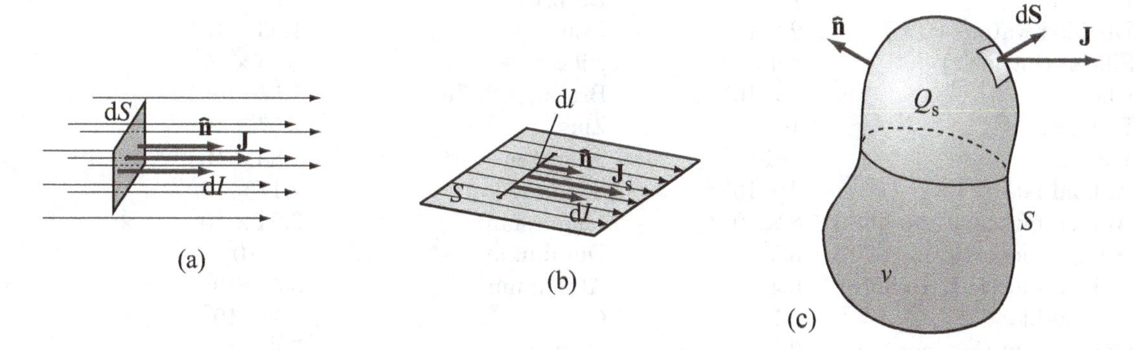

Figure 3.1 (a) Current density vector ($\mathbf{J}$). (b) Surface current density vector ($\mathbf{J}_\mathrm{s}$). (c) Arbitrary closed surface in a region with currents.

In many situations, current flow is localized in a very thin (theoretically infinitely thin) film over a surface, as shown in Fig.3.1(b). This is so-called surface current, described by the surface current density vector, $\mathbf{J}_s$, which is defined as

$$J_s = \frac{dI}{dl} \quad \text{(surface current density; unit: A/m)}, \tag{3.3}$$

where dI is the current flowing across a line element dl set normal to the current flow [Fig.3.1(b)]. Note that the surface current density vector is sometimes denoted as $\mathbf{K}$.

By the continuity equation, the total current of any time dependence leaving a domain v through a closed surface S, that is, the total outward flux of the current density vector through S, Fig.3.1(c), is equal to the negative derivative in time of the total charge enclosed by S,

$$\oint_S \mathbf{J} \cdot d\mathbf{S} = -\frac{dQ_S}{dt} \quad \text{(continuity equation, for currents of any time dependence)}. \tag{3.4}$$

For steady currents, the total current leaving v across S is zero, which is a generalization of Kirchhoff's current law in circuit theory.

Table 3.1 Conductivity of selected materials*

Material	σ (S/m)	Material	σ (S/m)
Quartz (fused)	$\sim 10^{-17}$	Carbon (graphite)	7.14×10^4
Wax	$\sim 10^{-17}$	Bismuth	8.70×10^5
Polystyrene	$\sim 10^{-16}$	Cast iron	$\sim 10^6$
Sulfur	$\sim 10^{-15}$	Nichrome	10^6
Mica	$\sim 10^{-15}$	Mercury (liquid)	1.04×10^6
Paraffin	$\sim 10^{-15}$	Stainless steel	1.1×10^6
Rubber (hard)	$\sim 10^{-15}$	Silicon steel	2×10^6
Porcelain	$\sim 10^{-14}$	Titanium	2.09×10^6
Carbon (diamond)	2×10^{-13}	Constantan (45% Ni)	2.26×10^6
Glass	$\sim 10^{-12}$	German silver	3×10^6
Polyethylene	1.5×10^{-12}	Lead	4.56×10^6
Wood	$10^{-11} - 10^{-8}$	Solder	7×10^6
Bakelite	$\sim 10^{-9}$	Niobium	8.06×10^6
Marble	10^{-8}	Tin	8.7×10^6
Granite	10^{-6}	Platinum	9.52×10^6
Dry soil	10^{-4}	Bronze	10^7
Distilled water	2×10^{-4}	Iron	1.03×10^7
Silicon (intrinsic)	4.4×10^{-4}	Nickel	1.45×10^7
Clay	5×10^{-3}	Brass (30% Zn)	1.5×10^7
Fresh water	10^{-2}	Zinc	1.67×10^7
Wet soil	$\sim 10^{-2}$	Tungsten	1.83×10^7
Animal fat**	4×10^{-2}	Sodium	2.17×10^7
Animal muscle ($\perp$ to fibre)**	8×10^{-2}	Magnesium	2.24×10^7
Animal, body (average)**	0.22	Duralumin	3×10^7
Animal muscle ($\parallel$ to fibre)**	0.4	Aluminium	3.5×10^7
Animal blood**	0.7	Gold	4.1×10^7
Germanium (intrinsic)	2.2	Copper	5.8×10^7
Seawater	3–5	Silver	6.17×10^7
Ferrite	10^2	Mercury (at $<$4.1 K)	∞
Tellurium	$\sim 5 \times 10^2$	Niobium (at $<$ 9.2 K)	∞
Silicon (doped)	1.18×10^3	$YBa_2Cu_3O_7$ (at $<$80 K)	∞

* For dc or low-frequency currents, at room temperature.

** Also for humans.

In linear conducting media,

$$\mathbf{J} = \sigma \mathbf{E} \quad \text{or} \quad \mathbf{E} = \rho \mathbf{J} \quad \text{(Ohm's law in local form)} , \tag{3.5}$$

where σ is the conductivity [unit: siemens per meter (S/m)] and ρ the resistivity [unit: ohm×meter (Ωm)] of the medium. These relationships are known as Ohm's law in local or point form. Almost always, the conductivity is a function of temperature, T. For metallic conductors, the conductivity decreases and resistivity increases with a temperature rise. Around a room temperature of $T_0 =$ 293 K (20°C), the resistivity varies almost linearly with T, and we can write

$$\rho(T) = \rho_0 \left[1 + \alpha (T - T_0)\right] , \tag{3.6}$$

where $\rho_0 = \rho(T_0)$. For most metals (copper, aluminum, silver, etc.), the temperature coefficient of resistivity, α, is approximately 0.4% per kelvin. Unlike the relative permittivity (ε_r), shown in Table 2.1, the conductivity of materials varies over an extremely wide range of values, as can be seen in Table 3.1.

MATLAB EXERCISE 3.1 **GUI for the conductivity table of materials.** Repeat MATLAB Exercise 2.1 but for the values of the conductivity (σ) of selected materials given in Table 3.1. *[folder ME3_1(GUI) on IR]*[1]

MATLAB EXERCISE 3.2 **Temperature dependence of resistivity.** Write a MATLAB program that calculates and plots the resistivity of copper, nickel, and constantan (55% copper, 45% nickel), $\rho(T)$, in a temperature range 0 – 100°C, based on Eq.(3.6). The temperature coefficient of resistivity (α) for copper, nickel, and constantan amounts to 0.0039 K^{-1}, 0.006 K^{-1}, and 0.000008 K^{-1}, respectively; use the GUI from the previous MATLAB exercise to get the corresponding values of σ at room temperature. *(ME3_2.m on IR)*

MATLAB EXERCISE 3.3 **2-D vector plots of volume current and field.** Figure 3.2(a) shows a parallel-plate capacitor with circular plates of radius a = 0.5 cm and a continuously inhomogeneous imperfect dielectric. The permittivity and conductivity of the dielectric are the following functions of the z coordinate: $\varepsilon(z) = 2(1+3z/d)\varepsilon_0$ and $\sigma(z) = \sigma_0/(1+3z/d)$, $0 \le z \le d$, where $\sigma_0 = 3 \times 10^{-8}$ S/m and d = 1 cm. The capacitor is connected to a time-constant voltage V = 1 V. Write a MATLAB code that, starting with the expression for the current density vector in the dielectric of the capacitor, $\mathbf{J} = 2\sigma_0 V\,\hat{\mathbf{z}}/(5d)$, obtained from the continuity equation, calculates the electric field intensity vector (**E**) and electric flux density vector (**D**) in the dielectric. The code also gives graphical representations of the permittivity and conductivity functions, $\varepsilon(z)$ and $\sigma(z)$, and distributions of vectors **J**, **E**, and **D**. *(ME3_3.m on IR)* **H**[2]

TUTORIAL:

We first define a **z** vector with coordinates z separated by an increment `dz` – for plotting (the

---

[1] *IR* = Instructor Resources (for the book).

[2] **H** = recommended to be done also "by hand," i.e., not using MATLAB.

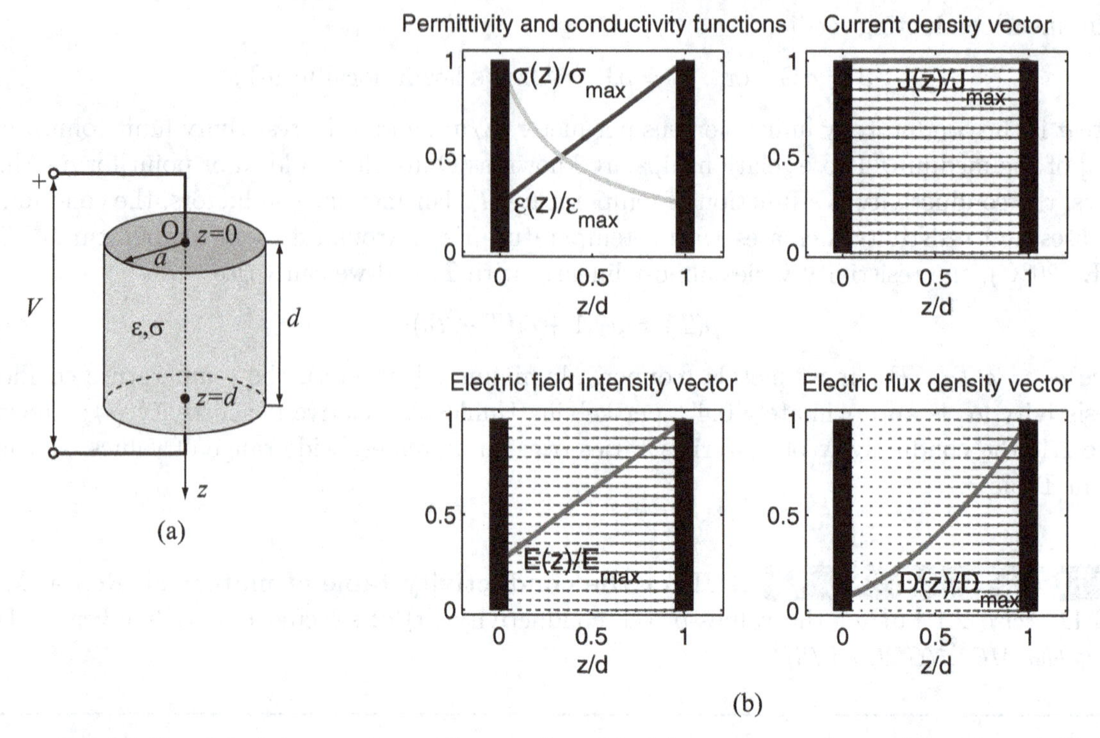

Figure 3.2 (a) Parallel-plate capacitor with a continuously inhomogeneous lossy dielectric and (b) MATLAB plots of the permittivity and conductivity functions, $\varepsilon(z)$ and $\sigma(z)$, and distributions of the current density vector (**J**), electric field intensity vector (**E**), and electric flux density vector (**D**) in the capacitor; for MATLAB Exercise 3.3. *(color figure on CW)*[3]

number of increments is N+1), and permittivity and conductivity functions,

```
N = 20;
dz = d/N;
z = [0:dz:d];
EPS = 2*EPS0*(1+3*z./d);
SIGMA = SIGMA0./(1+3*z/d);
```

Then, we compute magnitudes of vectors **J**, **E** [from Eq.(3.5)], and **D** [Eq.(2.2)], variables J, E, and D, respectively, all having the same length as the z vector,

```
J = 2*SIGMA0*V/(5*d)*ones(1,N+1);
E = J./SIGMA;
D = (EPS./SIGMA).*J;
```

To be able to create 2-D plots in the longitudinal cut of the capacitor in Fig.3.2(a) (in the plane containing the z-axis), we convert vectors z, J, E and D into matrices. Since $2a = d$, we can adopt the same increment for plotting along the y-axis (normal to the z-axis) in this plane, dy = dz. Of course, y-components of **J**, **E**, and **D** are zero, and none of the quantities changes its value in the y direction. The Z matrix contains z-coordinates of points (nodes of the 2-D mesh) at which

[3] CW = Companion Website (of the book).

the vectors are computed and plotted.

```
for i = 1:length(z);
EZ(i,:)  = E;
DZ(i,:)  = D;
JZ(i,:)  = J;
Z(i,:)  = z;
end;
EY = zeros(N+1,N+1);
DY = zeros(N+1,N+1);
JY = zeros(N+1,N+1);
```

The matrix `Y` (with y-coordinates of nodes) is the transpose of `Z` (`Y = Z'`). We also define a parameter **`scale`** to be used for the normalization of `Z` and `Y` in plotting, and fill vectors **`p1`** and **`p2`** with coordinates of capacitor plates ($z = 0$ and $z = d$, respectively), which will be used to draw the plates in 2-D plots.

```
Y = Z';
scale = max(z);
p1 = zeros(1,N+1);
p2 = d*ones(1,N+1);
```

For plots, we use MATLAB functions **`subplot`** (for division of the figure into subplots), **`quiver`** (see MATLAB Exercise 1.13), and **`axis equal`**. Function **`axis`** defines the limits of both axes on the plot. Note that plots need to be scaled. The first subplot visualizes $\varepsilon(z)$ and $\sigma(z)$:

```
subplot(2,2,1);
plot(z/scale,EPS/max(EPS),'b','linewidth',2); hold on;
plot(z/scale,SIGMA/max(SIGMA),'g','linewidth',2);
plot(p1/scale,z/scale,'k','linewidth',8);
plot(p2/scale,z/scale,'k','linewidth',8);
axis(1/scale*[-dz/2 d+dz, -dz/2 d+dz]); hold off;
axis equal;
```

The second subplot represents **J**,

```
subplot(2,2,2);
quiver(Z(:,1:N)/scale,Y(:,1:N)/scale,JZ(:,1:N),JY(:,1:N)); hold on;
plot(z/scale,J/max(J),'r','linewidth',2);
plot(p1/scale,z/scale,'k','linewidth',8);
plot(p2/scale,z/scale,'k','linewidth',8);
axis(1/scale*[-dz/2 d+dz, -dz/2 d+dz]);
hold off;
axis equal;
```

and the two remaining subplots, for **E** and **D**, are done in the analogous fashion. The plots are shown in Fig.3.2(b).

MATLAB EXERCISE 3.4 **3-D plot of surface currents over a spherical electrode.** A spherical capacitor, Fig.2.9(a), is filled with an imperfect homogeneous dielectric, so that the time-invariant current of intensity $I = 1$ A flows through the capacitor terminals. The inner electrode, of radius $a = 1$ m, is hollow, with a very thin wall, and hence the current through this electrode can be represented by a surface current density vector, $\mathbf{J}_s$ [Fig.3.1(b) and Eq.(3.3)]. In the spherical coordinate system shown in Fig.3.3, $\mathbf{J}_s$, as determined by the continuity equation, equals

$$\mathbf{J}_s(\theta) = \frac{I - J(a)S(\theta)}{2\pi a \sin\theta}\,\hat{\boldsymbol{\theta}}\,, \quad J(a) = \frac{I}{4\pi a^2}\,, \quad S(\theta) = 2\pi a^2(1 - \cos\theta)\,, \tag{3.7}$$

where $J(a)$ is the volume current density in the dielectric right at the surface of the inner electrode, while $S(\theta)$ is the surface area of the part of the electrode above the horizontal circle whose position on the electrode is defined by the angle θ in Fig.3.3, and whose circumference is $2\pi a \sin\theta$ [see also Fig.1.9 and Eq.(1.20)]. In MATLAB, plot the distribution of surface currents over the inner electrode. *(ME3_4.m on IR)*

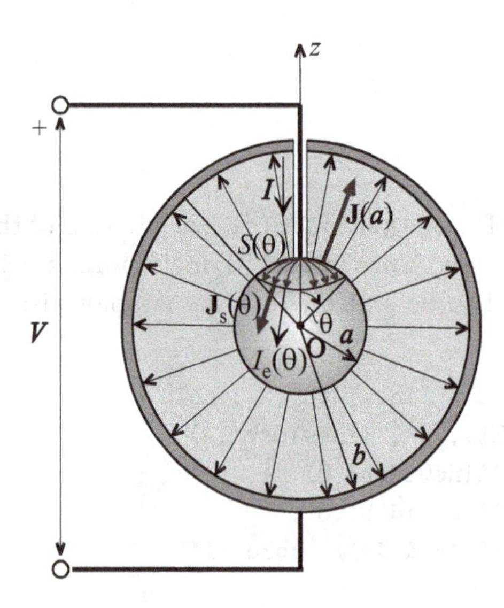

Figure 3.3 Surface current density vector ($\mathbf{J}_s$) through the thin wall of the inner electrode of a spherical capacitor with imperfect dielectric; for MATLAB Exercise 3.4.

TUTORIAL:

We define a unit sphere, for plotting, with the use of MATLAB function `sphere(N)`, whose output consists of three `(N+1)`× `(N+1)` matrices, with x-, y-, and z-coordinates `(x,y,z)`, respectively, of node points in a mesh (grid) over the sphere surface, in the Cartesian coordinate system associated with the spherical system in Fig.3.3 (we adopt `N = 25`, which results in a mesh of the sphere surface with 26 × 26 nodes).

```
[X Y Z]  = sphere(N);
```

After building the mesh, the magnitude of the surface current density vector, $\mathbf{J}_s$, at every node of the mesh is calculated, based on Eq.(3.7). This is implemented using two counters (indices), `i` and `j`, to access the Cartesian coordinates of the nodes, and function `car2Sph` from MATLAB Exercise 1.24 to convert Cartesian to spherical coordinates, and the result is a matrix `Js(i,j)`, suitable for plotting by means of MATLAB function **surf** – the graph is shown in Fig.3.4.

```
for i=1:N+1
for j=1:N+1
[r theta phi] = Car2Sph(X(i,j),Y(i,j),Z(i,j));
J = I/4/pi/a^2;
S = 2*pi*a^2*(1-cos(theta));
Js(i,j) = ((I-J*S)/2/pi/a/sin(theta));
end;
end;
surf(X,Y,Z,Js);
```

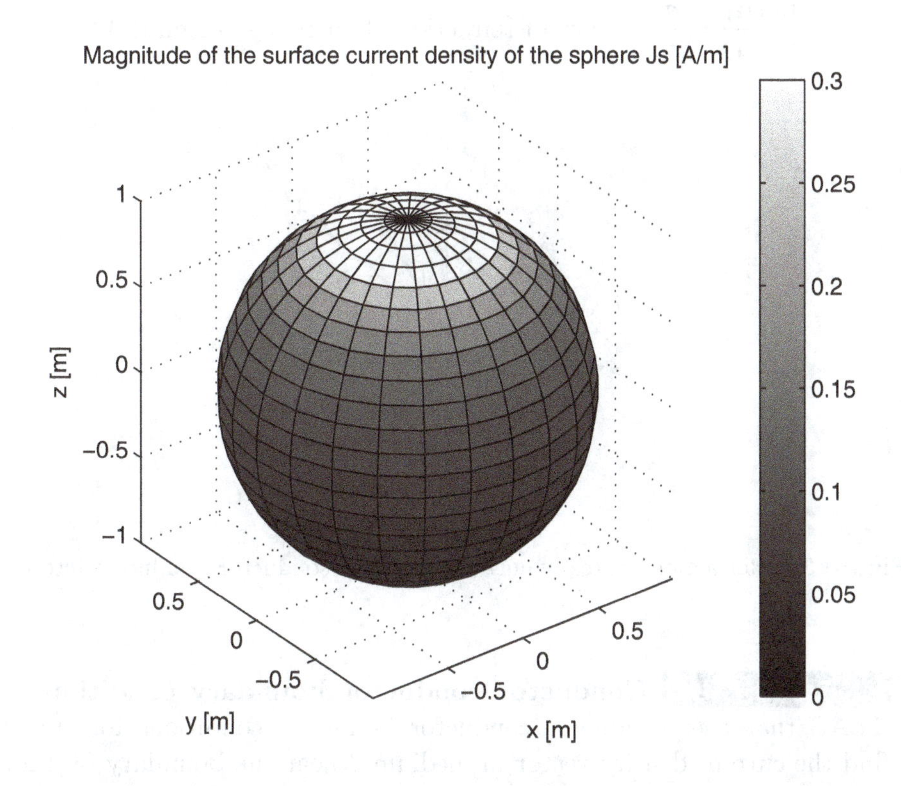

Figure 3.4 3-D plot – using MATLAB functions **sphere** and **surf** – of the distribution of surface currents over the inner electrode of the spherical capacitor in Fig.3.3; for MATLAB Exercise 3.4. *(color figure on CW)*

3.2 Boundary Conditions for Steady Currents

Comparing Eqs.(3.4) and (2.1), we conclude that the boundary condition for normal components of the vector **J** at interfaces between conducting media of different conductivity is of the same form as the boundary condition for the vector **D**, in Eqs.(2.4). The only difference is on the right-hand side of the equation, where ρ_s (the surface charge density that may exist on the surface) is replaced by $-\partial\rho_s/\partial t$. For steady currents, $-\partial\rho_s/\partial t = 0$, and the boundary condition that corresponds to Eq.(1.28) is the same as for the electrostatic field, so as in Eqs.(2.4); hence, the complete set of boundary conditions for steady currents is given by

$$E_{1t} = E_{2t} , \quad J_{1n} = J_{2n} \quad \text{(boundary conditions for steady currents)} . \tag{3.8}$$

With α_1 and α_2 denoting the angles that current lines in region 1 and region 2 make with the normal to the boundary interface, $\mathbf{n}$, as shown in Fig.3.5, we have, using Eq.(3.5), $\tan\alpha_1 = J_{1t}/J_{1n} = \sigma_1 E_{1t}/J_{1n}$ and $\tan\alpha_2 = J_{2t}/J_{2n} = \sigma_2 E_{2t}/J_{2n}$, which, combined with Eqs.(3.8), then results in the law of refraction of the current density lines at the boundary:

$$\frac{\tan\alpha_1}{\tan\alpha_2} = \frac{\sigma_1}{\sigma_2} \quad \text{(law of refraction of current streamlines)} . \tag{3.9}$$

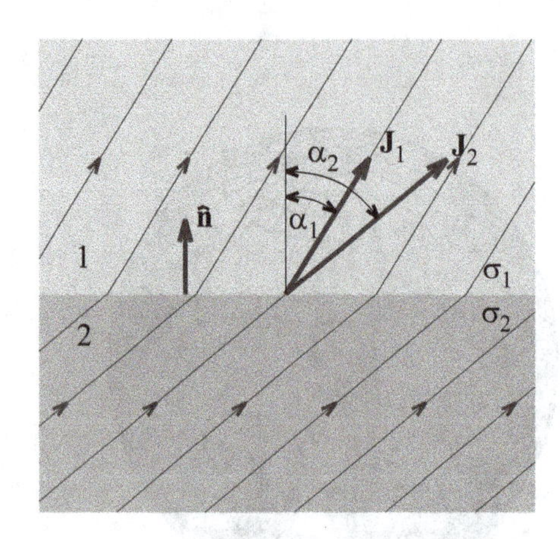

Figure 3.5 Refraction of steady current lines at a conductor–conductor interface.

MATLAB EXERCISE 3.5 **Conductor–conductor boundary conditions.** Write a program in MATLAB that uses conductor–conductor boundary conditions for steady currents in Eqs.(3.8) to find the current density vector in medium 2 near the boundary ($\mathbf{J}_2$) for a given current density vector in medium 1 near the boundary ($\mathbf{J}_1$), if the boundary plane coincides with the xy-plane ($z = 0$) of the Cartesian coordinate system and the unit normal directed from region 2 to region 1 (Fig.3.5) is $\hat{\mathbf{n}} = \hat{\mathbf{z}}$. The conductivities of the two materials amount to $\sigma_1 = 2 \times 10^{-3}$ S/m and $\sigma_2 = 8 \times 10^{-3}$ S/m, and $\mathbf{J}_1 = (\hat{\mathbf{x}} + 2\hat{\mathbf{y}} + 3\hat{\mathbf{z}})$ kA/m^2. Input data are entered from the keyboard. The result for the Cartesian components of $\mathbf{J}_2$ is displayed in the Command Window. The output graph shows vectors $\mathbf{J}_1$, $\mathbf{J}_2$, and $\hat{\mathbf{n}}$ (plotted by MATLAB function `quiver3`), as well as the boundary plane (using MATLAB function `surf`). *(ME3_5.m on IR)* **H**

HINT:
See MATLAB Exercises 2.4 and 2.6.

MATLAB EXERCISE 3.6 **Law of refraction of current streamlines.** With reference to Fig.3.5, let the angle that current streamlines in region 1 make with the normal to the interface be α_1, and let the magnitude of the current density vector in this region be J_1. In MATLAB, compute α_2 and J_2, on the other side of the boundary, for the given conductivities of the two media, σ_1 and σ_2, and graphically represent vectors $\mathbf{J}_1$, $\mathbf{J}_2$, and $\hat{\mathbf{n}}$ in a 2-D view (as in Fig.3.5). Using this MATLAB code, visualize refraction of current lines at interfaces between various combinations of media, taking the conductivity values from the GUI developed in MATLAB Exercise 3.1. *(ME3_6.m on IR)*

3.3 Relaxation Time

For a homogeneous conducting medium of conductivity σ and permittivity ε, a combination of Eqs.(2.1), (2.2), and (3.5) and the differential form of Eq.(3.4), $\nabla \cdot \mathbf{J} = -\partial\rho/\partial t$, gives the following first-order partial differential equation in time, t, for the charge density, ρ, in the medium and its solution:

$$\frac{\partial \rho}{\partial t} + \frac{\sigma}{\varepsilon}\rho = 0 \quad \longrightarrow \quad \rho = \rho_0\,\mathrm{e}^{-(\sigma/\varepsilon)\,t} = \rho_0\,\mathrm{e}^{-t/\tau}\,, \quad \tau = \frac{\varepsilon}{\sigma} \quad \text{(relaxation time)}\,, \tag{3.10}$$

where the time constant τ is referred to as the relaxation time (the unit is s) and equals the time required for the charge density at any point of the medium to decay to 1/e (36.8%) of its initial value (at $t = 0$), ρ_0.

Whether a material of parameters σ and ε is considered a good conductor or a good dielectric is decided on the basis of the relaxation time, as compared to times of interest in a given application. Thus, for a time-harmonic (sinusoidal) field of frequency f, the relaxation time, given by Eq.(3.10), is compared to the time period, $T = 1/f$. If $\tau \ll T$, the medium is classified as a good conductor. In particular, if $\tau = 0$ ($\sigma \to \infty$), the material is said to be the perfect electric conductor (PEC). On the other hand, the material is considered a good dielectric (insulator) if $\tau \gg T$. In a limit, $\tau \to \infty$ for perfect (lossless) dielectrics ($\sigma = 0$). For all other (intermediate) values of τ, the material is classified as a quasi-conductor.

MATLAB EXERCISE 3.7 **Relaxation time, good conductors and dielectrics.** Write a MATLAB code that calculates the relaxation time for the given conductivity and permittivity of a material. In addition, the code also determines – for the frequency f of a time-harmonic electric field in which the material resides – whether the material behaves like (1) a good conductor (use the criterion $\tau/T < 1/100$), (2) a quasi-conductor ($1/100 \le \tau/T \le 100$), or (3) a good dielectric ($\tau/T > 100$). The input data are entered from the keyboard, and the results are displayed in the Command Window. Using this code, evaluate relaxation times for various materials, taking the conductivity and permittivity values from GUI's written in MATLAB Exercises 3.1 and 2.1, respectively, and investigate their behavior, in terms of classifications (1)–(3), at different frequencies. *(ME3_7.m on IR)*

MATLAB EXERCISE 3.8 **Redistribution of charge in mica.** In MATLAB, compute and plot the exponential redistribution of charge (in %), based on Eq.(3.10), in mica ($\varepsilon_\mathrm{r} = 5.4$ and $\sigma = 10^{-15}$ S/m) during 6 days, assuming that $\rho_0 = 1$ C/m^3. *(ME3_8.m on IR)*

3.4 Resistance and Ohm's Law

A conductor, with two terminals and a (substantial) resistance, R, is usually referred to as a resistor. The relation between the voltage, current, and resistance of a resistor is known as Ohm's law:

$$V = RI \quad \text{or} \quad I = GV \quad \left(G = \frac{1}{R}\right) \quad \text{(Ohm's law)} . \tag{3.11}$$

The resistance is always nonnegative ($R \geq 0$), and the unit is the ohm (Ω), equal to V/A. The value of R depends on the shape and size of the conductor (resistor), and on the conductivity σ (or resistivity ρ) of the material. The reciprocal of resistance is called the conductance and symbolized by G. Its unit is the siemens (S), where S $= \Omega^{-1} =$ A/V. Note that sometimes the mho (ohm spelled backwards) is used instead of the siemens.

As an example, the resistance of a homogeneous resistor with a uniform cross section of an arbitrary shape and surface area S can be found as follows:

$$J = \frac{I}{S} \quad \longrightarrow \quad V = El = \frac{J}{\sigma} l = \frac{Il}{\sigma S} \quad \longrightarrow \quad R = \frac{V}{I} = \frac{l}{\sigma S} , \tag{3.12}$$

where I, J, E, l, and V are the current intensity, current density, electric field intensity, length, and voltage of the resistor and σ is the conductivity of its material.

MATLAB EXERCISE 3.9 **Resistances of resistors with uniform cross sections.** Write a MATLAB function `resistance()` that finds the resistances of multiple homogeneous resistors with uniform cross sections. Input data are arrays of conductivities, lengths, and cross-sectional surface areas of resistors, where ith element of each array corresponds to the ith resistor, and the output is an array of resulting resistances. *(resistance.m on IR)*

MATLAB EXERCISE 3.10 **Multiple resistors in series and parallel.** Write MATLAB functions `resistorsInSeries()` and `resistorsInParallel()` that calculate the equivalent resistance of an arbitrary number of resistors connected in series and parallel, respectively, for their given resistances. *(resistorsInSeries.m and resistorsInParallel.m on IR)*

MATLAB EXERCISE 3.11 **Two resistors with two cuboidal parts.** Write a program in MATLAB that calls functions from the previous two MATLAB exercises to find the resistances of the two resistors shown in Figs.3.6(a) and (b), respectively, formed from two rectangular cuboids of the same size, with sides $a = 8$ mm, $b = 2$ mm, and $c = 4$ mm. The cuboids are made out from different resistive materials, with conductivities $\sigma_1 = 10^5$ S/m and $\sigma_2 = 4 \times 10^5$ S/m. Input data are read from a text file `resistorsSpecification.txt`, with numerical values for σ_1 and σ_2 appearing in the first line of the file, followed by those for a, b, and c in the second, third, and fourth line, respectively. The results are displayed in the Command Window. *(ME3_11.m on IR)* **H**

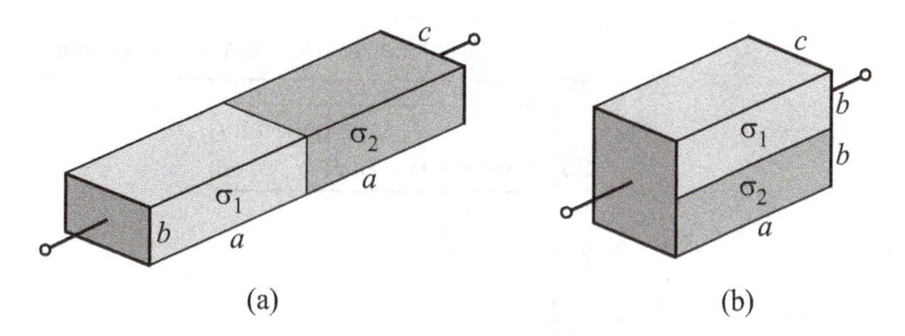

Figure 3.6 Two rectangular cuboids made out from different resistive materials connected in (a) series and (b) parallel; for MATLAB Exercise 3.11.

HINT:
Here, we explain the data input from a text file – in MATLAB. First, create a file with extension `.txt`, `resistorsSpecification.txt` in our case, and fill it with numerical data – for σ_1 and σ_2 (first line) and a, b, and c (in the following lines) – as prescribed above (we duplicate data for each of the dimensions, so that they can be different for the two resistors in a more general case),

```
1e5 4e5
8e-3 8e-3
2e-3 2e-3
4e-3 4e-3
```

(data in the same row are separated by empty spaces). The use of the file and its data in a MATLAB program is enabled by MATLAB function `load`,

```
load resistorsSpecification.txt;
```

The data, for the conductivity array `sigma` and dimensions of resistors `a`, `b`, and `c` as variables in the program, are read from the file calling the name of the file and specifying the appropriate rows of the file, as follows:

```
sigma = resistorsSpecification(1,:);
a = resistorsSpecification(2,:);
b = resistorsSpecification(3,:);
c = resistorsSpecification(4,:);
```

MATLAB EXERCISE 3.12 **Graphical and numerical solutions for a nonlinear circuit.** Consider a circuit with a voltage generator connected to a nonlinear resistor, whose resistance depends on the applied voltage, $R = R(V)$, shown in Fig.3.7(a). Solve this circuit in MATLAB. Input data are the electromotive force (emf) and internal resistance of the generator, $\mathcal{E}$ (note that V_{emf} is also used to denote emf) and R_{g}, respectively, and the function $R(V)$ in a range of voltages $0-30$ V (data entry is from the keyboard). The output consists of the results for the resistance, voltage, and current of the nonlinear resistor, as well as the graphical representation of the solution in the current-voltage diagram (plane). Test the code for $\mathcal{E}=10$ V, $R_{\mathrm{g}}=100\ \Omega$, and $R(V)=10(V+5)\ \Omega$ (V in V). *(ME3_12.m on IR)* **H**

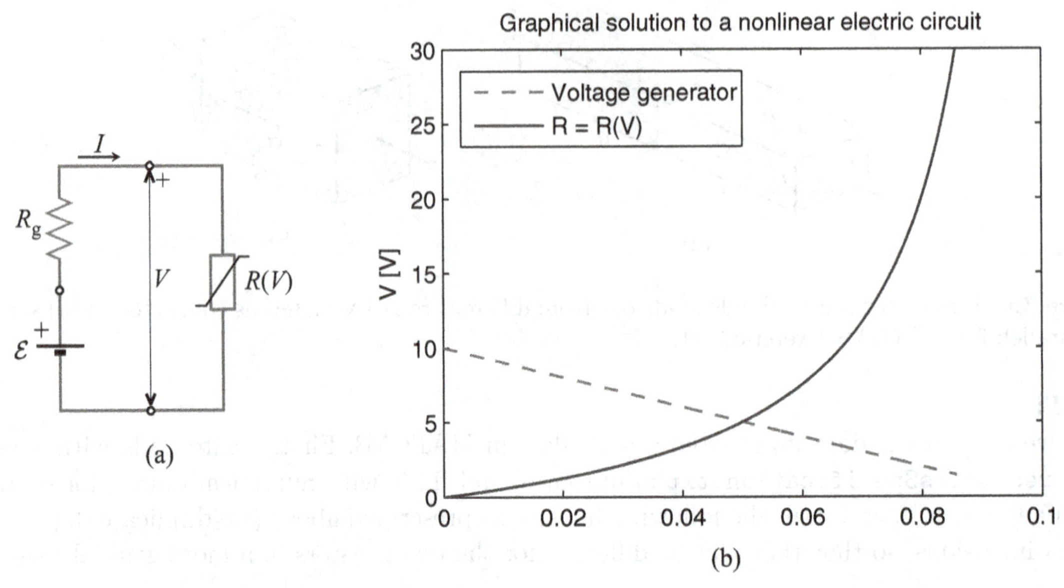

Figure 3.7 (a) Circuit with a voltage generator and a nonlinear resistor and (b) analysis of the nonlinear electric circuit in MATLAB: graphical representation of Eqs.(3.13) and (3.14) for the circuit; for MATLAB Exercise 3.12. *(color figure on CW)*

TUTORIAL:

Combining Eqs.(3.11) and $R=R(V)$, we can express the voltage V in Fig.3.7(a) as

$$V=R(V)I\ . \tag{3.13}$$

It, on the other side, also equals the voltage of the generator,

$$V=\mathcal{E}-R_{\mathrm{g}}I\ . \tag{3.14}$$

The solution for V and I of these two equations, one of which being nonlinear, is the solution of the circuit.

First, it is necessary to declare the voltage variable `V` as a symbolic one, because of the dependence of `R` on `V` at input. Then, the voltage range is defined, and input data are entered from the keyboard.

```
syms V
V = 0:0.01:30;
R = input('Enter R [Ohm] dependence on V[V]: ');
Rg = input('Enter Rg [Ohm]:  ');
E = input('Enter E [V]: ');
```

Next, we make the resistance variable `R` a numerical one, by means of MATLAB function `double`; the numerical variable is named `R1`, to distinguish it from its symbolic version. The current through the resistor is calculated from Eq.(3.13).

```
R1 = double(R);
I = V./R1;
```

Using MATLAB function `plot`, we graphically represent Eqs.(3.13) and (3.14), i.e., the current-voltage characteristic of the nonlinear resistor and the equation of the generator, as depicted in Fig.3.7(b). The latter graph is the load line for the circuit in Fig.3.7(a) (the locus of all possible combinations of values `V` and `I`, not yet taking into account the characteristic of the load).

```
plot(I,E-Rg.*I,'g--',I,V,'b');
```

(`'g'` and `'b'` stand for green and blue, respectively – for colors of curves, and, more specifically, `'g--'` makes the green curve dashed, while `'b'` results in a solid blue curve).

The intersection of the two curves represents the operating point for the circuit, and we can find the abscissa and ordinate of this point either graphically (reading the values as precisely as possible from the figure) or numerically, which is more accurate. In the numerical solution, we find the point (locus) of the minimum difference (ideally, zero difference) between the two functions, in Eqs.(3.13) and (3.14), using MATLAB function `min`. This function searches for the minimum value of an array of values and returns the minimum as output. It also returns as output the index (position) of the minimum value in the array. In the case of multiple minima, the function returns the index of the first one.

```
[Vmin,n]  = min(abs(E-Rg.*I-V));
```

Once the index (`n`) of the operating point is found, we display in the Command Window (by MATLAB function `disp`) the associated solutions for `R`, `V`, and `I`,

```
disp('Solution for R is:  ');
fprintf('%g ohm.\n',R1(n));
disp('Solution for V is:');
fprintf('%g V.\n',V(n));
disp('Solution for I is:');
fprintf('%g A.\n',I(n));
```

and here is what we get in the Command Window:

```
Solution for R is:
100 ohm.
Solution for V is:
5 V.
Solution for I is:
0.05 A.
```

4 MAGNETOSTATIC FIELD IN FREE SPACE

Introduction:

We now introduce a series of new phenomena associated with steady electric currents, which are essentially the consequence of a new simple experimental fact – that conductors with currents exert forces on one another. These forces are called magnetic forces, and the field due to one current conductor in which the other conductor is situated and which causes the force on it is called the magnetic field. Any motion of electric charges and any electric current are followed by the magnetic field. The magnetic field due to steady electric currents is termed the steady (static) magnetic field or magnetostatic field. The theory of the magnetostatic field, the magnetostatics, restricted to a vacuum and nonmagnetic media is the subject of this chapter. The magnetic materials will be studied in the following chapter.

4.1 Magnetic Force and Magnetic Flux Density Vector

To quantitatively describe the magnetic field, we introduce a vector quantity called the magnetic flux density vector, $\mathbf{B}$. It is defined analogously to the electric field intensity vector, $\mathbf{E}$, in electrostatics [Eq.(1.7)] through the force, magnetic force, on a small probe point charge Q_p moving at a velocity $\mathbf{v}$ (note that symbol $\mathbf{u}$ is also often used to denote velocity) in the field, which equals the cross product[1] of vectors $Q_\mathrm{p}\mathbf{v}$ and $\mathbf{B}$,

$$\mathbf{F}_\mathrm{m} = Q_\mathrm{p}\mathbf{v} \times \mathbf{B} \quad (Q_\mathrm{p} \to 0) \quad \text{(definition of the magnetic flux density } \mathbf{B}\text{; unit: T)} . \tag{4.1}$$

The unit for $\mathbf{B}$ is tesla (abbreviated T).

The magnetic equivalent of Coulomb's law, Eq.(1.1), states that the magnetic force on a point charge Q_2 that moves at a velocity $\mathbf{v}_2$ in the magnetic field due to a point charge Q_1 moving with a velocity $\mathbf{v}_1$ in a vacuum (or air), Fig.4.1(a), is given by

$$\mathbf{F}_\mathrm{m12} = \frac{\mu_0}{4\pi} \frac{Q_2\mathbf{v}_2 \times \left(Q_1\mathbf{v}_1 \times \hat{\mathbf{R}}_{12}\right)}{R^2} \quad \text{(magnetic force)} , \tag{4.2}$$

where μ_0 is the permeability of a vacuum (free space),

$$\mu_0 = 4\pi \cdot 10^{-7}\ \mathrm{H/m} \quad \text{(permeability of a vacuum)} . \tag{4.3}$$

---

[1]The cross product of vectors $\mathbf{a}$ and $\mathbf{b}$, $\mathbf{a}\times\mathbf{b}$, is a vector whose magnitude is given by $|\mathbf{a}\times\mathbf{b}| = |\mathbf{a}||\mathbf{b}|\sin\alpha$, where α is the angle between the two vectors in the product. It is perpendicular to the plane defined by the vectors $\mathbf{a}$ and $\mathbf{b}$, and its direction (orientation) is determined by the right-hand rule when the first vector ($\mathbf{a}$) is rotated by the shortest route toward the second vector ($\mathbf{b}$). In this rule, the direction of rotation is defined by the fingers of the right hand when the thumb points in the direction of the cross product.

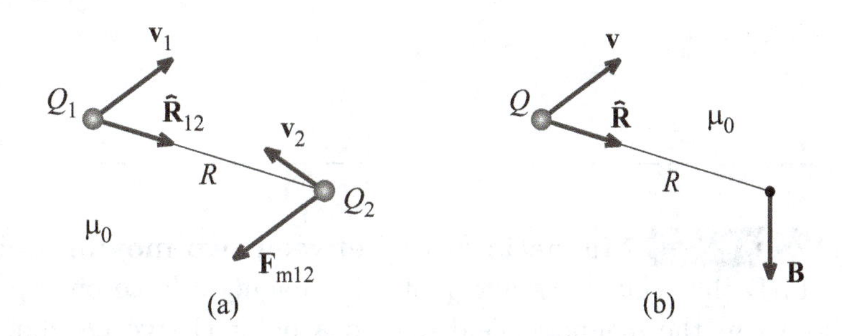

Figure 4.1 (a) Magnetic force between two point charges moving in a vacuum, given by Eq.(4.2). (b) Magnetic flux density vector due to a moving point charge, Eq.(4.4).

Combining Eqs.(4.1) and (4.2), we can identify the expression for the magnetic flux density vector of a point charge Q moving with a velocity $\mathbf{v}$ [Fig.4.1(b)]:

$$\mathbf{B} = \frac{\mu_0}{4\pi} \frac{Q\mathbf{v} \times \hat{\mathbf{R}}}{R^2} \quad (\mathbf{B} \text{ due to a moving point charge in free space}) . \tag{4.4}$$

MATLAB EXERCISE 4.1 **Cross product of two vectors.** Write a function `crossProduct()` in MATLAB that computes the cross product of two vectors in the Cartesian coordinate system. Test the function for vectors $\mathbf{a} = \hat{\mathbf{x}} + \hat{\mathbf{y}} + 2\hat{\mathbf{z}}$ and $\mathbf{b} = 3\hat{\mathbf{x}} + 2\hat{\mathbf{y}} + 4\hat{\mathbf{z}}$. *(crossProduct.m on IR)*[2]

TUTORIAL:

For two vectors given by their Cartesian components,

$$\mathbf{a} \times \mathbf{b} = (a_x\,\hat{\mathbf{x}} + a_y\,\hat{\mathbf{y}} + a_z\,\hat{\mathbf{z}}) \times (b_x\,\hat{\mathbf{x}} + b_y\,\hat{\mathbf{y}} + b_z\,\hat{\mathbf{z}})$$

$$= (a_y b_z - a_z b_y)\,\hat{\mathbf{x}} + (a_z b_x - a_x b_z)\,\hat{\mathbf{y}} + (a_x b_y - a_y b_x)\,\hat{\mathbf{z}} \,. \tag{4.5}$$

In an `if-else` branching, we check the lengths of the input vectors. The function returns the components of the vector $\mathbf{c} = \mathbf{a} \times \mathbf{b}$, computed based on Eq.(4.5), as follows:

```
function [CX,CY,CZ] = crossProduct(A,B)
if length(A) == 3 && length(B) == 3
C = [A(2)*B(3)-A(3)*B(2), A(3)*B(1)-A(1)*B(3), A(1)*B(2)-A(2)*B(1)];
CX = C(1);
CY = C(2);
CZ = C(3);
else
disp('Error - one or more vector components missing!');
CX = 0;
CY = 0;
```

[2] *IR* = Instructor Resources (for the book).

```
CZ = 0;
end;
```

MATLAB EXERCISE 4.2 **Magnetic force between two moving point charges.** Write a program in MATLAB that calculates and plots the magnetic force on a point charge Q_2 that moves at a velocity $\mathbf{v}_2$ in the magnetic field due to a point charge Q_1 moving with a velocity $\mathbf{v}_1$ in free space, at an instant when the charge Q_1 is at the coordinate origin and Q_2 is at the point defined by its coordinates (x, y, z) in a Cartesian coordinate system. The velocity vectors are given by their Cartesian components. Test the program for $Q_1 = 10$ pC, $Q_2 = -10$ pC, $(x, y, z) = (20\ \text{mm}, 10\ \text{mm}, -5\ \text{mm})$, $\mathbf{v}_1 = (\hat{\mathbf{x}} + 3\,\hat{\mathbf{y}} + 4\,\hat{\mathbf{z}})$ m/s, and $\mathbf{v}_2 = (6\,\hat{\mathbf{x}} + 3\,\hat{\mathbf{y}} + 2\,\hat{\mathbf{z}})$ m/s. *(ME4_2.m on IR)* **H**[3]

TUTORIAL:

We input data using MATLAB function `input`, as in MATLAB Exercise 1.4, for instance. Upon forming the velocity vectors, `V1` and `V2`, from their Cartesian components, we find their magnitudes, `V1mag` and `V2mag`, by means of function `vectorMag` from MATLAB Exercise 1.1, and normalize vectors `V1` and `V2`, to be able to plot them. Similarly to the computation in MATLAB Exercise 1.4, we also calculate the distance between the charges, `r`, and the unit vector ($\hat{\mathbf{R}}_{12}$) along the line connecting them [Fig.4.1(a)].

```
V1 = [V1x,V1y,V1z];
V2 = [V2x,V2y,V2z];
V1mag = vectorMag(V1);
V2mag = vectorMag(V2);
Vmag = max(V1mag,V2mag);
V1norm = V1./Vmag;
V2norm = V2./Vmag;
r = vectorMag([x,y,z]);
ur = [x,y,z]/r;
```

Then, we implement the magnetic force expression in Eq.(4.2), with the use of function `crossProduct` (from the previous MATLAB exercise),

```
Q1V1 = Q1.*V1;
[cP1x,cP1y,cP1z] = crossProduct(Q1V1,ur);
cP1 = [cP1x,cP1y,cP1z];
Q2V2 = Q2.*V2;
[cP12x,cP12y,cP12z] = crossProduct(Q2V2,cP1);
cP12 = [cP12x,cP12y,cP12z];
Fm12 = MU0/4/pi.*cP12./r^2;
Fmmag = vectorMag(Fm12);
Fm12norm = Fm12./Fmmag;
```

where `MU0` is the permeability of free space (μ_0), given in Eq.(4.3).

---

[3]**H** = recommended to be done also "by hand," i.e., not using MATLAB.

The result for the magnitude of the force is displayed in the Command Window as:

```
Magnetic force between two point charges equals:  6.40767e-025 N.
```

Figure 4.2 shows the associated graphical 3-D view of the system of charges, obtained as in MATLAB Exercise 1.4, with MATLAB function `plot3` being employed to plot the charge points and to draw a line connecting them, while `vecPlot3D` (MATLAB Exercise 1.3) is used to draw vectors (`V1`, `V2`, and `Fm12`).

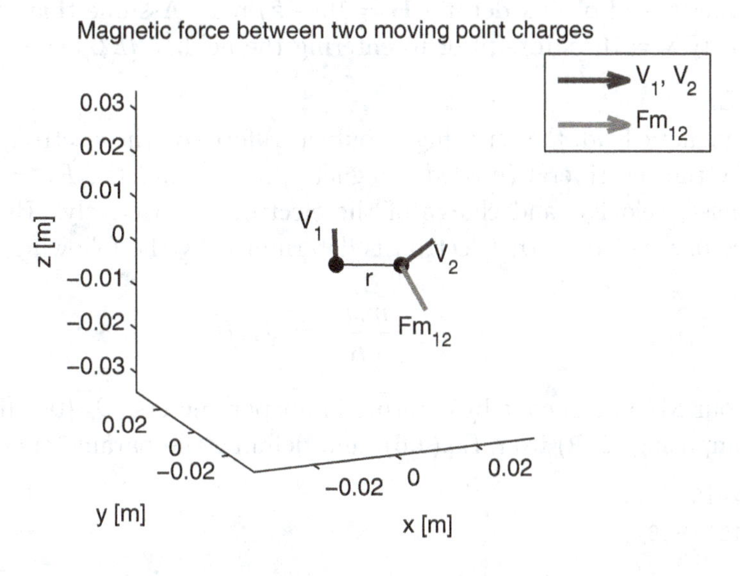

Figure 4.2 3-D visualization of a MATLAB solution for the magnetic force between two moving point charges in free space; for MATLAB Exercise 4.2. *(color figure on CW)*[4]

MATLAB EXERCISE 4.3 **Magnetic flux density vector due to a moving charge.** Expand the code developed in the previous MATLAB exercise to also calculate and plot the magnetic force on the charge Q_1 (due to Q_2), as well as the magnetic flux density vector $\mathbf{B}_1$ due to the moving charge Q_1 at the location of the charge Q_2, based on Eq.(4.4), and vice versa (vector $\mathbf{B}_2$ due to Q_2). The plot should contain the charges and vectors $\mathbf{v}_1$, $\mathbf{v}_2$, $\mathbf{B}_1$, $\mathbf{B}_2$, $\mathbf{F}_{\rm m12}$, and $\mathbf{F}_{\rm m21}$. *(ME4_3.m on IR)*

MATLAB EXERCISE 4.4 **Magnetic field due to multiple moving charges.** Using Eq.(4.4) and the superposition principle, write a MATLAB program that calculates and plots, at an arbitrary point in space, the resultant magnetic flux density vector due to N point charges, Q_1, Q_2, …, Q_N, placed at arbitrary locations in a Cartesian coordinate system, in a vacuum, and moving with given velocities, $\mathbf{v}_1$, $\mathbf{v}_2$, …, $\mathbf{v}_N$ (vectors), at these locations. *(ME4_4.m on IR)*

[4] *CW* = Companion Website (of the book).

HINT:
See MATLAB Exercises 1.7 and 1.4.

MATLAB EXERCISE 4.5 **Electron travel in a uniform magnetic field – movie.** Create a movie in MATLAB that traces the path traveled by an electron as it enters a uniform time-constant magnetic field of flux density $\mathbf{B} = 10(-\hat{\mathbf{z}})$ mT. Assume that the electron moves with a constant velocity $\mathbf{v} = 10^5\,\hat{\mathbf{x}}$ m/s prior to entering the field. *(ME4_5.m on IR)*

TUTORIAL:
Once in the magnetic field, the centrifugal (outward) force on the electron, given by $F_{\text{c.f.}} = m_\text{e}v^2/R$, is balanced by the centripetal (inward) magnetic force [Eq.(4.1)], $F_\text{m} = Q_\text{e}vB$, where m_e, v, and Q_e are the mass, velocity, and charge of the electron, respectively. Hence, the radius R of the circular trajectory traced by the electron is determined by the following equation:

$$\frac{m_\text{e}v^2}{R} = Q_\text{e}vB\,. \tag{4.6}$$

We start our MATLAB code by entering input parameters, Q_e (`Qe`) [Eq.(1.3)], m_e (`me`), B (`B`), and v (`V`), computing R (`R`) from Eq.(4.6), and defining the parameters of the mesh for plotting,

```
Qe = -1.6022*10^(-19);
me = 9.103*10^(-28);
B = 0.01;
V = 10^5;
R = abs(me*V/(Qe*B));
dtheta = 0.01*pi;
dl = R*dtheta;
l = -1.5*R:dl:0;
```

To aid the movie realization, we create a function `draw()` to plot the magnetic field vector at the mesh nodes, using MATLAB function `meshgrid`:

```
function draw(dl,R)
hold off;
[x,y] = meshgrid(0:10*dl:1.5*R,-1.5*R:10*dl:1.5*R);
[M,N]=size(x);
for k = 1:M
for t = 1:N
plot(x(k,t),y(k,t),'bx'); hold on;
end;
end;
title('Electron travel in a magnetic field');
text(x(M),y(N),'B','FontSize',12);
```

In the first part of the movie, we show the travel of the electron – prior to entering the magnetic field – in the positive x direction, from the starting position `[x,y] = [-1.5R,R]` toward the region with the magnetic field, which is visualized by function **draw**. The current position of the electron is marked by MATLAB function `plot` and its velocity vector is drawn using MATLAB function `quiver`. Also, MATLAB function **getframe** is used to merge pictures into a movie (see MATLAB Exercise 1.20).

```
for k = 1:length(l)
draw(dl,R);
plot(l(k),R,'ro','MarkerFaceColor','r');
scale = 0.25*R/V;
quiver(l(k),R,real(V),imag(V),scale,'Color','m');
axis equal;
ylim([-1.5*R 1.5*R]); xlim([-1.5*R 1.5*R]);
axis off;
pause(0.05);
M(k) = getframe;
end;
```

Figure 4.3(a) shows a snapshot of the movie.

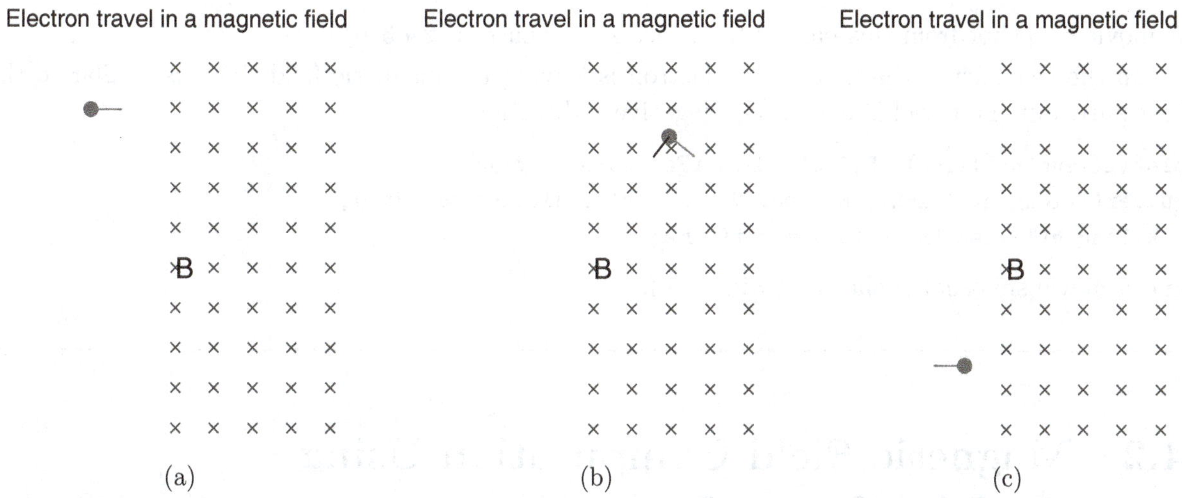

Figure 4.3 MATLAB movie tracing the path traveled by an electron in a uniform time-constant magnetic field of flux density **B**: (a) electron movement prior to entering the field, (b) snapshot of the second part of the movie, showing electron travel around a circular trajectory, determined by Eq.(4.6), and (c) third part of the movie, with the electron tracing a straight line after leaving the magnetic field; for MATLAB Exercise 4.5. *(color figure on CW)*

The second part of the movie starts from the point of entrance of the electron into the magnetic field, when it starts tracing the circular trajectory, which is parametrically described in terms of an angle θ, $-\pi/2 \le \theta \le \pi/2$. For the visualization by `plot` and `quiver`, we compute the x- and y-coordinates of the current position of the electron and x- and y-components of the velocity vector at that point, and this is done by treating the geometrical (x, y) plane as a complex (Re, Im) plane, similarly to the situation in Fig.6.14 in Section 6.8. Note, however, that this computation can be done in an entirely geometrical fashion, which would, on the other hand, require more equations and more lines of MATLAB code.

```
theta = pi/2:-dtheta:-pi/2;
for k = 1:length(theta)
draw(dl,R);
pos = R*exp(i*theta(k));
Vv = V*exp(i*(-pi/2+theta(k)));
x = real(pos);
y = imag(pos);
plot(x,y,'ro','MarkerFaceColor','r');
scale = 0.25*R/V;
quiver(x,y,real(Vv),imag(Vv),scale,'Color','m');
quiver(x,y,-0.2*x,-0.2*y,0,'Color','k');
axis equal;
ylim([-1.5*R 1.5*R]); xlim([-1.5*R 1.5*R]);
axis off;
pause(0.05);
M(k+length(l)) = getframe;
end;
```

A movie snapshot from this part of the process is given in Fig.4.3(b).

In the last part of the movie, the electron is leaving the magnetic field. This is similar to the first part, but the travel is now in the negative x direction,

```
plot(l(length(l)+1-k),-R,'ro','MarkerFaceColor','r');
quiver(l(length(l)+1-k),-R,-real(V),imag(V),scale,'Color','m');
M(k+length(l)+length(theta)) = getframe;
```

and a movie snapshot is shown in Fig.4.3(c).

4.2 Magnetic Field Computation Using the Biot–Savart Law

Generalizing, by the principle of superposition, the expression in Eq.(4.4), we obtain the expression for the resultant magnetic flux density vector due to a current of intensity I flowing along a line (wire) l, Fig.4.4(a),

$$\mathbf{B} = \frac{\mu_0}{4\pi}\int_l \frac{I\,\mathrm{d}\mathbf{l}\times\hat{\mathbf{R}}}{R^2} \quad \text{(Biot–Savart law)}\,, \tag{4.7}$$

which is known as the Biot–Savart law.

Equation (4.7), along with its versions for surface and volume currents (see Fig.3.1), is a general means for evaluating (by superposition and integration) the field $\mathbf{B}$ due to given current distributions in free space (or any nonmagnetic medium). For example, $\mathbf{B}$ due to a circular current

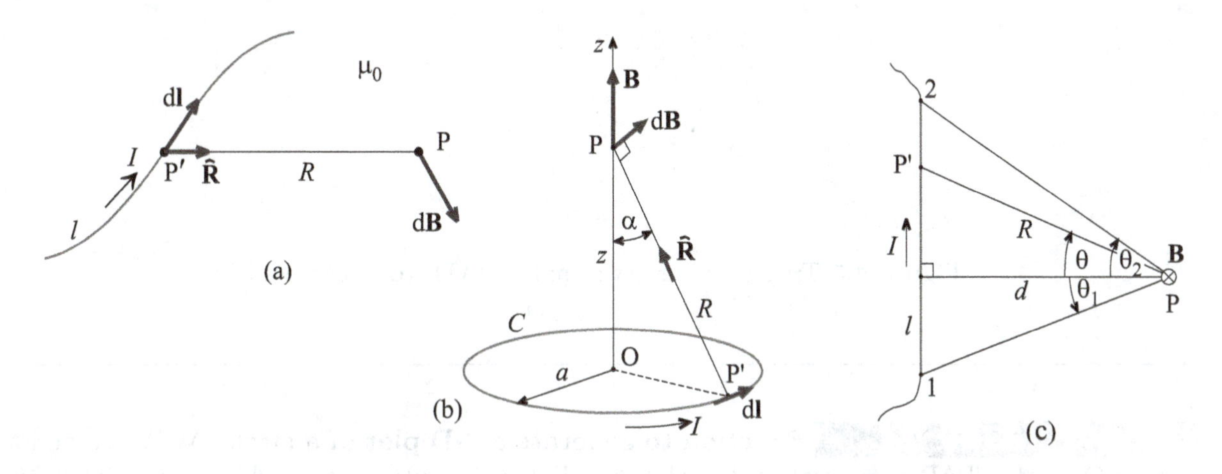

Figure 4.4 Evaluation of the magnetic flux density vector due to (a) an arbitrary line current, (b) a circular current loop, and (c) a finite straight wire conductor in free space.

loop at its axis normal to its plane, Fig.4.4(b), is found as follows [also see Eq.(1.15)]:

$$\mathbf{B} = \oint_C \mathrm{d}B_z \,\hat{\mathbf{z}} = \oint_C \mathrm{d}B \sin\alpha \,\hat{\mathbf{z}} = \oint_C \frac{\mu_0 I \,\mathrm{d}l}{4\pi R^2} \frac{a}{R} \,\hat{\mathbf{z}} = \frac{\mu_0 I a}{4\pi R^3} \,\hat{\mathbf{z}} \oint_C \mathrm{d}l = \frac{\mu_0 I a^2}{2(z^2 + a^2)^{3/2}} \,\hat{\mathbf{z}}$$

$$(\mathbf{B} \text{ due to a circular current loop}) \,. \quad (4.8)$$

Similarly, the magnetic flux density at an arbitrary point in space due to a steady current in a straight wire conductor of finite length, Fig.4.4(c), comes out to be

$$B = \frac{\mu_0 I}{4\pi d} (\sin\theta_2 - \sin\theta_1) \quad (B \text{ due to a straight wire conductor of finite length}) \,, \quad (4.9)$$

with the vector **B** shown in Fig.4.4(c). By taking $\theta_1 = -\pi/2$ and $\theta_2 = \pi/2$ in Fig.4.4(c), Eq.(4.9) gives the expression for B due to an infinitely long straight wire conductor carrying a current I, which, with a notation $d = r$, becomes

$$B = \frac{\mu_0 I}{2\pi r} \quad (B \text{ due to an infinite wire conductor}) \,. \quad (4.10)$$

MATLAB EXERCISE 4.6 **Magnetic field of a finite straight wire conductor.** Write a function `Bwireline()` in MATLAB to evaluate the magnetic field of a straight wire conductor of finite length, with a steady current, based on Eq.(4.9) and Fig.4.4(c). *(Bwireline.m on IR)*

MATLAB EXERCISE 4.7 **Triangular current loop.** A loop in the form of a triangle representing a half of a square of side $a = 5$ cm carries a steady current of intensity $I = 10$ mA, as shown in Fig.4.5. The medium is air. Calculate – in MATLAB, applying (three times) function `Bwireline` (from the previous MATLAB exercise) – the magnetic flux density vector at a point P located at the fourth vertex of the square. *(ME4_7.m on IR)* **H**

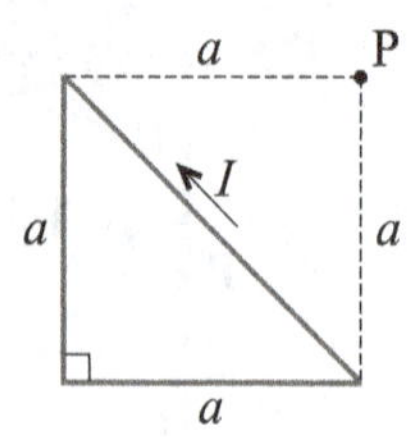

Figure 4.5 Triangular current loop; for MATLAB Exercise 4.7.

MATLAB EXERCISE 4.8 **Function to generate a 3-D plot of a circle.** Write a function `circle()` in MATLAB to generate a 3-D plot of a circle in the plane defined by $z = z_0$ (parallel to the xy-plane) of the Cartesian coordinate system, for the given radius r and coordinates (x_0, y_0, z_0) of the center of the circle. The input also includes the information on the color of the circle. (Note that this function is used in the next MATLAB exercise to visualize magnetic field lines of an infinite wire conductor.) *(circle.m on IR)*

TUTORIAL:

See Fig.4.6. The simplest way to create the circle function is to use a parametric equation of the circle given by

$$x = x_0 + r\cos\phi\,, \quad y = y_0 + r\sin\phi \qquad (0 \le \phi \le 2\pi)\,, \tag{4.11}$$

where ϕ is the azimuthal angle in the plane of the circle. An implementation in MATLAB reads

```
function circle(x0,y0,z0,r,color)
if (nargin == 4)
color = 'b';
end;
phi = 0:0.01*2*pi:2*pi;
x = r*cos(phi)+x0;
y = r*sin(phi)+y0;
```

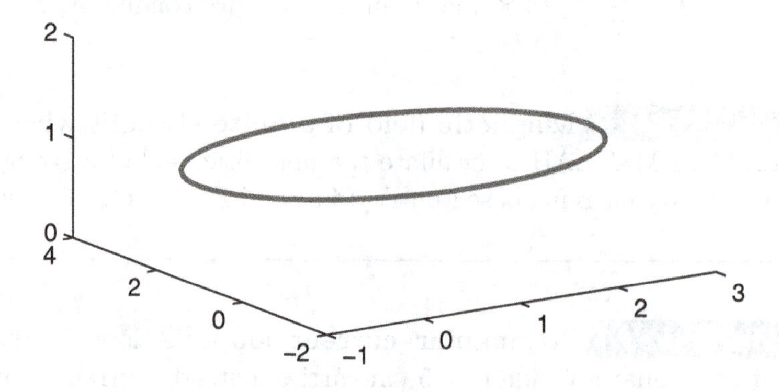

Figure 4.6 3-D plot of a circle using function `circle` in MATLAB written based on Eq.(4.11); for MATLAB Exercise 4.8. *(color figure on CW)*

```
z = z0*ones(1,length(phi));
plot3(x,y,z,'Color',color)
```

One can test the function from the Command Window as follows:

```
>> circle(1,1,1,2,'r')
```

and the resulting plot is shown in Fig.4.6.

MATLAB EXERCISE 4.9 **3-D visualization of magnetic field lines.** Applying function `circle` (from the previous MATLAB exercise), create a 3-D visualization of magnetic field lines due to an infinitely long straight wire conductor carrying a steady current in free space placed along the z-axis of a Cartesian coordinate system. *(ME4_9.m on IR)*

TUTORIAL:

From Eq.(4.10), the lines of the magnetic field due to the conductor current are circles centered at the conductor axis. For clarity of the graph, we choose only three equidistant points along the z-axis as centers of field lines (concentric circles) to be displayed. In plotting these circles, the difference between field intensities between any two adjacent lines is kept constant (step `delta` in our code), to have the density of field lines be representative of the magnitude of the field vector ($\mathbf{B}$). Equation (4.10) tells us that B decreases as $1/r$ from the z-axis, away from the conductor, and we accordingly compute the radii of circles to be shown (note that `N` is the total number of circles in one group in the horizontal plane). In addition, we specify the z-coordinates of the centers of groups of circles.

```
N = 20;
delta = 1/10;
for i = 1:N
r(i) = 1/delta/i;
end;
z = 1:6:13;
```

For plotting, we use function `circle` within two nested `for` loops. The first loop runs through radii of concentric circles, while the second one switches between the three centers along the z-axis,

```
L = length(r);
for k = 1:L
for t = 1:length(z)
circle(0,0,z(t),r(k),'r');
hold on;
end;
end;
```

Finally, we add some more detail to the graph, like the visualization of the wire conductor using MATLAB function `line`, the graph title, and settings of axes limits,

```
line([0 0],[0 0],[-2 15]);
hold off;
```

```
title('Magnetic field lines due to an infinite wire conductor');
axis equal;axis off;
xlim([-8 8]); ylim([-8 8]); zlim([-1 15]);
view(-36,18);
```

The plot is shown in Fig.4.7.

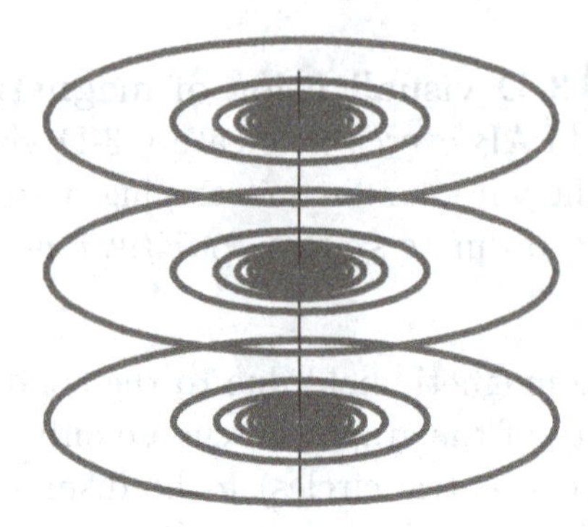
Magnetic field lines due to an infinite wire conductor

Figure 4.7 3-D visualization of magnetic field lines due to an infinitely long straight current conductor (vertical line) using function `circle`; for MATLAB Exercise 4.9.

MATLAB EXERCISE 4.10 Circular surface current distribution, symbolic integration. There is a surface current over a circular surface of radius b with a hole of radius a $(a < b)$ in free space. The surface lies in the plane $z = 0$ of a cylindrical coordinate system, with the coordinate origin coinciding with the surface center. The surface current density vector [see Fig.3.1(b)] is $\mathbf{J}_\mathrm{s} = J_{\mathrm{s}0}(a/r)\hat{\boldsymbol{\phi}}$ $(a \le r \le b)$, where $J_{\mathrm{s}0}$ is a constant. Compute the magnetic flux density vector along the z-axis – performing symbolic integration in MATLAB to add, according to the superposition principle, magnetic fields due to elementary rings over the circular surface. Specifically, assuming that $a = 1$ cm, $b = 2$ cm, and $J_{\mathrm{s}0} = 1$ A/m, plot the distribution of the resultant magnetic flux density along the z-axis, for $-a \le z \le a$. *(ME4_10.m on IR)* **H**

TUTORIAL:

We subdivide the circular surface with current into elemental rings of width $\mathrm{d}r$, as shown in Fig.4.8(a), where the current of a ring of radius r, which, using Eq.(3.3), equals $\mathrm{d}I = J_\mathrm{s}(r)\,\mathrm{d}r$ $(a \le r \le b)$, can be viewed as the current of an equivalent circular current loop of radius r [Fig.4.4(b)]. From Eq.(4.8), the magnetic flux density vector of this loop at an arbitrary point P at the z-axis (defined by a coordinate z) in Fig.4.8(a) is

$$\mathrm{d}\mathbf{B} = \frac{\mu_0\,\mathrm{d}I\,r^2}{2R^3}\,\hat{\mathbf{z}} = \frac{\mu_0 J_\mathrm{s}(r)\,r^2\,\mathrm{d}r}{2(z^2+r^2)^{3/2}}\,\hat{\mathbf{z}}\,, \tag{4.12}$$

and to find the total field $\mathbf{B}$ at the point P, we integrate $\mathrm{d}\mathbf{B}$ to sum the contributions of all equivalent loops over the surface S (superposition principle), in the range $a \le r \le b$, in MATLAB.

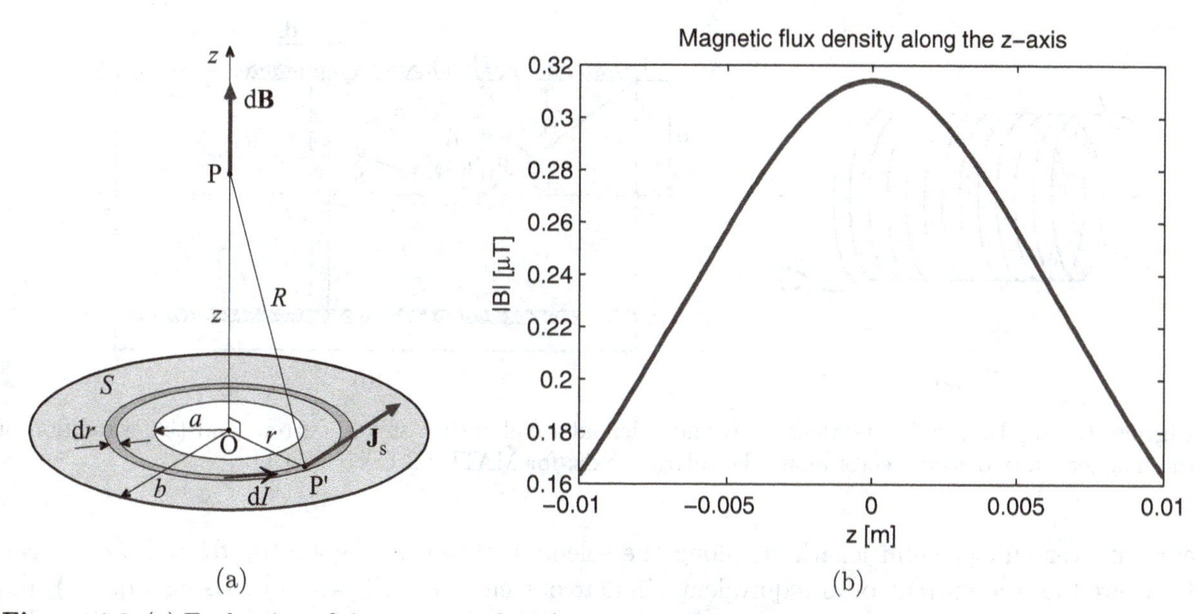

(a) (b)

Figure 4.8 (a) Evaluation of the magnetic field (at the z-axis) due to a circular surface current distribution given by $\mathbf{J}_s = J_{s0}(a/r)\hat{\boldsymbol{\phi}}$ $(a \leq r \leq b)$ and (b) plot of the magnitude of $\mathbf{B}$ along the z-axis – integration of d$\mathbf{B}$ performed symbolically in MATLAB; for MATLAB Exercise 4.10.

The integral is solved by symbolic integration, by means of MATLAB function `int` (note that `r` is first declared as a symbolic variable, using MATLAB function `syms`). However, the result of integration is then converted from the symbolic expression to a numerical value, by MATLAB function `double`, in order to plot it along the z-axis, as shown in Fig.4.8(b).

```
MU0 = 4*pi*10^(-7);
Js0 = 1;
a = 1*10^(-2);
b = 2*10^(-2);
syms r;
z = -a:a/100:a;
Js = Js0*a/r;
B = zeros(1,length(z));
for k = 1:length(z)
dB = MU0*Js*r^2/(2*(z(k)^2 + r^2)^1.5);
B(k) = double(int(dB,r,a,b));
end;
```

MATLAB EXERCISE 4.11 **Magnetic field of a finite solenoid.** Figure 4.9(a) shows a solenoid (cylindrical coil) consisting of N turns of an insulated thin wire wound uniformly and densely in one layer on a cylindrical nonmagnetic support with a circular cross section of radius a. The length of the solenoid is l, the current through the wire is I, and the medium is air. The

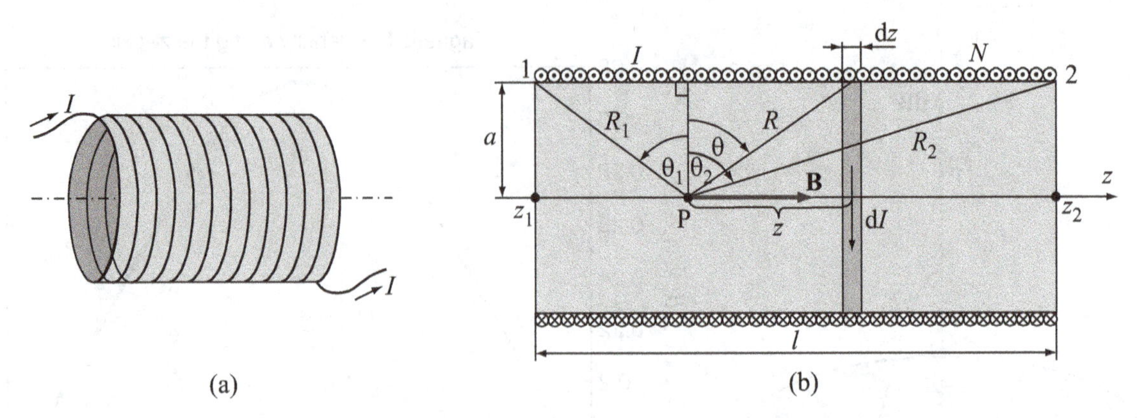

Figure 4.9 (a) Uniformly and densely wound solenoidal coil with a steady current and (b) evaluation of the magnetic flux density vector along the solenoid axis; for MATLAB Exercise 4.11.

current over an elemental length $\mathrm{d}z$ along the solenoid, shown in Fig.4.9(b), $\mathrm{d}I = NI\,\mathrm{d}z/l$, can be viewed as the current of an equivalent circular current loop [Fig.4.4(b)]. Using Eq.(4.8) and superposition (integration), the total magnetic flux density vector at an arbitrary point P at the solenoid axis [Fig.4.9(b)] comes out to be

$$\mathbf{B} = \frac{\mu_0 NI}{2l}(\sin\theta_2 - \sin\theta_1)\,\hat{\mathbf{z}} \quad (\mathbf{B} \text{ along the axis of a finite solenoid}) \,, \tag{4.13}$$

where the position of the point is defined by angles θ_1 and θ_2. Based on this expression, write a function `BzFiniteSolenoid()` in MATLAB to compute **B** along the z-axis. *(BzFiniteSolenoid.m on IR)*

MATLAB EXERCISE 4.12 **Field plots for different length-to-diameter ratios.** Consider a solenoid (Fig.4.9) of length $l = 50$ cm, with a nonmagnetic core and $N = 1000$ tightly wound turns of wire carrying a steady current $I = 1$ A. In MATLAB, calculate the **B** field at the solenoid center, at the center of the 250th (or 750th) wire turn, and at the center of the first (or last) wire turn, and plot the function $B(z)$ along the solenoid axis ($-\infty < z < \infty$) for the solenoid radius equal to (a) $a = 25$ cm and (b) $a = 2$ cm, respectively. Use function `BzFiniteSolenoid` (from the previous MATLAB exercise) and MATLAB function `plot`. *(ME4_12.m on IR)* **H**

4.3 Ampère's Law in Integral Form

In magnetostatics, the law that helps us evaluate the magnetic field due to highly symmetrical current distributions in free space more easily than the Biot–Savart law is Ampère's law. It states that the line integral (circulation) of the magnetic flux density vector around any contour (C) in a vacuum (free space), Fig.4.10(a), is equal to μ_0 times the total current enclosed by that contour, I_C,

$$\oint_C \mathbf{B}\cdot\mathrm{d}\mathbf{l} = \mu_0 I_C \,, \quad \oint_C \mathbf{B}\cdot\mathrm{d}\mathbf{l} = \mu_0\int_S \mathbf{J}\cdot\mathrm{d}\mathbf{S} \quad (\text{Ampère's law}) \,, \tag{4.14}$$

with S being a surface of arbitrary shape spanned over (bounded by) C. The reference direction of the current flow, that is, the orientation of the surface S, is related to the reference direction of the contour by means of the right-hand rule: the current is in the direction defined by the thumb of the right hand when the other fingers point in the direction of the contour, as shown in Fig.4.10(a). Equation (4.14) represents Maxwell's second equation for static fields in free space.

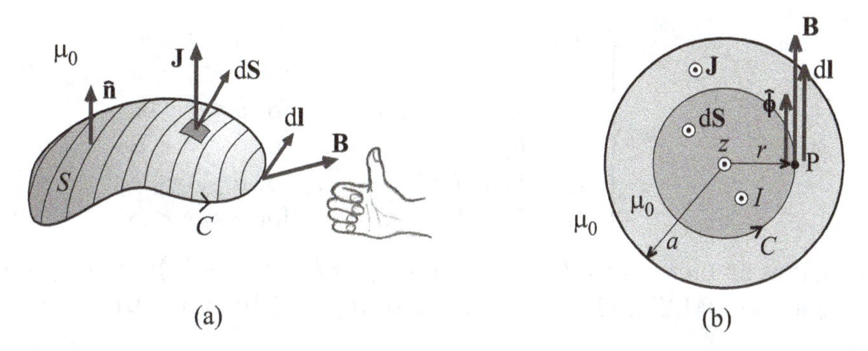

Figure 4.10 (a) Arbitrary contour in a magnetostatic field – for the formulation of Ampère's law. (b) Cross section of a cylindrical conductor with a steady current I: application of Ampère's law to find the field $\mathbf{B}$.

As an illustration of the application of Ampère's law, consider an infinitely long cylindrical copper conductor of radius a in air, carrying a steady current of intensity I. Because of symmetry, the lines of the magnetic field due to the conductor current are circles centered at the conductor axis, i.e., the vector $\mathbf{B}$ at an arbitrary point P either inside or outside the conductor is of the form $\mathbf{B} = B(r)\,\hat{\boldsymbol{\phi}}$, as shown in Fig.4.10(b). From Eq.(4.14) and Fig.4.10(b),

$$\oint_C \mathbf{B}\cdot \mathrm{d}\mathbf{l} = \oint_C B(r)\,\mathrm{d}l = B(r)\oint_C \mathrm{d}l = B(r)\,l = B(r)\,2\pi r = \mu_0 \begin{cases} J\pi r^2 & \text{for } r < a \\ I & \text{for } r \geq a \end{cases}, \tag{4.15}$$

where $J = I/(\pi a^2)$ is the current density of the conductor, and hence

$$B = \begin{cases} \mu_0 I r/(2\pi a^2) & \text{for } r < a \\ \mu_0 I/(2\pi r) & \text{for } r \geq a \end{cases} \quad (B \text{ due to a thick cylindrical conductor})\,. \tag{4.16}$$

MATLAB EXERCISE 4.13 **Magnetic field of a cylindrical conductor.** In MATLAB, plot the dependence on r of the magnetic flux density, $B(r)$, inside and outside the infinitely long cylindrical copper conductor carrying a steady current in Fig.4.10(b), based on Eq.(4.16), assuming that $a = 5$ mm and $I = 1$ A. *(ME4_13.m on IR)*

MATLAB EXERCISE 4.14 **Magnetic field of a triaxial cable.** Figure 4.11(a) shows a cross section of a triaxial cable, having three coaxial cylindrical conductors. The radius of the inner conductor of the cable is $a = 1$ mm, the inner and outer radii of the middle conductor are $b = 2$ mm and $c = 2.5$ mm, and those of the outer conductor $d = 5$ mm and $e = 5.5$ mm. The cable conductors and dielectric, as well as the surrounding medium, are all nonmagnetic. Assuming that steady currents of intensities $I_1 = 2$ A, $I_2 = -1$ A, and $I_3 = -1$ A flow through the inner, middle, and outer conductor, respectively, all given with respect to the same reference direction, Ampère's law, applied to a circular contour C of radius r ($0 \leq r < \infty$) to find the magnetic flux densities in the six characteristic regions in Fig.4.11(a), in a way similar to that in Eq.(4.15), results in the following:

$$B\,2\pi r = \mu_0 \begin{cases} J_1 \pi r^2 & \text{for } 0 \leq r \leq a \\ I_1 & \text{for } a < r < b \\ I_1 + J_2\pi(r^2 - b^2) & \text{for } b \leq r \leq c \\ I_1 + I_2 & \text{for } c < r < d \\ I_1 + I_2 + J_3\pi(r^2 - d^2) & \text{for } d \leq r \leq e \\ I_1 + I_2 + I_3 & \text{for } e < r < \infty \end{cases}, \tag{4.17}$$

with current densities amounting to $J_1 = I_1/(\pi a^2)$, $J_2 = I_2/[\pi(c^2 - b^2)]$, and $J_3 = I_3/[\pi(e^2 - d^2)]$. Based on this, generate a MATLAB plot of $B(r)$, for $0 \leq r \leq 10$ mm. *(ME4_14.m on IR)*

HINT:
The plot obtained in MATLAB is shown in Fig.4.11(b).

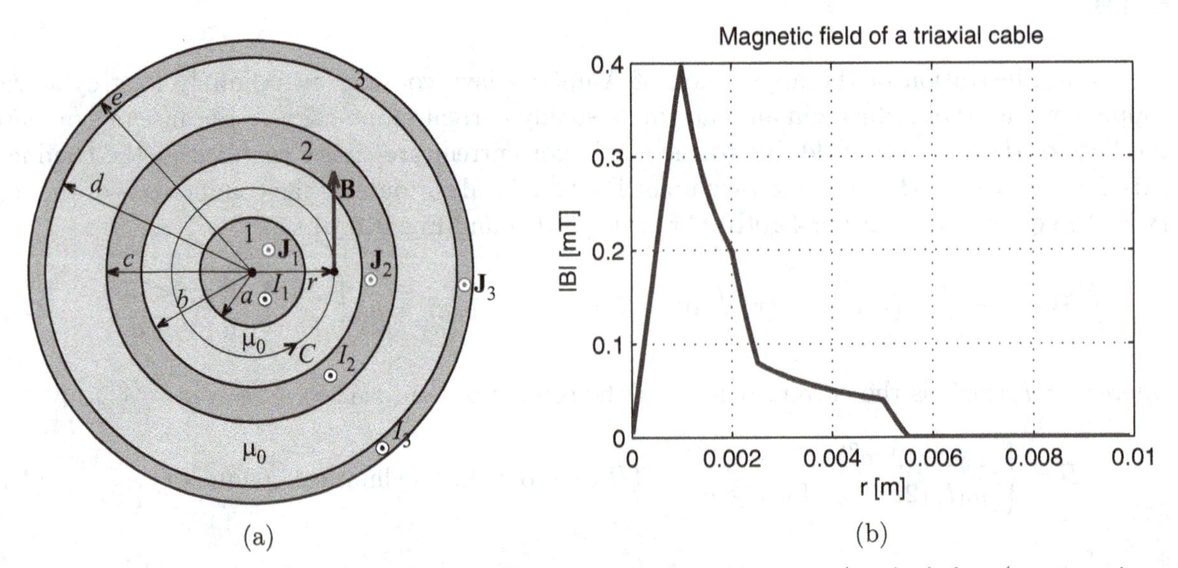

(a) (b)

Figure 4.11 (a) Evaluation of the magnetic field of a triaxial cable using Ampère's law (cross section of the structure) and (b) MATLAB plot of the dependence of the magnetic flux density of the cable on the radial distance from the cable axis; for MATLAB Exercise 4.14.

MATLAB EXERCISE 4.15 **Visualization of the B-vector using `quiver`.** Use MATLAB function `quiver` to visualize the magnetic field distribution (vector **B**) inside and outside the infinite cylindrical conductor with a steady current in Fig.4.10(b), for $a = 5$ mm and $I = 1$ A. *(ME4_15.m on IR)*

TUTORIAL:
First, we specify the input to the code and define the mesh of points (nodes) at which the magnetic

field is calculated:

```
MU0 = 4*pi*10^(-7);
a = 5*10^(-3);
I = 1;
R = 7*10^(-3);
dr = (R-a)/10;
dphi = 2*pi/50;
[r,phi] = meshgrid(0:dr:R,0:dphi:2*pi);
```

where R ($R = 7$ mm) is the radius of the boundary of the plot ($R > a$). We then convert the coordinates of the nodes from cylindrical (polar) to Cartesian coordinates, using function `cyl2Car` (MATLAB Exercise 1.24), which is needed for plotting **B** at the nodes (with MATLAB function `quiver`),

```
[x,y,z]=cyl2Car(r,phi,0);
```

The unit vector of **B** inside and outside the conductor in Fig.4.10(b), $\hat{\boldsymbol{\phi}}$ (in the cylindrical coordinate system), can be decomposed into its x- and y-components (in the associated Cartesian coordinate system) as:

$$\hat{\boldsymbol{\phi}} = -\sin\phi\,\hat{\mathbf{x}} + \cos\phi\,\hat{\mathbf{y}}\ . \tag{4.18}$$

With this, and the expression for the magnitude of **B** in Eq.(4.16), we obtain the x- and y-components of **B**, `Bx` and `By`, via the following lines of MATLAB code:

```
Bx = zeros(size(r));
By = zeros(size(r));
Const = MU0/(2*pi);
for i = 1:length(phi(:,1));
for j = 1:length(r(1,:));
if r(i,j) < a;
Bx(i,j) = Const*I/a^2*(-y(i,j));
By(i,j) = Const*I/a^2*x(i,j);
else
Bx(i,j) = Const*I*(-y(i,j))/r(i,j)^2;
By(i,j) = Const*I*x(i,j)/r(i,j)^2;
end;
end;
end;
```

Finally, we plot the vector field distribution of the conductor using `quiver` (for instance, see MATLAB Exercise 1.13). To enhance the clarity and readability of the plot, we also draw a circle representing the conductor contour ($r = a$) applying function `circle` written in MATLAB Exercise 4.8.

```
figure(1);
quiver(x*10^3,y*10^3,Bx,By); hold on;
circle(0,0,0,a*10^3,'r'); hold off; axis equal;
title('Magnetic field of a cylindrical conductor (cross section)');
```

```
xlabel('x [mm]'); ylabel('y [mm]');
```

The plot is shown in Fig.4.12.

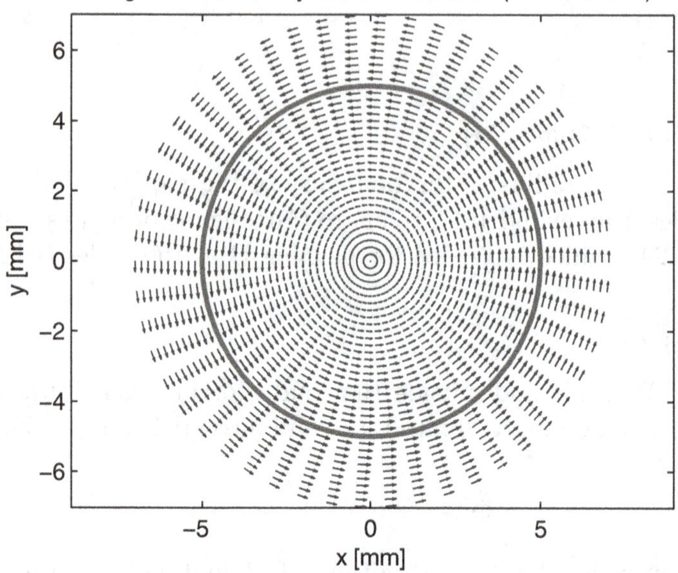

Figure 4.12 Visualization of the magnetic field distribution (vector **B**) inside and outside the cylindrical conductor with a steady current in Fig.4.10(b) using MATLAB function `quiver` (the circle representing the contour of the conductor is plotted by function `circle`); for MATLAB Exercise 4.15. *(color figure on CW)*

MATLAB EXERCISE 4.16 **Field visualization by `quiver` for a triaxial cable.** Repeat the previous MATLAB exercise but for the triaxial cable from MATLAB Exercise 4.14 (the plot should span $0 \leq r \leq 6$ mm). *(ME4_16.m on IR)*

4.4 Differential Form of Ampère's Law, Curl

The counterpart of Eq.(4.14) in differential notation reads

$$\left(\frac{\partial B_z}{\partial y} - \frac{\partial B_y}{\partial z}\right)\hat{\mathbf{x}} + \left(\frac{\partial B_x}{\partial z} - \frac{\partial B_z}{\partial x}\right)\hat{\mathbf{y}} + \left(\frac{\partial B_y}{\partial x} - \frac{\partial B_x}{\partial y}\right)\hat{\mathbf{z}} = \mu_0 \mathbf{J} \,, \tag{4.19}$$

with the expression on the left-hand side of the equation being the so-called curl of a vector function ($\mathbf{B}$), written as $\operatorname{curl}\mathbf{B}$. We notice that applying formally the formula for the cross product of two vectors in the Cartesian coordinate system, Eq.(4.5), to $\nabla \times \mathbf{B}$, where the del operator is given by Eq.(1.37), we obtain exactly $\operatorname{curl}\mathbf{B}$. Hence,

$$\operatorname{curl}\mathbf{B} = \nabla \times \mathbf{B} = \left(\frac{\partial B_z}{\partial y} - \frac{\partial B_y}{\partial z}\right)\hat{\mathbf{x}} + \left(\frac{\partial B_x}{\partial z} - \frac{\partial B_z}{\partial x}\right)\hat{\mathbf{y}} + \left(\frac{\partial B_y}{\partial x} - \frac{\partial B_x}{\partial y}\right)\hat{\mathbf{z}}$$

$$\text{(curl in Cartesian coordinates)} \,, \tag{4.20}$$

and the differential Ampère's law can be written in a short form as

$$\operatorname{curl}\mathbf{B} = \nabla \times \mathbf{B} = \mu_0 \mathbf{J} \quad \text{(Ampère's law in differential form)} . \tag{4.21}$$

In cylindrical coordinates [Fig.1.18(a)], the curl is computed as

$$\operatorname{curl}\mathbf{B} = \nabla \times \mathbf{B} = \left(\frac{1}{r}\frac{\partial B_z}{\partial \phi} - \frac{\partial B_\phi}{\partial z}\right)\hat{\mathbf{r}} + \left(\frac{\partial B_r}{\partial z} - \frac{\partial B_z}{\partial r}\right)\hat{\boldsymbol{\phi}}$$
$$+\frac{1}{r}\left[\frac{\partial}{\partial r}(rB_\phi) - \frac{\partial B_r}{\partial \phi}\right]\hat{\mathbf{z}} \quad \text{(curl in cylindrical coordinates)} , \tag{4.22}$$

while the corresponding formula for the spherical coordinate system [Fig.1.18(b)] is

$$\operatorname{curl}\mathbf{B} = \nabla \times \mathbf{B} = \frac{1}{r\sin\theta}\left[\frac{\partial}{\partial \theta}(\sin\theta B_\phi) - \frac{\partial B_\theta}{\partial \phi}\right]\hat{\mathbf{r}} + \frac{1}{r}\left[\frac{1}{\sin\theta}\frac{\partial B_r}{\partial \phi} - \frac{\partial}{\partial r}(rB_\phi)\right]\hat{\boldsymbol{\theta}}$$
$$+\frac{1}{r}\left[\frac{\partial}{\partial r}(rB_\theta) - \frac{\partial B_r}{\partial \theta}\right]\hat{\boldsymbol{\phi}} \quad \text{(curl in spherical coordinates)} . \tag{4.23}$$

MATLAB EXERCISE 4.17 **Symbolic curl in different coordinate systems.** (a) Applying symbolic programming in MATLAB, write a function `curlCar()` that takes as input symbolic expressions for `fx(x,y,z)`, `fy(x,y,z)`, and `fz(x,y,z)` representing the x-, y-, and z-components, respectively, of a vector function in the Cartesian coordinate system and returns a symbolic expression for the curl of the function, obtained based on Eq.(4.20). Repeat (a) but for (b) a function `curlCyl()` that returns the symbolic curl in the cylindrical coordinate system, based on Eq.(4.22), and (c) the curl in spherical coordinates [function `curlSph()`], using Eq.(4.23). *(curlCar.m, curlCyl.m, and curlSph.m on IR)*

MATLAB EXERCISE 4.18 **Ampère's law in differential form.** In a certain region, the magnetic field is given by $\mathbf{B} = [4(z-1)^2\,\hat{\mathbf{x}} + 2x^3\,\hat{\mathbf{y}} + xy\,\hat{\mathbf{z}}]$ mT (x, y, z in m). The medium is air. (a) Find the current density, applying function `curlCar` (from the previous MATLAB Exercise). (b) From the result in (a), find the total current enclosed by a square contour lying in the xy-plane, with the center at the coordinate origin and sides, of length 2 m, parallel to the x- and y-axes, by symbolic integration in MATLAB. *(ME4_18.m on IR)* **H**

4.5 Magnetic Vector Potential

In analogy to the computation of the electric scalar potential (V) in Eq.(1.31) and the electric field intensity vector ($\mathbf{E}$) from V in Eq.(1.38), in magnetostatics we have

$$\mathbf{A} = \frac{\mu_0}{4\pi}\int_l \frac{I\,\mathrm{d}\mathbf{l}}{R} \quad \longrightarrow \quad \mathbf{B} = \nabla \times \mathbf{A} \quad \text{(magnetic flux density from potential)} , \tag{4.24}$$

where the vector $\mathbf{A}$ is called the magnetic vector potential, and its unit is $\mathrm{T}\cdot\mathrm{m}$.

MATLAB EXERCISE 4.19 **Magnetic flux density from vector potential.** In a certain region, the magnetic vector potential is given as the following function in a cylindrical coordinate system: $\mathbf{A} = 2r^2\hat{\boldsymbol{\phi}}$ T · m (r in m). Find the magnetic flux density vector in this region, using Eq.(4.24) and function `curlCyl` (from MATLAB Exercise 4.17). *(ME4_19.m on IR)* **H**

4.6 Magnetic Dipole

A small loop with a steady current I constitutes the magnetic equivalent of the electric dipole of Fig.1.19, and is referred to as a magnetic dipole. With reference to Fig.4.13, the expression for the magnetic vector potential due to a dipole at large distances compared with the loop dimensions is given by

$$\mathbf{A} = A_\phi\hat{\boldsymbol{\phi}} = \frac{\mu_0 m \sin\theta}{4\pi r^2}\,\hat{\boldsymbol{\phi}}, \quad \mathbf{m} = I\mathbf{S} \quad (m = IS) \quad \text{(magnetic dipole potential)}, \tag{4.25}$$

where $\mathbf{m}$ is the magnetic moment of the dipole (the unit is $\mathrm{A\cdot m^2}$) and $\mathbf{S} = S\hat{\mathbf{n}}$ is the loop surface area vector, oriented in accordance to the right-hand rule with respect to the reference direction of the loop current. Combining Eqs.(4.23)–(4.25), the magnetic flux density vector of the dipole comes out to be

$$\mathbf{B} = \nabla\times\mathbf{A} = \frac{\mu_0 m}{4\pi r^3}\left(2\cos\theta\,\hat{\mathbf{r}} + \sin\theta\,\hat{\boldsymbol{\theta}}\right) \quad \text{(magnetic dipole field)}. \tag{4.26}$$

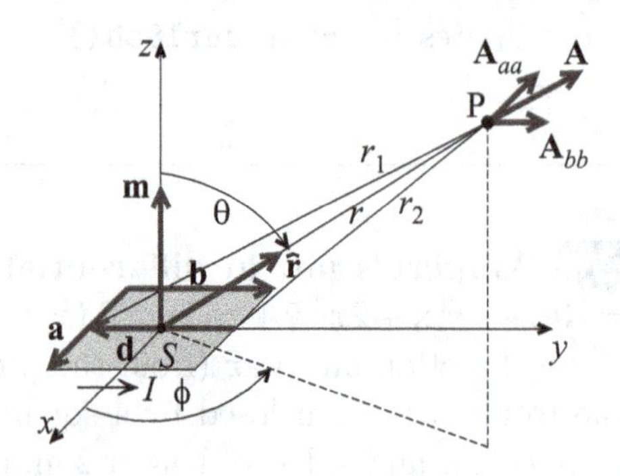

Figure 4.13 Magnetic dipole.

MATLAB EXERCISE 4.20 **Magnetic dipole potential and field functions.** Based on Eqs.(4.25) and (4.26), write functions `magDipoleA()` and `magDipoleB()` in MATLAB that compute the magnetic vector potential, $\mathbf{A}$, and flux density vector, $\mathbf{B}$, at a point defined by (r, θ, ϕ) due to a magnetic dipole of moment $\mathbf{m} = m\,\hat{\mathbf{z}}$ placed at the origin of the spherical coordinate system (Fig.4.13). *(magDipoleA.m and magDipoleB.m on IR)*

MATLAB EXERCISE 4.21 **A and B computation for a magnetic dipole.** A magnetic dipole with a moment $\mathbf{m} = 400\ \mu\text{Am}^2\,\hat{\mathbf{z}}$ is located at the origin of a spherical coordinate system. Using functions from the previous MATLAB exercise, calculate the magnetic potential $\mathbf{A}$ and flux density $\mathbf{B}$ at the following points defined by spherical coordinates: (a) $(1\text{ m}, 0, 0)$, (b) $(1\text{ m}, \pi/2, \pi/2)$, (c) $(1\text{ m}, \pi, 0)$, (d) $(1\text{ m}, \pi/4, 0)$, (e) $(10\text{ m}, \pi/4, 0)$, and (f) $(100\text{ m}, \pi/4, 0)$. The dipole dimensions are much smaller than 1 m. *(ME4_21.m on IR)* **H**

MATLAB EXERCISE 4.22 **B from A for a magnetic dipole, symbolic differentiation.** For a magnetic dipole (Fig.4.13), from the potential expression in Eq.(4.25), obtain, using the symbolic curl function `curlSph` (MATLAB Exercise 4.17), the corresponding field expression [Eq.(4.26)]. *(ME4_22.m on IR)* **H**

MATLAB EXERCISE 4.23 **Visualization of the magnetic dipole potential using `quiver`.** A square loop of edge length $a = 1$ cm is placed in the xy-plane of the Cartesian coordinate system, with its center coinciding with the coordinate origin. The loop carries a steady current of intensity $I = 0.1$ A. Use MATLAB function `quiver` to visualize the distribution of the magnetic vector potential ($\mathbf{A}$) due to the loop current in the plane defined by $z = 1$ m. *(ME4_23.m on IR)*

HINT:
See MATLAB Exercise 4.15.

---

5 MAGNETOSTATIC FIELD IN MATERIAL MEDIA

Introduction:

In analysis of the magnetostatic field in the presence of magnetic materials, many basic concepts, physical laws, and mathematical techniques are entirely analogous to the corresponding concepts, laws, and techniques in electrostatics (Chapter 2). Magnetized materials can be represented by vast collections of tiny atomic current loops, i.e., magnetic dipoles, while the concept of permeability of a medium allows for macroscopic characterization of materials and their fields. The most important difference, however, with respect to the analysis of dielectric materials is the inherent nonlinear behavior of the most important class of magnetic materials, called ferromagnetics. This is a class of materials with striking magnetic properties (many orders of magnitude stronger than in other materials), with iron as a typical example. Analysis of magnetic circuits (consisting of ferromagnetic cores of different shapes with current-carrying windings), which is a culmination of our study of the magnetostatic field in material media, thus essentially resembles the dc analysis of nonlinear electric circuits.

5.1 Permeability of Magnetic Materials

In a magnetostatic system that includes arbitrary media, inclusion of magnetization properties of materials in Eqs.(4.14) and (4.21) leads to the corresponding forms of the generalized Ampère's law:

$$\oint_C \mathbf{H} \cdot \mathrm{d}\mathbf{l} = I_C \,, \quad \nabla \times \mathbf{H} = \mathbf{J} \quad \text{(Generalized Ampère's law)} \,, \tag{5.1}$$

where $\mathbf{H}$ is called the magnetic field intensity vector and is measured in A/m. For linear magnetic materials, it is related to the magnetic flux density vector, $\mathbf{B}$, through the permeability of the medium, μ, as

$$\mathbf{B} = \mu \mathbf{H} \,, \quad \mu = \mu_\mathrm{r} \mu_0 \quad \text{(permeability)} \,, \tag{5.2}$$

with the unit for μ being henry per meter (H/m) and μ_r denoting the relative permeability of the material (dimensionless). Shown in Table 5.1 are values of the relative permeability of an illustrative set of selected materials. Note, however, that ferromagnetic materials often exhibit permanent magnetization, highly nonlinear behavior, and hysteresis effects. Typical ferromagnetics are iron, nickel, and cobalt, and their alloys. The name ferromagnetic comes from the Latin word for iron, "ferrum." In such materials, the function $B(H)$ is in general nonlinear and has multiple branches. The magnetization properties of the material depend on the applied magnetic field intensity, H, and also on the history of magnetization of the material, i.e., on its previous states. In other words, the value of μ (or μ_r), in Eq.(5.2), for a ferromagnetic material generally is not unique, but is a function of H and the previous history of the material.

Table 5.1 Relative permeability of selected materials

Material	μ_r	Material	μ_r
Bismuth	0.999833	Titanium	1.00018
Gold	0.99996	Platinum	1.0003
Mercury	0.999968	Palladium	1.0008
Silver	0.9999736	Manganese	1.001
Lead	0.9999831	Cast iron	150
Copper	0.9999906	Cobalt	250
Water	0.9999912	Nickel	600
Paraffin	0.99999942	Nickel–zinc ferrite (Ni–Zn–Fe_2O_3)	650
Wood	0.9999995	Manganese–zinc ferrite (Mn–Zn–Fe_2O_3)	1200
Vacuum	1	Steel	2000
Air	1.00000037	Iron (0.4% impurity)	5000
Beryllium	1.0000007	Silicon iron (4% Si)	7000
Oxygen	1.000002	Permalloy (78.5% Ni, 21.5% Fe)	7×10^4
Magnesium	1.000012	Mu-metal (75% Ni, 14% Fe, 5% Cu, 4% Mo, 2% Cr)	10^5
Aluminum	1.00002	Iron (purified – 0.04% impurity)	2×10^5
Tungsten	1.00008	Supermalloy (79.5% Ni, 15% Fe, 5% Mo, 0.5% Mn)	10^6

MATLAB EXERCISE 5.1 **GUI for the permeability table of materials.** Repeat MATLAB Exercise 2.1 but for the values of the relative permeability (μ_r) of selected materials given in Table 5.1. *[folder ME5_1(GUI) on IR]*[1]

MATLAB EXERCISE 5.2 **Permeability tensor of an anisotropic medium.** Some magnetic materials, such as ferrites, are anisotropic – accordingly, Eq.(5.2) becomes a matrix equation, with a permeability tensor, $[\mu]$, analogous to the permittivity tensor, $[\varepsilon]$, defined by Eq.(2.3). Based on such an equation, write a MATLAB program that calculates the magnetic flux density vector, **B** (`B`), in an anisotropic magnetic material if the magnetic field intensity vector and the relative-permeability tensor are given by `H = [1,1,1]` A/m and `mur = [7500 0 0; 0 2500 0; 0 0 1]`, respectively. In addition to displaying the result in the Command Window, the program should generate a 3-D plot of normalized vectors **H** and **B** – using MATLAB function `quiver3` (see MATLAB Exercise 1.3). *(ME5_2.m on IR)* **H**[2]

MATLAB EXERCISE 5.3 **Inverse of the permeability tensor.** Repeat the previous MATLAB exercise but for an input consisting of `B = [1,2,3]` μT and `mur = [1400 470 560; 250 80 370; 0 0 1]`, and the result for `H` and 3-D plot of **B** and **H** as output. *(ME5_3.m on IR)*

---

[1] *IR* = Instructor Resources (for the book).

[2] **H** = recommended to be done also "by hand," i.e., not using MATLAB.

HINT:
Use MATLAB function `inv` to obtain the inverse of the `mur` matrix (see MATLAB Exercise 1.39, for instance).

5.2 Boundary Conditions for the Magnetic Field

From the generalized Ampère's law in integral form, Eq.(5.1), and the law of conservation of magnetic flux (Gauss' law for the magnetic field), $\oint_S \mathbf{B} \cdot \mathrm{d}\mathbf{S} = 0$, we obtain the following set of boundary conditions for the magnetic field on the boundary surface between two arbitrary media, analogous to those in Eqs.(2.4):

$$H_{1\mathrm{t}} - H_{2\mathrm{t}} = J_\mathrm{s} \,, \quad B_{1\mathrm{n}} = B_{2\mathrm{n}} \quad \text{(magnetic–magnetic boundary conditions)} \,, \tag{5.3}$$

where J_s is the density of a surface current that may exist on the boundary. In vector form,

$$\hat{\mathbf{n}} \times \mathbf{H}_1 - \hat{\mathbf{n}} \times \mathbf{H}_2 = \mathbf{J}_\mathrm{s} \,, \quad \hat{\mathbf{n}} \cdot \mathbf{B}_1 - \hat{\mathbf{n}} \cdot \mathbf{B}_2 = 0 \quad (\hat{\mathbf{n}} \text{ directed from region 2 to region 1}) \,. \tag{5.4}$$

MATLAB EXERCISE 5.4 **Magnetic–magnetic boundary conditions, oblique plane.** Write a MATLAB code that, based on magnetic–magnetic boundary conditions in Eqs.(5.3) or (5.4), computes the magnetic field intensity vector in medium 2 near the boundary ($\mathbf{H}_2$) for a given magnetic field intensity vector in medium 1 near the boundary, $\mathbf{H}_1 = H_{1x}\,\hat{\mathbf{x}} + H_{1y}\,\hat{\mathbf{y}} + H_{1z}\,\hat{\mathbf{z}}$, if no surface conduction current exists on the boundary surface ($\mathbf{J}_\mathrm{s} = 0$). The program should apply to an arbitrarily positioned (oblique) boundary plane between magnetic media 1 and 2, which does not necessarily coincide with one of the coordinate planes of a Cartesian coordinate system, but contains the coordinate origin. Assume that relative permeabilities $\mu_{\mathrm{r}1}$ and $\mu_{\mathrm{r}2}$ of the media are also known, as well as the normal on the boundary, $\hat{\mathbf{n}} = n_x\,\hat{\mathbf{x}} + n_y\,\hat{\mathbf{y}} + n_z\,\hat{\mathbf{z}}$, directed from region 2 to region 1. *(ME5_4.m on IR)*

HINT:
See MATLAB Exercise 2.4

MATLAB EXERCISE 5.5 **Horizontal current-free boundary plane.** Repeat the previous MATLAB exercise but for a current-free boundary plane coinciding with the xy-plane ($z = 0$) of the Cartesian coordinate system. In specific, write a separate MATLAB program specialized for this particular boundary plane (and no other plane). The program creates a 3-D plot of vectors $\mathbf{H}_1$, $\mathbf{H}_2$, and $\hat{\mathbf{n}}$, along with the boundary plane. Compare the results for $\mathbf{H}_2$ (and various inputs $\mathbf{H}_1$) with those obtained by the general program for an oblique plane run with `n = [0 0 1]`. *(ME5_5.m on IR)*

MATLAB EXERCISE 5.6 **Horizontal boundary plane with surface current.** Repeat the previous MATLAB exercise but for a horizontal boundary plane ($z = 0$) with surface conduction current, given by $\mathbf{J}_s = J_{sx}\,\hat{\mathbf{x}} + J_{sy}\,\hat{\mathbf{y}}$. *(ME5_6.m on IR)*

MATLAB EXERCISE 5.7 **MATLAB computations of magnetic boundary conditions.** Assume that the plane $z = 0$ separates medium 1 ($z > 0$) and medium 2 ($z < 0$), with relative permeabilities $\mu_{r1} = 600$ and $\mu_{r2} = 250$, respectively. The magnetic field intensity vector in medium 1 near the boundary (for $z = 0^+$) is $\mathbf{H}_1 = (5\,\hat{\mathbf{x}} - 3\,\hat{\mathbf{y}} + 2\,\hat{\mathbf{z}})$ A/m. In MATLAB, calculate the magnetic field intensity vector in medium 2 near the boundary (for $z = 0^-$), $\mathbf{H}_2$, if (a) no conduction current exists on the boundary ($\mathbf{J}_s = 0$) and (b) there is a surface current of density $\mathbf{J}_s = 3\,\hat{\mathbf{y}}$ A/m on the boundary. *(ME5_7.m on IR)* **H**

HINT:
Use MATLAB Exercises 5.5 and 5.6.

MATLAB EXERCISE 5.8 **Law of refraction of magnetic field lines.** At an interface between two linear magnetic media of permeabilities μ_1 and μ_2 with no surface conduction current ($\mathbf{J}_s = 0$), the law of refraction of the magnetic field lines can be derived from boundary conditions in Eqs.(5.3) that is entirely analogous to the law in Eq.(3.9), so with σ_1 and σ_2 replaced by μ_1 and μ_2, respectively. With this in mind, let the angle that magnetic field lines in region 1 make with the normal to the interface be α_1, and let the magnitude of the magnetic flux density vector in this region be B_1. In MATLAB, compute α_2 and B_2, on the other side of the boundary, for the given relative permeabilities of the two media, μ_{r1} and μ_{r2}, and graphically represent vectors $\mathbf{B}_1$, $\mathbf{B}_2$, and $\hat{\mathbf{n}}$ in a 2-D view (as in Fig.3.5). Using this MATLAB code, visualize refraction of magnetic field lines at interfaces between various combinations of media, taking the relative permeability values from the GUI developed in MATLAB Exercise 5.1. *(ME5_8.m on IR)*

5.3 Magnetic Circuits

Magnetic circuit in general is a collection of bodies and media that form a way along which the magnetic field lines close upon themselves, i.e., it is a circuit of the magnetic flux flow. The name arises from the similarity to electric circuits. In practical applications, including transformers, generators, motors, relays, magnetic recording devices, etc., magnetic circuits are formed from ferromagnetic cores of various shapes, that may or may not have air gaps, with current-carrying windings wound about parts of the cores. Figure 5.1(a) shows a typical magnetic circuit. With assumptions that the field is restricted to the branches of the magnetic circuit (flux leakage and fringing are negligible) and is uniform in every branch, we now apply the law of conservation of magnetic flux to a closed surface S placed about a node (junction of branches) and the generalized Ampère's law to a contour C placed along a closed path of flux lines in a magnetic circuit, as indicated in Fig.5.1(b), to obtain

$$\oint_S \mathbf{B}\cdot d\mathbf{S} = 0 \quad \longrightarrow \quad \sum_{i=1}^{M} B_i S_i = 0\,, \quad \oint_C \mathbf{H}\cdot \mathrm{d}\mathbf{l} = I_C \quad \longrightarrow \quad \sum_{j=1}^{P} H_j l_j = \sum_{k=1}^{Q} N_k I_k$$

(Kirchhoff's laws for magnetic circuits) , (5.5)

where B_i ($i = 1, 2, \ldots, M$) and H_j ($j = 1, 2, \ldots, P$) are the magnetic flux densities and field intensities in the branches. Equations (5.5) are referred to as Kirchhoff's laws for magnetic circuits. In addition to these circuital laws, we need the "element laws," namely the relationships $B = B(H)$, called the magnetization curves, for the branches of the circuit, which are most frequently nonlinear; shown in Fig.5.1(c) is a typical initial (the material is completely demagnetized and both B and H are zero before a field is applied) magnetization curve for a ferromagnetic sample.

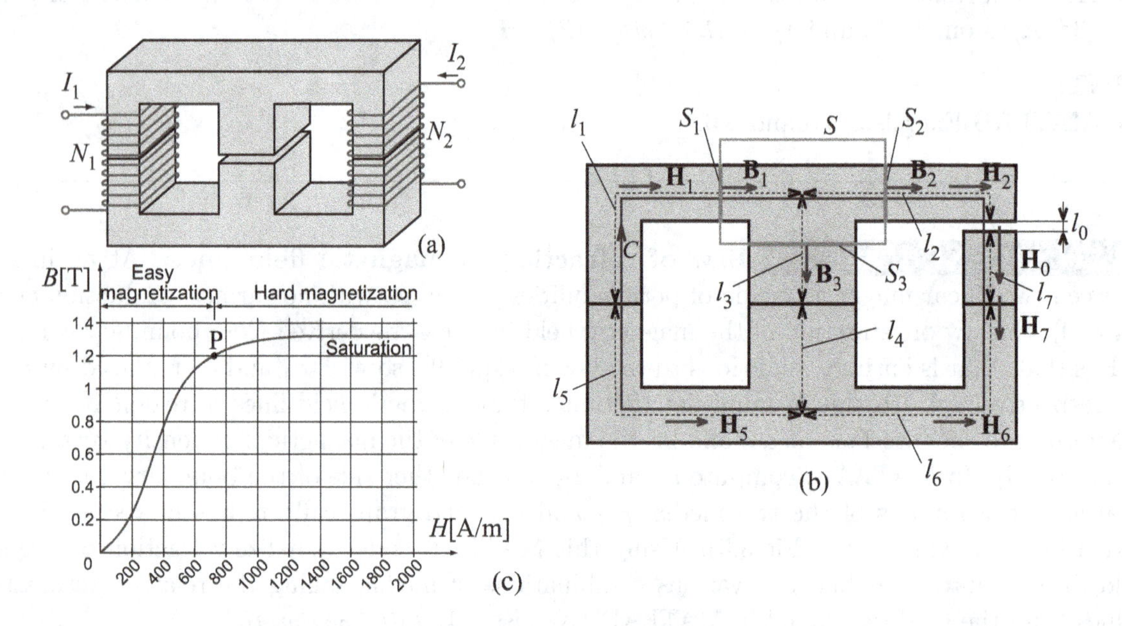

Figure 5.1 (a) Typical magnetic circuit. (b) A closed surface S about a node and a closed path C along the axes of branches of a circuit – for the formulation of Kirchhoff's laws for magnetic circuits. (c) Typical initial magnetization curve for a ferromagnetic material.

MATLAB EXERCISE 5.9 Generation of a linearized initial magnetization curve. Write a function `magCurveSat(H,Hk,Bm)` in MATLAB that generates points – values (H, B) – of a piece-wise linear initial magnetization curve (of a ferromagnetic core) with a linear part and a saturation part, like the one in Fig.5.2(b) (in the next MATLAB exercise). The input to the function consists of an array `H` of values of the magnetic field intensity, H, and the "knee" value `Hk` (H_k) up to which the material is in the linear regime and above which is in saturation, as well as the magnetic flux density of the core in saturation, `Bm` (B_m), and the output is an array of the B values, on the curve, corresponding to the H values at input. *(magCurveSat.m on IR)*

MATLAB EXERCISE 5.10 **Numerical solution for a complex nonlinear magnetic circuit.** Dimensions of the magnetic circuit shown in Fig.5.2(a) are $l_1 = l_3 = 2l_2 = 20$ cm and $S_1 = S_2 = S_3 = 1$ cm^2. The mmf in the first branch of the circuit is $NI = 400$ ampere-turns.[3] The initial magnetization curve of the core can be linearized as in Fig.5.2(b). In MATLAB, calculate the magnetic flux densities and field intensities in the three branches of the circuit. *(ME5_10.m on IR)* **H**

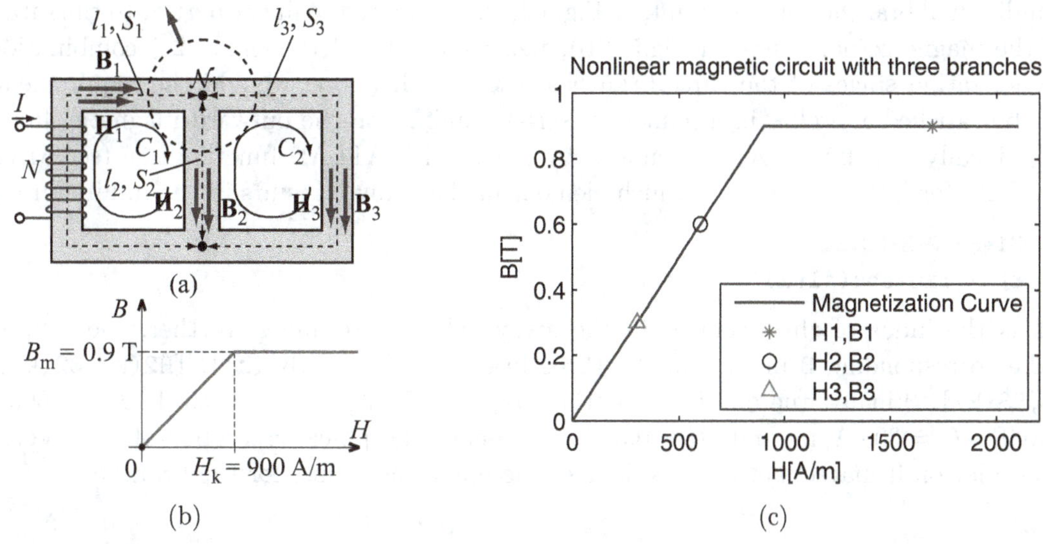

Figure 5.2 Analysis of a nonlinear magnetic circuit with three branches: (a) circuit geometry, with the adopted independent node and closed paths in the circuit, (b) idealized initial magnetization curve of the material, and (c) MATLAB solution of the circuit; for MATLAB Exercise 5.10. *(color figure on CW)*[4]

TUTORIAL:

After specifying the input data for the circuit, we generate arrays of values `(H,B)` describing the magnetization curve in Fig.5.2(b) using function `magCurveSat` (from the previous MATLAB exercise), with $H_k = 900$ A/m and $B_m = 0.9$ T,

```
H = [0:10:2100];
B = MagCurveSat(H,900,0.9);
```

Orienting the branches of the circuit as in Fig.5.2(a), we then apply Kirchhoff's "current" and "voltage" laws for magnetic circuits, Eqs.(5.5), to the node $\mathcal{N}_1$,

$$-B_1S_1 + B_2S_2 + B_3S_3 = 0 \quad \longrightarrow \quad B_1 = B_2 + B_3 \,, \tag{5.6}$$

and closed paths C_1 and C_2,

$$H_1l_1 + H_2l_2 = NI \,, \quad -H_2l_2 + H_3l_3 = 0 \,. \tag{5.7}$$

To solve the equations, we implement in MATLAB Eqs.(5.7) first,

```
H2 = H;
```

[3]The product N_kI_k in Eqs.(5.5), expressed in ampere-turns, is termed a magnetomotive force (mmf), in analogy to an electromotive force (emf) of a voltage generator in electric circuits.

[4]*CW* = Companion Website (of the book).

```
B2 = B;
H3 = H2*12/13;
B3 = MagCurveSat(H3,900,0.9);
H1 = (NI-H2*12)/11;
B1 = MagCurveSat(H1,900,0.9);
```

Because of the nonlinearity of the problem, we do not know in advance whether the operating points for the individual branches of the circuit, in Fig.5.2(a), belong to the linear part or to the saturation part of the magnetization curve in Fig.5.2(b), i.e., which one of the eight (2^3) combinations for the magnetization stages of the three branches is true. Therefore, we add the requirement that Eq.(5.6) be satisfied as well – in a numerical sense, minimizing the outward magnetic flux of the node $\mathcal{N}_1$ (ideally, this flux is zero). Namely, by means of MATLAB function `min` (see MATLAB Exercise 3.12, for instance), we find which element in the array `H` results in the minimum flux,

```
flux = -B1*S+B2*S+B3*S;
[error,k] = min(abs(flux));
```

where `k` is the index of that element in the array, which determines, further, the solution for `H` and the corresponding `B` in each of the three branches, `[H1(k),B1(k)]`, `[H2(k),B2(k)]`, and `[H3(k),B3(k)]`. Finally, the obtained solution, $H_1 = 1700$ A/m, $B_1 = 0.9$ T, $H_2 = 600$ A/m, $B_2 = 0.6$ T, $H_3 = 300$ A/m, and $B_3 = 0.3$ T, is presented graphically: we plot the magnetization curve and plot on it markers of the positions of the operating points for the branches,

```
plot(H,B);
hold on;
plot(H1(k),B1(k),'r*',H2(k),B2(k),'ko',H3(k),B3(k),'m^');
```

as shown in Fig.5.2(c) (note that `'r*'`, `'ko'`, and `'m^'` result in red star, black circle, and magenta upward-pointing-triangle markers in the graph).

MATLAB EXERCISE 5.11 **General numerical solution for the operating point.** Write a function `magCurveSolution()` in MATLAB that plots an arbitrary magnetization curve of a ferromagnetic core of a magnetic circuit and the load line of the circuit [see Fig.5.3(b) (in the next MATLAB exercise), for instance], and finds numerically the coordinates (H, B) of their intersection – the operating point of the circuit, which is also plotted. The input to the function consists of arrays of values `H`, `B1`, and `B2`, with `B1` and `B2` representing the dependences on `H` of the magnetization curve and the load line, respectively, as well as information, `c1` and `c2`, on the colors of curves, and a parameter `star` determining whether the operating point, once found, is plotted in the graph. *(magCurveSolution.m on IR)*

TUTORIAL:

First, we plot the magnetization curve and the load line in the same graph,

```
function [error,k] = magCurveSolution(H,B1,B2,c1,c2,star)
plot(H,B1,c1,H,B2,c2);
legend('Magnetization curve','Load line');
```

Next, we find the index k (which is an output parameter of the function) of the element H (and thus B1 or B2) that determines the solution for the operating point, i.e., that minimizes **error**, which is the other output parameter – the minimum difference (ideally, zero) between B1 and B2, and a measure of the error in solution due to the finite steps between array elements (sample points). In other words, H(k) is the closest of all points constituting the H array to the intersection of the magnetization curve with the load line of the circuit. Finally, we plot the solution (Hsol,Bsol) in a graph – if **star** equals unity.

```
[error,k]  = min(abs(B1-B2));
Bsol = B1(k);
Hsol = H(k);
hold on;
if star == 1
plot(Hsol,Bsol,'*k');
end;
xlabel('H [A/m]');
ylabel('B [T]');
text(1.3*Hsol,1.3*Bsol,'Operating point');
hold off;
```

MATLAB EXERCISE 5.12 **Simple nonlinear magnetic circuit with an air gap.** Consider a magnetic circuit consisting of a thin toroidal ferromagnetic core with a coil and an air gap shown in Fig.5.3(a). The coil has $N = 1000$ turns of wire with the total resistance $R = 50\ \Omega$. The length of the ferromagnetic portion of the circuit is $l = 1$ m, the thickness (width) of the gap is $l_0 = 4$ mm, the cross-sectional area of the toroid is $S = S_0 = 5\ \text{cm}^2$, and the emf of the generator in the coil circuit is $\mathcal{E} = 200$ V. The idealized initial magnetization curve of the material is given in Fig.5.3(b). With the use of MATLAB, find the magnetic field intensity (H) and flux density (B) in the core. In specific, write a MATLAB code that plots the magnetization curve of the core and the load line of the circuit (in the same graph), and computes numerically and plots the operating point of the circuit. *(ME5_12.m on IR)* **H**

HINT:
To find the equation of the load line for the magnetic circuit, we refer to Fig.5.3(a) and use Kirchhoff's laws for magnetic circuits, Eqs.(5.5), which give $BS = B_0S_0$, that is, $B = B_0$, and $Hl + H_0l_0 = NI$, where $I = \mathcal{E}/R$. These equations, combined with $B_0 = \mu_0 H_0$ (for the air gap), result in the following relationship between H and B – load line:

$$B = \frac{\mu_0}{l_0}(NI - Hl) \qquad \text{(load line for a simple magnetic circuit)} \tag{5.8}$$

[shown in Fig.5.3(b)].

Use function magCurveSolution (from the previous MATLAB exercise) and see the tutorial to MATLAB Exercise 5.10.

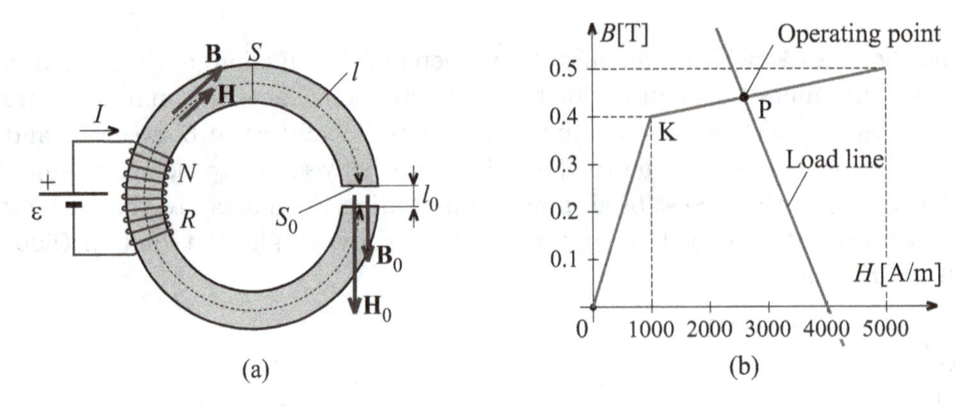

Figure 5.3 Analysis of a simple nonlinear magnetic circuit with an air gap: (a) circuit geometry, with flux density and field intensity vectors in the circuit, and (b) idealized initial magnetization curve of the material, with the load line and operating point for the circuit; for MATLAB Exercise 5.12.

MATLAB EXERCISE 5.13 **Another simple nonlinear magnetic circuit.** For the magnetic circuit in Fig.5.3(a), assume that $l = 40$ cm, $S = S_0 = 2.25$ cm^2, $l_0 = 0.25$ mm, $N = 800$, and $I = 1$ A. In addition, let the core be made from a nonlinear ferromagnetic material whose initial magnetization curve can be linearized in parts as in Fig.5.2(b) with $H_\mathrm{k} = 1000$ A/m and $B_\mathrm{m} = 1$ T. Solve this circuit in MATLAB: plot, in the same graph, B-H relationships in both the core and the air gap, and plot (mark) in the graph the obtained numerical solutions for (H, B) in both regions, in addition to displaying the results in the Command Window. *(ME5_13.m on IR)* **H**

HINT:

See the previous MATLAB exercise and make use of function `magCurveSat` from MATLAB Exercise 5.9.

MATLAB EXERCISE 5.14 **Magnetization–demagnetization – numerical solution and movie.** The ferromagnetic core of the circuit shown in Fig.5.4(a) has the cross-sectional area $S = 1$ cm^2, mean length $l = 20$ cm, and air-gap thickness $l_0 = 1$ mm. The number of wire turns of the coil is $N = 1000$, the resistance of the winding is $R = 20\ \Omega$, and the emf of the generator is $\mathcal{E} = 20$ V. The core does not have residual magnetization and the switch K is in the off position (open). By turning the switch on, the mmf is applied to the circuit and the magnetic flux through it rises following the initial magnetization curve of the material. This curve can be considered as linear, as shown in Fig.5.4(b), where the initial permeability is $\mu_\mathrm{a} = 0.001$ H/m $(B = \mu_\mathrm{a} H)$. The magnetic flux density in the core becomes $B = B_\mathrm{m}$ in the stationary state. The switch is then turned off (opened) and a new stationary state established in the circuit. The demagnetization curve for the material can be approximated by two straight-line segments [Fig.5.4(b)], where H_c (coercive force) equals H_m. In MATLAB, make a movie that simulates the magnetization–demagnetization of the circuit in the B-H diagram in Fig.5.4(b). *(ME5_14.m on IR)*

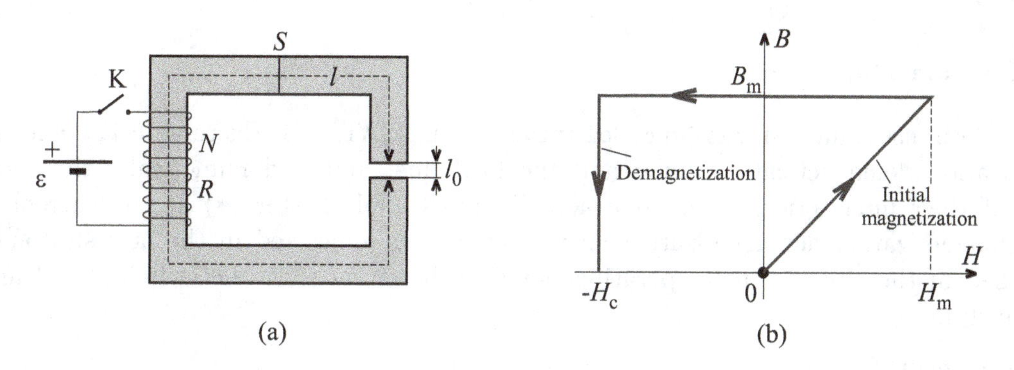

Figure 5.4 Magnetization and demagnetization in a magnetic circuit with an air gap: (a) circuit geometry and (b) idealized initial magnetization and demagnetization curves of the material; for MATLAB Exercise 5.14.

TUTORIAL:

After the switch K in Fig.5.4(a) is turned on (closed), the current in the winding is $I = \mathcal{E}/R$, and the magnetic flux density in the core reaches the value obtained from a combination of Eq.(5.8) and $B = \mu_a H$ [linear initial magnetization curve in Fig.5.4(b)], so it is given by

$$B = NI\left(\frac{l}{\mu_a} + \frac{l_0}{\mu_0}\right)^{-1} = B_m \quad \longrightarrow \quad H_m = \frac{B_m}{\mu_a}\,. \tag{5.9}$$

We implement this initial magnetization process in MATLAB as follows:

```
I = EPS/R;
Bmax = N*I*(l/MUa+l0/MU0)^(-1);
Hmax = Bmax/MUa;
H1 = Hmax*(0.05:0.05:1);
B1 = Bmax*(0.05:0.05:1);
```

The switch is then opened, and the demagnetization curve for the material, composed of two straight-line segments in Fig.5.4(b), is traced,

```
H2 = Hmax*(1:-0.05:-1);
B2 = Bmax*ones(1,length(H2));
B3 = Bmax*(1:-0.05:0);
H3 = -Hmax* ones(1,length(B3));
```

The arrays **Hmag** and **Bmag** contain data for all three parts (one magnetization part and two demagnetization parts) of the magnetization–demagnetization curve in Fig.5.4(b),

```
Hmag = [H1 H2 H3];
Bmag = [B1 B2 B3];
```

With the switch K open, Eq.(5.8) becomes

$$B = -\frac{\mu_0\, l}{l_0} H\,, \tag{5.10}$$

and this represents the equation of the load line for the magnetic circuit in this state, which must be satisfied as well,

```
B = -MU0*l*Hmag/10;
```

Next, we use function `magCurveSolution` (from MATLAB Exercise 5.11) and plot the magnetization–demagnetization curve and the load line, and find numerically the coordinates (H, B) of their intersection. The code actually plots a black star $(*)$ at the current position on the magnetization–demagnetization curve; the star becomes red in the new stationary state established in the circuit – new operating point of the circuit (for the switch closed and open, respectively).

```
for k = 1:length(Hmag)
[error,k1] = magCurveSolution(Hmag,Bmag,B,'k','g',0);
hold on;
if (k < k1)
if k == length(H1)
plot(Hmag(k),Bmag(k),'*r');
else
plot(Hmag(k),Bmag(k),'*k');
end;
else
plot(Hmag(k1),Bmag(k1),'*r');
end;
```

Finally, the following lines of the code write labels in the graph for different stages in the process, get the frame (snapshot) of the movie, and pause at the stationary state established with the switch closed,

```
if (k < length(H1))
text(0.55*Hmax,0.55*Bmax,'Initial magnetization');
end;
if (k == length(H1))
text(0.55*Hmax,0.55*Bmax,'Stationary state - switch closed');
end;
if (k > length(H1))
if (k < length(Hmag))
text(0.55*Hmax,0.55*Bmax,'Transient process - switch open');
else
text(0.55*Hmax,0.55*Bmax,'Stationary state - switch open');
end;
end;
M(k) = getframe();
if (k == length(H1))
pause(3);
end;
hold off;
end;
```

Figures 5.5(a)–(d) show four characteristic snapshots of the magnetization–demagnetization movie.

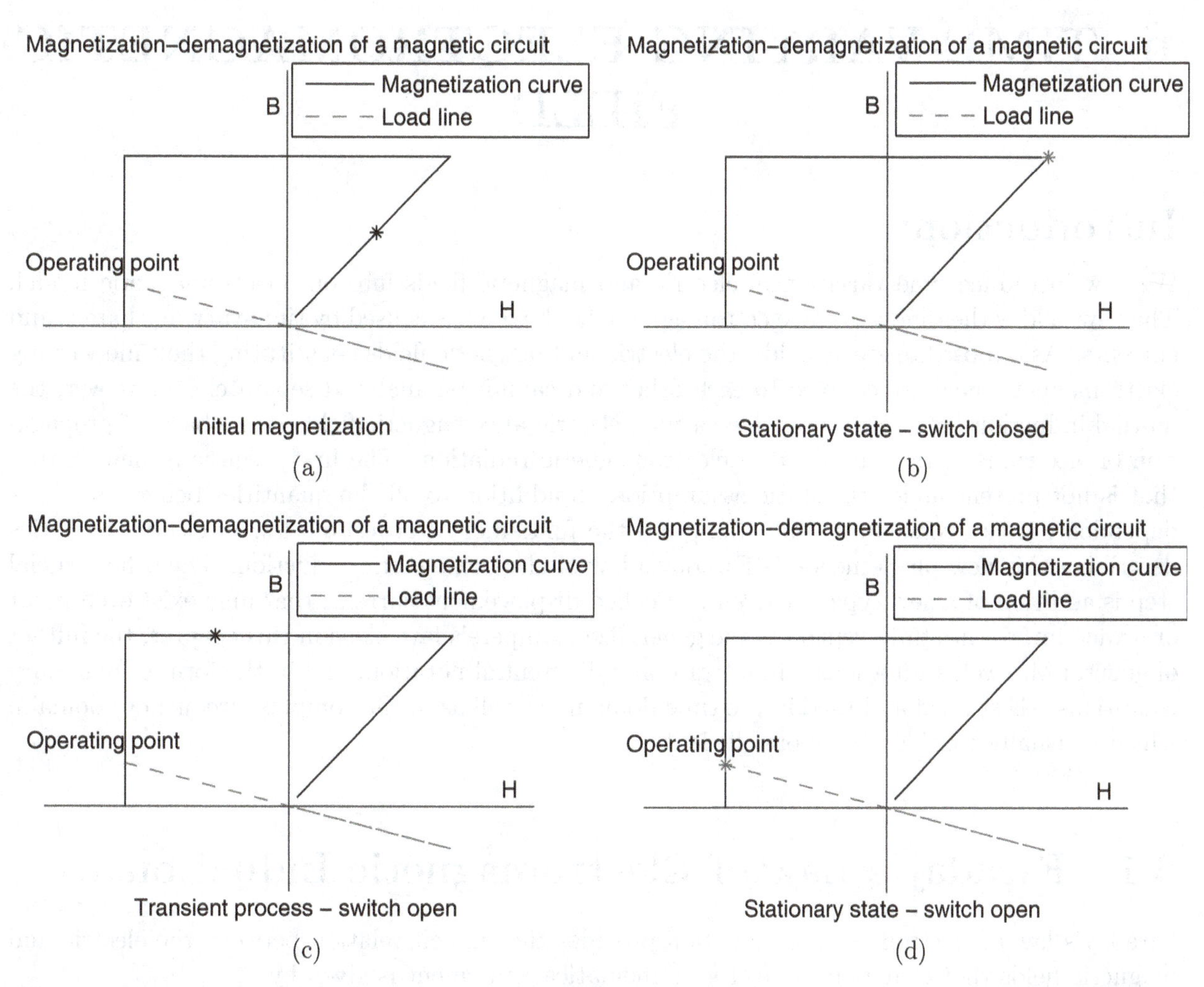

Figure 5.5 MATLAB movie simulating the magnetization–demagnetization of the magnetic circuit in Fig.5.4(a) in the *B*-*H* diagram in Fig.5.4(b): (a) snapshot of the initial magnetization of the core with the switch K closed, (b) stationary state (operating point of the circuit) with the switch closed, (c) movie snapshot of the demagnetization (transient process) with the switch open, and (d) stationary state (operating point) with the switch open; for MATLAB Exercise 5.14. *(color figure on CW)*

MATLAB EXERCISE 5.15 Movie with two magnetization–demagnetization curves. Repeat the previous MATLAB exercise but simulating and playing in the same movie, simultaneously, the magnetization–demagnetization processes of the circuit with two different emf's of the generator in Fig.5.4(a), amounting to $\mathcal{E} = 20$ V and $\mathcal{E} = 40$ V, respectively. *(ME5_15.m on IR)*

---

6 TIME-VARYING ELECTROMAGNETIC FIELD

Introduction:

We now introduce time variation of electric and magnetic fields into our electromagnetic model. The new field is the time-varying electromagnetic field, which is caused by time-varying charges and currents. As opposed to static fields, the electric and magnetic fields constituting the time-varying electromagnetic field are coupled to each other and cannot be analyzed separately. Moreover, the mutual induction (generation) of time-varying electric and magnetic fields is the basis of propagation of electromagnetic waves and of electromagnetic radiation. The first essentially new feature that is not present under the static assumption, in addition to all the quantities being now time-dependent, is electromagnetic induction, and the fundamental governing law of electromagnetics describing this new phenomenon is Faraday's law of electromagnetic induction. The other crucial step is addition of a new type of current, so-called displacement current, that may exist even in air or a vacuum, to the static version of the generalized Ampère's law. Also in this chapter, the full set of general Maxwell's equations – in integral and differential notation, and in the form of boundary conditions – is studied and used in the time domain, as well as in the complex (frequency) domain, which is usually considerably more efficient.

6.1 Faraday's Law of Electromagnetic Induction

Faraday's law of electromagnetic induction provides the explicit relation between the electric and magnetic fields that change in time. Its mathematical statement is given by

$$e_{\mathrm{ind}} = -\frac{\mathrm{d}\Phi}{\mathrm{d}t} \quad \text{or} \quad \oint_C \mathbf{E}_{\mathrm{ind}} \cdot \mathrm{d}\mathbf{l} = -\frac{\mathrm{d}}{\mathrm{d}t}\int_S \mathbf{B} \cdot \mathrm{d}\mathbf{S} \quad \text{(Faraday's law)} . \tag{6.1}$$

Namely, the induced electromotive force (emf), e_{ind} (also denoted as v_{emf} or V_{emf} in some texts), in an arbitrary contour is equal to the negative of the time rate of change of the magnetic flux, Φ (symbol Ψ is sometimes used for magnetic flux), through the contour, i.e., through a surface of arbitrary shape bounded by the contour and oriented in accordance to the right-hand rule with respect to the orientation of the contour. This rule tells us that the flux is in the direction defined by the thumb of the right hand when the other fingers point in the direction of the emf, as indicated in Fig.6.1. Faraday's law represents Maxwell's first equation for the time-varying electromagnetic field. It is essentially different from Maxwell's first equation for the time-invariant electromagnetic field in Eq.(1.28). In addition, the induced electric field intensity vector, $\mathbf{E}_{\mathrm{ind}}$, in Fig.6.1 and Eq.(6.1) can be expressed in terms of the magnetic vector potential in Eq.(4.24), due to a time-varying current, as $\mathbf{E}_{\mathrm{ind}} = -\partial\mathbf{A}/\partial t$.

The time change of the magnetic flux in Eq.(6.1) and thus the induced emf can be due to a time variation of the magnetic field, which gives rise to so-called transformer induction, and due to movement (translation, rotation, etc.) and/or deformation (changing shape and size) of the

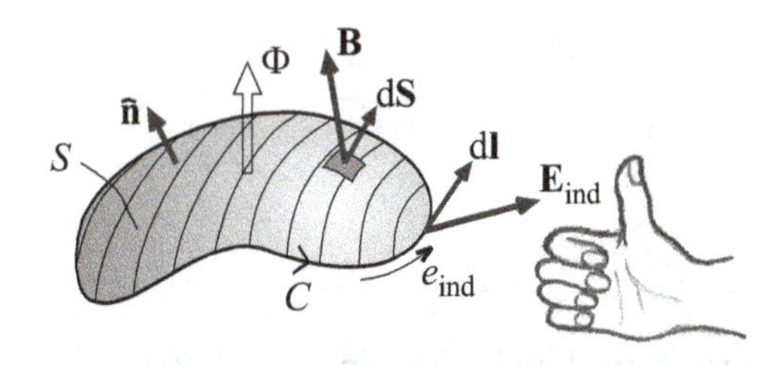

Figure 6.1 Arbitrary contour in a time-varying magnetic field – for the statement of Faraday's law of electromagnetic induction.

contour in a static magnetic field – motional induction, as well as due to a combination of the two mechanisms – total induction.

MATLAB EXERCISE 6.1 Transformer emf, symbolic integration and differentiation. An infinitely long straight wire carries a time-varying current of intensity $i(t)$. A rectangular contour of side lengths a and b lies in the same plane with the wire, with two sides parallel to it, as shown in Fig.6.2. The distance between the wire and the closer parallel side of the contour is c. The emf induced in the contour can be found from Eq.(6.1), with the magnetic flux through the contour being computed by integrating the magnetic flux density produced by the current in the wire, given by Eq.(4.10), across the flat surface spanned over the contour (Fig.6.2), namely,

$$B(x,t) = \frac{\mu_0 i(t)}{2\pi x} \quad \longrightarrow \quad \Phi(t) = \int_{x=c}^{c+a} B(x,t)\, \underbrace{b\,\mathrm{d}x}_{\mathrm{d}S} \quad \longrightarrow \quad e_{\mathrm{ind}}(t) = -\frac{\mathrm{d}\Phi}{\mathrm{d}t}\,. \tag{6.2}$$

Implement these equations symbolically in MATLAB – to obtain the expression for $e_{\mathrm{ind}}(t)$, assuming that $a = 4$ m, $b = 3$ m, $c = 1$ m, and $i(t) = \sin(377t)$ A (t in s). *(ME6_1.m on IR)*[1]
H[2]

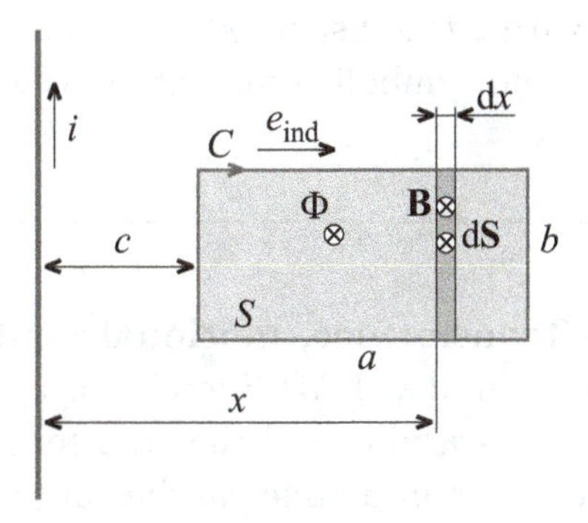

Figure 6.2 Evaluation of the emf in a rectangular contour in the vicinity of an infinitely long wire with a time-varying current; for MATLAB Exercise 6.1.

TUTORIAL:
We first declare symbolic variables for the computation, and define input quantities,

[1] *IR* = Instructor Resources (for the book).
[2] **H** = recommended to be done also "by hand," i.e., not using MATLAB.

```
syms MU0 x t pi a b c
a = 4;
b = 3;
c = 1;
i = sin(377*t);
```

Next, we calculate the magnetic flux density (B) in Eqs.(6.2),

```
B = MU0/(2*pi)*i/x;
```

and evaluate the magnetic flux Φ (Phi) through the contour by symbolic integration, according to Eqs.(6.2),

```
Phi = int(B*b,x,c,c+a);
```

Finally, we take the symbolic derivative of Φ to find e_{ind} (eind), Eqs.(6.2),

```
eind = -diff(Phi,t);
```

The result appears in the Command Window (using MATLAB function pretty) as follows:

$$\texttt{eind} = \frac{\texttt{-1131 MU0 cos(377t) ln(5)}}{\texttt{2 PI}}$$

MATLAB EXERCISE 6.2 **Motional emf, symbolic integration and differentiation.** Assume that the current in the straight wire conductor in Fig.6.2 is time-invariant, with intensity I, and that the contour moves away from the wire at a constant velocity v, such that the distance of the closer parallel side of the contour from the wire increases linearly in the course of time as $c+vt$. Since the magnetic field produced by the current in the wire is now time-invariant and the contour is moving, the system represents a motional induction version of the same geometry with transformer induction in Fig.6.2. The magnetic flux through the contour, $\Phi(t)$, can be obtained in the same way as in Eqs.(6.2) – with $i(t)$ substituted by I and c by $c+vt$. In MATLAB, perform symbolic integration to find $\Phi(t)$ and symbolic differentiation to determine the induced emf in the contour. *(ME6_2.m on IR)* **H**

MATLAB EXERCISE 6.3 **Transformer, motional, and total emf's in a moving contour.** As a combination of systems in MATLAB Exercises 6.1 and 6.2, assume that the current in the wire conductor in Fig.6.2 is time-varying and that the contour is moving away from the wire. Hence, we have a motion of a contour in a time-varying magnetic field, and the emf is induced in the contour due to combined (transformer plus motional) induction. With $c+vt$ in place of c, Eqs.(6.2) give the following expression for the total (combined) emf in the contour:

$$e_{\text{ind}}(t) = -\frac{\mathrm{d}\Phi}{\mathrm{d}t} = \underbrace{-\frac{\mu_0 b}{2\pi}\ln\frac{c+a+vt}{c+vt}\frac{\mathrm{d}i}{\mathrm{d}t}}_{\text{Transformer emf}} + \underbrace{\frac{\mu_0 i(t)abv}{2\pi(c+vt)(c+a+vt)}}_{\text{Motional emf}} . \tag{6.3}$$

Write a MATLAB code that plots $e_{\text{ind}}(t)$, as well as the parts of this emf corresponding to the transformer and motional induction, respectively, as indicated in Eq.(6.3), for $a = 4$ cm, $b = 1$ cm, $c = 10$ cm, $i(t) = \sin(377t)$ A (t in s), and $v = 50$ m/s. *(ME6_3.m on IR)*

HINT:
The resulting plots are given in Fig.6.3.

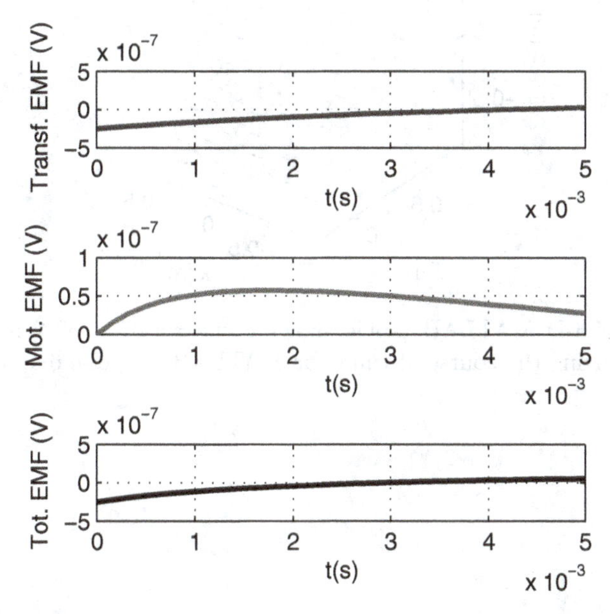

Figure 6.3 MATLAB plots of transformer, motional, and total emf's in a moving contour near a time-harmonic line current; for MATLAB Exercise 6.3. *(color figure on CW)*[3]

MATLAB EXERCISE 6.4 **Rotating loop in a time-harmonic magnetic field – 3-D movie.** A rectangular loop of edge lengths a and b rotates with a constant angular velocity ω_0 about its axis in a uniform time-harmonic magnetic field of flux density $B(t) = B_0 \cos \omega t$. At an instant $t = 0$, the loop lies in the yz-plane ($x = 0$) of a Cartesian coordinate system. In this coordinate system, the axis of loop rotation coincides with the y-axis and $\mathbf{B} = B(t)\,\hat{\mathbf{x}}$. Create a 3-D movie in MATLAB that shows the rotation of the loop and oscillation of the magnetic field during the course of time. Assume that $a = 40$ cm, $b = 80$ cm, $\omega_0 = 500$ rad/s, $\omega = 1000$ rad/s, and $B_0 = 0.1$ T. *(ME6_4.m on IR)*

TUTORIAL:
First, we enter the input data, and specify the duration time of the movie (`Tmovie`), and the time array and the spatial mesh in the yz-plane (see Fig.6.4) for time instants and nodes at which the field $\mathbf{B}$ is computed and plotted; of course, the same time instants are used to visualize the loop rotation.

```
a = 40*10^(-2);
b = 80*10^(-2);
```

[3] *CW* = Companion Website (of the book).

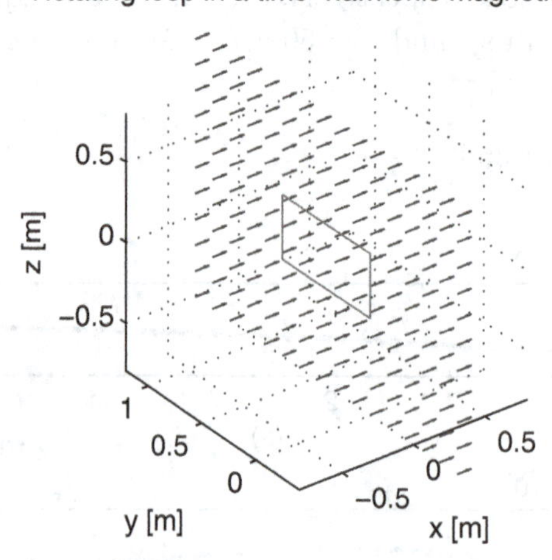

Figure 6.4 Snapshot of a 3-D MATLAB movie showing the rotation of a rectangular loop in a time-harmonic magnetic field, during the course of time; for MATLAB Exercise 6.4. *(color figure on CW)*

```
B0 = 0.1;
omega = 1000;
omega0 = 500;
T = 2*pi/omega;
T0 = 2*pi/omega0;
Tmovie = max(T,T0);
dt = Tmovie/200;
t = 0:dt:Tmovie;
[y1,z1] = meshgrid(-b:0.2*b:2*b,-2*a:0.4*a:2*a);
[L,N] = size(y1);
```

Then, in the movie `for` loop, we compute $B(t)$ (`B`) at a specified instant of time (from the time array) and Cartesian coordinates of loop vertices, defining the position of the loop as it rotates, at the same instant, and plot the **B** vectors by means of MATLAB function `quiver3` (see MATLAB Exercise 1.3) and the loop edges using MATLAB function `line`,

```
for i=1:length(t)
B = B0*cos(omega*t(i));
B1 = B*ones(L,N);
x = a/2*sin(omega0*t(i));
z = a/2*cos(omega0*t(i));
quiver3(zeros(L,N),y1,z1,B1,zeros(L,N),zeros(L,N),0);
hold on;
line([-x x],[b b],[-z z],'Color','r');
line([-x x],[0 0],[-z z],'Color','r');
line([x x],[0 b],[z z],'Color','r');
line([-x -x],[0 b],[-z -z],'Color','r');
axis equal;
xlim([-2*a,2*a]); ylim([-b/2,1.5*b]); zlim([-2*a,2*a])
```

```
M(i) = getframe();
hold off;
xlabel('x [m]'); ylabel('y [m]'); zlabel('z [m]');
title('Rotating loop in a time-harmonic magnetic field');
end;
```

Figure 6.4 shows a snapshot of the movie.

MATLAB EXERCISE 6.5 **Transformer, motional, and total emf's in a rotating loop.** Repeat MATLAB Exercise 6.3 but for the rotating loop in a time-harmonic magnetic field from the previous MATLAB exercise, assuming that $B(t) = B_0 \sin \omega t$ and $\omega_0 = \omega = 377$ rad/s. The total emf induced in the contour is given by

$$e_{\text{ind}}(t) = -\frac{\mathrm{d}\Phi}{\mathrm{d}t} = \underbrace{-ab\frac{\mathrm{d}B}{\mathrm{d}t}\cos \omega t}_{\text{Transformer emf}} + \underbrace{\omega abB(t)\sin \omega t}_{\text{Motional emf}} \ . \qquad (6.4)$$

(ME6_5.m on IR)

MATLAB EXERCISE 6.6 **Rotating magnetic field – movie.** A rectangular loop of edge lengths a and b is situated in the magnetic field produced by two mutually perpendicular large coils with time-harmonic currents. The field due to each coil can be considered to be uniform. The currents in the coils are of equal amplitudes and 90° out of phase, so that the magnetic flux densities they produce are given as $B_1(t) = B_0 \cos \omega t$ and $B_2(t) = B_0 \sin \omega t$, respectively, where $\mathbf{B}_1 \perp \mathbf{B}_2$. Vector $\mathbf{B}_1$ is perpendicular to the plane of the loop, while $\mathbf{B}_2$ lies in that plane. Create a movie in MATLAB that shows the time-harmonic oscillations of $\mathbf{B}_1(t)$ and $\mathbf{B}_2(t)$ and the resulting rotation of the vector $\mathbf{B}(t) = \mathbf{B}_1(t) + \mathbf{B}_2(t)$, assuming that $\omega = 377$ rad/s and $B_0 = 1$ T. *(ME6_6.m on IR)*

TUTORIAL:

The fist step is the entry of input data, and computation of the period of change of $B_1(t)$ and $B_2(t)$, $T = 2\pi/\omega$, and of a time array (vector) `t` of time instants for plotting (in the movie):

```
B0 = 1;
w = 377;
T = 2*pi/w;
dt = T/200;
t = 0:dt:T;
```

The next step defines the x- and y-axes of the graph and their bounds, using MATLAB function `linspace(a,b)`, which generates an array of 100 points equidistantly spaced between `a` and `b`. In our case, both `a` and `b` are determined relative to the amplitude B_0 of $B_1(t)$ and $B_2(t)$.

```
limit = 1.5*B0;
XAxis = linspace(-limit,limit);
```

```
YAxis = zeros(1,length(XAxis));
B1 = zeros(1,length(t));
B2 = zeros(1,length(t));
```

Within the `for` loop of the movie, we implement the component flux density expressions $B_1(t)$ and $B_2(t)$, and plot vectors $\mathbf{B}_1(t)$ and $\mathbf{B}_2(t)$, as well as the total vector $\mathbf{B}(t)$, using MATLAB function `quiver`, at a specified instant of time (ith element of the time array),

```
for i = 1:length(t);
B1(i) = B0*cos(w*t(i));
B2(i) = B0*sin(w*t(i));
quiver(0,0,B1(i),B2(i),'k','linewidth',2); hold on;
text(B1(i),B2(i),'B');
quiver(0,0,B1(i),0, 'b','linewidth',2);
text(B1(i),0,'B1');
quiver(0,0,0,B2(i), 'r','linewidth',2);
text(0,B2(i),'B2');
plot(XAxis,YAxis,'k');
plot(YAxis,XAxis,'k');
hold off;
axis equal; xlabel('B1 [T]'); ylabel('B2 [T]');
title('Rotating magnetic field');
M(i) = getframe;
end;
```

A snapshot of the movie is shown in Fig.6.5.

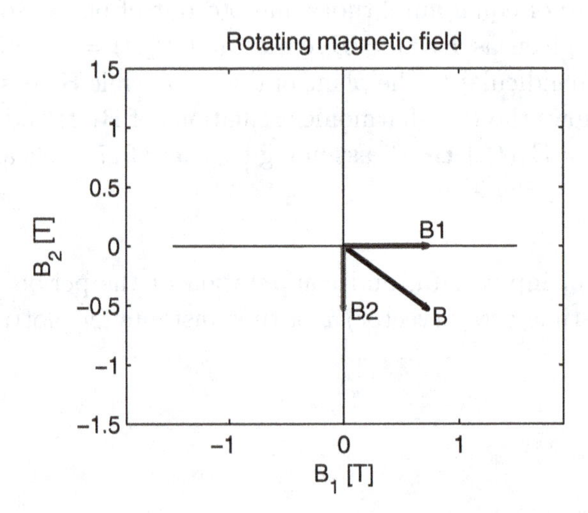

Figure 6.5 Snapshot of a 2-D MATLAB movie showing two time-harmonic magnetic fields ($\mathbf{B}_1$ and $\mathbf{B}_2$) of equal amplitudes and 90° out of phase, which superposed to each other represent a rotating magnetic field ($\mathbf{B}$); for MATLAB Exercise 6.6. *(color figure on CW)*

MATLAB EXERCISE 6.7 **Emf in a loop exposed to a rotating magnetic field.** Find the emf induced in the loop exposed to two mutually orthogonal time-harmonic magnetic fields of equal amplitudes and 90° out of phase from the previous MATLAB exercise. *(ME6_6.m on IR)* **H**

TUTORIAL:
From the movie developed and played in the previous MATLAB exercise, we conclude that the magnitude of the resultant magnetic flux density vector, $\mathbf{B}(t) = \mathbf{B}_1(t) + \mathbf{B}_2(t)$, is constant with respect to time, and equal to B_0. Also, $\mathbf{B}$ rotates at a constant angular velocity. Experimenting, in the movie, with different values for the angular frequency ω of the individual magnetic flux densities and currents in the coils, we realize that this velocity equals ω.

As the contour is stationary and vector $\mathbf{B}$ changes in time (its magnitude is constant, but its direction changes), this is a system based on transformer induction. On the other hand, for the generation of emf it is irrelevant whether $\mathbf{B}$ rotates around a stationary loop or a loop rotates (at the same rate) in a static $\mathbf{B}$, with this latter case being the system based on motional induction. Exploiting this equivalency, the emf in the loop is given by the expression for the motional part of the total emf in Eq.(6.4) with B substituted by B_0, so $e_{\text{ind}}(t) = \omega ab B_0 \sin \omega t$.

6.2 Self-Inductance

In general, inductance can be interpreted as a measure of transformer electromagnetic induction in a system of conducting contours (circuits) with time-varying currents in a linear magnetic medium. Briefly, self-inductance is a measure of the magnetic flux and induced emf in a single isolated contour (or in one of the contours in a system) due to its own current. Similarly, a current in one contour causes magnetic flux through another contour and induced emf in it, and mutual inductance is used to characterize this coupling between the contours.

Consider a stationary conducting wire contour (loop), C, in a linear, homogeneous or inhomogeneous, magnetic medium, and assume that a time-varying current of intensity i is established in the contour, as shown in Fig.6.6(a). This current produces a magnetic field whose flux density vector, $\mathbf{B}$, at any point of space and any instant of time is linearly proportional to i, as well as an induced electric field of intensity $\mathbf{E}_{\text{ind}}$ and an induced emf along C, e_{ind}, according to Eq.(6.1). The magnetic flux, Φ, through a surface S bounded by C [Fig.6.6(a)] is also linearly proportional to i, so we can write

$$L = \frac{\Phi}{i}\,, \quad e_{\text{ind}} = -\frac{\mathrm{d}\Phi}{\mathrm{d}t} = -L\frac{\mathrm{d}i}{\mathrm{d}t} \quad (L \text{ – self-inductance; unit: H})\,, \tag{6.5}$$

where L is termed the self-inductance or just inductance of the contour, and its unit is henry (H). More precisely, this is the external inductance, since it takes into account only the flux Φ of the magnetic field that exists outside the conductor of the loop.

As an example, let us find L per unit length of a thin symmetrical two-wire transmission line [Fig.2.9(c)]. The line can be considered as a wire loop that closes upon itself at both ends of the line at infinity. The fields $\mathbf{B}_1$ and $\mathbf{B}_2$ due to currents in individual conductors of the line are computed using Eq.(4.10), and Φ through the flat surface of length l spanned between the line

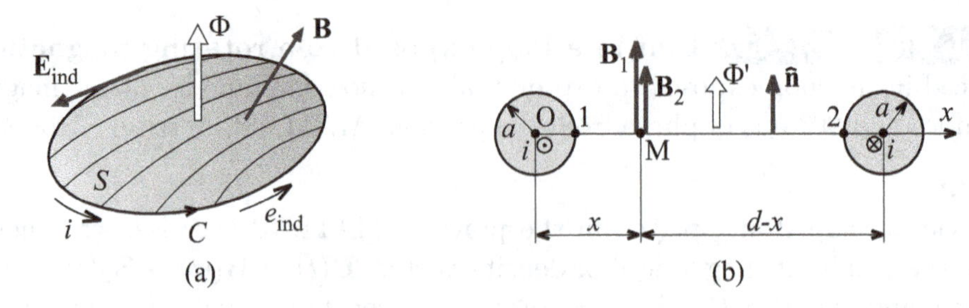

Figure 6.6 (a) Current contour in a linear magnetic medium – for the definition of self-inductance. (b) Evaluation of the external inductance per unit length of a thin two-wire transmission line (with $d \gg a$) – cross section of the structure.

conductors, Fig.6.6(b), is

$$\Phi = \int_{x=a}^{d-a} B \underbrace{l\,\mathrm{d}x}_{\mathrm{d}S} = \frac{\mu i l}{2\pi} \int_a^{d-a} \left(\frac{1}{x} + \frac{1}{d-x} \right) \mathrm{d}x = \frac{\mu i l}{\pi} \ln \frac{d-a}{a} \approx \frac{\mu i l}{\pi} \ln \frac{d}{a} \,. \tag{6.6}$$

Hence, the external inductance per unit length of the line comes out to be

$$L' = L_{\text{p.u.l.}} = \frac{\Phi'}{i} = \frac{\Phi}{il} = \frac{\mu}{\pi} \ln \frac{d}{a} \quad (L' \text{ – thin two-wire line}) \,. \tag{6.7}$$

Comparing the expression for L' in Eq.(6.7) and the expression in Eq.(2.27) for the capacitance per unit length of the same line, we note that the following relationship exists between the two parameters:

$$L'C' = \varepsilon\mu \quad (\text{duality of } L' \text{ and } C' \text{ of a transmission line}) \,. \tag{6.8}$$

As a matter of fact, the same duality relationship exists between L' and C' for any two-conductor transmission line with a homogeneous linear dielectric.

MATLAB EXERCISE 6.8 **Inductance calculator and GUI for transmission lines.** In MATLAB, create an inductance calculator in the form of a graphical user interface, with names of structures appearing in a pop-up menu, to calculate and display the external inductance per unit length of the following three types of transmission lines with air dielectric: thin symmetrical two-wire transmission line, Fig.6.6(b), coaxial cable, Fig.2.9(b), and wire-plane transmission line, consisting of a thin wire conductor of radius a placed at a height h ($h \gg a$) in parallel to a conducting ground plane. *[folder ME6_8(GUI) on IR]*

HINT:

See MATLAB Exercise 2.13. Use Eq.(6.7) for the two-wire line and Eqs.(2.26) and (6.8) for the coaxial cable; the inductance p.u.l. of the wire-plane line, obtained in a similar fashion to that in Eqs.(6.6) and (6.7), is given by $L' = \mu_0 \ln(2h/a)/(2\pi)$.

6.3 Mutual Inductance

The mutual inductance between two magnetically coupled contours (circuits) can be evaluated by assuming a current (i_1) to flow in one contour (primary circuit) and computing or measuring the resulting magnetic flux (Φ_2) or the induced emf (e_{ind2}) in the other contour (secondary circuit), as shown in Fig.6.7(a),

$$L_{21} = \frac{\Phi_2}{i_1}\,, \quad e_{\mathrm{ind2}} = -\frac{\mathrm{d}\Phi_2}{\mathrm{d}t} = -L_{21}\frac{\mathrm{d}i_1}{\mathrm{d}t} \quad (L_{21} \text{ – mutual inductance; unit: H})\,. \tag{6.9}$$

Because of reciprocity, $L_{12} = L_{21}$. Note that the symbol M is also used to denote mutual inductance.

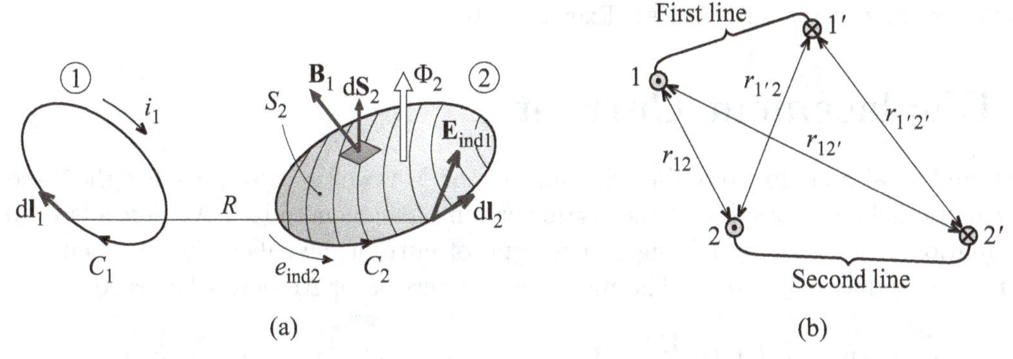

Figure 6.7 (a) Two magnetically coupled conducting contours – for the definition of mutual inductance. (b) Cross section of two parallel infinitely long thin two-wire lines.

As an example, the mutual inductance per unit length of two parallel infinitely long thin two-wire lines in air, Fig.6.7(b), obtained by a similar integration as in Eq.(6.6), amounts to

$$L'_{21} = (L_{21})_{\mathrm{p.u.l.}} = \frac{L_{21}}{l} = \frac{\mu_0}{2\pi}\ln\frac{r_{12'}r_{1'2}}{r_{12}r_{1'2'}} \quad (L'_{21} \text{ – two parallel two-wire lines})\,. \tag{6.10}$$

MATLAB EXERCISE 6.9 **Mutual inductance p.u.l. of two two-wire lines.** Write a function `mutualIndTwoLines()` in MATLAB that calculates the per-unit-length mutual inductance between two thin two-wire lines in Fig.6.7(b), based on Eq.(6.10). *(mutualIndTwoLines.m on IR)*

MATLAB EXERCISE 6.10 **Mutual inductance between phone and power lines.** Write a function `phoneLinePowerLine()` in MATLAB that returns the mutual inductance between the phone line and the power line shown in Fig.6.8, for given dimensions h, d_1, and d_2 in Fig.6.8 and length l of the system. Then use this function to calculate the amplitude (peak-value) $\mathcal{E}_0$ of the induced emf in the telephone line due to a time-harmonic current of amplitude I_0 and frequency f in the power line, based on Eq.(6.9), for different adopted values of parameters of the system. *(phoneLinePowerLine.m on IR)* **H**

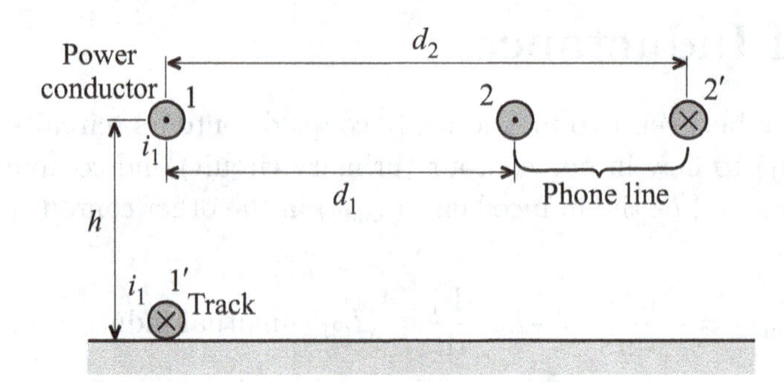

Figure 6.8 Evaluation of the mutual inductance between a two-wire telephone line and a nearby two-wire power line, approximating the transmission line formed by the power conductor of a cable car and the track (cross section of the system); for MATLAB Exercise 6.10.

6.4 Displacement Current

The next and final step in obtaining the full set of Maxwell's equations for the time-varying electromagnetic field is correction of the static version of the generalized Ampère's law (Maxwell's second equation), Eq.(5.1), by adding a new type of current, so-called displacement current, in parallel to the conduction current. The most general version of Ampère's law reads

$$\oint_C \mathbf{H} \cdot \mathrm{d}\mathbf{l} = \int_S \left(\mathbf{J} + \frac{\partial \mathbf{D}}{\partial t} \right) \cdot \mathrm{d}\mathbf{S} \quad \text{(corrected generalized Ampère's law)} , \tag{6.11}$$

where the expression $\partial \mathbf{D}/\partial t$ has the dimension of a current density (it is expressed in $\mathrm{A/m^2}$), and this is the density of a new type of current that may exist even in air or a vacuum. It is called the displacement current density and is denoted by $\mathbf{J}_\mathrm{d}$,

$$\mathbf{J}_\mathrm{d} = \frac{\partial \mathbf{D}}{\partial t} \quad \text{(displacement current density vector; unit: A/m}^2\text{)} . \tag{6.12}$$

MATLAB EXERCISE 6.11 **Time-harmonic magnetic field in a nonideal capacitor – 2-D movie.** The electrodes of a parallel-plate capacitor are circular plates of radius $a = 8$ cm, as shown in Fig.6.9. The dielectric between the plates is imperfect, with parameters $\varepsilon_\mathrm{r} = 12$, $\sigma = 4 \times 10^{-4}$ S/m, and $\mu_\mathrm{r} = 1$, and the plate separation is $d = 4$ mm. The capacitor is connected to a time-varying voltage $v(t) = V_0 \cos \omega t$, with $V_0 = 1$ V and $\omega = 10^6$ rad/s. As $d \ll a$, the fringing effects can be neglected. Equations (3.5), (6.12), and (2.2) give (Fig.6.9) $J(t) = \sigma E(t)$ and $J_\mathrm{d}(t) = \varepsilon\, \mathrm{d}E/\mathrm{d}t$, where $E(t) = v(t)/d$. Due to symmetry, the lines of the vector $\mathbf{H}$ in the dielectric are circles centered at the capacitor axis perpendicular to the plates, so that applying Eq.(6.11) to a circular contour C of radius r ($r < a$) in Fig.6.9, in exactly the same way as in Fig.4.10(b) and Eqs.(4.15) and (4.16), yields

$$\mathbf{H} = H(r,t)\, \hat{\boldsymbol{\phi}} = \frac{[J(t) + J_\mathrm{d}(t)]\, r}{2}\, \hat{\boldsymbol{\phi}} = \frac{V_0 r}{2d} (\sigma \cos \omega t - \omega \varepsilon \sin \omega t)\, \hat{\boldsymbol{\phi}} . \tag{6.13}$$

Create a 2-D movie in MATLAB that shows the distribution of the magnetic field intensity vector in the dielectric of the capacitor, during the course of time. *(ME6_11.m on IR)*

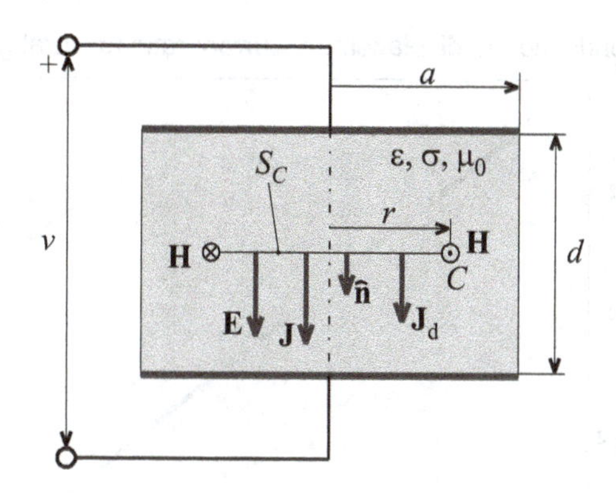

Figure 6.9 Evaluation of the magnetic field in a nonideal capacitor connected to a time-varying voltage; for MATLAB Exercise 6.11.

HINT:

Use Eq.(6.13) and see MATLAB Exercises 4.15 (for the visualization of the 2-D spatial distribution of **H** by MATLAB function `quiver`) and 6.6 (for the representation of the temporal variation of **H** in a movie employing MATLAB function `getframe`).

MATLAB EXERCISE 6.12 **Conduction to displacement current ratio, rural ground.** Write a function `condDispCurrentRatio()` in MATLAB that returns the ratio of the amplitudes of the conduction and displacement current densities in a given material. The input data to the function are the permittivity and conductivity of the material, and the vector (array) of frequencies. Then, with the use of this function, plot the conduction to displacement current ratio in a frequency range 10 kHz $\le f \le$ 10 GHz for a sample of rural ground with $\varepsilon_r = 14$ and $\sigma = 10^{-2}$ S/m that is occupied by a time-varying electric field of intensity $E(t) = E_0 \cos \omega t$. Assume no changes in the material parameters as a function of frequency. *(condDispCurrentRatio.m and ME6_12.m on IR)*

HINT:

From Eq.(6.13), we conclude that the amplitude ratio of current densities is given by

$$\frac{|J|_{\max}}{|J_d|_{\max}} = \frac{\sigma}{\omega\varepsilon} \quad \text{(conduction to displacement current ratio)} . \tag{6.14}$$

The graph is shown in Fig.6.10 (to present both the abscissa and the ordinate in log scale, MATLAB function `loglog` is used).

MATLAB EXERCISE 6.13 **Current ratio plots for fresh water and seawater.** Repeat the previous MATLAB exercise but for samples of (a) fresh water with $\varepsilon_r = 80$ and $\sigma = 10^{-3}$ S/m and (b) seawater with $\varepsilon_r = 80$ and $\sigma = 4$ S/m, respectively (on the same graph). *(ME6_13.m on IR)*

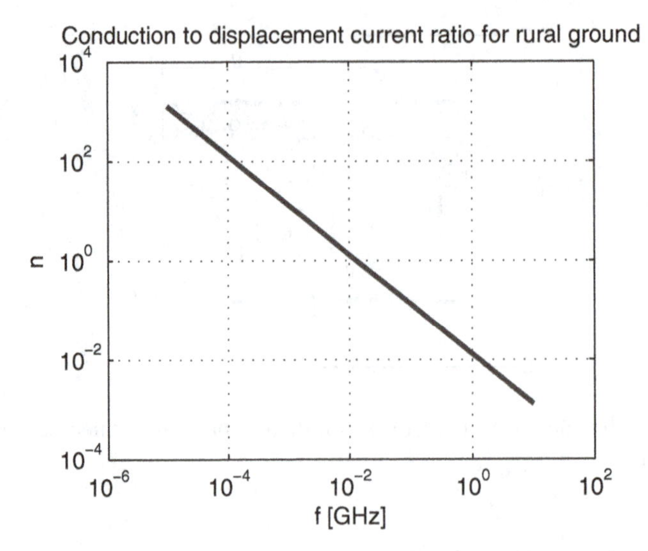

Figure 6.10 Conduction to displacement current ratio vs. frequency for rural ground; for MATLAB Exercise 6.12. *(color figure on CW)*

6.5 Maxwell's Equations for the Time-Varying Electromagnetic Field

Having now in place the corrected version of the generalized Ampère's law, Eq.(6.11), we are ready to summarize the full set of general Maxwell's equations for the time-varying electromagnetic field. We list them here in differential form, together with the three constitutive equations describing material properties of electromagnetic media:

$$\begin{cases} \nabla \times \mathbf{E} = -\frac{\partial \mathbf{B}}{\partial t} & \text{(Maxwell's first equation, differential)} \\ \nabla \times \mathbf{H} = \mathbf{J} + \frac{\partial \mathbf{D}}{\partial t} & \text{(Maxwell's second equation, differential)} \\ \nabla \cdot \mathbf{D} = \rho & \text{(Maxwell's third equation, differential)} \\ \nabla \cdot \mathbf{B} = 0 & \text{(Maxwell's fourth equation, differential)} \\ \mathbf{D} = \mathbf{D}(\mathbf{E}) \ [\mathbf{D} = \varepsilon \mathbf{E}] & \text{(constitutive equation for } \mathbf{D}) \\ \mathbf{B} = \mathbf{B}(\mathbf{H}) \ [\mathbf{B} = \mu \mathbf{H}] & \text{(constitutive equation for } \mathbf{B}) \\ \mathbf{J} = \mathbf{J}(\mathbf{E}) \ [\mathbf{J} = \sigma \mathbf{E}] & \text{(constitutive equation for } \mathbf{J}) \end{cases} . \tag{6.15}$$

MATLAB EXERCISE 6.14 Maxwell's equations, symbolic differentiation and integration. Write a function `diffMaxwellFirstEq()` in MATLAB to find, using Maxwell's equations in differential form, the symbolic expression for the time-varying magnetic field intensity vector (**H**) in free space from the associated electric field intensity vector (**E**) at the same point, if **E** is given symbolically as

$$\mathbf{E} = E_x(x, y, z, t)\,\hat{\mathbf{x}} + E_y(x, y, z, t)\,\hat{\mathbf{y}} + E_z(x, y, z, t)\,\hat{\mathbf{z}} \ , \tag{6.16}$$

in a Cartesian coordinate system. Specifically, find **H** for $\mathbf{E} = E_0 \cos(\omega t - \beta z)\,\hat{\mathbf{x}}$, where E_0 is a constant and $\beta = \omega\sqrt{\varepsilon_0\mu_0}$. *(diffMaxwellFirstEq.m and ME6_14.m on IR)* **H**

TUTORIAL:
With the use of Maxwell's first equation in differential form, Eqs.(6.15), we can write

$$\nabla \times \mathbf{E} = -\mu_0 \frac{\partial \mathbf{H}}{\partial t} \quad \longrightarrow \quad \mathbf{H} = -\frac{1}{\mu_0} \int \nabla \times \mathbf{E}\; \mathrm{d}t\,. \tag{6.17}$$

The input parameters to the function are **E** (E) in the form of a symbolic array (vector) of three elements (Cartesian components), symbolic time variable (for integration), t, and permeability of free space, μ_0 (MU0). For the symbolic curl in Cartesian coordinates, we apply function curlCar from MATLAB Exercise 4.17, while symbolic integration in Eq.(6.17) is performed by means of MATLAB function int,

```
function [H] = diffMaxwellFirstEq(E,t,MU0)
H = -int(curlCar(E(1),E(2),E(3)),t)/MU0;
return
```

The code calling the function to run the specified test case is as follows:

```
syms MU0 beta E0 omega t z;
Ex = E0*cos(omega*t - beta*z);
E = [Ex,0,0];
[H] = diffMaxwellFirstEq(E,t,MU0);
pretty(H);
```

The resulting symbolic expression for the vector **H**, as displayed in the Command Window, is shown in Fig.6.11.

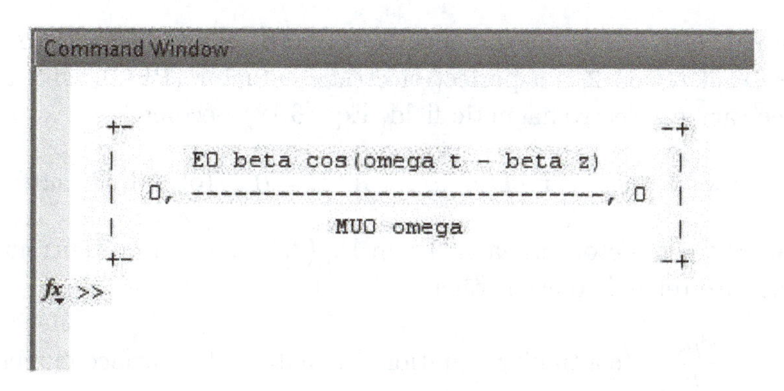

Figure 6.11 Display in the Command Window in MATLAB of the obtained symbolic expression for the magnetic field vector (**H**) accompanying the symbolically given time-varying electric field vector (**E**) in a Cartesian coordinate system, in free space; for MATLAB Exercise 6.14. *(color figure on CW)*

MATLAB EXERCISE 6.15 Magnetic from electric field of an antenna, symbolic computation. The electric field intensity vector radiated by an antenna placed at the coordinate origin of a spherical coordinate system is given, far away from the antenna, by $\mathbf{E}(r,\theta,t) =$

$E_0 \sin\theta \cos(\omega t - \beta r)\,\hat{\boldsymbol{\theta}}/r$, where E_0 is a constant and $\beta = \omega\sqrt{\varepsilon_0\mu_0}$. Using Maxwell's equations in differential form and symbolic differentiation and integration in MATLAB, find the magnetic field intensity vector of the antenna, at the same far point. *(ME6_15.m on IR)* **H**

HINT:
See the previous MATLAB exercise. For the symbolic curl, use function `curlSph` from MATLAB Exercise 4.17.

6.6 Boundary Conditions for the Time-Varying Electromagnetic Field

General electromagnetic boundary conditions – for the time-varying electromagnetic field, namely, for the tangential components of vectors **E** and **H** and normal components of vectors **D** and **B**, respectively, at a boundary surface between two electromagnetic media (regions 1 and 2) have the same form as in Eqs.(2.4) and (5.3); in vector notation, as in Eqs.(5.4), they are given by

$$\begin{cases} \hat{\mathbf{n}} \times \mathbf{E}_1 - \hat{\mathbf{n}} \times \mathbf{E}_2 = 0 & \text{(boundary condition for } \mathbf{E}_\mathrm{t}) \\ \hat{\mathbf{n}} \times \mathbf{H}_1 - \hat{\mathbf{n}} \times \mathbf{H}_2 = \mathbf{J}_\mathrm{s} & \text{(boundary condition for } \mathbf{H}_\mathrm{t}) \\ \hat{\mathbf{n}} \cdot \mathbf{D}_1 - \hat{\mathbf{n}} \cdot \mathbf{D}_2 = \rho_\mathrm{s} & \text{(boundary condition for } \mathbf{D}_\mathrm{n}) \\ \hat{\mathbf{n}} \cdot \mathbf{B}_1 - \hat{\mathbf{n}} \cdot \mathbf{B}_2 = 0 & \text{(boundary condition for } \mathbf{B}_\mathrm{n}) \end{cases}, \tag{6.18}$$

where $\hat{\mathbf{n}}$ is the normal unit vector on the surface, directed from region 2 to region 1. As an important special case, if region 2 is a perfect electric conductor (PEC), with $\sigma \to \infty$, in which there can be no time-varying electromagnetic field, Eqs.(6.18) become

$$\hat{\mathbf{n}} \times \mathbf{E} = 0\,, \quad \hat{\mathbf{n}} \times \mathbf{H} = \mathbf{J}_\mathrm{s}\,, \quad \hat{\mathbf{n}} \cdot \mathbf{D} = \rho_\mathrm{s}\,, \quad \hat{\mathbf{n}} \cdot \mathbf{B} = 0 \quad \text{(on a PEC surface)}\,, \tag{6.19}$$

with $\hat{\mathbf{n}}$ directed from the conductor outward. From Eq.(3.4), the surface currents and charges on the conductor surface are related to each other as

$$\nabla_\mathrm{s} \cdot \mathbf{J}_\mathrm{s} = -\frac{\partial \rho_\mathrm{s}}{\partial t} \quad \text{(continuity equation for plates – for surface currents)}\,. \tag{6.20}$$

This relationship is referred to as the continuity equation for (PEC) plates.

MATLAB EXERCISE 6.16 **PEC boundary conditions, plot of surface currents.** A surface of a perfect electric conductor in a time-varying electromagnetic field coincides with the xy-plane ($z = 0$) of a Cartesian coordinate system. The half-space $z > 0$ is air-filled and the electric field in this region is given by

$$\mathbf{E}(x, y, z, t) = E_0 \cos\left[\omega t - \beta(x + y)\right] \hat{\mathbf{z}}\,, \tag{6.21}$$

where $E_0 = 1$ V/m, $\omega = 10^9$ rad/s, and $\beta = \omega\sqrt{\varepsilon_0\mu_0}$. Using symbolic programming in MATLAB, find the surface current density vector, $\mathbf{J}_s$, on the PEC surface. Plot the distribution of these currents, in vector form, over a square area $2L$ on a side, where $L = \pi/\beta$, at an instant of time defined by $\omega t = \pi/2$. *(ME6_16.m on IR)* **H**

TUTORIAL:

We first specify input data for the program, including the normal unit vector on the PEC surface ($\hat{\mathbf{n}} = \hat{\mathbf{z}}$), NORMAL,

```
NORMAL = [0,0,1];
MU0 = 4*pi*10^(-7);
EPS0 = 8.8542*10^(-12);
MUR = 1;
EPSR = 1;
MU = MUR*MU0;
EPS = EPSR*EPS0;
E0 = 1;
w = 10^9;
beta = w*sqrt(EPS*MU);
```

Then we implement the symbolic expression for the electric field vector in air, E, in Eq.(6.21), and find the associated magnetic field vector, H, using function **diffMaxwellFirstEq** (from MATLAB Exercise 6.14). We also define a time instant **time** for plotting, equal to $t = T/4$ ($\omega t = \pi/2$), with T standing for the period of the time-harmonic variation of the field in Eq.(6.21), $T = 2\pi/\omega$.

```
x = sym('x','real');
y = sym('y','real');
z = sym('z','real');
t = sym('t','real');
Ex = 0;
Ey = 0;
Ez = E0*cos(w*t-beta*(x+y));
E = [Ex,Ey,Ez];
H = diffMaxwellFirstEq(E,t,MU0);
T = 2*pi/w;
time = T/4;
```

We next convert, by means of MATLAB function **subs**, the symbolic form of H to its numerical values at the instant **time** and coordinate **z = 0** (note that H still depends on symbolic variables x and y), and obtain a symbolic expression for $\mathbf{J}_s$ (Js) on the PEC surface applying the boundary condition for the magnetic field vector (namely, its tangential component) in Eqs.(6.19):

```
for k = 1:3
Hcurrent = H(k);
Hc = subs(Hcurrent,[t,z],[time,0]);
Hnum(k) = Hc;
end;
```

```
Js = cross(NORMAL,Hnum);
```

Finally, we define a mesh of nodes, coordinates `xi` and `yi`, for plotting, with steps $dl = \pi/(6\beta)$ between adjacent nodes in both directions (see Fig.6.12), convert `Js` to numerical values `JsNum` (needed for plotting) by function `subs` [substituting `x` by `xi(m)` and `y` by `yi(i)`] within two nested `for` loops, and plot `JsNum` using MATLAB function `quiver`, with a scaling factor (`scale`) adopted to be $\pi/(6\beta J_{\text{smax}})$,

```
xi = -pi/beta:pi/6/beta:pi/beta;
yi = -pi/beta:pi/6/beta:pi/beta;
for i = 1:length(yi)
for m = 1:length(xi)
JsNum(i,m,:)  = subs(Js,[x,y],[xi(m),yi(i)]);
JsNumMag(i,m) = sqrt(JsNum(i,m,1)^2 + JsNum(i,m,2)^2 + JsNum(i,m,3)^2);
end;
end;
scale = pi/6/beta/max(max(JsNumMag));
figure(1)
for i = 1:length(yi)
for m = 1:length(xi)
quiver(xi(m),yi(i),JsNum(i,m,1)*scale,JsNum(i,m,2)*scale,0,'r'); hold on;
end;
end;
axis equal;
title('Surface current density vector on a PEC plane at t = T/4');
xlabel('x [m]');
ylabel('y [m]');
```

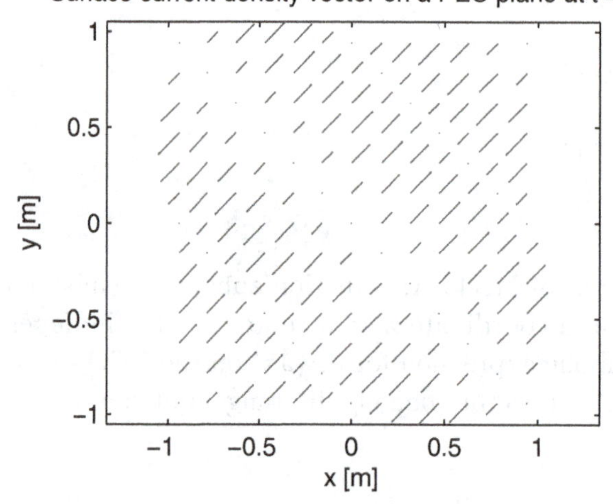

Figure 6.12 Plot of the distribution of the surface current density vector ($\mathbf{J}_s$) on a PEC surface associated with the electric field above the PEC given in Eq.(6.21), at an instant of time; for MATLAB Exercise 6.16. *(color figure on CW)*

```
hold off;
grid off;
```

The resulting graph is shown in Fig.6.12.

MATLAB EXERCISE 6.17 **PEC boundary conditions, movie of surface charges.** For the PEC surface and the electric field in air from the previous MATLAB exercise, create a 2-D movie in MATLAB that shows the spatial and temporal variations of the surface charge density, ρ_s, on the surface during one period of the time-harmonic oscillation of the field, $T = 2\pi/\omega$. *(ME6_17.m on IR)* **H**

TUTORIAL:

First, we specify input data and parameters of the system, as well as a time vector `t` for the movie and a mesh of nodes at which the electric field vector and surface charge density are computed:

```
EPS0 = 8.8542*10^(-12);
MU0 = 4*pi*10^(-7);
E0 = 1;
w = 10^9;
EPSR = 1;
MUR = 1;
EPS = EPSR*EPS0;
MU = MU0*MUR;
T = 2*pi/w;
beta = w*sqrt(EPS*MU);
t = 0:T/60:T;
[x,y] = meshgrid(-2*pi:pi/100:2*pi,-2*pi:pi/100:2*pi);
z = 0;
```

To determine ρ_s, we apply, in the plane $z = 0$, the boundary condition for the vector $\mathbf{D}$ (more precisely, for its normal component, $\mathbf{D}_n$) in Eqs.(6.19), where $\mathbf{D} = \varepsilon\mathbf{E}$ and $\hat{\mathbf{n}} = \hat{\mathbf{z}}$. In the `for` loop of the movie, we use MATLAB function `pcolor` to plot ρ_s over the PEC surface – at each (`j`th) instant (element) of the `t` vector, i.e., at each frame of the movie. In addition, to aid the continuity and smoothness of the movie, we introduce two scaling matrices (of the same size as the electric-field matrix `Ez`), `Escale1` and `Escale2`, that contain maximum and minimum values, respectively, of the electric field near the PEC boundary, and the corresponding scaling matrices for ρ_s, `RHOsScale1` and `RHOsScale2`.

```
for j = 1:length(t)
Ez = E0*cos(w*t(j)-beta*(x+y));
Escale1 = ones(size(Ez));
Escale2 = -1.*ones(size(Ez));
RHOs = Ez.*EPS;
RHOsScale1 = Escale1*EPS;
RHOsScale2 = Escale2*EPS;
```

```
pcolor(x,y,RHOsScale1); hold on;
pcolor(x,y,RHOsScale2);
pcolor(x,y,RHOs);hold off;
shading interp;
axis off;
xlim([-pi,pi]);
ylim([-pi,pi]);
title('Surface charge density on a PEC plane');
M(j)=getframe;
colorbar;
end;
```

Figure 6.13 shows a snapshot of the movie. Note that $\mathbf{J}_s$ computed in the previous MATLAB exercise and ρ_s found here are interconnected by the continuity equation in Eq.(6.20).

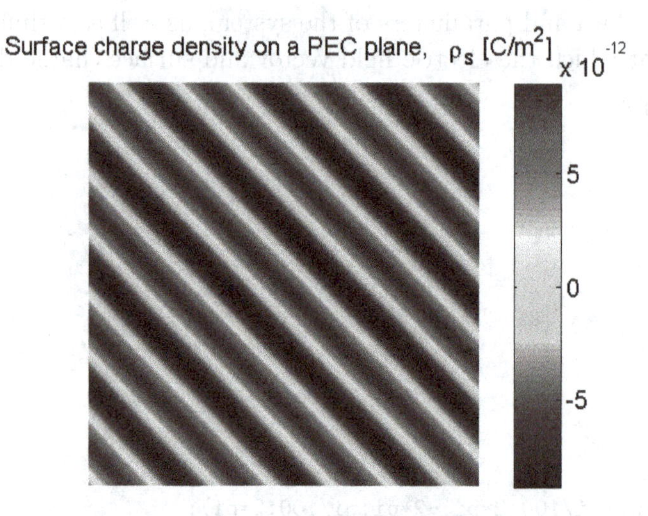

Figure 6.13 Snapshot of a 2-D MATLAB movie showing the distribution of surface charges (ρ_s) on a PEC surface associated with the electric field above the PEC given in Eq.(6.21), during the course of time; for MATLAB Exercise 6.17. *(color figure on CW)*

6.7 Time-Harmonic Electromagnetics

Consider a time-harmonic (steady-state sinusoidal) voltage of frequency (repetition rate) f and amplitude (peak-value) V_0. Its instantaneous value, that is, the value at an instant t, can be written as

$$v(t) = V_0 \cos(\omega t + \theta) = V\sqrt{2}\cos(\omega t + \theta) \quad (V = V_{\rm rms}) \,, \quad \omega = 2\pi f \,, \quad T = \frac{2\pi}{\omega} = \frac{1}{f}$$

$$\text{(time-harmonic voltage)} \,, \qquad (6.22)$$

where ω is the angular frequency (or radian frequency) and T is time period of time-harmonic oscillation [units are the hertz (Hz) for f (Hz = 1/s) and radian per second (rad/s) for ω], while θ

is the initial phase (phase at an instant $t = 0$) of the voltage. The root-mean-square (rms) value of $v(t)$, by definition, is found as

$$V_{\rm rms} = \sqrt{\frac{1}{T}\int_0^T v^2(t)\,{\rm d}t} \quad \text{(rms value)}\,, \tag{6.23}$$

so it amounts to[4] $V = V_{\rm rms} = V_0/\sqrt{2}$. In fact, it is more convenient to use rms values of time-harmonic quantities than their maximum values (amplitudes). Most instruments are calibrated to read rms values of measured quantities. In addition, all expressions for time-average powers, energies, and power and energy densities in the time-harmonic operation in circuit theory and electromagnetics can be computed just as for the time-invariant operation, if rms values of currents, voltages, field intensities, and other quantities are used. For example, the power of Joule's (ohmic) losses in a resistor of resistance R equals $P_{\rm J} = RI^2$ (dc), $P_{\rm J}(t) = Ri^2(t)$ (instantaneous), and $(P_{\rm J})_{\rm ave} = (1/T)\int_0^T P_{\rm J}(t)\,{\rm d}t = RI_0^2/2 = RI^2$ (time-average).

Time-harmonic expressions for electromagnetic quantities that vary also in space are written in a completely analogous way to that in Eq.(6.22), while keeping in mind that both the rms value and initial phase are, in general, functions of spatial coordinates. In addition, for a vector, separate expressions are written for each of its components. For instance, the Cartesian x-component of the time-harmonic electric field intensity vector, $\mathbf{E}$, can be expressed as $E_x(x,y,z,t) = E_x(x,y,z)\sqrt{2}\cos[\omega t + \theta_x(x,y,z)]$, and similarly for E_y and E_z.

MATLAB EXERCISE 6.18 **Symbolic rms value of a periodic signal.** Write a function `rmsValue()` in MATLAB to compute the root-mean-square (rms) value of an arbitrary periodic (not necessarily time-harmonic) signal $v(t)$, given by Eq.(6.23), using symbolic integration, i.e., MATLAB function `int`. The input to the function consists of the expression for $v(t)$, time (t) as the independent variable of integration, and the period T of the signal. Test the function for a time-harmonic signal (voltage) in Eq.(6.22), where V_0, ω, t, and θ need to be defined as symbolic quantities (variables) using MATLAB function `syms` (`syms V0 omega t theta`) in the main MATLAB program. *(rmsValue.m and ME6_18.m on IR)*

6.8 Complex Representatives of Time-Harmonic Field and Circuit Quantities

Time-harmonic quantities can be graphically represented as uniformly rotating vectors, called phasors, in the Cartesian xy-plane, as shown in Fig.6.14(a), where the projection on the x-axis of a vector of magnitude V_0 rotating with a constant angular velocity ω equals $v(t)$ in Eq.(6.22). In addition, we can formally proclaim the x- and y-axes of Fig.6.14(a) to be the real and imaginary axes of the complex plane, as indicated in Fig.6.14(b), and use complex numbers to represent

[4] As rms quantities will be used regularly throughout the rest of this text, we drop the subscripts ('rms') identifying them. With such convention, $V_{\rm rms}$ will be denoted simply as V, $H_{\rm rms}$ as H, and so on.

time-harmonic quantities. A complex number[5] $\underline{c}$ is a number composed of two real numbers, a and b, and it corresponds to a point, (a, b), or to a vector [position vector of the point (a, b) with respect to the coordinate origin] in the complex plane, as illustrated in Fig.6.14(c). Its rectangular (algebraic) and polar (exponential) forms read

$$\underline{c} = a + \mathrm{j}b = c\mathrm{e}^{\mathrm{j}\phi} , \quad a = \mathrm{Re}\{\underline{c}\} , \quad b = \mathrm{Im}\{\underline{c}\} , \quad \mathrm{j} = \sqrt{-1} , \quad c = |\underline{c}| = \sqrt{a^2 + b^2} , \tag{6.24}$$

so a and b represent the real and imaginary parts, respectively, of $\underline{c}$, and j stands for the imaginary unit, while c is the magnitude (or modulus) and ϕ the phase angle (argument) of $\underline{c}$. The argument (arg) function equals

$$\phi = \arg(a, b) = \begin{cases} \arctan(b/a) & \text{for } a > 0 \\ \pi/2 & \text{for } a = 0 \text{ and } b > 0 \\ -\pi/2 & \text{for } a = 0 \text{ and } b < 0 \\ \arctan(b/a) + \pi & \text{for } a < 0 \text{ and } b \geq 0 \\ \arctan(b/a) - \pi & \text{for } a < 0 \text{ and } b < 0 \\ \text{not defined} & \text{for } a = b = 0 \end{cases} \qquad \left(\arctan \equiv \tan^{-1}\right) . \tag{6.25}$$

From the right-angled triangle with arms $|a|$ and $|b|$ in Fig.6.14(c) and Eqs.(6.24),

$$\mathrm{e}^{\mathrm{j}\phi} = \cos\phi + \mathrm{j}\sin\phi \quad \text{(Euler's identity)} , \tag{6.26}$$

and this relation is known as Euler's identity. It is now clear that the projection of the rotating vector in Fig.6.14(a) on the real axis in Fig.6.14(b) equals the real part of the complex number.

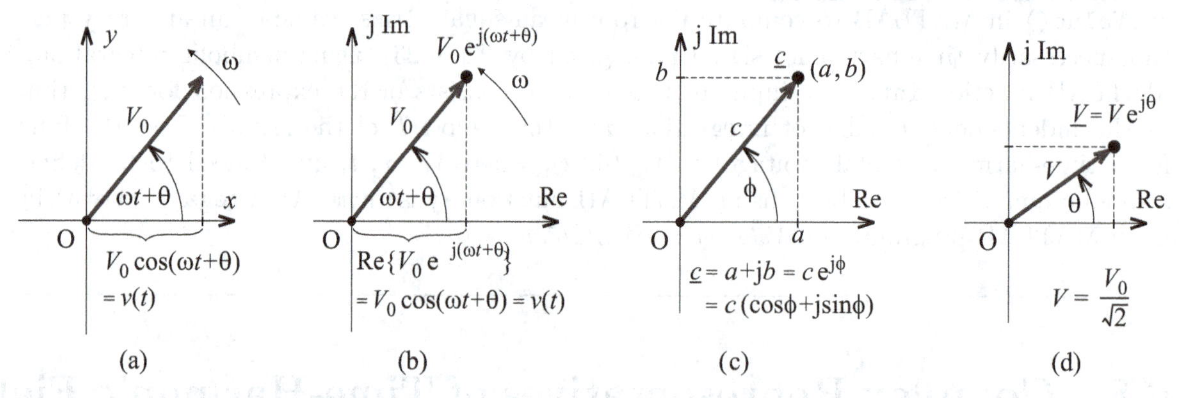

Figure 6.14 Representing time-harmonic quantities by phasors and complex numbers: (a) a phasor (rotating vector) whose magnitude and angular velocity equal the amplitude (peak-value) and angular frequency, respectively, of the instantaneous quantity [see Eq.(6.22)], (b) a complex number with magnitude (modulus) and phase angle (argument) equal to the amplitude and instantaneous phase of the instantaneous quantity, (c) different forms of a complex number, in general, and (d) the final adopted complex root-mean-square (rms) representative of a time-harmonic quantity with the time factor $\mathrm{e}^{\mathrm{j}\omega t}$ suppressed.

As all the phasors representing time-harmonic quantities in a system rotate with the same angular velocity, the picture with all the vectors frozen at instant $t = 0$ contains all relevant

[5]In this text, letters denoting complex numbers and complex variables are underlined, which is in compliance with the recommendation of the International Electrotechnical Commission (IEC).

data. We can, therefore, disregard rotation of phasors, or drop the time factor $e^{j\omega t}$ from the associated equations, and use the complex representation shown in Fig.6.14(d), where the complex rms representative of a time-harmonic quantity is defined as

$$v(t) = V\sqrt{2}\cos(\omega t + \theta) \quad \longleftrightarrow \quad \underline{V} = V\,e^{j\theta}\,, \quad v(t) = \mathrm{Re}\left\{\underline{V}\sqrt{2}\,e^{j\omega t}\right\}$$

(time-complex conversion; $\underline{V}$ – complex rms value) . (6.27)

The magnitude of the complex quantity is represented with the rms value, $|\underline{V}| = V = V_{\mathrm{rms}}$, rather than with the amplitude of the corresponding instantaneous quantity because, as already mentioned and illustrated on two examples (reading of instruments and computation of time-average power) in the previous section, it is more convenient to deal with rms quantities. Note, however, that many electromagnetics texts do use the latter representation, namely, complex amplitude representatives, in the form $\underline{V} = V_0\,e^{j\theta}$ ($V_0 = V\sqrt{2}$). Note also that alternative notations to $\underline{V}$ for complex (phasor) quantities include $\tilde{V}$ (tilde over the letter), V_s (subscript s), etc. A notable feature of the time-complex conversion in Eq.(6.27), given by $dv/dt \leftrightarrow j\omega\underline{V}$, allows us to replace all time derivatives in field/circuit equations by the factor $j\omega$, which enormously simplifies the analysis. For instance, the complex-domain equivalent of Maxwell's first equation in differential form, Eqs.(6.15), reads $\nabla \times \underline{\mathbf{E}} = -j\omega\underline{\mathbf{B}}$.

MATLAB EXERCISE 6.19 **Finding the phase of a complex number.** Write a function `phaseDeg()` in MATLAB that returns a phase angle (argument) ϕ in degrees of a complex number given as $\underline{c} = a + jb$, namely, as `a + i*b` in MATLAB, where a and b represent the real and imaginary parts, respectively, of $\underline{c}$, Eq.(6.24). *(phaseDeg.m on IR)*

HINT:

Implement Eq.(6.25) to find ϕ in radians, and then convert it to degrees using the relationship $\phi_{\mathrm{deg}} = \phi_{\mathrm{rad}} 180/\pi$.

Note that, while using `i` for the imaginary unit in MATLAB, one must be careful to avoid using the same symbol, `i`, for other quantities in the code, e.g., as an index (counter) in a `for` loop or otherwise. An alternative is to use `1i` as the imaginary unit, so to write `a + 1i*b` in place of `a + i*b`. Note also that `j` serves as the imaginary unit in MATLAB as well.

MATLAB EXERCISE 6.20 **Graphical representation of complex numbers.** Write a function `cplxNumPlot()` in MATLAB to represent (plot) an arbitrary complex number in the complex plane, as in Fig.6.14(c). The input arguments for the function are the complex number $\underline{c} = a + jb$ (`a + i*b`) and a color specification for plotting the vector $\underline{c}$. The graph also contains the textual information on the magnitude and phase angle of $\underline{c}$, placed near the vector $\underline{c}$. *(cplxNumPlot.m on IR)*

TUTORIAL:

We first compute the magnitude of $\underline{c}$, $|\underline{c}|$, using MATLAB function `abs` and the phase angle (ϕ) of $\underline{c}$ (in degrees) by function `phaseDeg` (from the previous MATLAB exercise). We also find the real and imaginary parts of $\underline{c}$, `a` and `b`, employing MATLAB functions `real` and `imag`, respectively.

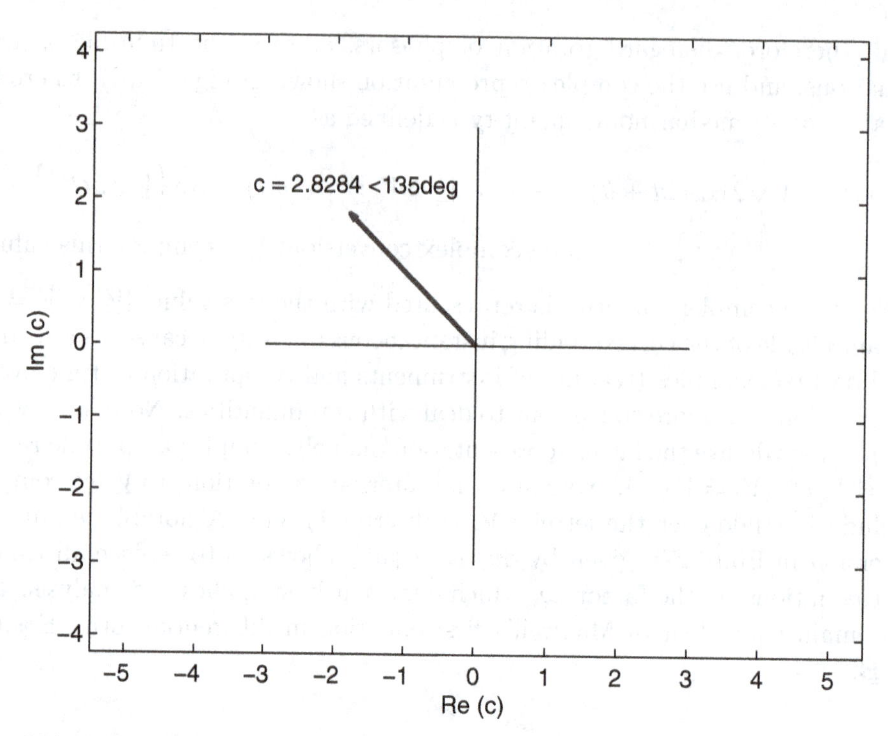

Figure 6.15 MATLAB graph showing an arbitrary complex number, $\underline{c}$, in the complex plane [see Fig.6.14(c)]; for MATLAB Exercise 6.20. *(color figure on CW)*

```
function cplxNumPlot(c,color)
r = abs(c);
phi = phaseDeg(c);
a = real(c);
b = imag(c);
```

Then, we introduce two vectors (arrays), `X` and `Y`, to represent the real and imaginary axes of the complex plane, in the graph (see Fig.6.15), with `X` being defined by means of MATLAB function **`linspace`**, which is explained in MATLAB Exercise 6.6. The bounds of the axes (`limit`) are determined as the greater of absolute values of `a` and `b`, multiplied by some `factor` (adopted to be 1.5).

```
factor = 1.5;
L = max(abs(a),abs(b));
limit = L*factor;
X = linspace(-limit,limit);
Y = zeros(1,length(X));
```

Finally, we plot the graph, with the complex number $\underline{c}$ in the complex plane. As we know, $\underline{c}$ corresponds to a vector representing the position vector of the point (a, b) with respect to the coordinate origin in the complex plane, Fig.6.14(c), and we draw this vector using MATLAB function **`quiver`** (with the first two arguments standing for the coordinates of the starting point of the vector and the other two for the coordinates of the ending point). The axes are plotted combining MATLAB functions **axis** and **plot**. Textual data, whose positions in the graph are determined based on **a** and **b**, are written using MATLAB function **text** [the first two arguments are coordinates of the text, and the last is a string (we convert numerical data into a string by MATLAB function **num2str**)].

```
figure(1);
quiver(0,0,a,b,'linewidth',2,'Color',color); hold on;
axis([-1 1 -1 1]*r*1.5); axis equal;
plot(X,Y,'k');
plot(Y,X,'k');
xlabel('Re (c)'); ylabel('Im (c)');
text(a-0.6*abs(a),1.1*b,['c = ' num2str(r),' <' num2str(phi) 'deg']);
```

We test the function as follows:

```
cplxNumPlot(-2 + i*2,'b');
```

and the resulting graph is shown in Fig.6.15. Note that, prior to running this test (e.g., from the Command Window), it is advisable to clear all previously used memory locations (type `clear all`), as it is possible that the symbol `i` was used and assigned a value (different from the imaginary unit) in previous runs of other MATLAB codes (see the corresponding note in the previous MATLAB exercise). An alternative is to type `cplxNumPlot(-2 + 1i*2,'b')`.

MATLAB EXERCISE 6.21 **Graphical representation of complex voltage and current.** Using function `cplxNumPlot` (developed in the previous MATLAB exercise), represent graphically in the complex plane, in separate subplots, the complex rms voltage $\underline{V} = (\sqrt{3} - \mathrm{j})$ V and current $\underline{I} = (-1 + \mathrm{j})$ A, respectively. *(ME6_21.m on IR)*

MATLAB EXERCISE 6.22 **Movie of voltage phasor rotation in complex plane.** In MATLAB, create a 2-D movie that visualizes the rotation of the voltage phasor $\underline{V} = V_0\,\mathrm{e}^{\mathrm{j}(\omega t+\theta)}$ in the complex plane, as indicated in Fig.6.14(b), assuming that $V_0 = 2$ V, $\omega = 377$ rad/s, and $\theta = \pi/6$. *(ME6_22.m on IR)*

TUTORIAL:

The first part of the MATLAB code specifies input data and a time vector (array) `t` for the movie, and computes the voltage phasor array, `V`, at each time instant, using MATLAB function **exp**:

```
V0 = 2;
w = 377;
theta = pi/6;
T = 2*pi/w;
dt = T/200;
t = 0:dt:T;
V = V0.*exp(i*(w.*t + theta));
```

where `T` is the time period of the time-harmonic variation of the voltage, Eqs.(6.22), that is, the time period for one rotation of the phasor in the complex plane [or in the xy-plane in Fig.6.14(a)]. The next step is to define axes of the complex plane as in MATLAB Exercise 6.20,

```
limit = V0;
X = linspace(-limit,limit);
Y = zeros(1,length(X));
```

Within the `for` loop of the movie (ending with MATLAB function `getframe`), we draw the phasor (rotating vector) $\underline{V}$ with the use of MATLAB function `quiver` (see MATLAB Exercise 6.20), based on the real and imaginary parts of $\underline{V}$. The axes of the complex plane are plotted by MATLAB function `plot`.

```
figure(1);
for i = 1:length(t);
Vreal = real(V(i));
Vimag = imag(V(i));
quiver(0,0,Vreal,Vimag,'linewidth',2); hold on;
plot(X,Y,'k');
plot(Y,X,'k');
hold off;
axis equal; xlabel('Re (V)');ylabel('Im (V)');
title('Voltage phasor rotation in complex plane');
M(i) = getframe;
end;
```

MATLAB EXERCISE 6.23 **Rotation of voltage and current phasors for an inductor.** Let the voltage phasor $\underline{V}$ specified in the previous MATLAB exercise represent the time-harmonic voltage of an inductor of inductance $L = 100$ nH. Find the corresponding current phasor, $\underline{I}$, for the inductor, using the element law (current-voltage characteristic) for an inductor, $\underline{V} = \mathrm{j}\omega L\underline{I}$ (the complex-domain equivalent of $v = L\,\mathrm{d}i/\mathrm{d}t$), and visualize the simultaneous rotation in the complex plane of both phasors, $\underline{V}$ and $\underline{I}$, in a 2-D movie. *(ME6_23.m on IR)*

MATLAB EXERCISE 6.24 **Rotation of voltage and current phasors for a capacitor.** Repeat the previous MATLAB exercise but now assuming that $\underline{V}$ represents the time-harmonic voltage of a capacitor of capacitance $C = 50$ pF; visualize the simultaneous rotation of both $\underline{V}$ and $\underline{I}$ for this capacitor ($\underline{I} = \mathrm{j}\omega C\underline{V}$), in a 2-D movie. *(ME6_24.m on IR)*

MATLAB EXERCISE 6.25 **Conversion from complex to time domain in symbolic form.** Write a function `cplx2TimeDomain()` in MATLAB that converts a symbolic expression of a time-harmonic signal in the complex domain given as the complex rms representative, Eq.(6.27), to its time-domain symbolic counterpart. *(cplx2TimeDomain.m on IR)*

TUTORIAL:

The instantaneous signal (e.g., voltage) can be obtained from the complex rms signal (voltage) applying the complex-time conversion in Eq.(6.27). However, if this equation is implemented directly, the symbolic programming option in MATLAB is not able to return clear expressions for signals in the time domain, because of the presence of the complex exponential function $e^{j\omega t}$, multiplying the complex expression for $\underline{V}$. Therefore, we first express $e^{j\omega t}$ via its real and imaginary parts, using Euler's identity, Eq.(6.26), and make use of the fact that

$$\underline{c} = a + jb \quad \longrightarrow \quad \text{Re}\{j\underline{c}\} = \text{Re}\{j(a+jb)\} = \text{Re}\{ja - b\} = -b = -\text{Im}\{\underline{c}\}\,, \tag{6.28}$$

to rewrite Eq.(6.27) into a form suitable for symbolic programming in MATLAB:

$$v(t) = \text{Re}\left\{\underline{V}\sqrt{2}\,e^{j\omega t}\right\} = \text{Re}\{\underline{V}\}\sqrt{2}\cos\omega t - \text{Im}\{\underline{V}\}\sqrt{2}\sin\omega t\,. \tag{6.29}$$

The input arguments of function `cplx2TimeDomain` are the symbolic complex signal, `Scplx`, symbolic angular frequency, `omega`, and symbolic time variable, `t`, while the output is a symbolic instantaneous signal, `Stime`. The expanded relationship for the complex-time conversion in Eq.(6.29) is implemented in MATLAB employing MATLAB functions `real` and `imag`, as follows:

```
function Stime = cplx2TimeDomain(Scplx,omega,t)
Stime = real(Scplx)*sqrt(2)*cos(omega*t)-imag(Scplx)*sqrt(2)*sin(omega*t);
return
```

Of course, numerical values of $v(t)$ at any instant of time obtained by this MATLAB function are the same as those resulting from a direct application of Eq.(6.27), and the same is true for the associated plots of $v(t)$.

MATLAB EXERCISE 6.26 **Transferring a complex E-vector to time domain.** Consider the complex rms electric field intensity vector given by $\underline{\mathbf{E}} = j\underline{E}_0 \sin\beta z\, e^{-j\beta x}\,\hat{\mathbf{x}} + \underline{E}_0\cos\beta z\, e^{-j\beta x}\,\hat{\mathbf{z}}$ ($\underline{E}_0 = E_0\, e^{j\theta_0}$), and obtain in MATLAB, symbolically, the instantaneous counterpart of $\underline{\mathbf{E}}$, assuming that the operating angular frequency is ω, as well as that $E_0 = 1$ V/m, $\theta_0 = 0$, and $x = 0$. *(ME6_26.m on IR)* **H**

TUTORIAL:

After declaring real symbolic variables (representing purely real quantities, they are defined as real symbolic and not just symbolic to help the symbolic programming feature in MATLAB generate nicer symbolic expressions), we implement the given complex vector $\underline{\mathbf{E}}$ as a three-element array (vector) `Etot`, with symbolic expressions for x-, y-, and z-components of $\underline{\mathbf{E}}$ (of course, $\underline{E}_y = 0$) as elements. The time-domain equivalent of the field, a symbolic vector `Etime`, is obtained using function `cplx2TimeDomain` (from the previous MATLAB exercise). The resulting symbolic expression is displayed in the Command Window by means of MATLAB function `pretty`.

```
t = sym('t','real');
beta = sym('beta','real');
z = sym('z','real');
```

```
omega = sym('omega','real');
E0 = 1;
theta0 = 0;
x = 0;
E = E0*exp(i*theta0);
Etot = [i*E*sin(beta*z)*exp(-i*beta*x);0;i*E*cos(beta*z)*exp(-i*beta*x)];
Etime = cplx2TimeDomain(Etot,omega,t);
pretty(Etime);
```

6.9 Instantaneous and Complex Poynting Vector

Consider an arbitrary electromagnetic field described by field vectors **E** and **H**. The associated Poynting vector, defined as

$$\boldsymbol{\mathcal{P}} = \mathbf{E} \times \mathbf{H} \quad \text{(Poynting vector; unit: W/m}^2\text{)} \tag{6.30}$$

(note that **S** is also widely used to denote the Poynting vector), has the dimension of a surface power density (power per unit area), and is expressed in W/m^2 (the unit for E, V/m, times the unit for H, A/m). Hence, the power transferred through any (open or closed) surface S (power flow) can be obtained as

$$P_{\mathrm{f}} = \int_S \boldsymbol{\mathcal{P}} \cdot \mathrm{d}\mathbf{S} \quad \text{(power flow through a surface } S\text{; unit: W)} . \tag{6.31}$$

In the case of a time-harmonic field, the complex Poynting vector is given by

$$\underline{\boldsymbol{\mathcal{P}}} = \underline{\mathbf{E}} \times \underline{\mathbf{H}}^* \quad \text{(complex Poynting vector)} , \tag{6.32}$$

where $\underline{\mathbf{E}}$ and $\underline{\mathbf{H}}$ are complex rms electric and magnetic field intensity vectors [Eq.(6.27)], so that the time average of the instantaneous Poynting vector, in Eq.(6.30), equals (see Section 6.7)

$$\boldsymbol{\mathcal{P}}_{\mathrm{ave}} = \frac{1}{T}\int_0^T \boldsymbol{\mathcal{P}}(t)\,\mathrm{d}t = \mathrm{Re}\{\underline{\boldsymbol{\mathcal{P}}}\} \quad \text{(time-average Poynting vector)} , \tag{6.33}$$

and represents the time-average (active) power flow per unit area.

MATLAB EXERCISE 6.27 **Complex Poynting vector, symbolic differentiation.** Having in mind Eqs.(6.1) and (1.38), the complex electric field intensity vector, $\underline{\mathbf{E}}$, can be expressed in terms of the complex magnetic vector potential, $\underline{\mathbf{A}}$, in the following way:

$$\underline{\mathbf{E}} = -\mathrm{j}\omega\underline{\mathbf{A}} - \nabla\underline{V} = -\mathrm{j}\omega\left[\underline{\mathbf{A}} + \frac{1}{\beta^2}\nabla(\nabla\cdot\underline{\mathbf{A}})\right] \quad (\nabla\cdot\underline{\mathbf{A}} = -\mathrm{j}\omega\varepsilon\mu\underline{V}) , \quad \beta = \omega\sqrt{\varepsilon\mu} , \tag{6.34}$$

where the differential relation between $\underline{\mathbf{A}}$ and the electric scalar potential, $\underline{V}$, analogous to the continuity equation interconnecting the current and charge densities, $\underline{\mathbf{J}}$ and $\underline{\rho}$, is called the Lorenz condition. Assuming that the magnetic potential in a region, in air, is given by $\underline{\mathbf{A}} = A_0(x^2\,\hat{\mathbf{x}} +$

$jy^2\,\hat{\mathbf{z}})/a^2$, where A_0 is a real constant, use Eqs.(6.32), (6.34), and (4.24) and symbolic programming in MATLAB to obtain the expression for the complex Poynting vector, $\underline{\boldsymbol{\mathcal{P}}}$, in this region. *(ME6_27.m on IR)* **H**

TUTORIAL:

First, we declare real symbolic variables (for the same reason as in the previous MATLAB exercise) and express the complex magnetic vector potential, $\underline{\mathbf{A}}$, as a symbolic array A of three elements (Cartesian components),

```
x = sym('x','real');
y = sym('y','real');
z = sym('z','real');
A0 = sym('A0','real');
omega = sym('omega','real');
MU0 = sym('MU0','real');
beta = sym('beta','real');
a = sym('a','real');
A = (A0/a^2)*[x^2 0 i*y^2];
```

Based on Eqs.(6.34) and (4.24), complex electric and magnetic field vectors in the region, E and H, are found applying functions divCar (MATLAB Exercise 1.33) gradCar (MATLAB Exercise 1.26), and curlCar (MATLAB Exercise 4.17), for the symbolic divergence, gradient, and curl, respectively, in Cartesian coordinates, to the vector A, as follows:

```
E = -i*omega*(A + 1/beta^2*gradCar(divCar(A(1),A(2),A(3))));
H = curlCar(A(1),A(2),A(3))/MU0;
```

The complex Poynting vector at the same point, P, is then computed using Eq.(6.32) and MATLAB function conj for the complex conjugate of H and function crossProduct (MATLAB Exercise 4.1) for the cross product of E and Hconj. To display the result for P in the Command Window, we employ MATLAB function pretty.

```
Hconj = conj(H);
[Px,Py,Pz] = crossProduct(E,Hconj);
P = [Px, Py, Pz];
pretty(P);
```

MATLAB EXERCISE 6.28 **Complex Poynting vector in spherical coordinates.** The complex magnetic vector potential in a region in free space is given by the following expression in a spherical coordinate system: $\underline{\mathbf{A}} = A_0(\cos\theta\,\hat{\mathbf{r}} - \sin\theta\,\hat{\boldsymbol{\theta}})/r$, where A_0 is a real constant ($\underline{\mathbf{A}}$ is purely real). Compute the complex Poynting vector ($\underline{\boldsymbol{\mathcal{P}}}$) in this region by symbolic programming in MATLAB. *(ME6_28.m on IR)* **H**

HINT:

Use Eqs.(6.32), (6.34), and (4.24) and symbolic divergence, gradient, and curl in spherical coordinates.

---

7 UNIFORM PLANE ELECTROMAGNETIC WAVES

Introduction:

Electromagnetic waves, i.e., traveling electric and magnetic fields, are the most important consequence of general Maxwell's equations, discussed in the preceding chapter. We now proceed with analysis of electromagnetic wave propagation, to describe the properties of waves as they propagate away from their sources – time-varying currents and charges in a source region (a transmitting antenna). Far away from it, the elementary spherical waves originated by the sources form a unified global spherical wavefront, which – if considered only over a receiving aperture (e.g., at the receiving end of a wireless link) – can be treated as if it were a part of a uniform plane wave. Such a wave has planar wavefronts and uniform (constant) distributions of fields over every plane perpendicular to the direction of wave propagation. Most importantly, we can completely remove the spherical wave from the analysis and assume that a uniform plane wave illuminating the aperture exists in the entire space. Once this model is established, we then deal with uniform plane waves only, and study their propagation not only in unbounded media with and without losses (this chapter) but also in the presence of planar interfaces between material regions with different electromagnetic properties (next chapter).

7.1 Time-Harmonic Uniform Plane Waves and Complex-Domain Analysis

We consider a time-harmonic uniform plane electromagnetic wave propagating in the positive z direction in an unbounded region filled with a linear, homogeneous, and lossless ($\sigma = 0$) material of permittivity ε and permeability μ. With reference to Fig.7.1, the electric and magnetic field intensities of the wave are

$$E_x = E_{\rm m} \cos(\omega t - \beta z + \theta_0) \,, \quad H_y = \frac{E_{\rm m}}{\eta} \cos(\omega t - \beta z + \theta_0) \quad \text{(uniform plane wave)} \,, \tag{7.1}$$

where $E_{\rm m}$ is the electric-field amplitude and θ_0 the initial (for $t = 0$) phase in the plane $z = 0$ of the wave, while ω is its angular frequency [see Eqs.(6.22)]. Since $H_y = E_x/\eta$, η has the unit of impedance, Ω [the units for E and H are V/m and A/m, respectively, and (V/m)/(A/m) = V/A = Ω]; it is called the intrinsic impedance of the medium and is computed as

$$\eta = \sqrt{\frac{\mu}{\varepsilon}} \quad \text{(intrinsic impedance of a medium; unit: } \Omega) \,. \tag{7.2}$$

For a vacuum or air (free space),

$$\eta_0 = \sqrt{\frac{\mu_0}{\varepsilon_0}} \approx 120\pi \ \Omega \approx 377 \ \Omega \quad \text{(intrinsic impedance of free space)} \,. \tag{7.3}$$

The constant β, in units of rad/m, is the phase coefficient or wavenumber (note that the symbol k is also used to denote the wavenumber), given by

$$\beta = \frac{\omega}{c} = \omega\sqrt{\varepsilon\mu} = \frac{2\pi f}{c} = \frac{2\pi}{\lambda} \quad \text{(phase coefficient or wavenumber; unit: rad/m)}, \tag{7.4}$$

with λ being the wavelength of the wave, measured in meters and defined as the distance traveled by a wave during one time period T [Eqs.(6.22)],

$$\lambda = cT = \frac{c}{f} \quad \text{(wavelength; unit: m)}. \tag{7.5}$$

The phase velocity of uniform plane waves (in a lossless unbounded medium) equals

$$v_{\mathrm{p}} = \frac{\omega}{\beta} = c = \frac{1}{\sqrt{\varepsilon\mu}} \quad \text{(phase velocity of uniform plane waves; unit: m/s)}. \tag{7.6}$$

If the medium is air (vacuum),

$$c_0 = \frac{1}{\sqrt{\varepsilon_0\mu_0}} = 299{,}792{,}458 \text{ m/s} \approx 3 \times 10^8 \text{ m/s} \quad \text{(velocity of waves in free space)}. \tag{7.7}$$

This constant is commonly referred to as the speed of light.

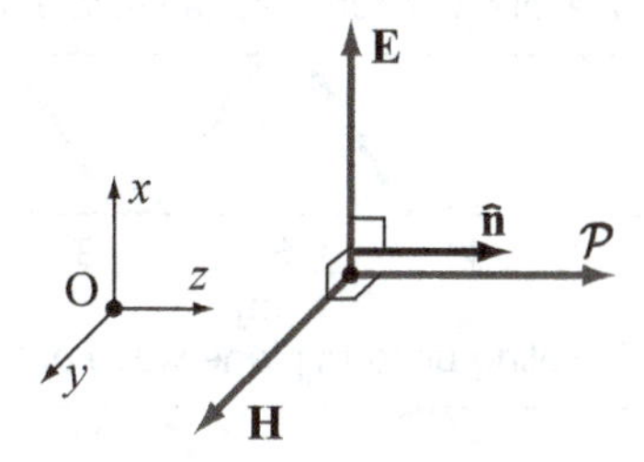

Figure 7.1 Electric field intensity vector (**E**), magnetic field intensity vector (**H**), propagation unit vector ($\hat{\mathbf{n}}$), and Poynting vector ($\boldsymbol{\mathcal{P}}$), Eq.(6.30), of a uniform plane electromagnetic wave propagating in an unbounded medium.

Applying the time-complex conversion in Eq.(6.27) to the expressions for the instantaneous field intensities in Eqs.(7.1), we obtain the following expressions for complex rms field intensities of the wave in Fig.7.1:

$$\underline{E}_x = \underline{E}_0\,\mathrm{e}^{-\mathrm{j}\beta z}, \quad \underline{H}_y = \frac{\underline{E}_0}{\eta}\,\mathrm{e}^{-\mathrm{j}\beta z}, \quad \underline{E}_0 = E_0\,\mathrm{e}^{\mathrm{j}\theta_0} \quad \text{(complex rms fields)}, \tag{7.8}$$

where E_0 is the rms electric field intensity of the wave ($E_0 = E_{\mathrm{m}}/\sqrt{2}$).

MATLAB EXERCISE 7.1 **Propagation parameters in a lossless medium.** A uniform plane time-harmonic electromagnetic wave of frequency f travels through a lossless medium of relative permittivity ε_{r} and relative permeability μ_{r}. Write a function `propParamLossless()` that returns the following propagation parameters of the wave and the medium: angular (radian) frequency, ω, time period, T, phase coefficient (wavenumber), β, wavelength, λ, and phase velocity, $v_p = c$, of the wave and intrinsic impedance, η, of the propagation medium. *(propParamLossless.m on IR)*[1]

HINT:
Use Eqs.(6.22), (7.4)–(7.6), and (7.2) for ω, T, β, λ, v_p, and η, respectively.

MATLAB EXERCISE 7.2 **Visualization of traveling-wave snapshots in space.** Write a MATLAB code that calculates and plots the electric field intensity of a time-harmonic uniform plane wave propagating in air, given by $E_x(z,t) = 2\cos(\omega t - \beta z)$ V/m [see Eqs.(7.1)], at time instants $t = 0$, $t = T/2$, and $t = T$ (T is the time period of the wave) in a region of space defined by $0 \le z \le 12$ m, if the frequency of the wave is $f = 100$ MHz. *(ME7_2.m on IR)*

HINT:
Use function `propParamLossless` (from the previous MATLAB exercise), to compute the necessary propagation parameters for the wave. The resulting E-field snapshots for the wave are shown in Fig.7.2, where the three subplots, for the three time instants, are obtained by MATLAB function `subplot` (see MATLAB Exercise 3.3).

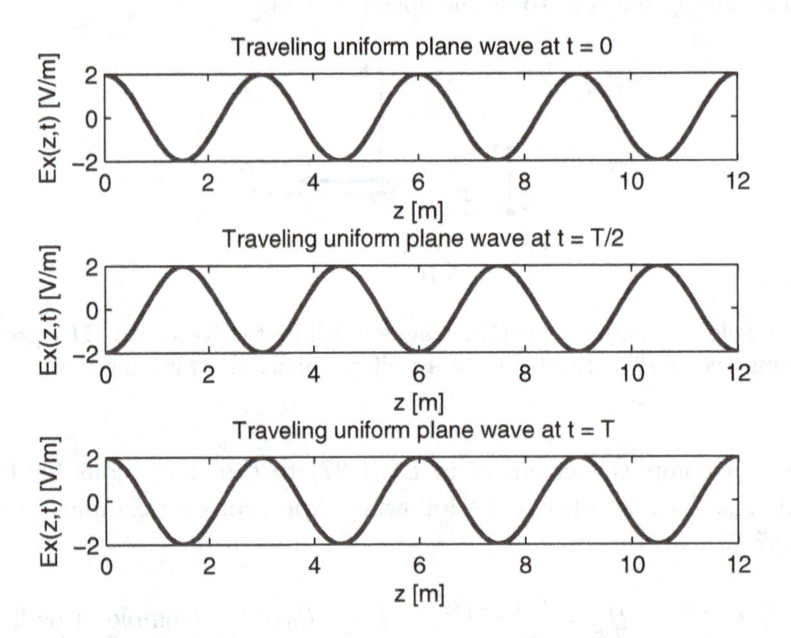

Figure 7.2 Snapshots (scans) at three characteristic instants of time, in a region of space, of the electric field intensity of a time-harmonic uniform plane wave traveling in air; for MATLAB Exercise 7.2. *(color figure on CW)*[2]

---

[1] *IR* = Instructor Resources (for the book).
[2] *CW* = Companion Website (of the book).

MATLAB EXERCISE 7.3 **Two plane waves traveling in opposite directions – movie.** Write a MATLAB code that plays a 2-D movie visualizing the spatial and temporal variations of the electric fields of two time-harmonic uniform plane electromagnetic waves that propagate in free space in the positive (forward) and negative (backward) x directions, respectively, approaching one another. The fields are given by

$$E_{\text{forward}} = E_{\text{m}} \sin(\omega t - \beta x) \quad \text{and} \quad E_{\text{backward}} = E_{\text{m}} \sin(\omega t + \beta x) \,, \tag{7.9}$$

where the field amplitude is $E_{\text{m}} = 1$ V/m (for both waves), and the operating frequency amounts to $f = 100$ MHz. The movie lasts two time periods of the waves, $2T$, and spans a range of two wavelengths, 2λ, along the x-axis, with the two waves meeting at the center of this range. At the beginning of the movie ($t = 0$), the waves appear at the opposite sides of the graph. *(ME7_3.m on IR)*

TUTORIAL:

After entering input data, propagation parameters of the waves are calculated by function `propParamLossless` (from MATLAB Exercise 7.1). Also, we define the time step and the step in the x-coordinate, as well as the corresponding vectors (arrays) `t` and `x`.

```
[w,T,beta,lambda,vp,eta] = propParamLossless(f,EPSR,MUR);
dt = T/200;
dx = lambda/50;
t = [0:dt:2*T];
x = [-lambda:dx:lambda];
lx = length(x);
```

Within the `for` loop of the movie, the E-fields of the forward and backward propagating waves, E_{forward} (`Eforward`) and E_{backward} (`Ebackward`), are computed based on Eqs.(7.9). In particular, E_{forward} is evaluated at an instant `t(i)` at all points reached by the forward wave, implementing the first expression in Eqs.(7.9), with the last point at the x-axis reached by the wave, past which there is no field, being determined using MATLAB functions `min` and `heaviside`. Namely, function `min` returns the index of the first minimum in the vector (array) `heaviside(w*t(i)-beta.*(x+lambda))` – the first point where `heaviside(w*t(i)-beta.*(x+lambda))` becomes zero, and the wave is plotted in the graph only up to that point. Similarly, E_{backward} at the time `t(i)`, obtained from the second expression in Eqs.(7.9), is plotted only up to the point the backward wave actually reaches at that time. Since the propagation of the wave is in the negative x direction, we are looking for the first point where `heaviside(w*t(i)+beta.*(x-lambda))` becomes unity, and find it as the first maximum in the vector `heaviside(w*t(i)+beta.*(x-lambda))`, using MATLAB function `max`. Note that `Eforward` and `Ebackward` are plotted in blue (solid line) and red (dashed line), respectively.

```
for i = 1:length(t);
Eforward = Em*sin(w*t(i)-beta.*x).*heaviside(w*t(i)-beta.*(x+lambda));
Ebackward = Em*sin(w*t(i)+beta.*x).*heaviside(w*t(i)+beta.*(x-lambda));
[errorf, nf] = min(heaviside(w*t(i)-beta.*(x+lambda)));
[errorb, nb] = max(heaviside(w*t(i)+beta.*(x-lambda)));
plot(x(1:nf), Eforward(1:nf),'b','linewidth',2);hold on;
plot(x(nb-1:lx), Ebackward(nb-1:lx),'r--', 'linewidth',2);
```

```
hold off;
axis ([-lambda lambda -3/2*Em 3/2*Em]);
title('Two plane waves traveling in opposite directions');
xlabel('x [m]');
ylabel('E [V/m]');
M(i) = getframe;
end;
```

A snapshot (frame) of the movie is shown in Fig.7.3.

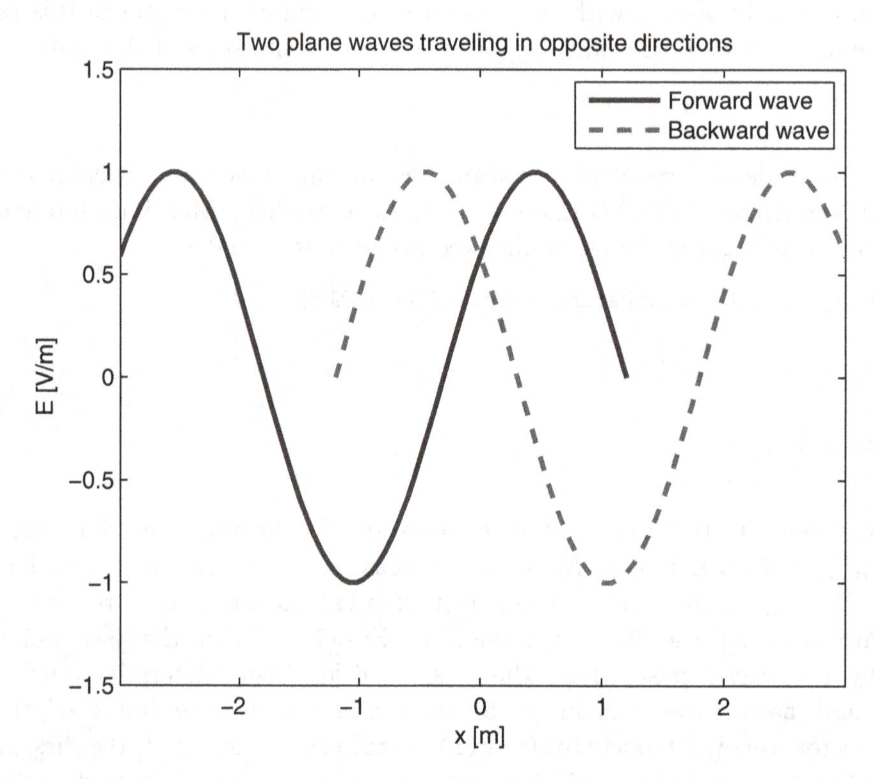

Figure 7.3 Snapshot of a 2-D movie visualizing the electric fields of two time-harmonic uniform plane waves that travel in opposite directions; for MATLAB Exercise 7.3. *(color figure on CW)*

MATLAB EXERCISE 7.4 **Superposition of two traveling waves – movie.** Modify the movie created in the previous MATLAB exercise by showing the sum of electric fields of the two traveling waves; the movie lasts $8T$ (two waves start their travel toward the meeting point at $t = 0$) and covers a range of 4λ along the x-axis. *(ME7_4.m on IR)*

HINT:
Figure 7.4 shows a snapshot of the resulting movie.

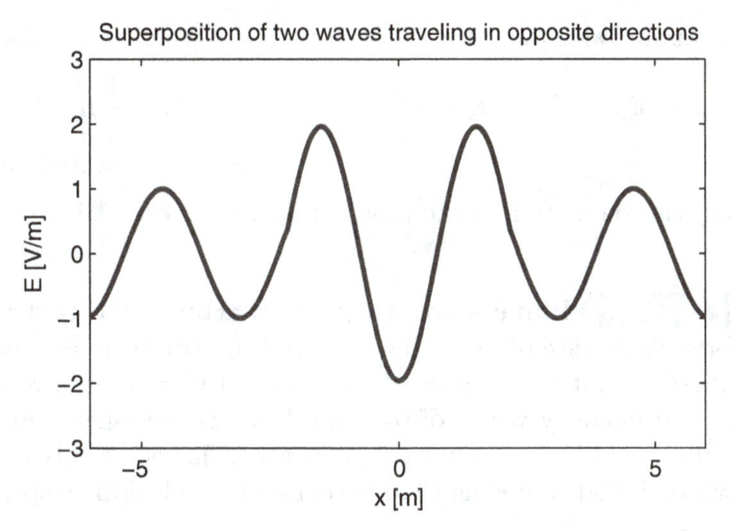

Figure 7.4 Snapshot of a 2-D movie visualizing the superposition (sum of electric fields) of the two traveling waves from Fig.7.3; for MATLAB Exercise 7.4. *(color figure on CW)*

MATLAB EXERCISE 7.5 **Movie of field pictures of a plane wave using `imagesc`.** Write a program in MATLAB that plays a 2-D movie visualizing the electric field of a uniform plane time-harmonic wave of frequency $f = 300$ MHz in air given by $\mathbf{E} = E(x,t)\,\hat{\mathbf{z}} = \cos(\omega t - \beta x)\,\hat{\mathbf{z}}$ V/m, for $0 \le t \le 2T$, with T being the time period of the wave. In specific, each frame of the movie shows E in the xy-plane, for $0 \le x, y \le \lambda$ (and $z = 0$), where λ is the wavelength of the wave, using MATLAB function `imagesc` (which provides a 2-D view with color representing the value of the quantity – field intensity in our case). *(ME7_5.m on IR)*

MATLAB EXERCISE 7.6 **GUI for plane waves in free space.** In MATLAB, create a graphical user interface that displays the waveform of a time-harmonic uniform plane wave in free space given by $E_x(z,t) = \cos(\omega t - \beta z + \theta_0)$ V/m [see Eqs.(7.1)] for $0 \le z \le 2\lambda$ (λ is the wavelength of the wave). Input data are the frequency of the wave (f), the instant of time for display (t), and the initial phase of E in the plane $z = 0$ (θ_0). *[folder ME7_6(GUI) on IR]*

HINT:
See MATLAB Exercise 2.13 for the GUI development and use function `propParamLossless` from MATLAB Exercise 7.1 to obtain the necessary propagation parameters of the wave.

7.2 Arbitrarily Directed Uniform Plane Waves

The complex rms field intensity vectors of a uniform plane wave whose propagation direction is completely arbitrary with respect to a given coordinate system, i.e., it does not coincide with an axis (x, y, or z) of the system and is defined by the unit vector $\hat{\mathbf{n}} = n_x\,\hat{\mathbf{x}} + n_y\,\hat{\mathbf{y}} + n_z\,\hat{\mathbf{z}}$, are, in place of Eqs.(7.8), computed at a point the position vector of which is $\mathbf{r} = x\,\hat{\mathbf{x}} + y\,\hat{\mathbf{y}} + z\,\hat{\mathbf{z}}$ as

$$\underline{\mathbf{E}} = \underline{\mathbf{E}}_0\,\mathrm{e}^{-\mathrm{j}\beta\mathbf{r}\cdot\hat{\mathbf{n}}} = \underline{\mathbf{E}}_0\,\mathrm{e}^{-\mathrm{j}\beta(xn_x+yn_y+zn_z)}\,, \quad \underline{\mathbf{H}} = \frac{1}{\eta}\,\hat{\mathbf{n}}\times\underline{\mathbf{E}}$$

(arbitrarily directed plane wave) . (7.10)

Analogous expressions can be written in the time domain, as in Eqs.(7.1).

MATLAB EXERCISE 7.7 **Plane wave travel in an arbitrary direction – 3-D movie.** A time-harmonic uniform plane wave of frequency $f = 300$ MHz travels in free space in the direction defined by the propagation unit vector $\hat{\mathbf{n}} = \sqrt{2}\,(\hat{\mathbf{x}} + \hat{\mathbf{y}})\,/2$ ($|\hat{\mathbf{n}}| = 1$) in a rectangular coordinate system. The electric field intensity vector of the wave has a z-component only, and its amplitude is $E_\mathrm{m} = 1$ V/m. Write a MATLAB code that plays a 3-D movie of this field, for $0 \le t \le 2T$ and $0 \le x, y \le \lambda$, where T and λ are the time period and wavelength, respectively, of the wave. *(ME7_7.m on IR)*

TUTORIAL:

From Eq.(7.10), since the Cartesian components of the vector $\hat{\mathbf{n}}$ are $n_x = n_y = \sqrt{2}/2$ and $n_z = 0$, the instantaneous electric field intensity vector of the wave is given by

$$\mathbf{E}(x, y, t) = E_\mathrm{m}\cos\left[\omega t - \beta\frac{\sqrt{2}}{2}\,(x + y)\right]\,\hat{\mathbf{z}}\ \mathrm{V/m}\,. \tag{7.11}$$

The first part of the MATLAB code is similar to that in MATLAB Exercise 7.3,

```
[w,T,beta,lambda,vp,eta] = propParamLossless(f,EPSR,MUR);
dt = T/50;
dx = lambda/50;
dy = dx;
[x,y] = meshgrid(0:dx:lambda,0:dy:lambda);
t = [0:dt:2*T];
```

Within the movie `for` loop (ending with MATLAB function `getframe`), we implement Eq.(7.11), and visualize the spatial and temporal variations of E with the use of MATLAB function `surf` (see MATLAB Exercise 1.39). The viewpoint (MATLAB function `view`) is defined by azimuthal and elevation angles of 60° and 30°, respectively.

```
for i = 1:length(t);
E = Em*cos(w*t(i) - sqrt(2)/2*beta.*(x + y));
surf(x,y,E);
view([60 30]);
xlabel('x [m]'); ylabel('y [m]'); zlabel('E(x,y,t) [V/m]');
M(i) = getframe;
end;
```

Figure 7.5 shows a snapshot of the movie.

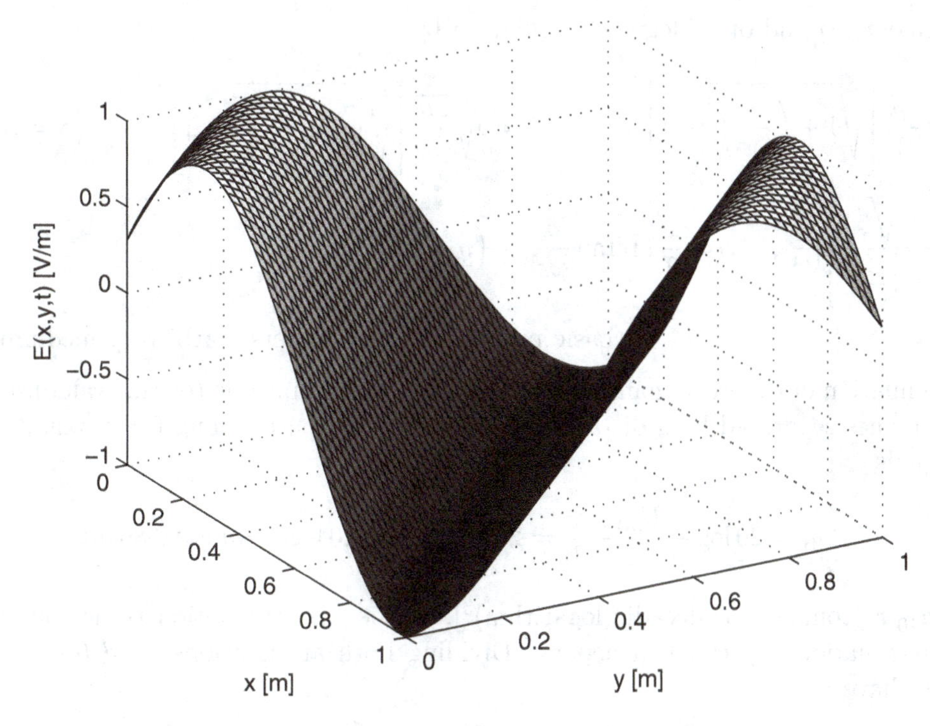

Figure 7.5 Snapshot of a 3-D movie – by means of MATLAB function `surf` – visualizing the electric field of a uniform plane wave given by Eq.(7.11); for MATLAB Exercise 7.7. *(color figure on CW)*

MATLAB EXERCISE 7.8 **Another 3-D movie of an arbitrarily directed plane wave.** Repeat the previous MATLAB exercise but for a plane wave traveling in the direction defined by the propagation unit vector $\hat{\mathbf{n}} = \sqrt{5}\,(-\,\hat{\mathbf{y}} + 2\,\hat{\mathbf{z}})\,/5$, with the vector $\mathbf{E}$ having an x-component only ($f = 300$ MHz and $E_{\mathrm{m}} = 1$ V/m). Play the movie for $0 \le t \le 2T$, $-\lambda \le y \le 0$, and $0 \le z \le 2\lambda$. *(ME7_8.m on IR)*

7.3 Theory of Time-Harmonic Waves in Lossy Media

For a uniform plane time-harmonic electromagnetic wave in a linear and homogeneous medium that exhibits losses ($\sigma \neq 0$), Eqs.(7.1) become

$$E_x = E_{\mathrm{m}}\,\mathrm{e}^{-\alpha z}\cos(\omega t - \beta z + \theta_0)\,, \quad H_y(z,t) = \frac{E_{\mathrm{m}}}{|\underline{\eta}|}\,\mathrm{e}^{-\alpha z}\cos(\omega t - \beta z + \theta_0 - \phi)$$

$$\text{(plane wave in a lossy medium)}\,, \qquad (7.12)$$

where the attenuation coefficient, α [unit is Np/m (neper per meter)], and phase coefficient, β (rad/m), which combined give the complex propagation coefficient, $\underline{\gamma}$ (expressed in m^{-1}), of the wave, and the magnitude of the complex intrinsic impedance of the medium, $|\underline{\eta}|$ (Ω), and its phase angle (argument), ϕ [rad or ° (degrees)], are given by

$$\alpha = \omega\sqrt{\frac{\varepsilon\mu}{2}}\left[\sqrt{1+\left(\frac{\sigma}{\omega\varepsilon}\right)^2}-1\right]^{1/2}, \quad \beta = \omega\sqrt{\frac{\varepsilon\mu}{2}}\left[\sqrt{1+\left(\frac{\sigma}{\omega\varepsilon}\right)^2}+1\right]^{1/2} \quad \left(\underline{\gamma} = \alpha + \mathrm{j}\beta\right),$$

$$|\underline{\eta}| = \frac{\sqrt{\frac{\mu}{\varepsilon}}}{\left[1+\left(\frac{\sigma}{\omega\varepsilon}\right)^2\right]^{1/4}}, \quad \phi = \frac{1}{2}\arctan\frac{\sigma}{\omega\varepsilon} \quad \left(\underline{\eta} = |\underline{\eta}|\,\mathrm{e}^{\mathrm{j}\phi}\right)$$

(basic propagation parameters – arbitrary medium) . (7.13)

The attenuation of the wave, namely, of its electric field amplitude (or rms value), between two transversal planes separated by a distance d along the wave path (along the z-axis), measured in decibels, equals

$$A_{\mathrm{dB}} = 20\log\frac{E_{\mathrm{m}}(z)}{E_{\mathrm{m}}(z+d)} = 20\log\mathrm{e}^{\alpha d} = (20\log e)\,\alpha d = 8.686\,\alpha d \quad (7.14)$$

[$\log x \equiv \log_{10} x$ (common or decadic logarithm)[3]], so the dB attenuation comes out to be 8.686 times the attenuation expressed in nepers. Dividing both attenuations by d (or assuming that $d = 1$ m), we have

$$\alpha \text{ in dB/m} = 8.686 \times (\alpha \text{ in Np/m}) \quad \text{(conversion from Np/m to dB/m)} . \quad (7.15)$$

The magnitude of the time-average Poynting vector [Eq.(6.33)] of the wave ($\mathcal{P}$) is proportional to the rms electric field intensity squared; however, the attenuation in dB is the same whether decibels are defined for power ratios or the corresponding field intensity ratios,

$$A_{\mathrm{dB}} = 10\log\frac{\mathcal{P}_1}{\mathcal{P}_2} = 20\log\frac{E_1}{E_2} \quad \text{(decibel attenuation via } \mathcal{P} \text{ or } E\text{)} . \quad (7.16)$$

MATLAB EXERCISE 7.9 Basic propagation parameters in an arbitrary medium. Write a function `basicPropParam()` in MATLAB that computes basic propagation parameters for a uniform plane time-harmonic wave in an arbitrary linear and homogeneous medium. Namely, based on the electromagnetic parameters of the medium, ε_{r}, μ_{r}, and σ, as well as the operating frequency of the wave, f, the function returns the values of the attenuation and phase coefficients of the wave, α and β, and of the magnitude and phase angle of the complex intrinsic impedance of the medium, $|\underline{\eta}|$ and ϕ. *(basicPropParam.m on IR)*

HINT:
Use Eqs.(7.13) to compute the propagation parameters of the wave in Eqs.(7.12).

MATLAB EXERCISE 7.10 GUI for basic propagation parameters. Create a graphical user interface in MATLAB to calculate and display propagation parameters α, β, $|\underline{\eta}|$, and ϕ of a

[3]Whereas "log" in our mathematical notation means "$\log_{10}$," MATLAB uses "log" in place of "ln," to denote the natural logarithm, and "log10" for "$\log_{10}$."

uniform plane time-harmonic wave in a lossy medium, for the medium parameters, ε_r, μ_r, and σ, and wave frequency, f, as input. *[folder ME7_10(GUI) on IR]*

HINT:
Use function `basicPropParam` (from the previous MATLAB exercise). See MATLAB Exercise 2.13 for the GUI development.

MATLAB EXERCISE 7.11 **E and H fields of an attenuated traveling wave – 3-D movie.** Write a code in MATLAB that plays a movie of a time-harmonic uniform plane wave in a lossy medium, based on the expressions for instantaneous electric and magnetic field intensities of the wave in Eqs.(7.12). Assume that $\varepsilon_r = 81$, $\mu_r = 1$, $\sigma = 0.4$ S/m, $f = 835$ MHz, $E_m = 1$ V/m, and $\theta_0 = 0$. The movie shows the wave in a region defined by $0 < z < 4\lambda$ and lasts $4T$ (λ and T are the wavelength and time period, respectively, of the wave). *(ME7_11.m on IR)*

TUTORIAL:
We use function `basicPropParam` (from MATLAB Exercise 7.9) to find the necessary propagation parameters of the wave, and specify the time array and the spatial array of nodes along the z-axis for the computation and plotting of fields,

```
[alpha,beta,absEta,phiEta] = basicPropParam(EPSR,MUR,SIGMA,f);
Hm = Em/absEta;
T = 1/f;
dt = T/50;
lambda = 2*pi/beta;
vp = C0/sqrt(EPSR*MUR);
dz = vp*dt;
[z,t] = meshgrid(0:dz:4*lambda, 0:dt:4*T);
z1 = 0:dz:4*lambda;
[k,N] = size(t);
```

with `k` and `N` standing for respective dimensions of the `[z,t]` mesh matrix, which are equal in this case.

Then, we implement Eqs.(7.12), to calculate the electric and magnetic field intensities of the wave,

```
phi = W.*t-beta.*z;
Ex = Em * exp(-alpha*z).*cos(phi).*heaviside(W.*t-beta.*z);
Hy = Hm * exp(-alpha*z).*cos(phi - phiEta).*heaviside(W.*t-beta.*z);
```

where we assume that the fields are zero for $t < 0$ (and $0 < z < 4\lambda$) (and are nonzero for $0 \leq t \leq 4T$), and accordingly use MATLAB function `heaviside` in a similar way as in MATLAB Exercise 7.3. Within the `for` loop of the movie (number of frames is `k`), the first four lines plot four points with coordinates given by `(0,-Em,-Hm)`, `(z1(N),-Em,-Hm)`, `(0,Em,-Hm)`, and `(0,-Em,Hm)` – in white (so they are not visible) – to set and maintain the boundaries of the 3-D plot, namely, to ensure that every frame of the movie covers and shows the same range along coordinate axes [$z1(N)$ is the last element of the array $z1$, that is, the largest coordinate z to be covered in the movie]. The next two lines of the code plot the electric and magnetic field intensities, and the following ones specify the viewpoint (MATLAB function `view`) for the movie, labels of axes, and graph (movie) title. As usual, the movie loop ends with MATLAB function **getframe** (making a frame of the movie for every pass through the loop). In each movie frame, the fields are plotted only up to the point that is actually reached by the wave at that time.

```
for j = 1:k-1
[error, n] = min(heaviside(W*t(j)-beta.*z1));
plot3(0,-Em,-Hm,'w');hold on;
plot3(z1(N),-Em,-Hm,'w');
plot3(0,Em,-Hm,'w');
plot3(0,-Em,Hm,'w');
plot3(z1(1:n-1),Ex(j,1:n-1),zeros(1,n-1),'r','linewidth',2);
plot3(z1(1:n-1),zeros(1,n-1),Hy(j,1:n-1),'b','linewidth',2);
grid on;
hold off;
view([60,30,45]);
xlabel('z [m]');
ylabel('Ex [V/m]');
zlabel('Hy [A/m]');
title('E and H fields of an attenuated traveling plane wave');
M(j) = getframe;
end;
```

A frame of the movie is shown in Fig.7.6.

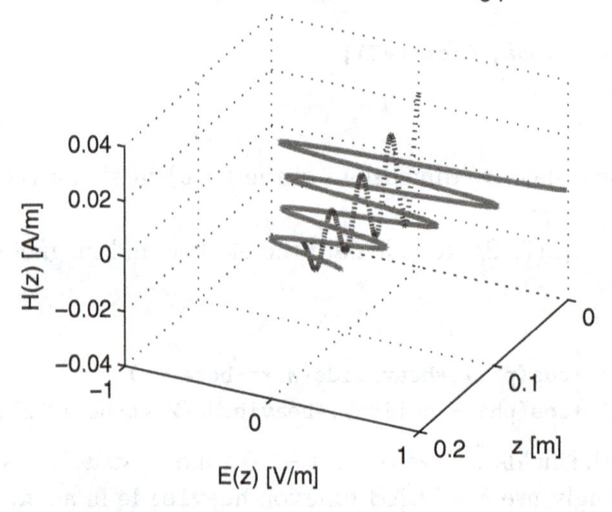

Figure 7.6 Snapshot of a 3-D movie visualizing the electric and magnetic fields of an attenuated traveling uniform plane wave (in a lossy medium), based on Eqs.(7.12); for MATLAB Exercise 7.11. *(color figure on CW)*

MATLAB EXERCISE 7.12 **Another 3-D movie of an attenuated wave.** Repeat the previous MATLAB exercise but for instantaneous electric and magnetic field intensities of a wave propagating in a lossy medium given by $\mathbf{E} = 486.7\,\mathrm{e}^{-1.26x}\cos(2.4\times 10^8 t - 2x + 24.1°)\,\hat{\mathbf{y}}$ V/m and $\mathbf{H} = 5\,\mathrm{e}^{-\alpha x}\cos(2.4\times 10^8 t - 2.83x)\,\hat{\mathbf{z}}$ A/m (t in s; x in m). *(ME7_12.m on IR)*

MATLAB EXERCISE 7.13 **E and H fields of an unattenuated wave – 3-D movie.** Repeat MATLAB Exercise 7.11 but for a wave propagating in a lossless medium, with instantaneous electric and magnetic field intensities given in Eqs.(7.1), assuming that $\varepsilon_r = 1$, $\mu_r = 1$, $f =$ 300 MHz, $E_m = 1$ V/m, and $\theta_0 = 0$. In specific, do this exercise in the following two ways: (1) write a separate MATLAB program specialized for the lossless case [based on Eqs.(7.1)] and (2) run the program for the lossy case (from MATLAB Exercise 7.11) for the new set of data. *(ME7_13.m on IR)*

HINT:
A snapshot of the resulting movie, obtained in the way (1) above, is shown in Fig.7.7 (note a different arrangement of coordinate axes when compared to that in Fig.7.6).

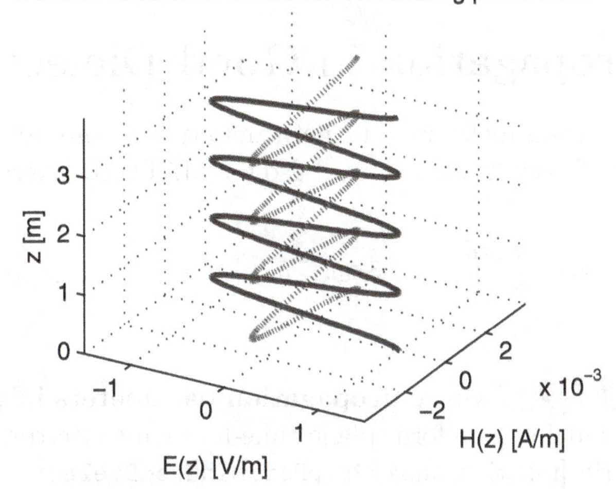

Figure 7.7 Snapshot of a 3-D movie visualizing the electric and magnetic fields of an unattenuated traveling uniform plane wave (in a lossless medium); for MATLAB Exercise 7.13. *(color figure on CW)*

MATLAB EXERCISE 7.14 **Neper to decibel conversion.** Write a function `Np2dB()` in MATLAB that converts the wave attenuation coefficient (α) expressed in Np/m to the one expressed in dB/m. *(Np2dB.m on IR)*

HINT:
To convert Np/m to dB/m, use the relationship in Eq.(7.15).

MATLAB EXERCISE 7.15 **Power ratio of two waves in decibels.** The rms electric field intensities of two time-harmonic uniform plane waves are E_1 and E_2, respectively, and the corresponding time-average Poynting vector magnitudes are $\mathcal{P}_1$ and $\mathcal{P}_2$. Write a function `dBPowerRatio()` in MATLAB that returns the power ratio for the two waves in decibels ($A_{\rm dB}$), Eq.(7.16), for the given field ratio E_1/E_2 or power ratio $\mathcal{P}_1/\mathcal{P}_2$. Input data to the function are the value of one of the ratios and one character to signify which ratio is used: the letter `P` means the power ratio, and any other letter (character), for instance, `F`, indicates the field ratio. Using this function, compute $A_{\rm dB}$ for the following values of E_1/E_2: 100, 10, 2, 1.41, 1, 0.707, 0.5, 0.1, and 0.01. *(dBPowerRatio.m on IR)* **H**[4]

MATLAB EXERCISE 7.16 **Conversion from dB power ratios to natural numbers.** Write a function `dB2naturalNum()` in MATLAB that performs a conversion reverse to that in the previous MATLAB exercise: it returns both ratios E_1/E_2 and $\mathcal{P}_1/\mathcal{P}_2$ for the given dB ratio $A_{\rm dB}$ at input. Use this function to find E_1/E_2 and $\mathcal{P}_1/\mathcal{P}_2$ for the following values of $A_{\rm dB}$: 60 dB, 14 dB, 6 dB, 1 dB, 0 dB, −3 dB, −14 dB, and −100 dB. *(dB2naturalNum.m on IR)* **H**

7.4 Wave Propagation in Good Dielectrics

For wave propagation in good dielectrics, namely, materials whose permittivity and conductivity at a given frequency satisfy condition $\sigma \ll \omega\varepsilon$ (also see MATLAB Exercise 3.7), Eqs.(7.13) can be approximated to

$$\sigma \ll \omega\varepsilon \quad \longrightarrow \quad \alpha \approx \frac{\sigma}{2}\sqrt{\frac{\mu}{\varepsilon}}, \quad \beta \approx \omega\sqrt{\varepsilon\mu}\,, \quad \underline{\eta} \approx \sqrt{\frac{\mu}{\varepsilon}} \quad \text{(good dielectrics)}\,. \qquad (7.17)$$

MATLAB EXERCISE 7.17 **Basic propagation parameters in good dielectrics.** Repeat MATLAB Exercise 7.9 but for a uniform plane time-harmonic electromagnetic wave propagating through a good dielectric [function `basicPropParamInGoodDielectrics()`]. *(basicPropParamIn-GoodDielectrics.m on IR)*

HINT:
Make use of Eqs.(7.17).

7.5 Wave Propagation in Good Conductors

In the case of good conductors, for which $\sigma \gg \omega\varepsilon$, Eqs.(7.13) are simplified to

$$\sigma \gg \omega\varepsilon \quad \longrightarrow \quad \alpha \approx \beta \approx \sqrt{\frac{\omega\mu\sigma}{2}}, \quad |\underline{\eta}| \approx \sqrt{\frac{\omega\mu}{\sigma}}, \quad \phi \approx 45^\circ \quad \text{(good conductors)}\,. \qquad (7.18)$$

[4]**H** = recommended to be done also "by hand," i.e., not using MATLAB.

MATLAB EXERCISE 7.18 **Basic propagation parameters in good conductors.** Repeat MATLAB Exercise 7.9 but for a plane wave in a good conductor [function `basicPropParamInGoodConductors()`]. *(basicPropParamInGoodConductors.m on IR)*

HINT:
Use Eqs.(7.18).

MATLAB EXERCISE 7.19 **Various combinations of material parameters and frequency.** Consider the following combinations of material parameters and frequency: (a) glass ($\varepsilon_\mathrm{r} = 5$ and $\sigma = 10^{-12}$ S/m), fresh water ($\varepsilon_\mathrm{r} = 80$ and $\sigma = 10^{-3}$ S/m), and copper ($\varepsilon_\mathrm{r} = 1$ and $\sigma = 58$ MS/m) at a frequency of $f = 100$ kHz, and (b) rural ground from MATLAB Exercise 6.12 at frequencies of $f_1 = 12.84$ GHz, $f_2 = 12.84$ MHz, and $f_3 = 12.84$ kHz, respectively. For each of them, determine if the material is a good dielectric [use the criterion $\sigma/(\omega\varepsilon) < 1/100$], good conductor [$\sigma/(\omega\varepsilon) > 100$], or quasi-conductor [$1/100 \le \sigma/(\omega\varepsilon) \le 100$], and calculate, in MATLAB, the attenuation and phase coefficients, dB attenuation per meter traveled, wavelength, and phase velocity of a uniform plane wave propagating in the material, as well as the complex intrinsic impedance of the material. Use approximate (simpler) expressions for good dielectrics or good conductors. *(ME7_19.m on IR)* **H**

HINT:
For good dielectrics, α (in Np/m), β, $\underline{\eta}$, α in dB/m, λ, and v_p should be computed using Eqs.(7.17), (7.15), (7.4), and (7.6), respectively. For quasi-conductors, these parameters should be evaluated based on Eqs.(7.13), (7.15), (7.4), and (7.6). Finally, Eqs.(7.18), (7.15), (7.4), and (7.6) should be invoked if the material at a given frequency is classified as a good conductor.

7.6 Skin Effect

The losses in conductors are considerable, especially at higher frequencies, and an electromagnetic wave incident on the surface of a conductor attenuates rapidly with distance from the surface, quickly reaching rather negligible field intensities. To quantitatively express the degree to which the wave can penetrate into the conductor (with still substantial intensity), we introduce a simple parameter defined as the depth into the conductor (distance from the conductor surface) at which the amplitude of the electric field of the wave is attenuated to 1/e (or about 36.8%) of its initial value, i.e., value at the surface. The same applies to the amplitude of the current density, $\mathbf{J} = \sigma\mathbf{E}$ [Eq.(3.5)], in the conductor. From Eqs.(7.12), it is a simple matter to conclude that this parameter, designated by δ, equals

$$\delta = \frac{1}{\alpha} \quad \text{(skin depth; unit: m)} . \tag{7.19}$$

It is termed the skin depth, to emphasize that the substantial wave penetration is confined to a very thin layer near the conductor surface ("skin" of the conductor). By the same token, ac currents of higher frequencies flow practically only through the skin region below the surface of metallic cylindrical conductors (with circular or any other cross-sectional shape), and the thickness of the layer that carries most of the current is on the order of the skin depth, δ. In general, the phenomenon of predominant localization of fields, currents, and power in the skin of a conducting body is referred to as the skin effect. For good conductors, and, in particular, for copper (Cu), for which $\sigma = 58$ MS/m (Table 3.1) and $\mu = \mu_0$, Eqs.(7.18) give

$$\delta \approx \frac{1}{\sqrt{\pi\mu f\sigma}} \quad \text{(for good conductors)}, \quad \delta_{\rm Cu} \approx \frac{66}{\sqrt{f}}\ \text{mm} \quad (f \text{ in Hz}) \quad \text{(for copper)}. \tag{7.20}$$

In a limit, $\delta = 0$ for a perfect electric conductor, at any frequency, that is, an electromagnetic wave cannot penetrate at all into a PEC.

MATLAB EXERCISE 7.20 **Skin depth for some materials at different frequencies.** Write a program in MATLAB to compute the skin depth (δ) for four materials, namely, copper, iron ($\sigma_{\rm Fe} = 10$ MS/m and $\mu_{\rm Fe} = 1000\mu_0$), seawater with $\sigma = 4$ S/m and $\varepsilon_{\rm r} = 81$, and rural ground with $\sigma = 10^{-2}$ S/m and $\varepsilon_{\rm r} = 14$, at frequencies of 60 Hz, 1 kHz, 100 kHz, 1 MHz, 100 MHz, and 1 GHz, respectively. The results should be tabulated in a previously created text file `skindepth.txt`. *(ME7_20.m on IR)* **H**

HINT:
For copper, use the formula for $\delta_{\rm Cu}$ in Eqs.(7.20). For iron, compute $\delta = \delta_{\rm Fe}$ from the first expression in Eqs.(7.20). Since seawater and rural ground do not behave like good conductors at some of the higher frequencies in the list [i.e., when $\sigma/(\omega\varepsilon) \leq 100$], δ in such cases should be computed using Eqs.(7.19) and (7.13), rather than the expression for δ in Eqs.(7.20).

MATLAB EXERCISE 7.21 **Decibel attenuation of an aluminum foil.** Write a MATLAB program that calculates the wave attenuation in decibels of a 3-mm thick aluminum ($\sigma_{\rm Al} = 38$ MS/m and $\varepsilon_{\rm r} = \mu_{\rm r} = 1$) foil at frequencies $f_1 = 1$ kHz, $f_2 = 10$ kHz, $f_3 = 100$ kHz, and $f_4 = 1$ MHz. The results are tabulated in a text file. *(ME7_21.m on IR)* **H**

TUTORIAL:
First, we create a file `attenuation.txt` and define all input parameters. The four frequencies for computation are stored in the following frequency vector:

```
f = [f1, f2, f3, f4];
```

Then, the attenuation of the foil is computed as αd ($d = 3$ mm) using the expression for α in Eqs.(7.18), the corresponding dB attenuation is obtained by means of function `Np2dB` (from MATLAB Exercise 7.14), and the two rows of a matrix `TAB` are filled with frequencies and the resulting attenuations, respectively,

```
A = sqrt(pi*MU0*MUR*f.*SIGMA)*d;
a = Np2dB(A);
TAB = [f;a];
```

The text file is opened by means of MATLAB function `fopen`, with a restriction to writing in only, printing of text and data to the file is done by MATLAB function `fprintf`, and the file is closed using MATLAB function `fclose`,

```
fid = fopen('attenuation.txt', 'w');
fprintf (fid,'Decibel attenuation of an aluminum foil \n \n');
fprintf (fid, 'f(kHz)    %.4g   %.4g   %.4g  %.4g \n', TAB(1,:)/1000);
fprintf (fid, ' A(dB)  %.4g   %.4g   %.4g   %.4g \n', TAB(2,:));
fclose(fid);
```

Upon running the program, the content of the file `attenuation.txt` appears to be:

```
Decibel attenuation of an aluminum foil

f(kHz)      1      10      100     1000
A(dB)   10.09   31.92   100.9    319.2
```

MATLAB EXERCISE 7.22 **Decibel attenuation of a microwave oven wall.** Repeat the previous MATLAB exercise but for the 1-mm thick stainless steel ($\sigma = 1.2$ MS/m, $\mu_r = 500$, and $\varepsilon_r = 1$) wall of a microwave oven, and frequencies $f_1 = 2$ GHz, $f_2 = 2.2$ GHz, $f_3 = 2.45$ GHz (standard frequency for microwave cooking), $f_4 = 2.7$ GHz, and $f_5 = 3$ GHz. In addition to f and A_{dB}, calculate and print-to-file (to file `microwave.txt`) the one-percent depth of penetration ($\delta_{1\%}$) into the wall as well, where $\delta_{1\%}$ is defined as the depth into a conductor at which the electric field intensity decreases to 1% of the initial value and it comes out to be $\delta_{1\%} = \delta \ln 100 \approx 4.6\,\delta$ [δ is given by the first expression in Eqs.(7.20)]. *(ME7_22.m on IR)* **H**

7.7 Wave Propagation in Plasmas

Plasmas are ionized gases which, in addition to neutral atoms and molecules, include a large enough number of ionized atoms and molecules and free electrons that macroscopic electromagnetic effects caused by Coulomb forces between charged particles are notable. The phase coefficient of a uniform plane electromagnetic wave of frequency f propagating through a plasma medium is given by

$$\beta = \frac{2\pi f}{c_0}\sqrt{1 - \frac{f_p^2}{f^2}}\,, \quad f_p = 9\sqrt{N} \quad (N \text{ in m}^{-3};\ f_p \text{ in Hz}) \quad \text{(plasma medium)}\,, \tag{7.21}$$

where c_0 is the free-space wave velocity, Eq.(7.7), f_p is the so-called plasma frequency, and N is the concentration of free electrons in the gas. For $f < f_p$, β becomes purely imaginary and thus effectively acts like a large attenuation coefficient α in Eqs.(7.12), so that the wave does not propagate (at $f = f_p$, $\beta = 0$, meaning that f_p is also a non-propagating frequency) – plasma apparently behaves like a high-pass filter.

An important example of a plasma medium is the upper region of the earth's atmosphere, from about 50 to 500 km altitude above the earth's surface, called the ionosphere. It consists of a highly rarefied gas that is ionized by the sun's radiation and plays an essential role in a number of radio-wave applications.

MATLAB EXERCISE 7.23 **Wave propagation in a parabolic ionospheric slab.** The ionosphere can approximately be represented by a parabolic profile $N(h)$, of the concentration of free electrons (N) as a function of the altitude (h) above the earth's surface, within a plasma slab of thickness $2d$ as follows:

$$N(h) = N_{\mathrm{m}}\left[1 - \frac{(h-h_{\mathrm{m}})^2}{d^2}\right] , \quad h_{\mathrm{m}} - d \leq h \leq h_{\mathrm{m}} + d , \tag{7.22}$$

where h_{m} is the altitude at the middle of the slab and N_{m} is the maximum electron concentration – at this altitude, which is illustrated in Fig.7.8. A uniform plane time-harmonic electromagnetic wave of frequency f is incident vertically (normally) from the earth's surface onto the lower boundary of the ionosphere, as shown in Fig.7.8, or from space onto the upper boundary. From Eqs.(7.21), maximum plasma frequency of the ionosphere is $(f_{\mathrm{p}})_{\max} = 9\sqrt{N_{\mathrm{m}}}$ Hz (N_{m} in m^{-3}). If $f > (f_{\mathrm{p}})_{\max}$, the propagating condition for a plasma ($f > f_{\mathrm{p}}$) is satisfied at every layer of the ionosphere (100 km $\leq h \leq$ 500 km), and the wave will pass through the entire slab in Fig.7.8. If $f < (f_{\mathrm{p}})_{\max}$, on the other hand, there is an altitude in the ionosphere at which $f = f_{\mathrm{p}}$, and the wave will bounce back off that layer. Combining Eqs.(7.21) and (7.22), the bounce-off altitude (h_{b}) is thus determined as

$$f < (f_{\mathrm{p}})_{\max} \quad \longrightarrow \quad f = f_{\mathrm{p}}(h_{\mathrm{b}}) = 9\sqrt{N(h_{\mathrm{b}})} = 9\sqrt{N_{\mathrm{m}}\left[1 - \frac{(h_{\mathrm{b}}-h_{\mathrm{m}})^2}{d^2}\right]}$$

$$\longrightarrow \quad \text{wave bounces at } h_{\mathrm{b}} = h_{\mathrm{m}} \pm d\sqrt{1 - \frac{f^2}{81N_{\mathrm{m}}}} \quad (N \text{ in m}^{-3};\ f \text{ in Hz}) , \tag{7.23}$$

where the solution with the minus sign in the expression for h_{b} corresponds to the incidence from the earth's surface onto the lower boundary of the slab, while the solution with the plus sign is for the incidence from space onto the upper boundary. Assuming that $h_{\mathrm{m}} = 250$ km, $d = 100$ km, and $N_{\mathrm{m}} = 10^{12}$ m^{-3}, write a MATLAB code that, based on Eqs.(7.23), calculates and plots the

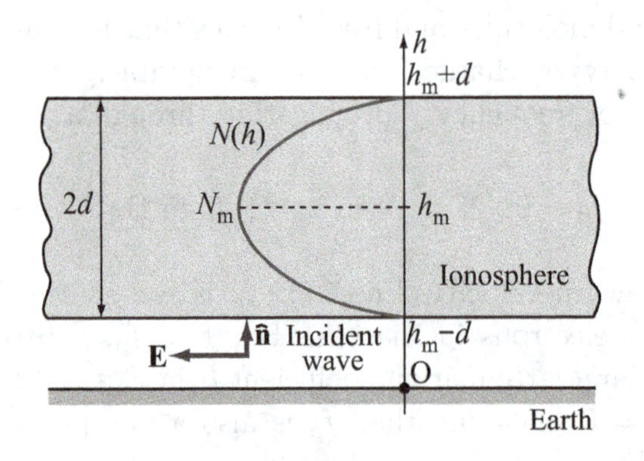

Figure 7.8 Normal incidence of a uniform plane wave on a parabolic ionospheric slab with the concentration of free electrons (N) given by Eq.(7.22); for MATLAB Exercise 7.23.

altitude h_b for each of the frequencies separated by steps of $\Delta f = 0.5$ MHz within a frequency range 8 MHz $\leq f \leq$ 12 MHz. In a case of the wave passing through the ionosphere, indicate so on the graph. *(ME7_23.m on IR)* **H**

7.8 Polarization of Electromagnetic Waves

All time-harmonic electromagnetic waves considered so far in this chapter exhibit so-called linear polarization (LP), with the tip of the electric field intensity vector of the wave (and the same is true for the magnetic one) at a given point in space tracing a straight line in the course of time. For instance, we see that the tip of the vector $\mathbf{E} = E(z,t)\,\hat{\mathbf{x}}$ given by Eqs.(7.1) for a fixed z oscillates, in the course of time, along the x-axis of the Cartesian coordinate system, as shown in Fig.7.9, and similarly for $\mathbf{E}$ in Eqs.(7.12), for a wave propagating in a lossy medium.

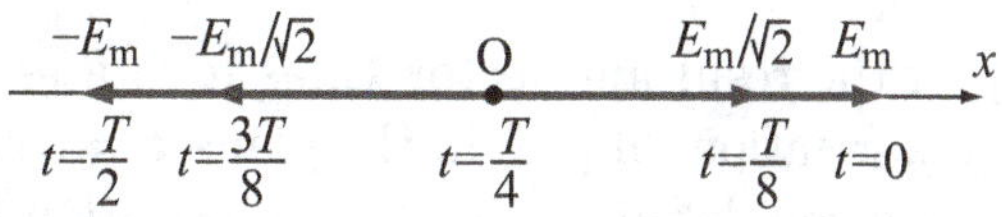

Figure 7.9 Linearly polarized time-harmonic uniform plane wave: the electric field vector in Eqs.(7.1) for $z = 0$ and $\theta_0 = 0$.

However, if two waves with mutually orthogonal linear polarizations at the same frequency co-propagate in the same direction, the polarization of the resultant wave depends on the relative amplitudes and phases of its individual LP components. For instance, if the two transverse components have the same amplitudes but are out of phase by 90° (i.e., they are in time-phase quadrature),

$$E_x = E_{\mathrm{m}} \cos(\omega t - \beta z)\,, \quad E_y = E_{\mathrm{m}} \cos(\omega t - \beta z \mp 90°) = \pm E_{\mathrm{m}} \sin(\omega t - \beta z)$$
$$\text{(circular polarization)}\,, \tag{7.24}$$

the wave is circularly polarized (CP). Namely, the resultant vector $\mathbf{E}$ for $z =$ const rotates with an angular velocity equal to ω and its tip describes a circle (so-called polarization circle), of radius E_{m}, as a function of time in a transversal plane defined by the fixed coordinate z, as in the example of a rotating magnetic field in Fig.6.5. With δ denoting the relative phase of $\mathbf{E}_y$ with respect to $\mathbf{E}_x$, Fig.7.10(a) shows the rotation of the total field vector for the case $\delta = -90°$ $[E_y = E_{\mathrm{m}} \sin(\omega t - \beta z)]$. Such CP wave is said to be right-hand circularly polarized (RHCP), given that, when the thumb of the right hand points into the direction of the wave travel, the other fingers curl in the direction of rotation of $\mathbf{E}$. Conversely, the case with $\delta = 90°$ in Eqs.(7.24) $[E_y = -E_{\mathrm{m}} \sin(\omega t - \beta z)]$ gives rise to a left-hand circularly polarized (LHCP) wave, Fig.7.10(b).

If we change one of the amplitudes of the transverse components of the wave in Eqs.(7.24) so that they are no longer the same,

$$E_x = E_1 \cos(\omega t - \beta z)\,, \quad E_y = \pm E_2 \sin(\omega t - \beta z) \quad \text{(elliptical polarization)} \tag{7.25}$$

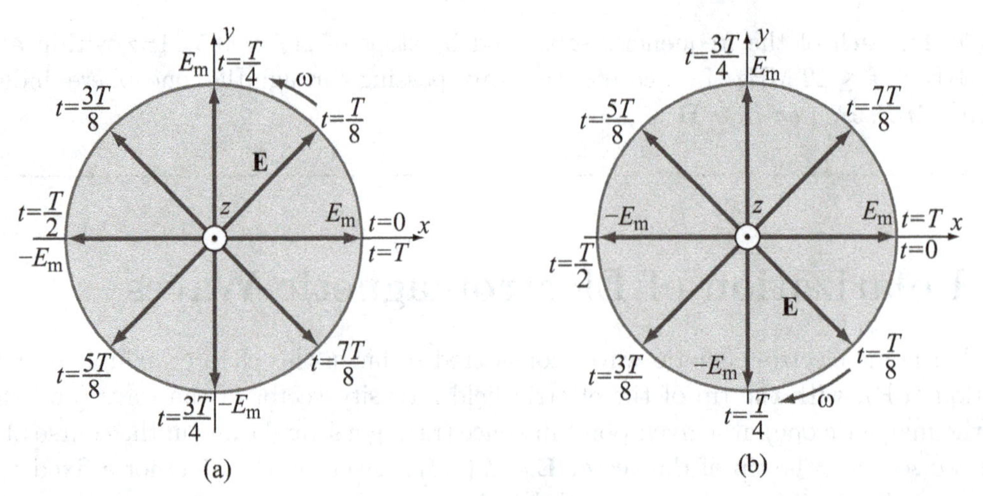

Figure 7.10 Right-hand (a) and left-hand (b) circularly polarized time-harmonic uniform plane wave (propagating in the positive z direction), Eqs.(7.24) with $\delta = -90°$ (RHCP) or $\delta = 90°$ (LHCP).

($E_1 \neq E_2$, $\delta = \mp 90°$), the tip of the resultant vector $\mathbf{E}$, as it rotates in time with angular velocity equal to ω, traces an ellipse (polarization ellipse) in the plane $z =$ const, and the wave is said to be elliptically polarized (EP). The same definition for the polarization handedness applies as for the CP waves: for $\delta = -90°$, the wave is right-hand elliptically polarized (RHEP), whereas $\delta = 90°$ yields a left-hand elliptically polarized (LHEP) wave, as shown in Fig.7.11.

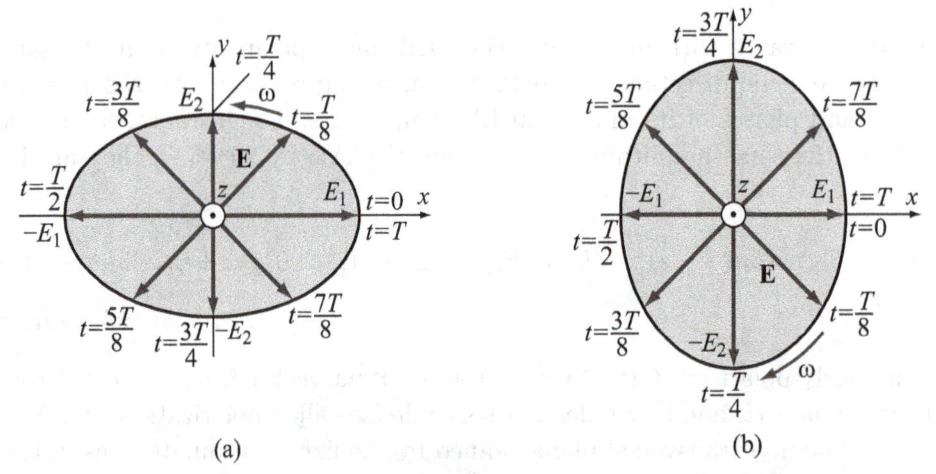

Figure 7.11 Elliptical polarization of a time-harmonic uniform plane wave (propagating out of the page), Eqs.(7.25): (a) RHEP ($\delta = -90°$) with the major axis of the polarization ellipse adopted to be in the x direction ($E_1 > E_2$) and (b) LHEP ($\delta = 90°$) with a y-directed major axis ($E_2 > E_1$).

In the most general case, the phase difference between the two orthogonal transverse components of $\mathbf{E}(z,t)$ is arbitrary ($-180° < \delta \leq 180°$), in addition to an arbitrary ratio between their amplitudes,

$$E_x = E_1 \cos(\omega t - \beta z) , \quad E_y = E_2 \cos(\omega t - \beta z + \delta) \quad \text{(the most general EP wave)} . \quad (7.26)$$

This wave is also elliptically polarized, with $\delta > 0$ implying a left-handed polarization and $\delta < 0$

the right-handed one, and the polarization ellipse being tilted with respect to the x and y axes. So, elliptical polarization is the most general polarization of time-harmonic vectors.

MATLAB EXERCISE 7.24 **Right-hand circularly polarized wave – 2-D movie.** Write a code in MATLAB that plays a 2-D movie in the xy-plane ($z = 0$) of a right-hand circularly polarized (RHCP) time-harmonic uniform plane wave propagating in the positive z direction, Fig.7.10(a), assuming that $E_\mathrm{m} = 1$ V/m, $T = 300$ ns, and the duration of the movie is $3T$. *(ME7_24.m on IR)*

TUTORIAL:

First, we specify input data and a time vector (array) `t` for the movie,

```
N = 3;
Em = 1;
T = 300*10^(-9);
W = 2*pi/T;
dt = T/300;
t = 0:dt:N*T;
```

Then, we compute components E_x and E_y of the RHCP wave – from Eqs.(7.24) with $z = 0$ and $\delta = -90°$ $[E_x = E_\mathrm{m}\cos\omega t$ and $E_y = E_\mathrm{m}\sin\omega t]$, and coordinates x and y of the points determining the polarization circle in Fig.7.10(a), that the tip of the resultant vector $\mathbf{E}$ traces during the course of time,

```
Ex = Em*cos(W*t);
Ey = Em*sin(W*t);
x = Em*cos(0:2*pi/100:2*pi);
y = Em*sin(0:2*pi/100:2*pi);
```

Finally, we play the movie in a usual fashion. In specific, the first code line within the movie `for` loop plots the polarization circle in blue, the next two plot the coordinate axes, by means of MATLAB function `line`, then we plot vectors $\mathbf{E} = E_x\,\hat{\mathbf{x}} + E_y\,\hat{\mathbf{y}}$, $E_y\,\hat{\mathbf{y}}$, and $E_x\,\hat{\mathbf{x}}$, in this order, using function `vecPlot2D` from MATLAB Exercise 1.2, do some more formatting and explanation of the graph and movie, and the last line calls MATLAB function `getframe`. Note that it is essential to insert MATLAB command `hold on` between calls of `vecPlot2D` to keep both plotted vectors in the graph (rather than just the latter one).

```
for i=1:length(t)
plot(x,y,'b'); hold on;
line([0 0],[-Em Em]*1.2,'Color','k');
line([-Em Em]*1.2,[0 0],'Color','k');
vecPlot2D([0 0],[Ex(i) Ey(i)],0,'r',0);
hold on;
vecPlot2D([0 0],[0 Ey(i)],0,'r',0);
hold on;
vecPlot2D([0 0],[Ex(i) 0],0,'r',0);
hold off;
```

```
axis equal;
title('Circular polarization for z=0');
xlabel('x');
ylabel('y');
text(Em,0,'Ex');
text(0,Em,'Ey');
text(Ex(i)*1.1,Ey(i)*1.1,'Etot');
xlim([-Em*1.2 Em*1.2]);
ylim([-Em*1.2 Em*1.2]);
M(i) = getframe;
end;
```

Figure 7.12 shows a frame of the movie.

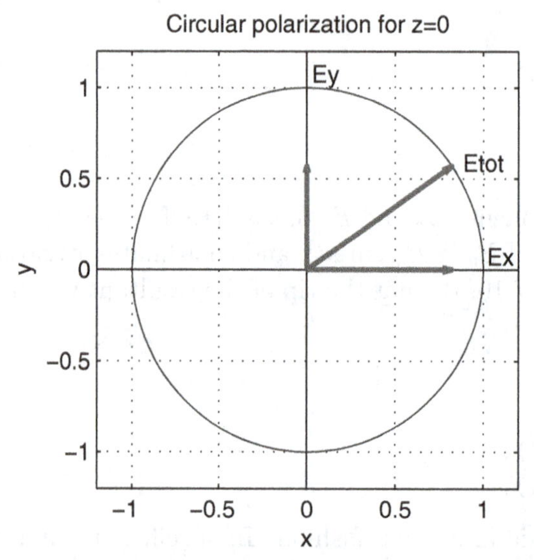

Figure 7.12 Snapshot of a 2-D movie visualizing a right-hand circularly polarized (RHCP) time-harmonic uniform plane wave propagating in the positive z direction [Fig.7.10(a)]; for MATLAB Exercise 7.24. *(color figure on CW)*

MATLAB EXERCISE 7.25 **Sum of two counter-rotating CP waves.** Two oppositely rotating, right- and left-handed, circularly polarized time-harmonic uniform plane electromagnetic waves of the same frequency, $f = 1$ GHz, travel in the same direction, the positive z direction, and the amplitudes of their electric field intensity vectors are $E_{\text{m1}} = 2$ V/m and $E_{\text{m2}} = 5$ V/m, respectively. Write a MATLAB program that plays a 2-D movie in the xy-plane ($z = 0$) of the two waves, as well as of the resultant wave, obtained by superposition of the two CP waves, during the time $0 \leq t \leq 2T = 2/f$. Show the three waves in three separate subplots (use MATLAB functions `subplot`). Arrange the subplots such that the resultant wave is in between the two CP waves. *(ME7_25.m on IR)*

MATLAB EXERCISE 7.26 **Elliptically polarized wave – 2-D movie.** Repeat MATLAB Exercise 7.24 but for an elliptically polarized time-harmonic uniform plane wave traveling in the positive z direction – described by Eqs.(7.25), for $E_1 = 1$ V/m, $E_2 = 2$ V/m, and $f = 750$ MHz. The movie plays either a right-hand wave (RHEP), Fig.7.11(a), or a left-hand one (LHEP), Fig.7.11(b), as specified at the input. *(ME7_26.m on IR)*

MATLAB EXERCISE 7.27 **Tilted polarization ellipse – 2-D movie.** Repeat MATLAB Exercise 7.24 but for a left-hand elliptically polarized time-harmonic uniform plane wave with a tilted polarization ellipse – given by Eqs.(7.26) for $\delta = \pi/3$. Assume that $E_1 = 1$ V/m, $E_2 = 2$ V/m, and $f = 750$ MHz. Also, play the movie with $\delta = -\pi/3$ (right-handed polarization). *(ME7_27.m on IR)*

MATLAB EXERCISE 7.28 **Circularly polarized wave – 3-D plot.** Write a MATLAB program that creates a 3-D plot for $0 \leq z \leq 2\lambda = 4\pi/\beta$ and $t = 0$ of the right-hand circularly polarized wave from Fig.7.10(a), assuming that $E_\text{m} = 0.5$ V/m and $f = 500$ MHz. *(ME7_28.m on IR)*

TUTORIAL:

Vectors (arrays) `z` and `z1`, to be used for plotting the electric field vector, **E**, along the z-axis and its "envelope," respectively, are defined as

```
W = 2*pi*f;
beta = W/Vp;
z = 0:2*pi/beta/16:4*pi/beta;
z1 = 0:2*pi/beta/100:4*pi/beta;
```

with a finer mesh for `z1` so that the "envelope" appears continuous in the graph.

Field components E_x and E_y, of the RHCP wave, are computed from Eqs.(7.24) with $\delta = -90°$ and $t = 0$ for every point of the `z` vector, and plotted in a `for` loop using function `vecPlot3D` from MATLAB Exercise 1.3. Argument `1/length(z)*[1 i length(z)-i/4]` of the function enables plotting vectors in different colors, for better clarity of the graph. MATLAB command `hold on` keeps all plotted vectors in the graph.

```
t=0;
phi = W.*t - beta.*z;
Ex = Em*cos(phi);
Ey = Em*sin(phi);
for i = 1:length(z)
vecPlot3D([0 0 z(i)],[Ex(i) Ey(i) z(i)],0,1/length(z)*[1 i length(z)-i/4],0);
hold on;
end;
```

The "envelope" of **E** vectors in the graph, namely, the helix containing the tips of **E** in space, according to Eqs.(7.24), is drawn (in red) using MATLAB functions `plot3` as follows:

```
phi1 = W.*t - beta.*z1;
Ex1 = Em*cos(phi1);
Ey1 = Em*sin(phi1);
plot3(Ex1,Ey1,z1,'r');
hold off;
```

The resulting 3-D plot of the wave is shown in Fig.7.13.

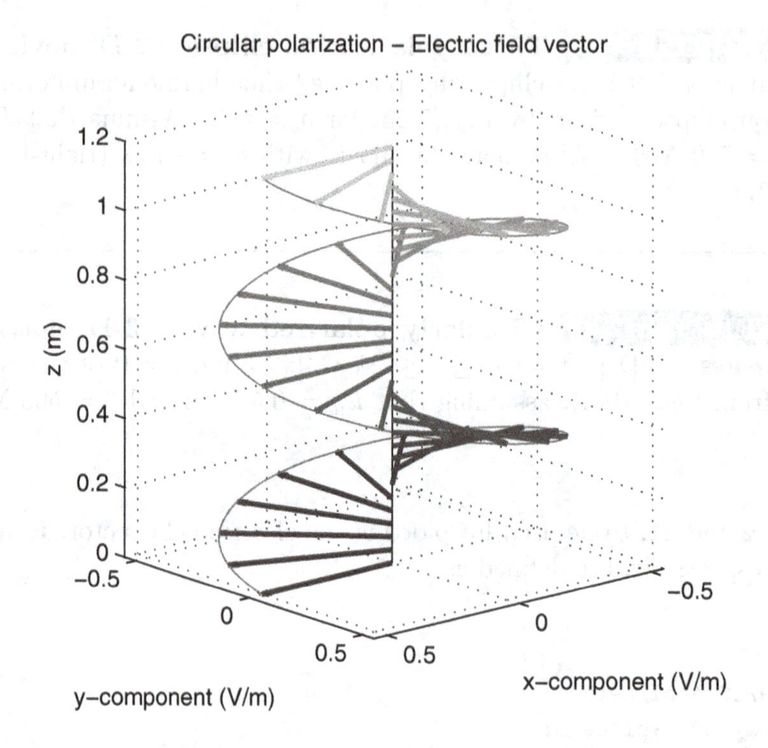

Figure 7.13 3-D visualization of the electric field vector (**E**) of a RHCP plane wave [Fig.7.10(a)]; for MATLAB Exercise 7.28. *(color figure on CW)*

MATLAB EXERCISE 7.29 **Elliptically polarized wave – 3-D plot.** Repeat the previous MATLAB exercise but for the left-hand elliptically polarized wave with a tilted polarization ellipse from MATLAB Exercise 7.27 ($\delta = \pi/3$), assuming that $E_1 = 1$ V/m, $E_2 = 4$ V/m, and $f = 100$ MHz. *(ME7_29.m on IR)*

MATLAB EXERCISE 7.30 **Change of EP wave handedness due to material anisotropy – 2-D movie.** Consider an anisotropic crystalline dielectric material whose relative permittivity has different values [Eq.(2.3)], $\varepsilon_{rx} = 2.2$ and $\varepsilon_{ry} = 2.1$, along two mutually orthogonal transverse directions in the crystal, namely, x and y directions, respectively. Assume that the right-hand elliptically polarized plane wave from MATLAB Exercise 7.26 (propagating in the positive z direction), but with the free-space wavelength of $\lambda_0 = 1$ μm, enters the crystal. Let the length of the crystal (along the z axis) be such that it results in a change of the phase difference between orthogonal (x and y) components of the electric field intensity vector of the wave, $\mathbf{E}$, equal to $\delta_{\rm a} = \pi$. Under these circumstances, create a 2-D movie in MATLAB that visualizes (simultaneously, in two subplots) the input wave, entering the crystal, and the output wave, emerging on the other side of it. *(ME7_30.m on IR)*

TUTORIAL:
Using Eq.(7.4), the phase coefficients in x and y directions in the crystal are

$$\beta_x = \omega\sqrt{\varepsilon_{{\rm r}x}\varepsilon_0\mu_0} = \frac{2\pi}{\lambda_0}\sqrt{\varepsilon_{{\rm r}x}}\,, \quad \beta_y = \omega\sqrt{\varepsilon_{{\rm r}y}\varepsilon_0\mu_0} = \frac{2\pi}{\lambda_0}\sqrt{\varepsilon_{{\rm r}y}} \quad (\mu_{\rm r} = 1)\,. \tag{7.27}$$

As a consequence of $\beta_x \neq \beta_y$, an additional relative phase $\delta_{\rm a}$ of the field component E_y with respect to E_x accumulated as the wave travels a length d of the crystal amounts to

$$\delta_{\rm a} = \phi_y - \phi_x = \omega t - \beta_y d - (\omega t - \beta_x d) = (\beta_x - \beta_y)\,d \quad \text{(due to material anisotropy)}\,. \tag{7.28}$$

From Eqs.(7.27) and (7.28), we compute β_x and β_y, and the shortest d that results in $\delta_{\rm a} = \pi$ (the result for d is displayed in the Command Window),

```
betax = 2*pi/LAMBDA0*sqrt(EPSRx);
betay = 2*pi/LAMBDA0*sqrt(EPSRy);
delta = pi;
d = delta/(betax - betay);
fprintf('d = %g m.\n',d);
```

Similarly to the computation in the respective code lines in MATLAB Exercise 7.24, we define a time vector `t` for the movie, and compute electric field components E_x and E_y, now using Eqs.(7.25) with $\delta = -90°$, of the input wave (with $z = 0$) and the output wave (with $z = d$), as well as coordinates x and y of the points determining the polarization ellipse,

```
W = 2*pi/T;
dt = T/300;
t = 0:dt:N*T;
Ex0 = EX*cos(W*t);
Ey0 = EY*sin(W*t);
Exd = EX*cos(W*t - betax*d);
Eyd = EY*sin(W*t - betay*d);
x = EX*cos(0:2*pi/100:2*pi);
y = EY*sin(0:2*pi/100:2*pi);
```

The movie is played, in two subplots (using MATLAB functions `subplot`), in a similar fashion to that in MATLAB Exercise 7.24:

```
for i=1:length(t)
subplot(1,2,1)
plot(x,y,'b'); hold on;
vecPlot2D([0 0],[Ex0(i) Ey0(i)],0,'r',0);
hold on;
```

```
vecPlot2D([0 0],[0 Ey0(i)],0,'r',0);
hold on;
vecPlot2D([0 0],[Ex0(i) 0],0,'r',0);
hold off;
M(i) = getframe;
subplot(1,2,2)
plot(x,y,'b'); hold on;
vecPlot2D([0 0],[Exd(i) Eyd(i)],0,'r',0);
hold on;
vecPlot2D([0 0],[0 Eyd(i)],0,'r',0);
hold on;
vecPlot2D([0 0],[Exd(i) 0],0,'r',0);
hold off;
M(i) = getframe;
end;
```

Figures 7.14 and 7.15 show two frames of the movie, where we see that, while the input wave (shown in subplots on the left-hand side of the figures) is a RHEP wave (as specified in the text of the exercise), the output wave (subplots on the right) is a LHEP one. In this case, therefore, the crystal reverses the handedness of the input wave, as expected, because of the addition of $\delta_a = \pi$

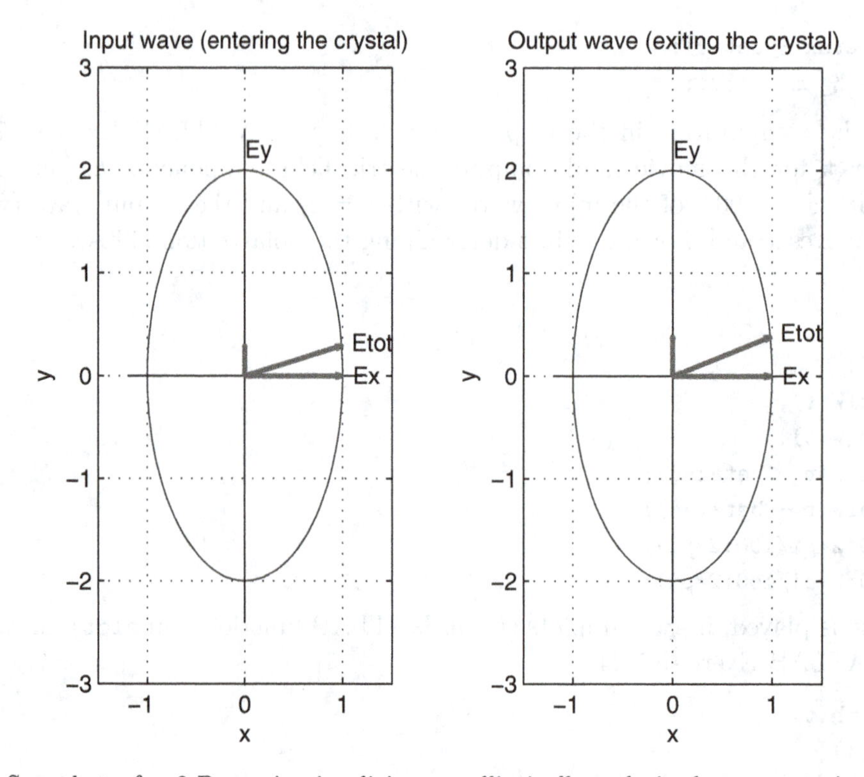

Figure 7.14 Snapshot of a 2-D movie visualizing an elliptically polarized wave entering an anisotropic crystal (subplot on the left) and the EP wave emerging on the other side of it (subplot on the right); for MATLAB Exercise 7.30. *(color figure on CW)*

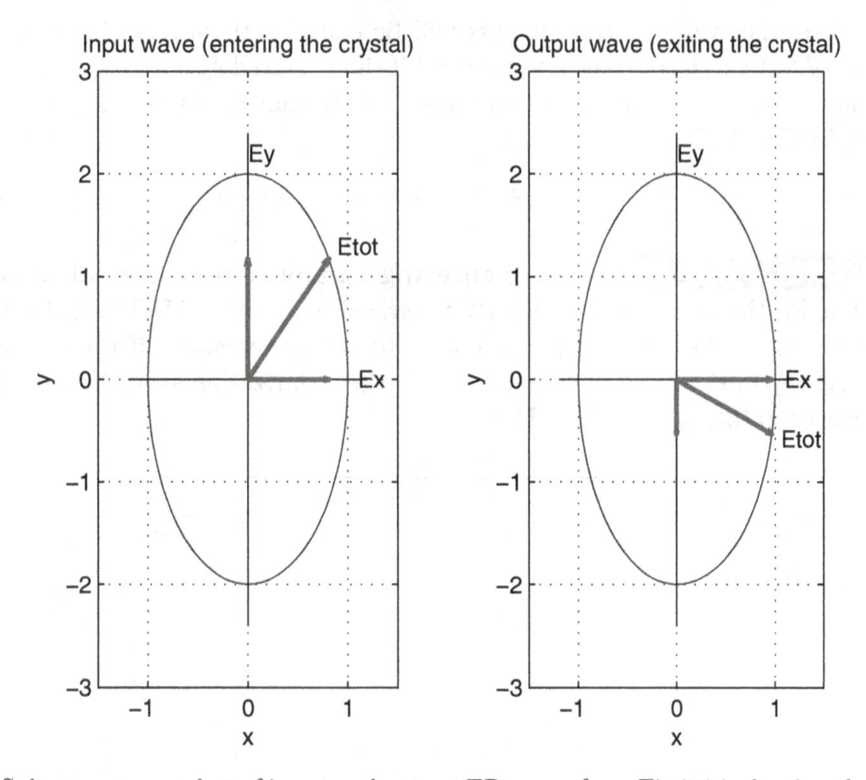

Figure 7.15 Subsequent snapshot of input and output EP waves from Fig.7.14, showing that the two waves are right- and left-handed, respectively; for MATLAB Exercise 7.30. *(color figure on CW)*

to the phase difference between components of the resultant vector **E** for the wave.

MATLAB EXERCISE 7.31 **EP to LP conversion due to material anisotropy.** Repeat the previous MATLAB exercise but assuming that the length of the crystal is such that the output wave is a linearly polarized (LP) one. *(ME7_31.m on IR)*

MATLAB EXERCISE 7.32 **LP to CP conversion by an anisotropic crystal.** Repeat MATLAB Exercise 7.30 but for a linearly polarized time-harmonic uniform plane electromagnetic wave propagating, with the free-space wavelength of $\lambda_0 = 1\ \mu$m, in the positive z direction and entering the crustal. The electric field intensity vector of the wave, **E**, is at 45° to x and y directions. The length of the crystal (along the z axis) is such that the wave emerging on the other side of it is circularly polarized, and out of many possible such lengths the shortest one is chosen. Also, determine the handedness (RH or LH) of the output CP wave. *(ME7_32.m on IR)* **H**

HINT:

The vector **E** of the input wave can be decomposed onto E_x and E_y components that have equal amplitudes (equal to the amplitude of **E** times $\cos 45°$) and are in phase with respect to each other

($\delta = 0$). Therefore, the two field components will be exactly 90° out of phase, which is circular polarization, Eq.(7.24), at the output of the crystal if the acquired δ_a due to material anisotropy, in Eq.(7.28), amounts to $\delta_a = \pi/2$ (the smallest value) – note that $\delta_a > 0$ for the given permittivities of the material, in Eqs.(7.27).

MATLAB EXERCISE 7.33 **CP wave entering an anisotropic crystal.** Repeat MATLAB Exercise 7.30 but for the right-hand circularly polarized wave as in MATLAB Exercise 7.24 (the free-space wavelength of the wave is $\lambda_0 = 1\ \mu\text{m}$) entering the crystal – if the crystal length from the previous MATLAB exercise is doubled. What is the polarization state (type and handedness) of the output wave? *(ME7_33.m on IR)* **H**

---

8 REFLECTION AND TRANSMISSION OF PLANE WAVES

Introduction:

Capitalizing on the concepts and techniques of the analysis of wave propagation in homogeneous and unbounded media of various electromagnetic properties from the previous chapter, we now proceed to develop the concepts and techniques for the analysis of wave interaction with planar boundaries between material regions. In general, as a wave encounters an interface separating two different media, it is partly reflected back to the incident medium (wave reflection) and partly transmitted to the medium on the other side of the interface (wave transmission), and hence the title of this chapter. The material will be presented as several separate cases of reflection and transmission (also referred to as refraction) of plane waves, in order of increasing complexity, from normal incidence (wave propagation direction is normal to the interface) on a perfectly conducting plane and normal incidence on a penetrable interface (between two arbitrary media), to oblique incidence (at an arbitrary angle) on these two types of interfaces, to wave propagation in multilayer media (with multiple interfaces). In all problems, however, the core of the solution will be the use of appropriate general electromagnetic boundary conditions, as a "connection" between the fields on different sides of the interfaces.

8.1 Normal Incidence on a Perfectly Conducting Plane

Consider a uniform plane linearly polarized time-harmonic electromagnetic wave of frequency f and rms electric field intensity $\underline{E}_{\mathrm{i0}}$ propagating through a lossless ($\sigma = 0$) medium of permittivity ε and permeability μ. Let the wave be incident normally on an infinite flat surface (the direction of wave propagation is normal to the surface) of a perfect electric conductor (PEC), with $\sigma \to \infty$, as shown in Fig.8.1(a). Complex electric and magnetic field intensity vectors of the wave, which we refer to as the incident (or forward) wave, can be written as [see Eqs.(7.8)]

$$\underline{\mathbf{E}}_{\mathrm{i}} = \underline{E}_{\mathrm{i0}}\,\mathrm{e}^{-\mathrm{j}\beta z}\,\hat{\mathbf{x}}\,, \quad \underline{\mathbf{H}}_{\mathrm{i}} = \frac{\underline{E}_{\mathrm{i0}}}{\eta}\,\mathrm{e}^{-\mathrm{j}\beta z}\,\hat{\mathbf{y}}\,, \quad \beta = \omega\sqrt{\varepsilon\mu}\,, \quad \eta = \sqrt{\frac{\mu}{\varepsilon}} \quad \text{(incident wave)} \tag{8.1}$$

($\omega = 2\pi f$). This wave excites currents to flow on the PEC surface, which, in turn, are sources of a reflected (or backward) wave, propagating in the negative z direction. Its field vectors are given by

$$\underline{\mathbf{E}}_{\mathrm{r}} = \underline{E}_{\mathrm{r0}}\,\mathrm{e}^{\mathrm{j}\beta z}\,\hat{\mathbf{x}}\,, \quad \underline{\mathbf{H}}_{\mathrm{r}} = \frac{\underline{E}_{\mathrm{r0}}}{\eta}\,\mathrm{e}^{\mathrm{j}\beta z}(-\hat{\mathbf{y}}) \quad \text{(reflected wave)}\,. \tag{8.2}$$

From the boundary condition for the vector $\mathbf{E}$ (more precisely, for its tangential component, $\mathbf{E}_{\mathrm{t}}$) in Eqs.(6.19) applied in the plane $z = 0$, we obtain

$$\underline{E}_{\mathrm{i0}} + \underline{E}_{\mathrm{r0}} = 0 \quad \longrightarrow \quad \underline{E}_{\mathrm{r0}} = -\underline{E}_{\mathrm{i0}} \quad \text{(boundary condition)}\,. \tag{8.3}$$

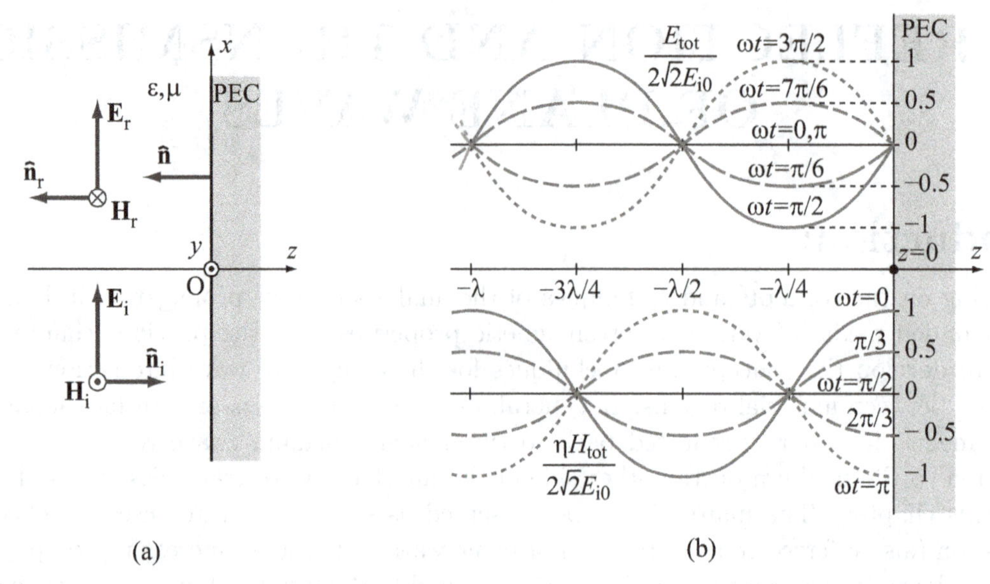

Figure 8.1 (a) Normal incidence of a uniform plane time-harmonic electromagnetic wave on a planar interface between a perfect dielectric and a perfect conductor. (b) Plots of normalized total electric and magnetic field intensities in Eqs.(8.5) against z at different instants of time.

Using Eqs.(8.1)–(8.3) and (6.26), the total complex electric and magnetic fields in the incident medium (for $z \leq 0$), i.e., the field vectors of the resultant wave, are

$$\underline{\mathbf{E}}_{\text{tot}} = -2\text{j}\underline{E}_{\text{i0}} \sin \beta z \, \hat{\mathbf{x}} \,, \quad \underline{\mathbf{H}}_{\text{tot}} = 2\frac{\underline{E}_{\text{i0}}}{\eta} \cos \beta z \, \hat{\mathbf{y}} \quad \text{(resultant wave)} \,, \tag{8.4}$$

and, by means of Eq.(6.27), their time-domain counterparts come out to be

$$\mathbf{E}_{\text{tot}}(t) = 2\sqrt{2}E_{\text{i0}} \sin \beta z \sin \omega t \, \hat{\mathbf{x}} \,, \quad \mathbf{H}_{\text{tot}}(t) = 2\sqrt{2}\,\frac{E_{\text{i0}}}{\eta} \cos \beta z \cos \omega t \, \hat{\mathbf{y}} \quad (\underline{E}_{\text{i0}} = E_{\text{i0}})$$

$$\text{(instantaneous fields of a standing wave)} \,. \tag{8.5}$$

Figure 8.1(b) shows snapshots at different time instants of the resultant field intensities as a function of z. We see that the fields do not travel as the time advances, but stay where they are, only oscillating in time between the stationary zeros. So, they do not represent a traveling wave in either direction. The resultant wave, which is a superposition of two traveling waves, is thus termed a standing wave.

MATLAB EXERCISE 8.1 **3-D plot of field vectors of a plane wave incident on a boundary plane.** Write a function `nEHgraphLabel()` in MATLAB that generates a 3-D plot of the electric field intensity vector ($\mathbf{E}$), magnetic field intensity vector ($\mathbf{H}$), and propagation unit vector ($\hat{\mathbf{n}}$), of a traveling uniform plane electromagnetic wave impinging a boundary plane at normal (or oblique) incidence. The wave propagates in the positive z direction, $\mathbf{E}$ and $\mathbf{H}$ are x- and y-directed vectors, respectively, and the boundary is in the plane defined by $z = 0$. The vectors are plotted at a point (x, y, z). The function places a label next to each of the vectors, like in Fig.8.1(a). It also plots (in 3-D) the boundary plane and labels it with a type of the boundary (e.g., PEC). Note that this function will be used in movies visualizing reflection and transmission of plane waves at both normal and oblique incidences on both PEC and dielectric boundaries. *(nEHgraphLabel.m on IR)*[1]

TUTORIAL:

At input to the function, coordinates of the field point are `X`, `Y`, and `Z`, `n`, `E`, and `H` are three-element arrays (vectors) with x-, y-, and z-components of vectors **n**, **E**, and **H**, and `name` is a string of characters describing the boundary type (e.g. `'PEC'`). Since the MATLAB default x-axis plays the role of the z-axis in our graph, the order of vector components at input is as follows: `n(1)` is n_z (and not n_x), `n(2)` equals n_x, and `n(3)` stands for n_y, and analogously for components of vectors **E**, and **H**. In addition, `minX` and `maxX` define the extent of the graph along the x- and y-axes, while `minZ` and `maxZ` do the same for the z-axis. Finally, parameters `disp1` and `disp2` convey information on whether or not some parts of the graph should be displayed (in a movie).

```
function nEHgraphLabel(Z,X,Y,n,E,H,name,minX,maxX,minZ,maxZ,disp1,disp2)
```

Vectors **n**, **E**, and **H** are plotted using MATLAB function `quiver3` (see MATLAB Exercise 1.3) – in red, green, and blue, respectively, and their labels are written next to the vectors by MATLAB function `text`. If the value of the parameter `disp1` is not unity, these plots are omitted.

```
if disp1 == 1
quiver3(Z,X,Y,n(1),n(2),n(3),0,'r');hold on;
text(Z+n(1)*9/10,X+n(2)*9/10,Y+n(3)*9/10,'n');
quiver3(Z,X,Y,H(1),H(2),H(3),0,'b');
text(Z+H(1)*9/10,X+H(2)*9/10,Y+H(3)*9/10,'H');
quiver3(Z,X,Y,E(1),E(2),E(3),0,'g');
text(Z+E(1)*9/10,X+E(2)*9/10,Y+E(3)*9/10,'E');
end;
```

Next, a mesh of four points defining the boundary plane (rectangular surface), for which $z = 0$, is generated, the plane is plotted using MATLAB function `surf` (see MATLAB Exercise 2.4), and the limits and labels of coordinate axes in the graph are defined (note, again, the order of coordinates, as explained for the components of input vectors); this is omitted if the value of the parameter `disp2` is not unity.

```
if disp2 == 1
[x,y] = meshgrid(minX:(maxX-minX):maxX,minX:(maxX-minX):maxX);
P = x.*0 + y.*0;
surf(P,x,y); alpha(0.3); colormap copper;
text(-1,1.5,1.5,name); hold off;
axis equal; grid off;
xlim([minZ-0.5,maxZ+0.5]);
ylim([minX,maxX]);
zlim([minX,maxX]);
xlabel('z [m]');
```

[1] *IR* = Instructor Resources (for the book).

```
ylabel('x [m]');
zlabel('y [m]');
end;
```

MATLAB EXERCISE 8.2 **3-D movie of normal incidence and reflection at a PEC plane.** Write a MATLAB code to play a 3-D movie of a uniform plane wave that travels through air in the positive z direction and is incident normally on and reflects from a perfectly conducting plane coinciding with the xy-plane ($z = 0$) of the Cartesian coordinate system, as in Fig.8.1(a). In particular, the movie shows the direction (polarization) and constant magnitudes of electric and magnetic field vectors of the incident wave as they move toward the PEC plane [theoretically, with a velocity c_0 in Eq.(7.7)], and of the reflected wave as it travels backward. *(ME8_2.m on IR)*

TUTORIAL:
First, we define the propagation unit vector as $\hat{\mathbf{n}} = \hat{\mathbf{z}}$, having in mind that default MATLAB coordinate axes x, y, and z (for plotting) correspond to z, x, and y in our graph (see the previous MATLAB exercise). Then we define vectors $\mathbf{E}$ and $\mathbf{H}$ as unit vectors $\hat{\mathbf{x}}$ and $\hat{\mathbf{y}}$, as constant-magnitude normalized traveling vectors, again conforming to the changed order of coordinates. As the propagation medium is lossless (air), the field magnitudes do not attenuate during the course of wave travel. We also compute the propagation unit vector and field vectors of the reflected wave, based on Eqs.(8.2) and (8.3).

```
NORMALi = [1,0,0];
Eio = [0,1,0];
Hio = [0,0,1];
NORMALr = -NORMALi;
Ero = -Eio;
Hro = Hio;
```

The spatial limit of movie frames in the z direction, `Zlim`, is specified next, and the time limit, `Tlim`, is introduced as the time that the wave needs to travel, with the velocity $v_p = c_0$ in Eq.(7.7), the distance `Zlim`. The time array of the movie and coordinates (x, y, z) of the propagation path of the wave, i.e., coordinates of points at which the propagation and field vectors are plotted, are defined [note that the z-coordinate in the incident medium in Fig.8.1(a) is negative].

```
Zlim = 2;
Vp = 1/sqrt(EPS0*MU0);
Tlim = Zlim/Vp;
t = 0:Tlim/20:Tlim;
z = Vp.*t-Zlim;
l = length(z);
X = 0;
Y = 0;
```

The movie is made in two `for` loops, one playing the travel of the wave before reflection from the PEC plane (incident wave) and the other showing the reflected wave. The plots in each frame of the movie are generated by function `nEHgraphLabel` (from the previous MATLAB exercise).

Both loop counters run from one to the length of vector `z`, where the points along the propagation path for the reflected wave are in the reversed order. After both loops are completed, one last frame of the movie shows the PEC plane without wave vectors, so the value of the parameter **`disp1`** is set to zero when calling function **`nEHgraphLabel`** in that frame.

```
figure(1)
for i = 1:length(t)
nEHgraphLabel(z(i),X,Y,NORMALi,Eio,Hio,'PEC',-2,2,min(z),max(z),1,1);
M(i) = getframe;
end;
for i = 1:length(t)
nEHgraphLabel(z(l+1-i),X,Y,NORMALr,Ero,Hro,'PEC',-2,2,min(z),max(z),1,1);
M(i) = getframe;
end;
nEHgraphLabel(0,0,0,0,0,0,'PEC',-2,2,min(z),max(z),0,1);
M(i) = getframe;
```

Figures 8.2(a) and (b) show two characteristic snapshots (frames) of the movie.

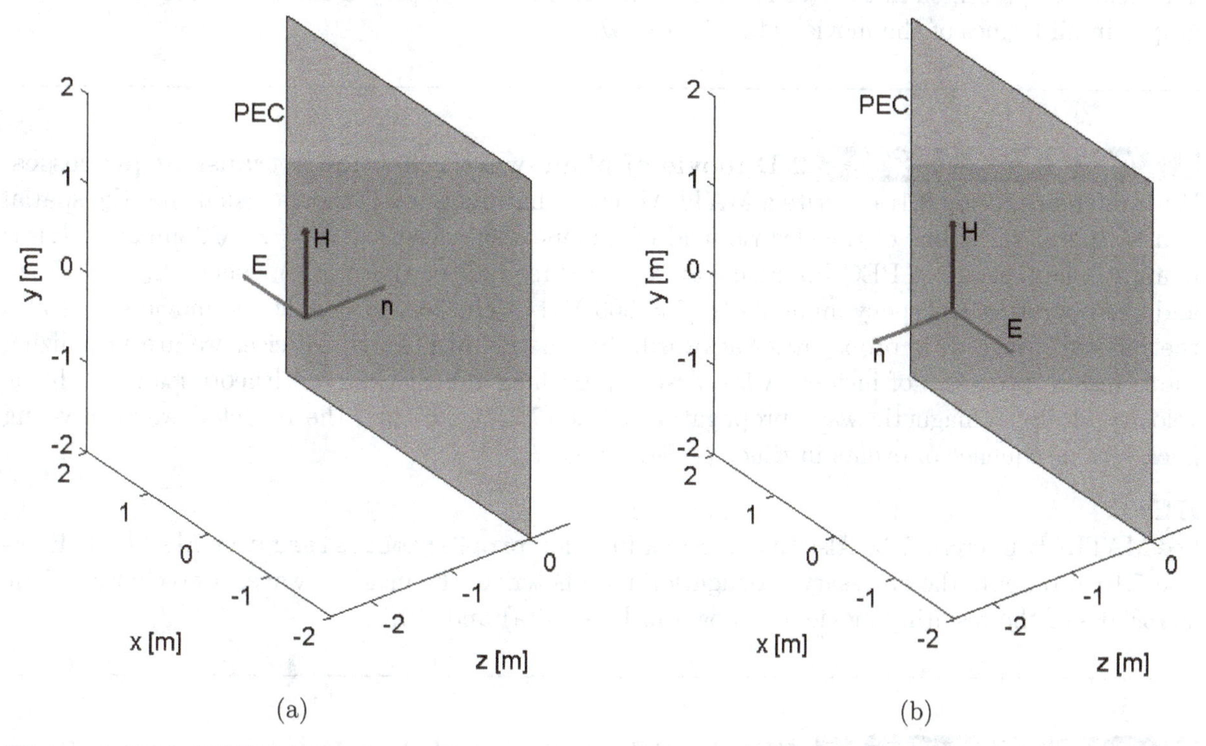

Figure 8.2 3-D MATLAB movie visualizing normal incidence and reflection of a uniform plane wave from a PEC boundary plane [Fig.8.1(a)]: (a) snapshot taken prior to the reflection (incident wave) and (b) snapshot taken upon the reflection (reflected wave); for MATLAB Exercise 8.2. *(color figure on CW)*[2]

[2] *CW* = Companion Website (of the book).

MATLAB EXERCISE 8.3 **2-D movie of fields of a standing wave in steady state.** Create a MATLAB movie that visualizes the instantaneous electric and magnetic field intensities of the resultant standing wave in the material medium in front of the PEC screen in Fig.8.1(a), according to Eqs.(8.5) – see Fig.8.1(b). In particular, the movie shows the wave in the steady state, that is, after all the initial transient processes have already occurred, and the resultant steady-state sinusoidal fields have been established in the entire domain considered. Assume that the incident medium is air, $f = 300$ MHz, and $E_{i0} = 0.1$ V/m. Plot normalized field intensities, $E_{tot}/(2\sqrt{2}E_{i0})$ and $H_{tot}\eta/(2\sqrt{2}E_{i0})$, as in Fig.8.1(b). The movie lasts two time periods of the incident wave, $2T$, and spans a range of two wavelengths along the z-axis, $-2\lambda \leq z \leq 0$. It shows the PEC boundary (vertical line), at $z = 0$. *(ME8_3.m on IR)*

MATLAB EXERCISE 8.4 **2-D movie of incident, reflected, and resultant E fields in steady state.** Write a MATLAB program that plays a 2-D movie of instantaneous electric field intensities of the incident, reflected, and resultant uniform plane waves in air in front of the PEC plane in Fig.8.1(a), for $f = 500$ MHz, $\underline{E}_{i0} = E_{i0}\,e^{j0} = E_{i0} = 1$ V/m, $0 \leq t \leq T$, and $-2\lambda \leq z \leq 0$. The fields are presented in the steady state, with all three waves played simultaneously in the same graph, in all frames of the movie. *(ME8_4.m on IR)*

MATLAB EXERCISE 8.5 **2-D movie of plane-wave reflection – transient processes.** With reference to Fig.8.1(a), write a MATLAB code that plays a 2-D movie visualizing the spatial and temporal variations of the electric field of a plane wave given by $E_i = E_{i0}\sqrt{2}\sin(\omega t - \beta z)$ as it approaches, in air, a PEC boundary at $z = 0$ (first half of the movie), where $E_{i0} = 1$ V/m and the operating frequency amounts to $f = 300$ MHz. The second half of the movie shows also the reflected wave as it propagates backward. So, in this MATLAB exercise, we are visualizing the transient processes of incident wave first approaching the PEC screen [theoretically, with the velocity of electromagnetic wave propagation in Eq.(7.7)], and then the reflected wave traveling back – as a sequence of events in time. *(ME8_5.m on IR)*

HINT:
See MATLAB Exercise 7.3. Also, make use of function `propParamLossless` from MATLAB Exercise 7.1 to compute the necessary propagation parameters of the incident wave. Two characteristic snapshots of the resulting movie are shown in Figs.8.3(a) and (b).

MATLAB EXERCISE 8.6 **GUI for field plots in a Fabry–Perot resonator.** We see in Fig.8.1(b) that there are planes in Fig.8.1(a), defined by $\sin\beta z = 0$, that is, $\beta z = -m\pi$ ($m = 0, 1, 2, \ldots$), in which $E_{tot}(t)$ is zero at all times, so we can insert another PEC surface (foil) in any of these planes, and nothing will change in the entire half-space, because boundary conditions in Eqs.(6.19) are automatically satisfied. We can now remove the field from the region on the left-hand side of the foil, and so obtain a self-contained structure, on the right-hand side of the foil, with a standing electromagnetic plane wave trapped between the two parallel PEC planes

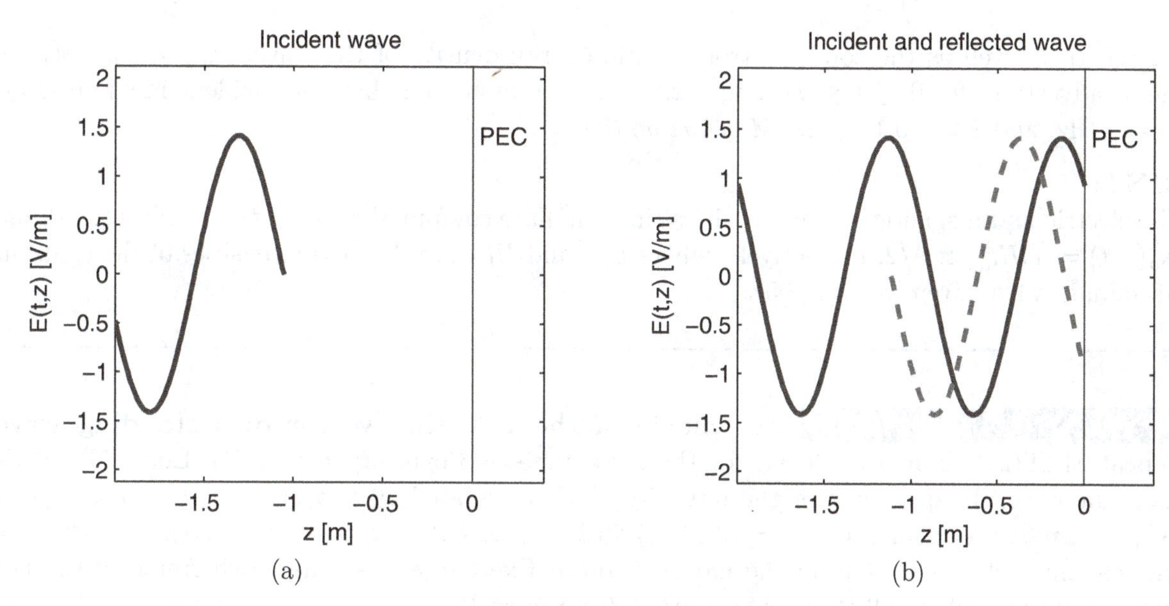

Figure 8.3 2-D MATLAB movie visualizing transient processes in normal incidence on a PEC screen [Fig.8.1(a)]: (a) snapshot taken as the incident wave approaches the screen and (b) snapshot taken as the reflected wave propagates backward; for MATLAB Exercise 8.5. *(color figure on CW)*

(like two mirrors). This structure behaves like an electromagnetic resonator, and is known as the Fabry–Perot resonator. Its resonant frequency, f_{res}, for its given length (separation between the PEC planes), a, is the frequency at which the zero-field condition at $z = -a$, $\beta a = m\pi$, is satisfied. Since $\beta = 2\pi f/c$ [Eq.(7.4)], where c is the velocity of (traveling) electromagnetic waves in the medium of parameters ε and μ, Eq.(7.6), we have

$$\sin\beta a = 0 \quad \longrightarrow \quad f_{\text{res}} = m\,\frac{c}{2a} \quad (m = 1, 2, \ldots) \quad \text{(Fabry–Perot resonator)}\,. \tag{8.6}$$

Create a graphical user interface (GUI) in MATLAB that visualizes the electric field distribution in a Fabry–Perot resonator. In specific, the GUI provides a 2-D plot of the total electric field intensity in the resonator [see Fig.8.1(b)] at a time instant $t = T/4$ (T is the time period of the incident wave), based on Eqs.(8.5), assuming that the dielectric in the resonator is air. Input parameters are the length of the resonator, a, and the mode m (positive integer) of the oscillating wave. *[folder ME8_6(GUI) on IR]*

HINT:
See MATLAB Exercise 2.1 – for the GUI development. Use Eq.(8.6) to compute the resonant frequency, f_{res}, of the resonator, and then use that frequency in the E-field expression in Eqs.(8.5). Display the value of f_{res} in the GUI as well.

MATLAB EXERCISE 8.7 **2-D movie of electric and magnetic energy densities of a standing wave.** Create a movie in MATLAB to visualize the instantaneous electric and magnetic energy densities, w_{e} and w_{m}, of the standing electromagnetic wave in front of the PEC screen in

Fig.8.1(a), as well as the total electromagnetic energy density of the wave, $w_{\rm em} = w_{\rm e} + w_{\rm m}$, in the steady state, for $0 \le t \le 2T$ and $-2\lambda \le z \le 0$, assuming that the incident medium is air, $f = 1$ GHz, and $E_{\rm i0} = 0.1$ V/m. *(ME8_7.m on IR)*

HINT:
The electric and magnetic energy densities (in J/m^3) are computed as $w_{\rm e}(z,t) = \varepsilon_0 E_{\rm tot}^2(z,t)/2$ and $w_{\rm m}(z,t) = \mu_0 H_{\rm tot}^2(z,t)/2$, respectively, where $E_{\rm tot}$ and $H_{\rm tot}$ are the field intensities of the resultant (standing) wave, given by Eqs.(8.5).

MATLAB EXERCISE 8.8 **2-D movie of the Poynting vector of a standing wave.** Repeat MATLAB Exercise 8.3 but for the instantaneous Poynting vector ($\mathcal{P}$), Eq.(6.30), of the resultant wave. In specific, plot the intensity of $\mathcal{P}$, calculated as $\mathcal{P}(z,t) = E_{\rm tot}(z,t)H_{\rm tot}(z,t) = (8E_{\rm i0}^2/\eta)\sin\beta z\cos\beta z\sin\omega t\cos\omega t = (2E_{\rm i0}^2/\eta)\sin 2\beta z\sin 2\omega t$ $(2\sin\alpha\cos\alpha = \sin 2\alpha)$, over the electric and magnetic field plots in the movie made in Exercise 8.3, so that each frame of the new movie contains plots of all three quantities. *(ME8_8.m on IR)*

8.2 Normal Incidence on a Penetrable Planar Interface

We now consider a more general case with the medium on the right-hand side of the interface in Fig.8.1(a) being penetrable for the (normally) incident wave. Moreover, let both media be lossy, as indicated in Fig.8.4. Having in mind Eqs.(8.1), (8.2), (7.12), (7.13), and (7.8), we can write – for the field vectors in Fig.8.4:

$$\underline{\mathbf{E}}_{\rm i} = \underline{E}_{\rm i0}\,{\rm e}^{-\underline{\gamma}_1 z}\,\hat{\mathbf{x}}\,, \quad \underline{\mathbf{H}}_{\rm i} = \frac{\underline{E}_{\rm i0}}{\underline{\eta}_1}\,{\rm e}^{-\underline{\gamma}_1 z}\,\hat{\mathbf{y}} \quad \text{(incident wave)}\,, \tag{8.7}$$

$$\underline{\mathbf{E}}_{\rm r} = \underline{E}_{\rm r0}\,{\rm e}^{\underline{\gamma}_1 z}\,\hat{\mathbf{x}}\,, \quad \underline{\mathbf{H}}_{\rm r} = \frac{\underline{E}_{\rm r0}}{\underline{\eta}_1}\,{\rm e}^{\underline{\gamma}_1 z}(-\hat{\mathbf{y}}) \quad \text{(reflected wave)}\,, \tag{8.8}$$

$$\underline{\mathbf{E}}_{\rm t} = \underline{E}_{\rm t0}\,{\rm e}^{-\underline{\gamma}_2 z}\,\hat{\mathbf{x}}\,, \quad \underline{\mathbf{H}}_{\rm t} = \frac{\underline{E}_{\rm t0}}{\underline{\eta}_2}\,{\rm e}^{-\underline{\gamma}_2 z}\,\hat{\mathbf{y}} \quad \text{(transmitted wave)}\,. \tag{8.9}$$

In Eqs.(8.8) and (8.9), we have two unknown field intensities at $z = 0$, $\underline{E}_{\rm r0}$ and $\underline{E}_{\rm t0}$, so we invoke two boundary conditions – those for tangential components of vectors **E** and **H** in Eqs.(6.18) – to solve for $\underline{E}_{\rm r0}$ and $\underline{E}_{\rm t0}$, for a given $\underline{E}_{\rm i0}$,

$$\underline{E}_{\rm i0} + \underline{E}_{\rm r0} = \underline{E}_{\rm t0}\,, \quad \frac{\underline{E}_{\rm i0}}{\underline{\eta}_1} - \frac{\underline{E}_{\rm r0}}{\underline{\eta}_1} = \frac{\underline{E}_{\rm t0}}{\underline{\eta}_2} \quad (\underline{\mathbf{J}}_{\rm s} = 0) \quad \text{(boundary conditions)}\,, \tag{8.10}$$

where the vector $\underline{\mathbf{J}}_{\rm s}$ is taken to be zero since surface currents in the plane $z = 0$ can only exist if one of the two media is a perfect conductor, like in Fig.8.1(a). The solution of Eqs.(8.10) is expressed in terms of the so-called reflection and transmission coefficients, $\underline{\Gamma}$ and $\underline{\tau}$, as follows:

$$\underline{\Gamma} = \frac{\underline{E}_{\rm r0}}{\underline{E}_{\rm i0}} = \frac{\underline{\eta}_2 - \underline{\eta}_1}{\underline{\eta}_1 + \underline{\eta}_2} \quad \text{(reflection coefficient; dimensionless)}\,, \tag{8.11}$$

$$\underline{\tau} = \frac{\underline{E}_{t0}}{\underline{E}_{i0}} = \frac{2\underline{\eta}_2}{\underline{\eta}_1 + \underline{\eta}_2} \quad \text{(transmission coefficient; dimensionless)}, \tag{8.12}$$

which completes the computation of the reflected and transmitted fields in Fig.8.4.

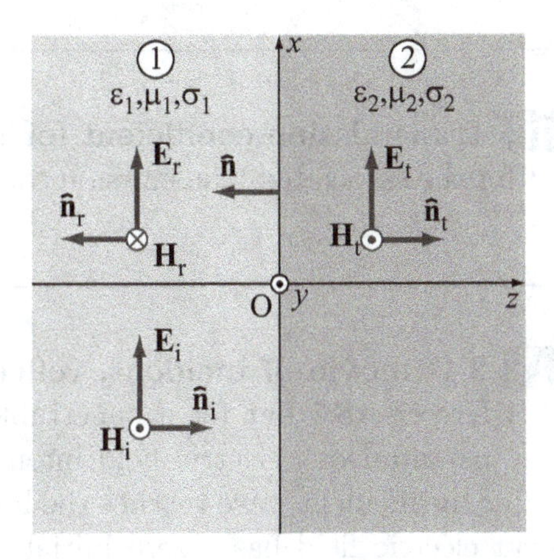

Figure 8.4 Normal incidence of a uniform plane time-harmonic electromagnetic wave on a planar interface between two linear homogeneous media with arbitrary electromagnetic parameters.

MATLAB EXERCISE 8.9 **Reflection coefficient for normal incidence on a penetrable interface.** Write a function `gammaReflCoef()` in MATLAB that computes the reflection coefficient, $\underline{\Gamma}$, for the normal incidence of a uniform plane time-harmonic electromagnetic wave on a planar interface between two media with arbitrary electromagnetic parameters, namely, for the normal incidence from a medium with complex intrinsic impedance $\underline{\eta}_1$ to a medium with impedance $\underline{\eta}_2$, as shown in Fig.8.4. Note that `gamma()` cannot be used for the name of this function, since it is reserved for an existing MATLAB function. In addition to $\underline{\eta}_1$ (`eta1`) and $\underline{\eta}_2$ (`eta2`), the input to the function contains a character `k` that specifies the format of the output $\underline{\Gamma}$: `'r'` for the rectangular (algebraic) form and `'p'` for the polar (exponential) form [Eqs.(6.24)], while `'d'` for $|\underline{\Gamma}|$ in decibels [Eq.(7.16)]. *(gammaReflCoef.m on IR)*

TUTORIAL:

Based on Eqs.(8.11), (6.24), and (7.16), the reflection coefficient $\underline{\Gamma}$ (or $\Gamma_{\text{dB}} = 20 \log |\underline{\Gamma}|$) is computed as follows (note that for `k` equaling `'r'` or `'d'`, the output parameter `result2` is not used, whereas it represents the phase angle ψ of $\underline{\Gamma} = |\underline{\Gamma}| e^{j\psi}$ if `k` equals `'p'`):

```
function [result1,result2] = gammaReflCoef(eta1,eta2,k)
if k == 'r';
result1 = (eta2 - eta1)/(eta1 + eta2);
result2 = ' ';
end;
if k == 'p';
result1 = abs((eta2 - eta1)/(eta1 + eta2));
result2 = phase((eta2 - eta1)/(eta1 + eta2));
end;
if k == 'd';
```

```
result1 = 20*log10(abs((eta2 - eta1)/(eta1 + eta2)));
result2 = ' ';
end;
```

MATLAB EXERCISE 8.10 **Transmission coefficient for normal incidence.** Repeat the previous MATLAB exercise but for the associated transmission coefficient, $\underline{\tau}$, in Eq.(8.12) [function `tau()`]. *(tau.m on IR)*

MATLAB EXERCISE 8.11 **2-D movie of incident, reflected, and transmitted transient fields.** Repeat MATLAB Exercise 8.5 but for a penetrable material interface, in Fig.8.4. In specific, the movie plays the instantaneous electric field intensities of incident, reflected, and transmitted waves before and after the incident wave reaches the boundary – as transient processes, for $\underline{E}_{\rm i0} = E_{\rm i0} = 1$ V/m (incident electric field has a zero initial phase at $z = 0$), $f = 300$ MHz, $\varepsilon_{\rm r1} = 4$, $\varepsilon_{\rm r2} = 9$, $\mu_1 = \mu_2 = \mu_0$, and $\sigma_1 = \sigma_2 = 0$. *(ME8_11.m on IR)*

TUTORIAL:
We first specify all input parameters:

```
EPS0 = 8.8542*10^(-12);
MU0 = 4*pi*10^(-7);
Ei0 = 1;
THETA0 = 0;
f = 300*10^(6);
w = 2*pi*f;
EPSR1 = 4;
MUR1 = 1;
SIGMA1 = 0;
EPSR2 = 9;
MUR2 = 1;
SIGMA2 = 0;
deg2rad = pi/180;
THETA0rad = THETA0*deg2rad;
```

We then calculate necessary propagation parameters of the waves and the two media, calling function `basicPropParam` (from MATLAB Exercise 7.9). We also find the reflection and transmission coefficients, using functions `gammaReflCoef` and `tau` (from the previous two MATLAB exercises), as well as the complex rms electric field intensities of the reflected and transmitted waves, respectively, in the plane $z = 0$, $\underline{E}_{\rm r0}$ and $\underline{E}_{\rm t0}$, based on Eqs.(8.11) and (8.12).

```
[alpha1,beta1,absEta1,phi1] = basicPropParam(EPSR1,MUR1,SIGMA1,f);
[alpha2,beta2,absEta2,phi2] = basicPropParam(EPSR2,MUR2,SIGMA2,f);
ETA1 = absEta1*exp(i*phi1);
ETA2 = absEta2*exp(i*phi2);
lambda1 = 2*pi/beta1;
```

```
lambda2 = 2*pi/beta2;
GAMMA = gammaReflCoef(ETA1,ETA2,'r');
TAU = tau(ETA1,ETA2,'r');
Er0 = GAMMA*Ei0*exp(i*THETA0rad);
Et0 = TAU*Ei0*exp(i*THETA0rad);
```

Next, we compute the spatial limit of movie frames, `Zlim`, as the maximum of the wavelengths in the two media, and from `Zlim`, the spatial mesh in the first medium, vector `z`. The mesh for the second medium is defined by `-z`. Constants `T1` and `T2` are introduced as times that the wave needs to travel distance `Zlim` trough the first and the second medium, respectively, and based on them, vectors `t1` and `t2` defining temporal meshes for playing the two parts of the movie, before and after the instant when the incident wave reaches the material interface, are filled. Also, amplitudes (peak-values) of incident, reflected, and transmitted electric fields are found.

```
Zlim = max(lambda1,lambda2);
dz = Zlim/100;
z = [-Zlim:dz:0];
T1 = Zlim*sqrt(EPS0*EPSR1*MU0*MUR1);
T2 = Zlim*sqrt(EPS0*EPSR2*MU0*MUR2);
dt = max(T1,T2)/200;
t1 = [0:dt:T1];
t2 = [0:dt:2*T2];
Eim = Ei0*sqrt(2);
Erm = abs(Er0)*sqrt(2);
Etm = abs(Et0)*sqrt(2);
Emax = max([Eim,Erm,Etm]);
Elim = Emax + Emax/2;
L = [-Elim:Elim];
boundary = zeros(1,length(L));
```

The movie is made in two `for` loops. The first loop plays the travel of the incident wave (in the first medium) before it reaches the boundary; in the loop, the electric field of the wave is evaluated at an instant `t1(i)` at all points reached by the wave, implementing the time-domain version of Eq.(8.7). The last point in the medium reached by the wave, past which there is no field, is determined using MATLAB functions `min` and `heaviside` in the same way as in MATLAB Exercise 7.3.

```
for i = 1:length(t1);
Ei = Eim.*exp(-alpha1*z).*cos(w*t1(i) - beta1.*z + THETA0rad).*heaviside(w*t1(i) - ...
beta1.*(z+Zlim));
[errori, ni] = min(heaviside(w*t1(i) - beta1.*(z+Zlim)));
plot(z(1:ni-2), Ei(1:ni-2), 'b','linewidth',2); hold on;
plot(boundary,L,'k'); hold off;
M(i) = getframe;
end;
```

Similarly, the second loop plays the travel of the incident and reflected waves in the first medium and the transmitted wave in the second medium, after the incident wave has reached the boundary, based on Eqs.(8.7)–(8.9), again plotting the fields of the reflected and transmitted waves only up to points they actually reach at an instant `t2(i)+T1` in the corresponding frame of the movie. These points are determined combining MATLAB functions `max` and `heaviside` again as in MATLAB Exercise 7.3.

```
for i = 1:length(t2);
Ei = Eim.*exp(-alpha1*z).*cos(w*(t2(i)+T1) - beta1.*z + THETA0rad);
Er = Erm.*exp(alpha1*z).*cos(w*(t2(i)+T1) + beta1.*z + phase(Er0)).*heaviside(w*t2(i)...
+ beta1.*(z));
Et = Etm.*exp(-alpha2*(-z)).*cos(w*(t2(i)+T1) - beta2.*(-z) + phase(Et0))...
.*heaviside(w*t2(i) - beta2.*(-z));
[errorr, nr] = max(heaviside(w*t2(i) + beta1.*(z)));
[errort, nt] = max(heaviside(w*t2(i) - beta2.*(-z)));
plot(z, Ei,'b','linewidth',2); hold on;
plot(z(nr:length(z)), Er(nr:length(z)),'r--','linewidth',2);
plot(-z(nt:length(z)), Et(nt:length(z)),'g-.','linewidth',2);
plot(boundary,L,'k'); hold off;
M(i) = getframe;
end;
```

(note that line specifiers `'b'`, `'r--'`, and `'g-.'` make `Ei`, `Er`, and `Et` plotted as blue solid, red dashed, and green dash-dot lines, respectively).

Figures 8.5 and 8.6 show two characteristic frames of the movie.

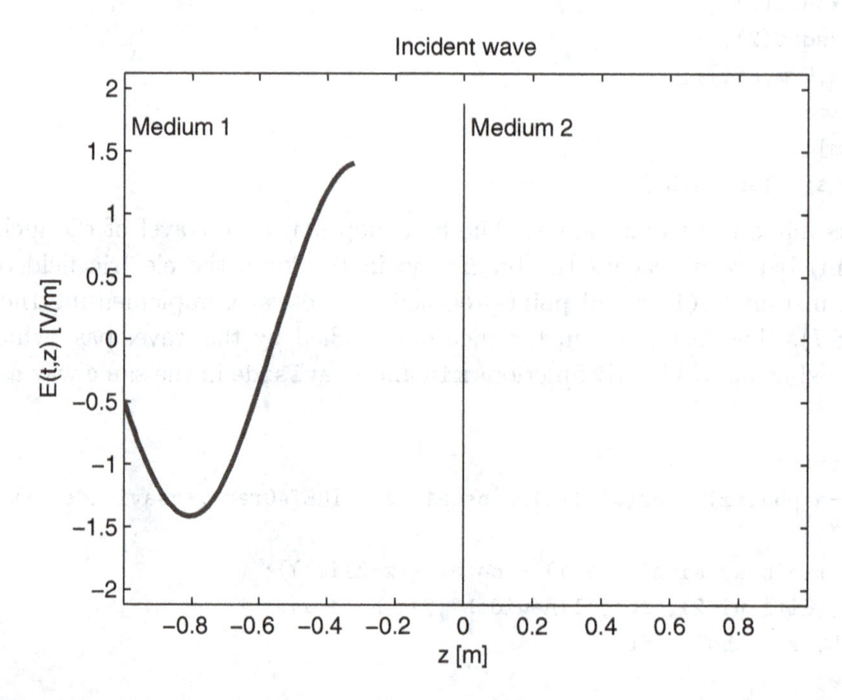

Figure 8.5 2-D MATLAB movie visualizing transient processes in normal incidence on an interface between two lossless dielectric media (Fig.8.4): snapshot taken as the incident wave approaches the interface; for MATLAB Exercise 8.11. *(color figure on CW)*

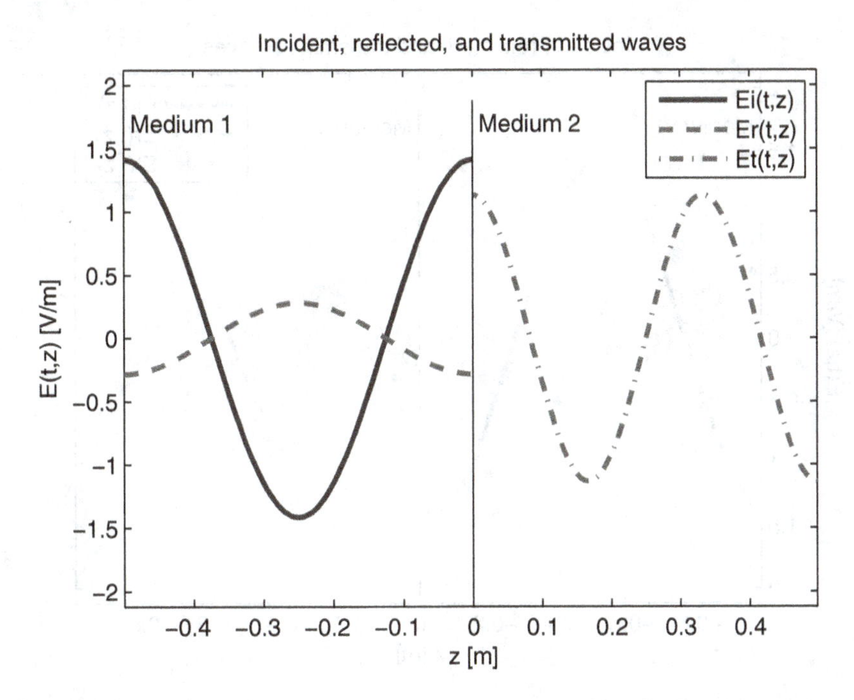

Figure 8.6 2-D MATLAB movie visualizing normal incidence on a dielectric–dielectric interface: snapshot taken after all three waves have reached all points in the domain considered; for MATLAB Exercise 8.11. *(color figure on CW)*

MATLAB EXERCISE 8.12 **Normal incidence on the interface of a lossy medium.** Repeat the previous MATLAB exercise but for the following combination of two media in Fig.8.4: medium 1 is air and medium 2 is wet soil with $\varepsilon_{r2} = 5$ and $\sigma_2 = 10^{-2}$ S/m ($\mu_2 = \mu_0$). *(ME8_12.m on IR)*

HINT:

Figure 8.7 shows a snapshot of the resulting movie taken after incident, reflected, and transmitted waves have reached all points in the domain considered.

MATLAB EXERCISE 8.13 **3-D MATLAB movie of normal incidence, reflection, and transmission.** Repeat MATLAB Exercise 8.2 but to visualize in a 3-D movie the propagation and field vectors of incident, reflected, and transmitted waves in a two-media structure in Fig.8.4 with air as the incident medium and a lossless nonmagnetic dielectric with relative permittivity $\varepsilon_{r2} = 5$ as the second medium, for an operating frequency of $f = 3$ GHz. *(ME8_13.m on IR)*

HINT:

The propagation unit vectors can be determined as in MATLAB Exercise 8.2. The propagation parameters, reflection and transmission coefficients, and complex rms electric field intensities of the reflected and transmitted waves in the plane $z = 0$ ($\underline{E}_{r0}$ and $\underline{E}_{t0}$) can be calculated as in MATLAB

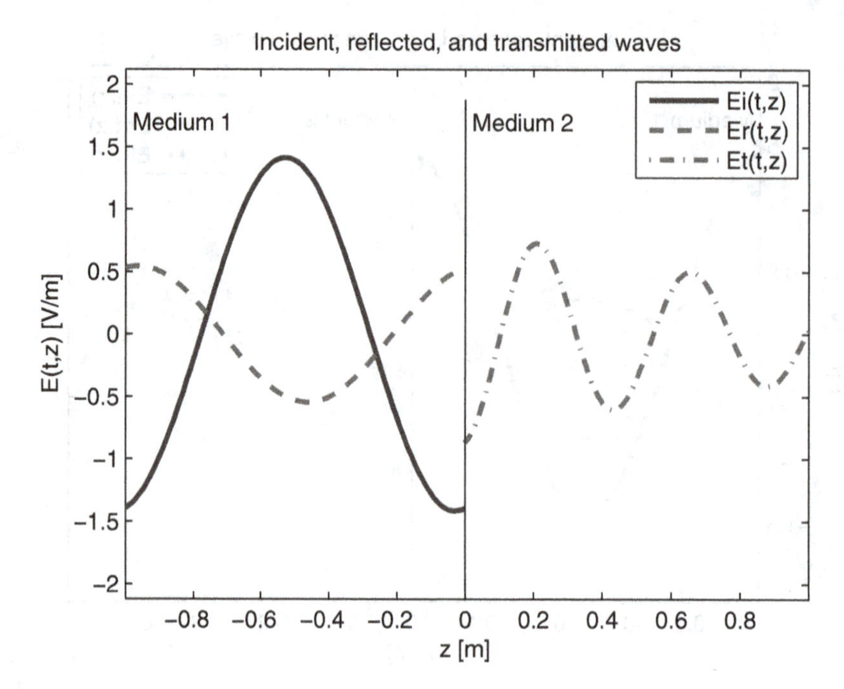

Figure 8.7 The same as in Fig.8.6 but for an interface between air and lossy dielectric (wet soil); for MATLAB Exercise 8.12. *(color figure on CW)*

Exercise 8.11. The z-coordinates of the propagation path of the wave, i.e., z-coordinates of points at which the propagation and field vectors are plotted (see MATLAB Exercise 8.2), should be calculated taking into account different phase velocities of the waves in the two media, `Vp1` and `Vp2`, as follows:

```
Zlim = 3;
Tlim = Zlim/Vp1;
t = 0:Tlim/20:Tlim;
z1 = Vp1.*t-Zlim;
z2 = Vp2.*t;
```

Figures 8.8(a) and (b) show two characteristic frames of the resulting movie.

MATLAB EXERCISE 8.14 **Array factor for waves in the complex plane – movie.** A combination of Eqs.(8.7), (8.8), (8.11), and $\underline{\Gamma} = |\underline{\Gamma}|\,\mathrm{e}^{\mathrm{j}\psi}$ gives the following expression for the total electric field intensity vector in the incident medium (medium 1) in Fig.8.4, assuming that it is lossless ($\alpha_1 = 0$, $\gamma_1 = \mathrm{j}\beta_1$):

$$\underline{\mathbf{E}}_{\mathrm{tot1}} = \underline{\mathbf{E}}_{\mathrm{i}} + \underline{\mathbf{E}}_{\mathrm{r}} = \underline{E}_{\mathrm{i0}}\,\hat{\mathbf{x}}\left(\mathrm{e}^{-\mathrm{j}\beta_1 z} + \underline{\Gamma}\,\mathrm{e}^{\mathrm{j}\beta_1 z}\right) = \underbrace{\underline{E}_{\mathrm{i0}}\,\mathrm{e}^{-\mathrm{j}\beta_1 z}\,\hat{\mathbf{x}}}_{\text{Incident wave}}\underbrace{\left(1 + |\underline{\Gamma}|\,\mathrm{e}^{\mathrm{j}(2\beta_1 z+\psi)}\right)}_{\text{Array factor, }\underline{F}_{\mathrm{a}}}, \tag{8.13}$$

in which the field intensity of the incident wave is multiplied by a factor that we symbolize by $\underline{F}_{\mathrm{a}}$ and refer to as the array factor for waves, in analogy with the analysis of antenna arrays. Write a

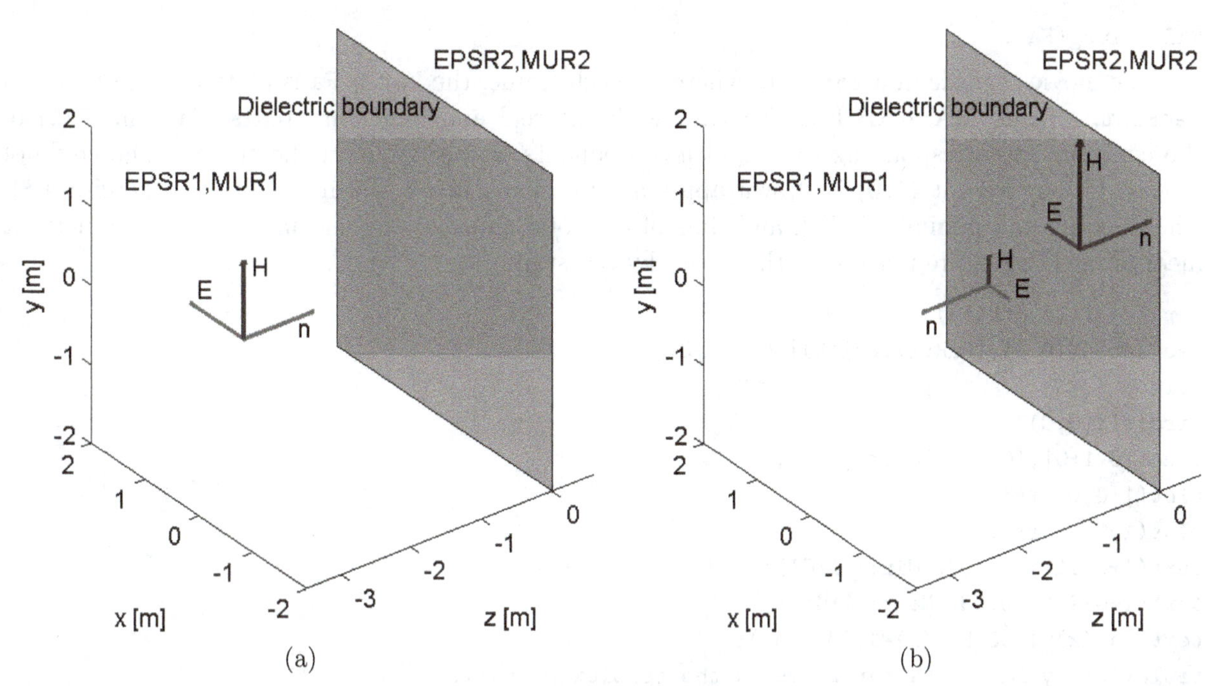

Figure 8.8 3-D MATLAB movie visualizing normal incidence, reflection, and transmission of a uniform plane wave in a two-media structure (Fig.8.4): (a) snapshot taken prior to the reflection (incident wave) and (b) snapshot taken upon the reflection/transmission (reflected and transmitted waves); for MATLAB Exercise 8.13. *(color figure on CW)*

MATLAB code that plays a movie visualizing $\underline{F}_{\mathrm{a}}$ in the complex plane. The movie demonstrates that, as the coordinate z varies from $z = 0$ to, theoretically, $z \to -\infty$, the tip of the vector $\underline{F}_{\mathrm{a}}$ traces a circle of radius $|\underline{\Gamma}|$, centered at $(1, 0)$ in the complex plane; the tip of $\underline{F}_{\mathrm{a}}$ (point representing complex number $\underline{F}_{\mathrm{a}}$) rotates in the mathematically negative (clockwise) direction. In specific, assume that the incident medium is air, $\beta_1 = 0.2\pi$ rad/m, $|\underline{\Gamma}| = 1/3$, and $\psi = \pi$. Also, the movie shows two full rotations of point $\underline{F}_{\mathrm{a}}$ on the circle. *(ME8_14.m on IR)*

TUTORIAL:

After specifying the input parameters, the array of coordinates **z** is defined such that the tip of the vector $\underline{F}_{\mathrm{a}}$ traces the circle in the complex plane twice. The complex quantity $\underline{F}_{\mathrm{a}}$ (`Fa`) is computed based on Eq.(8.13).

```
G = 1/3;
psi = pi;
BETA = 0.2*pi;
LAMBDA = 2*pi/BETA;
Vp = 1/sqrt(EPS0*MU0);
Tlim = LAMBDA/Vp;
t = 0:Tlim/500:Tlim;
z = Vp*t-LAMBDA;
Fa = 1+G*exp(i*(2*BETA*z+psi));
FaR = real(Fa);
```

```
FaI = imag(Fa);
```

The movie is made in a `for` loop, where, in each frame, the vector `Fa` is plotted using function `vecPlot2D` (from MATLAB Exercise 1.2), with the real and imaginary parts (`FaR` and `FaI`) of the complex `Fa` corresponding to x- and y-components, respectively, of the vector. The circle of radius $|\underline{\Gamma}|$ centered at $(1,0)$ is drawn applying function `circle` (from MATLAB Exercise 4.8). The maxima and minima of $|\underline{F}_{\text{a}}|$, and thus of the total rms electric field intensity in the incident medium in Fig.8.4, are marked in the graph by red stars.

```
for k = 1:length(Fa)
vecPlot2D([0 0],[FaR(k),FaI(k)],0,'r',0);
hold on;
circle(1,0,G);
line([0 1+G],[0 0],'Color','k');
plot(1-G,0,'r*');
plot(1+G,0,'r*');
text(1-G*11/10,0.03,'Min:  1-G');
text(1+G*9/10,0.03,'Max:  1+G');
text(FaR(k)*1.03,FaI(k)*1.03,'Fa');
title('Array factor Fa for waves in the complex plane');
xlabel('Re');
ylabel('j*Im');
hold off;
M(k) = getframe;
end;
```

Figure 8.9 shows a frame of the movie.

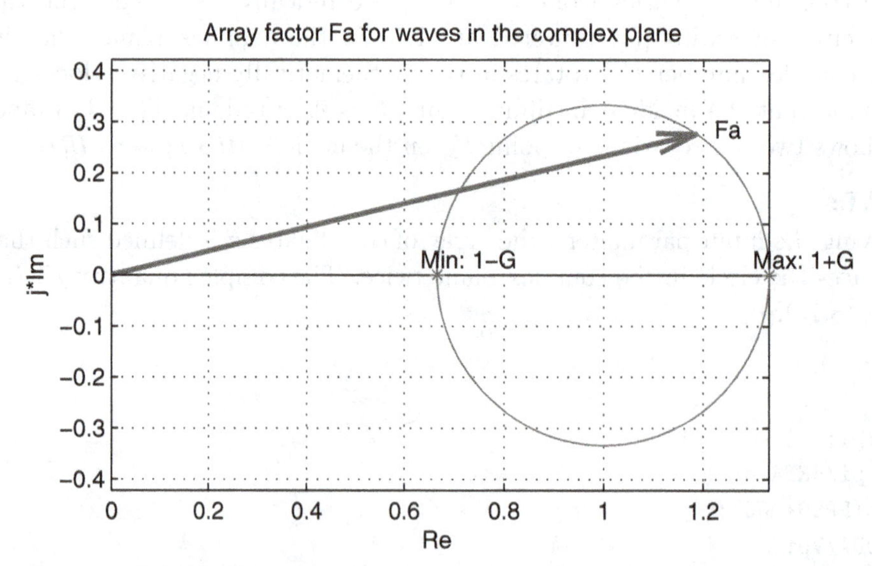

Figure 8.9 Snapshot of a MATLAB movie visualizing – in the complex plane – the dependence on the coordinate z in Fig.8.4 of the array factor ($\underline{F}_{\text{a}}$) for waves in Eq.(8.13); for MATLAB Exercise 8.14. *(color figure on CW)*

MATLAB EXERCISE 8.15 **Standing wave ratio.** From Eq.(8.13) and Fig.8.9, the ratio of maxima to minima of the total electric field in the incident medium in Fig.8.4, called the standing wave ratio (SWR), equals

$$s = \frac{|\mathbf{E}_{\mathrm{tot1}}|_{\max}}{|\mathbf{E}_{\mathrm{tot1}}|_{\min}} = \frac{|\underline{F}_{\mathrm{a}}|_{\max}}{|\underline{F}_{\mathrm{a}}|_{\min}} = \frac{1+|\underline{\Gamma}|}{1-|\underline{\Gamma}|} \quad \text{(standing wave ratio – SWR)} . \tag{8.14}$$

In MATLAB, compute and plot s for the case of both media in Fig.8.4 being lossless ($\underline{\Gamma}$ being purely real) – for the range of Γ given by $-1 < \Gamma < 1$. *(ME8_15.m on IR)* **H**[3]

MATLAB EXERCISE 8.16 **Wave impedance.** This MATLAB exercise introduces a new quantity, equal to the ratio of the total electric to magnetic complex field intensities at a point in space. It is called the wave impedance and is denoted by $\underline{\eta}_{\mathrm{w}}$. With this definition and the use of Eqs.(8.7) and (8.8), $\underline{\eta}_{\mathrm{w}}$ in the incident medium in Fig.8.4 comes out to be

$$\underline{\eta}_{\mathrm{w}} = \frac{\underline{E}_{\mathrm{tot1}}}{\underline{H}_{\mathrm{tot1}}} = \frac{\underline{E}_{\mathrm{i}}+\underline{E}_{\mathrm{r}}}{\underline{H}_{\mathrm{i}}+\underline{H}_{\mathrm{r}}} = \underline{\eta}_1 \frac{\mathrm{e}^{-\underline{\gamma}_1 z}+\underline{\Gamma}\,\mathrm{e}^{\underline{\gamma}_1 z}}{\mathrm{e}^{-\underline{\gamma}_1 z}-\underline{\Gamma}\,\mathrm{e}^{\underline{\gamma}_1 z}} \quad \text{(wave impedance; unit: } \Omega) . \tag{8.15}$$

Write a function `waveImpedance()` in MATLAB that calculates $\underline{\eta}_{\mathrm{w}}$. Then, use this function to find and plot the magnitude and phase angle of $\underline{\eta}_{\mathrm{w}}$ for $\varepsilon_{r1} = 2.25$, $\varepsilon_{r2} = 4$, $\sigma_1 = 1.5 \times 10^{-12}$ S/m, $\sigma_2 = 10^{-4}$ S/m, $\mu_{r1} = \mu_{r2} = 1$, and $f = 300$ MHz in the range $-\lambda_1 \le z \le 0$ ($\lambda_1 = 2\pi/\beta_1$). *(waveImpedance.m on IR)*

8.3 Oblique Incidence on a Perfect Conductor

In this section, we generalize the analysis of plane-wave reflections upon perfectly conducting surfaces for the normal incidence on the surface [Fig.8.1(a)] to the case of an arbitrary, oblique, incidence. Namely, we now let an incident uniform plane time-harmonic electromagnetic wave approach the PEC boundary at an arbitrary angle, so-called incident angle, θ_{i} ($0 \le \theta_{\mathrm{i}} \le 90°$) with respect to the normal on the boundary (for the normal incidence, $\theta_{\mathrm{i}} = 0$). Furthermore, let the incident electric field intensity vector, $\mathbf{E}_{\mathrm{i}}$, be normal to the plane of incidence, defined by the direction of incident-wave propagation (incident ray) and normal on the boundary, as shown in Fig.8.10(a). An obliquely incident wave with such orientation of $\mathbf{E}_{\mathrm{i}}$ is said to be normally (or perpendicularly) polarized. The other characteristic case, with $\mathbf{E}_{\mathrm{i}}$ lying in the plane of incidence (and, of course, being perpendicular to the direction of wave travel), is depicted in Fig.8.10(b). It is referred to as the parallel polarization of the incident wave. Note that the normal (perpendicular) and parallel polarizations of an incident wave, in Fig.8.10(a) and (b), are sometimes referred to as the horizontal and vertical polarizations, respectively. This comes from a frequent situation where the reflection plane is the earth's surface, with the case in Fig.8.10(a) corresponding to a horizontal $\mathbf{E}_{\mathrm{i}}$ and the case in Fig.8.10(b) to $\mathbf{E}_{\mathrm{i}}$ in a vertical plane. In addition, these two cases are sometimes

---

[3]**H** = recommended to be done also "by hand," i.e., not using MATLAB.

also labeled, respectively, as s- and p-polarization ("s" being an abbreviation for the German *senkrecht*, meaning normal, and "p" for the German word for parallel, which is, also, *parallel*). Finally, some texts use TE versus TM polarization terminology, which refers to the electric field, in case (a), versus magnetic field, in case (b), being transverse to the observation direction parallel to the PEC interface in Fig.8.10. Using the field expression for an arbitrarily directed plane wave in Eqs.(7.10), for both the incident and reflected waves in Fig.8.10, and applying, in the plane $z = 0$, the boundary condition for the vector $\mathbf{E}$ in Eqs.(6.19), we obtain

$$\theta_\mathrm{r} = \theta_\mathrm{i} \,, \quad \underline{E}_{\mathrm{r}0} = -\underline{E}_{\mathrm{i}0} \quad \text{(oblique incidence on a PEC)} \,, \tag{8.16}$$

for both wave polarizations, where the first relationship is known as Snell's law of reflection.

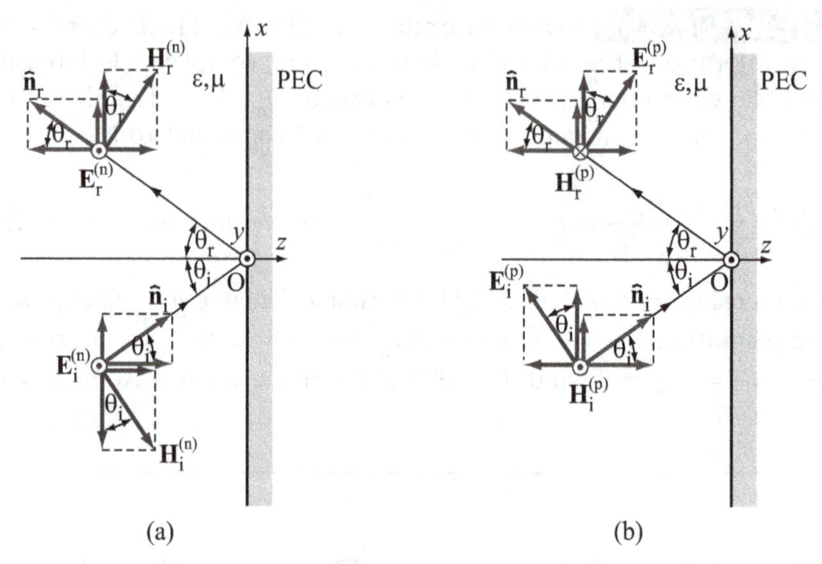

Figure 8.10 Oblique incidence, on a planar perfect electric conductor, of a uniform plane time-harmonic electromagnetic wave with (a) normal (perpendicular) polarization and (b) parallel polarization. The incident medium is a perfect dielectric, and the plane of drawing is also the plane of incidence.

MATLAB EXERCISE 8.17 **3-D movie of oblique incidence, normal polarization, on a PEC plane.** Repeat MATLAB Exercise 8.2 but for an oblique incidence on a PEC plane of a uniform plane wave with normal polarization, as in Fig.8.10(a), assuming that the incident medium is air and the incident angle is $\theta_\mathrm{i} = 60°$. *(ME8_17.m on IR)*

TUTORIAL:

The reflected angle in Fig.8.10(a), θ_r, is obtained from Snell's law of reflection, in Eqs.(8.16). The Cartesian x- and z-components of propagation unit vectors $\hat{\mathbf{n}}_\mathrm{i}$ and $\hat{\mathbf{n}}_\mathrm{r}$ of the incident and reflected waves (their y-components are zero) are computed based on the decompositions shown in Fig.8.10(a) and the fact that the order of coordinates is changed to z, x, and y, as in MATLAB Exercise 8.2.

```
THETAi = 60*pi/180;
THETAr = THETAi;
```

```
NORMAL = 1;
NORMALiz = NORMAL*cos(THETAi);
NORMALrz = -NORMALiz;
NORMALix = NORMAL*sin(THETAi);
NORMALrx = NORMALix;
NORMALi = [NORMALiz,NORMALix,0];
NORMALr = [NORMALrz,NORMALrx,0];
```

Cartesian components of the electric and magnetic field vectors of the incident and reflected waves, $\underline{\mathbf{E}}_{\mathrm{i}}^{(\mathrm{n})}$, $\underline{\mathbf{H}}_{\mathrm{i}}^{(\mathrm{n})}$, $\underline{\mathbf{E}}_{\mathrm{r}}^{(\mathrm{n})}$, and $\underline{\mathbf{H}}_{\mathrm{r}}^{(\mathrm{n})}$, are obtained from Fig.8.10(a) and the second relationship in Eqs.(8.16),

```
Eio = 1;
Hio = 1;
Hioz = Hio*sin(THETAi);
Hiox = -Hio*cos(THETAi);
Ero = -Eio;
Hroz = -Hioz;
Hrox = Hiox;
Ei = [0,0,Eio];
Hi = [Hioz,Hiox,0];
Er = [0,0,Ero];
Hr = [Hroz,Hrox,0];
```

Next, similarly to the computation in MATLAB Exercise 8.2, we specify a spatial limit of movie frames along the z-axis, `Zlim`, a temporal mesh for playing the movie, `t`, and coordinates (x, y, z) of propagation paths of the incident and reflected waves in Fig.8.10(a),

```
Zlim = 1;
Vpz = cos(THETAi)/sqrt(EPS0*MU0);
Tlim = Zlim/Vpz;
t = 0:Tlim/20:Tlim;
z = Vpz.*t-Zlim;
l = length(z);
Xi = z.*tan(THETAi);
Xr = 0-z.*tan(THETAr);
Y = 0;
```

The movie is implemented in a similar fashion to that in MATLAB Exercise 8.2:

```
figure(1)
for i = 1:length(z)
nEHgraphLabel(z(i),Xi(i),Y,NORMALi,Ei,Hi,'PEC',-2,2,min(z),max(z),1,1);
M(i) = getframe;
end;
for i = 1:length(z)
nEHgraphLabel(z(l+1-i),Xr(l+1-i),Y,NORMALr,Er,Hr,'PEC',-2,2,min(z),max(z),1,1);
M(i) = getframe;
end;
```

```
nEHgraphLabel(0,0,0,0,0,0,'PEC',-2,2,min(z),max(z),0,1);
M = getframe;
```

MATLAB EXERCISE 8.18 **3-D movie of oblique incidence on a PEC – parallel polarization.** Repeat the previous MATLAB exercise but for parallel polarization of the incident wave, as in Fig.8.10(b). *(ME8_18.m on IR)*

8.4 Oblique Incidence on a Dielectric Boundary

If the medium on the right-hand side of the interface in Fig.8.10 is another lossless dielectric, a part of the incident energy will be transmitted through the interface, like in Fig.8.4 for the normal incidence case. In Fig.8.11, $\theta_\mathrm{r} = \theta_\mathrm{i}$, as in Eqs.(8.16), while the transmitted (refracted) angle, θ_t, is determined by Snell's law of refraction:

$$\frac{\sin\theta_\mathrm{i}}{\sin\theta_\mathrm{t}} = \frac{c_1}{c_2} = \frac{\beta_2}{\beta_1} = \sqrt{\frac{\varepsilon_2\mu_2}{\varepsilon_1\mu_1}} = \frac{n_2}{n_1} \quad \text{(Snell's law of refraction)}\,, \tag{8.17}$$

with c_1 and c_2 being the velocities [Eq.(7.6)] and β_1 and β_2 the phase coefficients [Eq.(7.4)] of waves in media 1 and 2, respectively, and $n_1 = c_0/c_1 = \sqrt{\varepsilon_{1\mathrm{r}}\mu_{1\mathrm{r}}}$ and $n_2 = c_0/c_2 = \sqrt{\varepsilon_{2\mathrm{r}}\mu_{2\mathrm{r}}}$ standing for the indices of refraction of the two media.

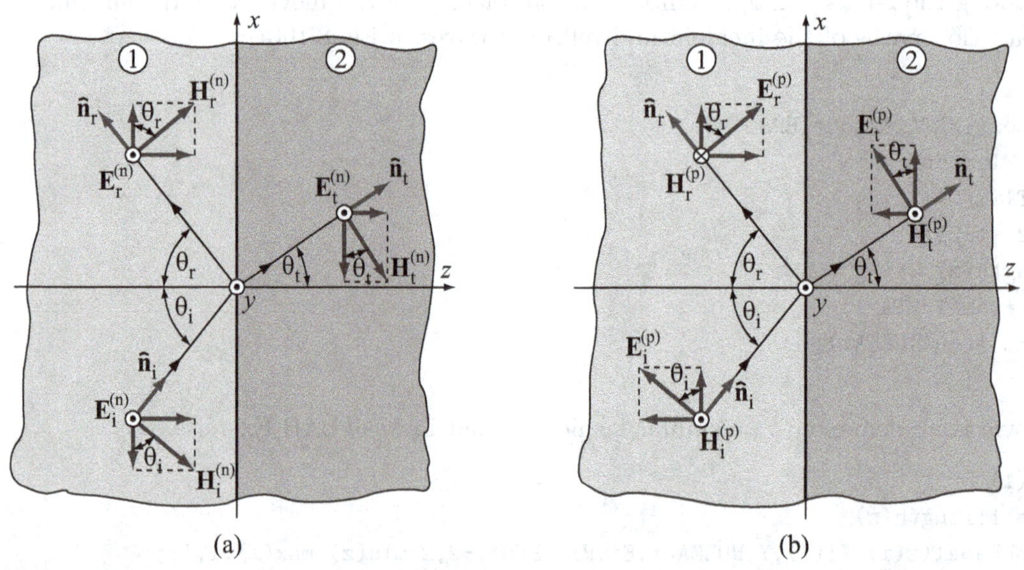

Figure 8.11 Reflection and refraction of an obliquely incident uniform plane time-harmonic electromagnetic wave at a planar dielectric boundary: electric and magnetic field vectors of the incident, reflected, and transmitted waves for normal (a) and parallel (b) polarizations of waves.

To find the unknown complex rms electric field intensities of the reflected and transmitted waves in the plane $z = 0$, $\underline{E}_{\mathrm{r}0}$ and $\underline{E}_{\mathrm{t}0}$, for a given intensity $\underline{E}_{\mathrm{i}0}$ at $z = 0$ of the incident wave, we apply the boundary conditions for tangential components of both electric and magnetic field vectors, as in Eqs.(8.10). Here, however, we need to distinguish between the normal and parallel polarizations of the incident wave, which are depicted, respectively, in Fig.8.11(a) and (b), and in place of Eqs.(8.11) and (8.12) we obtain

$$\Gamma_\mathrm{n} = \left(\frac{\underline{E}_{\mathrm{r}0}}{\underline{E}_{\mathrm{i}0}}\right)_\mathrm{n} = \frac{\eta_2\cos\theta_\mathrm{i} - \eta_1\cos\theta_\mathrm{t}}{\eta_1\cos\theta_\mathrm{t} + \eta_2\cos\theta_\mathrm{i}}, \quad \tau_\mathrm{n} = \left(\frac{\underline{E}_{\mathrm{t}0}}{\underline{E}_{\mathrm{i}0}}\right)_\mathrm{n} = \frac{2\eta_2\cos\theta_\mathrm{i}}{\eta_1\cos\theta_\mathrm{t} + \eta_2\cos\theta_\mathrm{i}}$$

$$\text{(Fresnel's coefficients for normal polarization)}, \quad (8.18)$$

$$\Gamma_\mathrm{p} = \left(\frac{\underline{E}_{\mathrm{r}0}}{\underline{E}_{\mathrm{i}0}}\right)_\mathrm{p} = \frac{\eta_2\cos\theta_\mathrm{t} - \eta_1\cos\theta_\mathrm{i}}{\eta_1\cos\theta_\mathrm{i} + \eta_2\cos\theta_\mathrm{t}}, \quad \tau_\mathrm{p} = \left(\frac{\underline{E}_{\mathrm{t}0}}{\underline{E}_{\mathrm{i}0}}\right)_\mathrm{p} = \frac{2\eta_2\cos\theta_\mathrm{i}}{\eta_1\cos\theta_\mathrm{i} + \eta_2\cos\theta_\mathrm{t}}$$

$$\text{(Fresnel's coefficients for parallel polarization)}. \quad (8.19)$$

These coefficients are known as Fresnel's (reflection and transmission) coefficients.

MATLAB EXERCISE 8.19 **Fresnel's coefficients for normal and parallel polarizations.** (a) Write a function `gammaFresnel()` in MATLAB that calculates Fresnel's reflection coefficients for normal and parallel polarizations, Γ_n and Γ_p, respectively, given by Eqs.(8.18) and (8.19), assuming that both media in Fig.8.11 are lossless. The input to the function consists of intrinsic impedances of the two media, η_1 and η_2, incident angle, θ_i, and transmitted angle, θ_t, as well as a character that specifies the polarization type of the incident wave (`'n'` for normal and `'p'` for parallel polarization). (b) Repeat (a) but for Fresnel's transmission coefficients for normal and parallel polarizations, τ_n and τ_p, respectively, in Eqs.(8.18) and (8.19) [function `tauFresnel()`]. *(gammaFresnel.m and tauFresnel.m on IR)*

MATLAB EXERCISE 8.20 **3-D movie of oblique incidence, parallel polarization, on a dielectric interface.** Repeat MATLAB Exercise 8.2 but for an oblique incidence on a dielectric–dielectric interface of a uniform plane wave with parallel polarization, as in Fig.8.11(b), assuming that medium 1 is air and medium 2 is glass with $\varepsilon_\mathrm{r} = 9$ and $\sigma = 0$, the operating frequency amounts to $f = 300$ MHz, and incident angle is $\theta_\mathrm{i} = 30°$. *(ME8_20.m on IR)*

TUTORIAL:

See MATLAB Exercises 8.13 and 8.17. The propagation parameters of the waves and the two media are computed calling function `propParamLossless` (from MATLAB Exercise 7.1). The reflected and transmitted (refracted) angles, θ_r and θ_t, are obtained from Snell's laws of reflection and refraction, in Eqs.(8.16) and (8.17), respectively. Fresnel's reflection and transmission coefficients for parallel polarization are calculated by means of functions `gammaFresnel` and `tauFresnel` (from the previous MATLAB exercise).

```
[w1,T1,BETA1,lambda1,Vp1,ETA1] = propParamLossless(f,EPSR1,MUR1);
[w2,T2,BETA2,lambda2,Vp2,ETA2] = propParamLossless(f,EPSR2,MUR2);
THETAi = 30*pi/180;
THETAr = THETAi;
THETAt = asin(sin(THETAi)*BETA1/BETA2);
```

```
GAMMA = gammaFresnel(ETA1,ETA2,THETAi,THETAt,'p');
TAU = tauFresnel(ETA1,ETA2,THETAi,THETAt,'p');
```

Cartesian components of propagation unit vectors, $\hat{\mathbf{n}}_{\mathrm{i}}$, $\hat{\mathbf{n}}_{\mathrm{r}}$, and $\hat{\mathbf{n}}_{\mathrm{t}}$, components of electric and magnetic field vectors, $\underline{\mathbf{E}}_{\mathrm{i}}^{(\mathrm{p})}$, $\underline{\mathbf{H}}_{\mathrm{i}}^{(\mathrm{p})}$, $\underline{\mathbf{E}}_{\mathrm{r}}^{(\mathrm{p})}$, $\underline{\mathbf{H}}_{\mathrm{r}}^{(\mathrm{p})}$, $\underline{\mathbf{E}}_{\mathrm{t}}^{(\mathrm{p})}$, and $\underline{\mathbf{H}}_{\mathrm{t}}^{(\mathrm{p})}$, and coordinates (x, y, z) of propagation paths of the incident, reflected, and transmitted waves are obtained – from Fig.8.11(b) – similarly to computations in MATLAB Exercise 8.17,

```
NORMAL = 1;
NORMALiz = NORMAL*cos(THETAi);
NORMALrz = -NORMALiz;
NORMALtz = NORMAL*cos(THETAt);
NORMALix = NORMAL*sin(THETAi);
NORMALrx = NORMALix;
NORMALtx = NORMAL*sin(THETAt);
ni = [NORMALiz,NORMALix,0];
nr = [NORMALrz,NORMALrx,0];
nt = [NORMALtz,NORMALtx,0];
Eio = 1;
Hio = 1;
Eioz = -Eio*sin(THETAi);
Eiox = Eio*cos(THETAi);
Hro = -Hio*GAMMA;
Ero = Eio*GAMMA;
Eroz = Ero*sin(THETAr);
Erox = Ero*cos(THETAr);
Hto = Hio*TAU*ETA1/ETA2;
Eto = Eio*TAU;
Etoz = -Eto*sin(THETAt);
Etox = Eto*cos(THETAt);
Hi = [0,0,Hio];
Ei = [Eioz,Eiox,0];
Hr = [0,0,Hro];
Er = [Eroz,Erox,0];
Ht = [0,0,Hto];
Et = [Etoz,Etox,0];
Vp1z = Vp1*cos(THETAi);;
Vp2z = Vp2*cos(THETAr);;
Zlim = 1;
Tlim = Zlim/Vp1z;
t = 0:Tlim/20:Tlim;
z1 = Vp1z.*t-Zlim;
z2 = Vp2z.*t;
Xi = z1.*tan(THETAi);
Xr = 0-z1.*tan(THETAi);
Xt = z2.*tan(THETAt);
```

```
Y = 0;
l1 = length(z1);
```

The movie is made as in MATLAB Exercises 8.2, 8.13, and 8.17, in two **for** loops, one playing the travel of the incident wave before reflection/refraction at the dielectric–dielectric interface and the other showing the reflected and transmitted waves [see Figs.8.12(a) and (b)], with plots in each frame of the movie being generated by function **nEHgraphLabel** (from MATLAB Exercise 8.1),

```
figure(1)
for i = 1:length(z1)
nEHgraphLabel(z1(i),Xi(i),Y,ni,Ei,Hi,'boundary',-2,2,min(z1),max(z2),1,1);
view([-50,20]);
text(min(z1),2,2,' medium 1');
text(max(z2),2,2,' medium 2');
M(i) = getframe;
end;
for i = 1:length(z2)
nEHgraphLabel(z1(l1+1-i),Xr(l1+1-i),Y,nr,Er,Hr,'boundary',-2,2,min(z1),max(z2),1,1);
hold on;
nEHgraphLabel(z2(i),Xt(i),Y,nt,Et,Ht,'boundary',-2,2,min(z1),max(z2),1,0);
hold off;
```

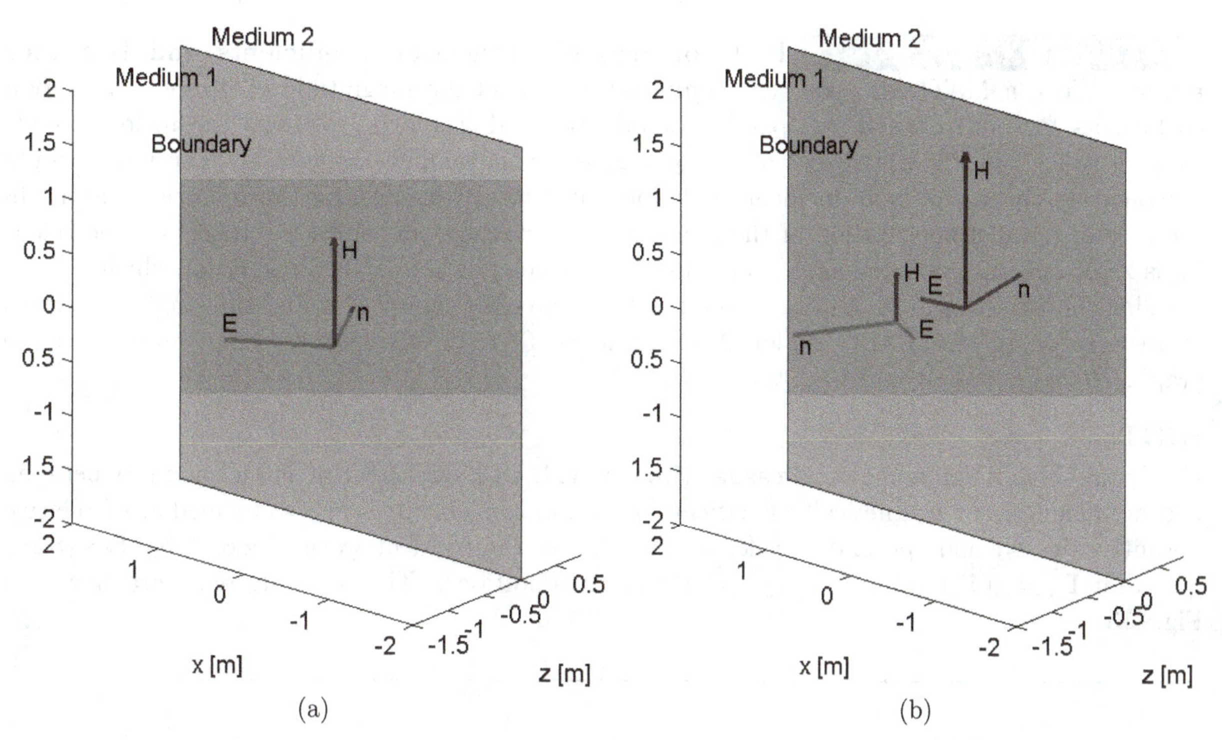

Figure 8.12 3-D MATLAB movie visualizing an oblique incidence of a uniform plane wave with parallel polarization on a dielectric–dielectric interface and its reflection and refraction [Fig.8.11(b)]: (a) snapshot taken prior to the reflection/refraction (incident wave) and (b) snapshot taken upon the reflection/refraction (reflected and transmitted waves); for MATLAB Exercise 8.20. *(color figure on CW)*

```
text(min(z1),2,2,' medium 1');
text(max(z2),2,2,' medium 2');
view([-50,20]);
M(i) = getframe;
end;
nEHgraphLabel(0,0,0,0,0,0,'boundary',-2,2,min(z1),max(z2),0,1);
text(min(z1),2,2,' medium 1');
text(max(z2),2,2,' medium 2');
view([-50,20]);
M = getframe;
```

Two characteristic frames of the movie are shown in Figs.8.12(a) and (b).

MATLAB EXERCISE 8.21 Oblique incidence, normal polarization, on a dielectric interface. Repeat the previous MATLAB exercise but for normal polarization of the incident wave, as in Fig.8.11(a), and relative permittivity of the second medium $\varepsilon_r = 2.7$ (PVC) (other numerical data are unchanged). *(ME8_21.m on IR)*

MATLAB EXERCISE 8.22 Plots of Fresnel's reflection coefficients and Brewster angle. Write a MATLAB code that computes and plots the magnitude of Fresnel's reflection coefficients for both normal and parallel polarizations, $|\Gamma_n|$ and $|\Gamma_p|$, versus the incident angle ($0 \le \theta_i \le 90°$) in Fig.8.11, if $\varepsilon_{r1} = 1$, $\varepsilon_{r2} = 2$, $\mu_1 = \mu_2$, and $\sigma_1 = \sigma_2 = 0$. The plots should demonstrate that there is an incident angle for which total transmission (no reflection) occurs in the case of parallel polarization of the incident wave, namely, for which the reflection coefficient Γ_p is zero, and that no such angle exists for normal polarization. The angle θ_i for which $\Gamma_p = 0$ is called the Brewster angle and is denoted as θ_{iB}. The code should also find this angle in degrees [write a separate MATLAB function `BrewsterAngle()` that calculates θ_{iB}] and mark it in the graph. *(BrewsterAngle.m and ME8_22.m on IR)*

HINT:

Use Eq.(8.17) and function `gammaFresnel` (from MATLAB Exercise 8.19). For a boundary between two nonmagnetic or magnetically identical ($\mu_1 = \mu_2$) lossless ($\sigma_1 = \sigma_2 = 0$) media, of relative permittivities ε_{r1} and ε_{r2}, and parallel polarization of the incident wave, Fig.8.11(b), Eqs.(8.19) give that $\Gamma_p = 0$ if $\tan\theta_{iB} = \sqrt{\varepsilon_{r2}/\varepsilon_{r1}}$ (Brewster condition). The resulting plots are shown in Fig.8.13.

MATLAB EXERCISE 8.23 Prism function – for computation of light beams in a glass prism. Write a function `prismFunc()` in MATLAB that determines the coordinates of the intersection of two straight lines in the xy-plane of the Cartesian coordinate system. This function is specially developed for computation and plotting of incident and refracted beams associated

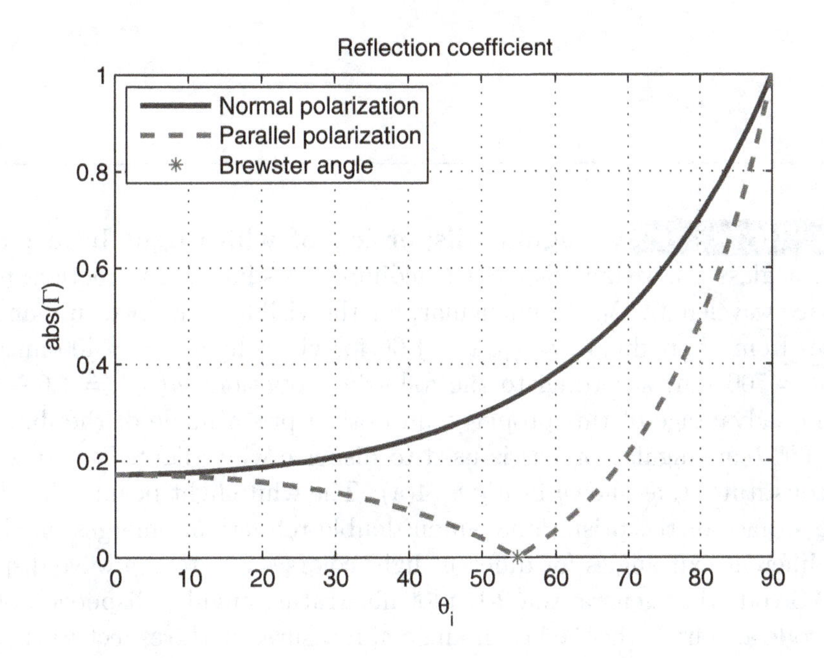

Figure 8.13 Plots of the magnitude of Fresnel's reflection coefficients against the incident angle (θ_i) in Fig.8.11 for incidence from air onto a perfect-dielectric material with relative permittivity $\varepsilon_r = 2$ (paraffin); for MATLAB Exercise 8.22. *(color figure on CW)*

with angular dispersion of white light by a glass prism, analyzed in a cross section of the prism, to be done in the next MATLAB exercise. One of the two lines is defined by one point and an angle that the line makes with the x-axis, while the other line is defined by two points, with the intersection point being between these two points. *(prismFunc.m on IR)*

TUTORIAL:

At input to the function, the coordinates (`x0,y0`) and angle `beta` (β) determine the first line. The second line is defined by coordinates (`x1,y1`) and (`x2,y2`) of the two points.

```
function [x3 y3] = prismFunc(x0,y0,beta,x1,y1,x2,y2)
```

We introduce a dense mesh of points (`x,y`) along the second line, bounded by points (`x1,y1`) and (`x2,y2`), as follows:

```
dx = 0.0001*(x2-x1);
dy = 0.0001*(y2-y1);
x = x1:dx:x2;
y = y1:dy:y2;
```

Then we compute slopes (array `k`) of all lines starting at (`x0,y0`) and ending at each of the points (`x,y`) of the mesh on the second line. By means of MATLAB function `min` (see MATLAB Exercise 3.12, for instance), we find which of slopes `k` is the closest to the slope of the first given line, $\tan\beta$, and from that, the coordinates (`x3,y3`) of the intersection point (output of the function).

```
k = (y-y0)./(x-x0);
[error,l] = min(abs(k-tan(beta)));
```

```
x3 = x(1);
y3 = y(1);
```

MATLAB EXERCISE 8.24 **Angular dispersion of white light by a glass prism.** At optical frequencies, glass is a weakly dispersive medium, as its index of refraction, n, slightly varies with the free-space wavelength, λ_0. In particular, for the visible light spectrum and a type of flint glass, n decreases from approximately $n_{\text{violet}} = 1.66$ for violet light ($\lambda_0 = 400$ nm) to $n_{\text{red}} = 1.62$ for red light ($\lambda_0 = 700$ nm) according to the following equation: $n(\lambda_0) = 1.6 + 9.5 \times 10^{-15}/\lambda_0^2$ (λ_0 in m). Taking advantage of this property, an optical prism made of the flint glass and with apex angle $\alpha = 60°$ (equiangular prism) is used to disperse white light, i.e., to separate in space the colors that constitute it, as shown in Fig.8.14(a). The white light beam is incident at an angle $\theta_{\text{i}} = 65°$ on one surface of the prism, and, upon double refraction, emerges on the other side of the prism with different exit angles for different light colors – due to the wave dispersion in glass. Write a MATLAB code that generates a 2-D plot illustrating angular dispersion of white light by the prism. The code also finds the deviation angle δ, measured with respect to the incident beam, of the outgoing beam for each of the colors in the visible spectrum [violet, blue ($\lambda_0 = 470$ nm), green ($\lambda_0 = 540$ nm), yellow ($\lambda_0 = 590$ nm), orange ($\lambda_0 = 610$ nm), and red], as well as the total angular dispersion of the prism, defined as $\gamma = \delta_{\text{violet}} - \delta_{\text{red}}$ [Fig.8.14(a)]; these results are displayed in the Command Window. In the graph, refracted beams are plotted in colors that correspond to their wavelengths (components of visible light). *(ME8_24.m on IR)* **H**

TUTORIAL:
We first specify the input data for the prism apex angle, α (`alpha`), and the incident angle of the white light beam at the first (front) surface of the prism in Fig.8.14(a), θ_{i} (`thetai`). In addition, we define an array `lambda0` with free-space wavelengths, λ_0, for colors in the visible light spectrum, array `n` with the corresponding values of the index of refraction of the flint glass, according to the given function $n(\lambda_0)$, and array `color` with codes for colors in the visible spectrum (violet, blue, green, yellow, orange, and red).

```
alpha = 60/180*pi;
thetai = 65/180*pi;
lambda0 = [400 470 540 590 610 700]*10^(-9);
n = 1.6 + 9.5*10^(-15)./lambda0.^2;
color = [[1,0,1]; [0,0,1]; [0,1,0]; [1,1,0]; [1,0.5,0]; [1,0,0]];
```

With reference to Fig.8.14(b), Snell's law of refraction, Eq.(8.17), applied at the first surface of the prism, at the point A, results in the following expression for the transmitted (refracted) angle θ_{t}:

$$\frac{\sin\theta_{\text{i}}}{\sin\theta_{\text{t}}} = n = n(\lambda_0) \quad (n_{\text{air}} = 1) \quad \longrightarrow \quad \theta_{\text{t}} = \arcsin\frac{\sin\theta_{\text{i}}}{n(\lambda_0)} \quad \left(\arcsin \equiv \sin^{-1}\right), \tag{8.20}$$

and in the associated lines of MATLAB code to fill an array `thetat`,

```
sinThetat = sin(thetai)./n;
thetat = asin(sinThetat);
```

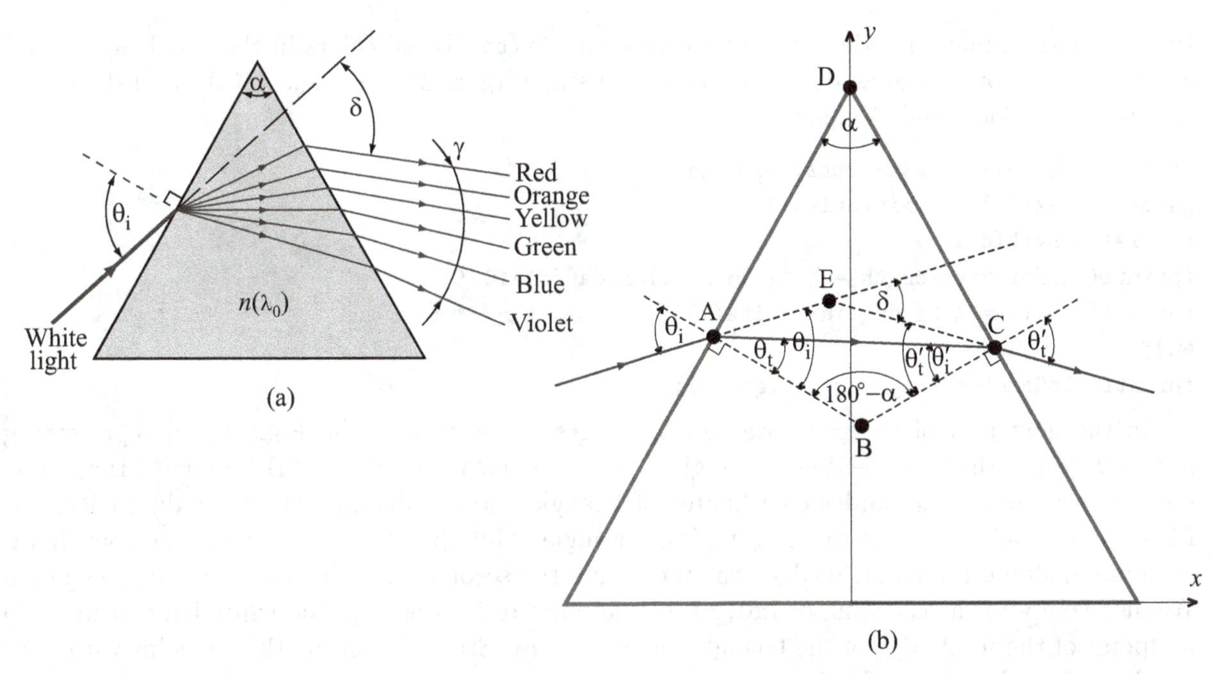

Figure 8.14 (a) Angular dispersion of a beam of white light into its constituent colors in the visible spectrum using a glass prism, and (b) geometrical considerations for the prism to find, in conjunction with Snell's law of refraction, the deviation angle δ of the outgoing beam (relative to the incident beam) in terms of the incident angle θ_i and the index of refraction of the glass; for MATLAB Exercise 8.24.

The sum of angles of the quadrilateral ABCD in Fig.8.14(b) being 360°, and the three angles of this quadrilateral being 90°, α, and 90°, the fourth angle (at the vertex B) must be $180° - \alpha$, and then – from the triangle ABC – we obtain the incident angle θ_i' on the other side of the prism, at the point C,

$$\theta_\mathrm{i}' = 180° - (180° - \alpha + \theta_\mathrm{t}) = \alpha - \theta_\mathrm{t} , \tag{8.21}$$

and hence the array `thetaip`:

```
thetaip = alpha - thetat;
```

Another application of Snell's law of refraction, now at the point C, gives the exit angle θ_t' (array `thetatp`):

$$\frac{\sin\theta_\mathrm{i}'}{\sin\theta_\mathrm{t}'} = \frac{1}{n} \quad \longrightarrow \quad \theta_\mathrm{t}' = \arcsin(n\sin\theta_\mathrm{i}') , \tag{8.22}$$

implemented as

```
sinThetatp = sin(thetaip).*n;
thetatp = asin(sinThetatp);
```

Finally, from the quadrilateral ABCE in Fig.8.14(b), we obtain the expression for the deviation angle δ, relative to the incident beam, of the outgoing beam [also see Fig.8.14(a)]:

$$\theta_\mathrm{i} + (180° - \alpha) + \theta_\mathrm{t}' + (180° - \delta) = 360° \quad \longrightarrow \quad \delta = \theta_\mathrm{i} + \theta_\mathrm{t}' - \alpha . \tag{8.23}$$

In the following lines of code, we compute an array **delta** (for all colors in the visible spectrum), as well as the total angular dispersion of the prism, γ (**gamma**) [see Fig.8.14(a)], and display all results in the Command Window,

```
delta = (thetai+thetatp-alpha).*180/pi;
gamma = (max(delta)- min(delta));
for i=1:length(delta)
fprintf('\nFor wavelength = %.4g nm\n', lambda0(i)*10^9);
fprintf('delta = %.5g deg.\n',delta(i));
end;
fprintf('\nGamma = %.4g deg.\n',gamma);
```

In the next part of the code, we plot the edges of the triangle [in Fig.8.14(a)] representing, in a 2-D graph, the cross section of the glass prism, by means of MATLAB function `line`, whose respective inputs are x- and y-coordinates of triangle vertices defining the individual edges [see Fig.8.14(b)]. Note that `h` is the height of the triangle, while `(x1,y1)` and `(x2,y2)` are coordinates of points defining the triangle edge that represents the second (exit) interface of the glass prism. We also compute the coordinates `(x0,y0)` of the point of incidence of the white light beam – the midpoint of the front edge of the triangle [point A in Fig.8.14(b)], and mark this point with a red star in the graph. We call function `prismFunc` (written in the previous MATLAB exercise) to find the coordinates `(xb,yb)` of the starting point (for plotting) of the incident white light beam, and plot this beam (in black) by function `line`, as a line connecting points `(xb,yb)` and `(x0,y0)`.

```
h = 1;
x1 = 0;
y1 = h;
x2 = h*tan(alpha/2);
y2 = 0;
line([-x2 x2],[y2 y2]); hold on;
line([-x2 x1],[y2 y1]);
line([x1 x2],[y1 y2]);
x0 = -x2/2;
y0 = y1/2;
plot(x0,y0,'r*');
[xb,yb] = prismFunc(x0,y0,-alpha/2+thetai+pi,-h,0,-1.1*h,h);
line([xb x0],[yb y0],'Color','k');
```

Then, we compute an array of angles `beta` that individual dispersed beams inside the prism make with the x-axis and corresponding array of angles `betaout` for the outgoing dispersed beams, outside the prism. By function `prismFunc`, we determine, in a `for` loop, the array of coordinates `(x3,y3)` of points where each of the dispersed beams in the prism hits the exit edge of the triangle (interface of the prism) and array of coordinates `(x4,y4)` of ending points (for plotting) of outgoing dispersed beams. These coordinates are used to plot the interior dispersed beams, as lines connecting the point `(x0,y0)` and the array of points `(x3,y3)`, and outgoing dispersed beams, as lines connecting the array of points `(x3,y3)` and the corresponding array of points `(x4,y4)`. Both sets of beams are plotted in proper colors, using the array of color codes `color`, as shown in Fig.8.15.

```
beta = thetat - alpha/2;
```

```
betaout = -thetatp + alpha/2;
for i = 1:length(beta)
[x3 y3] = prismFunc(x0,y0,beta(i),x1,y1,x2,y2);
[x4 y4] = prismFunc(x3,y3,betaout(i),0,0,5*h,0.5*h);
line([x0 x3],[y0 y3],'Color',color(i,:));
line([x3 x4],[y3 y4],'Color',color(i,:));
hold on;
end;
```

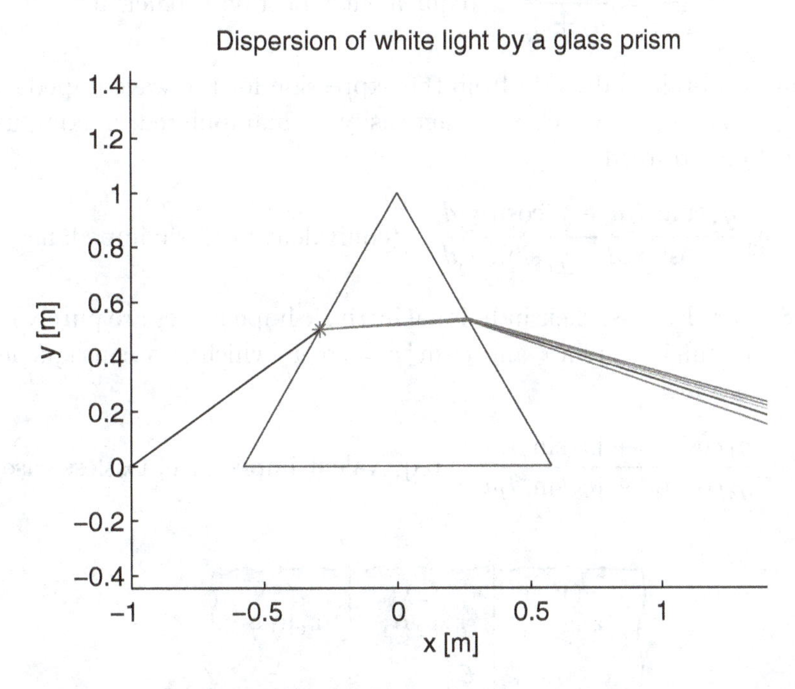

Figure 8.15 MATLAB plot and computation of angular dispersion of white light by a glass prism; for MATLAB Exercise 8.24. *(color figure on CW)*

MATLAB EXERCISE 8.25 Variations of the light dispersion experiment with a glass prism. Repeat the previous MATLAB exercise but with the following combinations of the prism apex angles, α, and incident angles of the white light beam at the front surface of the prism, $\theta_{\rm i}$, in Fig.8.14(a): (i) $\alpha = 45°$ and $\theta_{\rm i} = 30°$, (ii) $\alpha = 50°$ and $\theta_{\rm i} = 50°$, and (iii) $\alpha = 30°$ and $\theta_{\rm i} = 0$. Also, for $\alpha = 60°$ and $\theta_{\rm i} = 65°$, explore how changes in the function describing the index of refraction of glass, $n(\lambda_0) = n_0 + k \times 10^{-15}/\lambda_0^2$ (λ_0 in m), affect the plot of the outgoing dispersed beams and the total angular dispersion of the prism (γ); namely, try the following constants in the function: (iv) $n_0 = 1.5$ and $k = 9$ and (v) $n_0 = 1.8$ and $k = 10$. Modify the program from the previous exercise such that the input values of α, $\theta_{\rm i}$, n_0, and k are entered from the keyboard. *(ME8_25.m on IR)* **H**

8.5 Wave Propagation in Multilayer Media

In this section, we study plane-wave propagation in a three-layer medium, where a layer of thickness d is placed between two semi-infinite regions, as shown in Fig.8.16, and a uniform plane time-harmonic electromagnetic wave is incident from medium 1 normally on interface 1-2 (at $z = -d$); generalization to an arbitrary number of layers is straightforward. Let us introduce the equivalent intrinsic impedance of the combination of media 2 and 3, $\underline{\eta}_e$, so that the corresponding equivalent reflection coefficient [Eq.(8.11)] at the first interface in Fig.8.16 is given by

$$\underline{\Gamma}_e = \frac{\underline{E}_{1r}}{\underline{E}_{1i}} = \frac{\underline{\eta}_e - \underline{\eta}_1}{\underline{\eta}_1 + \underline{\eta}_e} \quad \text{(equivalent reflection coefficient)} . \tag{8.24}$$

This impedance can be obtained directly from the expression for the wave impedance in Eq.(8.15), as $\underline{\eta}_e = \underline{E}_2/\underline{H}_2 = \underline{\eta}_{w2}(z = -d^+)$, which can then easily be manipulated, making use of hyperbolic sine and cosine functions, to read

$$\underline{\eta}_e = \underline{\eta}_2 \frac{\underline{\eta}_2 \sinh \underline{\gamma}_2 d + \underline{\eta}_3 \cosh \underline{\gamma}_2 d}{\underline{\eta}_2 \cosh \underline{\gamma}_2 d + \underline{\eta}_3 \sinh \underline{\gamma}_2 d} \quad \text{(equivalent intrinsic impedance)} . \tag{8.25}$$

If all media in Fig.8.16 are lossless, their individual intrinsic impedances are purely real and $\underline{\gamma}_2 = \mathrm{j}\beta_2$ ($\alpha_2 = 0$). In addition, $\sinh \mathrm{j}x = \mathrm{j}\sin x$ and $\cosh \mathrm{j}x = \cos x$ [which is a consequence of Eq.(6.26)], and hence Eq.(8.25) becomes

$$\underline{\eta}_e = \eta_2 \frac{\eta_3 \cos \beta_2 d + \mathrm{j}\eta_2 \sin \beta_2 d}{\eta_2 \cos \beta_2 d + \mathrm{j}\eta_3 \sin \beta_2 d} \quad \text{(equivalent impedance, lossless case)} . \tag{8.26}$$

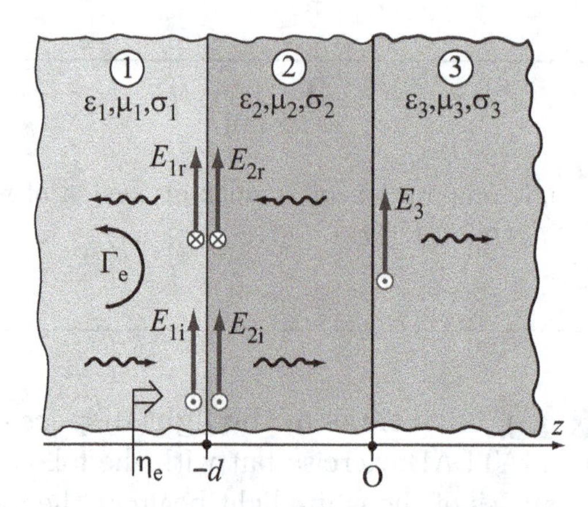

Figure 8.16 Normal incidence, from medium 1, of a uniform plane time-harmonic electromagnetic wave on two parallel interfaces separating three different homogeneous material regions with arbitrary electromagnetic parameters.

MATLAB EXERCISE 8.26 **Lossless three-media structure.** Write a function `threeMediaLossless()` in MATLAB that computes the equivalent input impedance $\underline{\eta}_e$ seen at interface 1-2 looking to the right in Fig.8.16, based on Eq.(8.26), and the associated equivalent reflection coefficient, $\underline{\Gamma}_e$, by means of Eq.(8.24), assuming that media 2 and 3 are lossless and that the incident medium is air. *(threeMediaLossless.m on IR)*

MATLAB EXERCISE 8.27 **Lossy three-media structure.** Repeat the previous MATLAB exercise but now assuming that media 2 and 3 in Fig.8.16 are lossy [function `threeMediaLossy()`], so that $\underline{\eta}_e$ is given by Eq.(8.25). Show that in the lossless case this function gives the same results as function `threeMediaLossless` (from the previous exercise). *(threeMediaLossy.m on IR)*

---

9 FIELD ANALYSIS OF TRANSMISSION LINES

Introduction:

In addition to wireless links, which use free (unbounded) electromagnetic waves propagating in free space or material media (previous two chapters), electromagnetic signals and energy can be transported to a distance also using guided electromagnetic waves. Such waves are channeled through a guiding system composed of conductors and dielectrics. Guiding systems normally have a uniform cross section, and are classified into transmission lines, having two or more separate conductors, and waveguides, consisting of a single conductor or only dielectrics. In this chapter, we present a field analysis of two-conductor transmission lines, which is important for understanding physical processes that constitute the propagation and attenuation along a line of a given geometry and material composition. The principal result of the analysis are the parameters of a circuit model of an arbitrary line, in the form of a network of many cascaded equal small cells with lumped elements. This network is then solved, in the next chapter, using circuit-theory concepts and equations, as the starting point of the frequency-domain (complex-domain) and transient (time-domain) analysis of transmission lines as circuits with distributed parameters (circuit analysis of two-conductor transmission lines).

9.1 Field Analysis of Lossless Transmission Lines

Consider a transmission line consisting of two perfect conductors of arbitrary cross section, shown in Fig.9.1, in a homogeneous perfect dielectric of permittivity ε and permeability μ. We assume a time-harmonic variation of the electromagnetic field in the line, of frequency f (and angular or radian frequency $\omega = 2\pi f$), and perform the analysis in the complex domain. The voltage $\underline{V}_{12}$ between the line conductors, that is, the difference between their potentials, $\underline{V}_1$ and $\underline{V}_2$, and the current $\underline{I}$ through the conductors are waves propagating along the z-axis in Fig.9.1,

$$\underline{V}_{12} = \underline{V}_{12}(z) = \underline{V}_{12}(0)\,\mathrm{e}^{-\mathrm{j}\beta z}\,, \quad \underline{I} = \underline{I}(z) = \underline{I}(0)\,\mathrm{e}^{-\mathrm{j}\beta z}\,, \quad \beta = \omega\sqrt{\varepsilon\mu}\,, \tag{9.1}$$

with the phase coefficient, β, being the same [see Eq.(7.4)] as for a uniform plane wave traveling in an unbounded medium with the same parameters as the dielectric of the transmission line. The phase velocity, v_p, and wavelength, $\lambda_z = 2\pi/\beta$, along the line (along the z-axis) are thus also the same as in Eqs.(7.4)–(7.6).

The electric and magnetic field lines in a cross section of the system (Fig.9.1) are as in electrostatics and magnetostatics, respectively, and both the electric and magnetic field vectors, $\underline{\mathbf{E}}$ and $\underline{\mathbf{H}}$, are transverse to the direction of wave propagation (i.e., to the line axis), constituting a TEM (transverse electromagnetic) wave. In addition, the field vectors are perpendicular to each other, and the ratio of their complex rms intensities equals a real constant, denoted by Z_TEM and called the wave impedance of TEM waves [it has the same value as the intrinsic impedance (η) in Eq.(7.2) of the medium of parameters ε and μ],

$$\underline{\mathbf{E}} \perp \underline{\mathbf{H}}\,, \quad \underline{\mathbf{E}}_z = 0\,, \quad \underline{\mathbf{H}}_z = 0\,, \quad Z_\mathrm{TEM} = \frac{\underline{E}}{\underline{H}} = \sqrt{\frac{\mu}{\varepsilon}} \quad \text{(TEM wave)}\,. \tag{9.2}$$

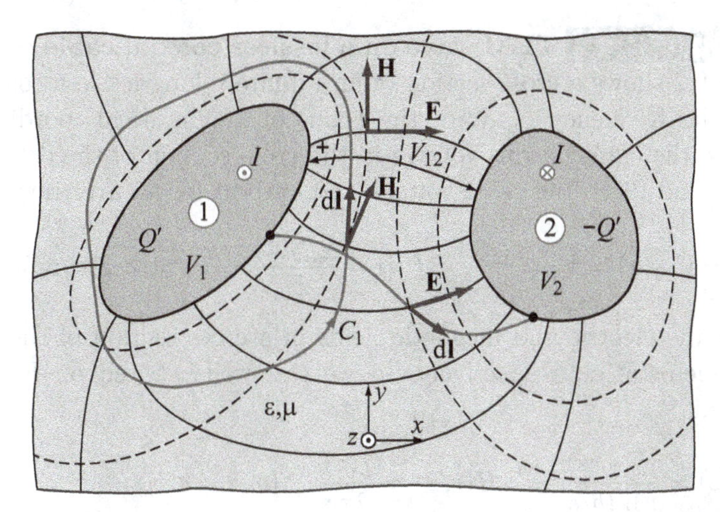

Figure 9.1 Cross section of a two-conductor transmission line with a homogeneous dielectric.

With reference to Fig.9.1, the voltage, current, and charge per unit length ($\underline{Q}'$) of the transmission line can be computed as

$$\underline{V}_{12} = \underline{V}_1 - \underline{V}_2 = \int_1^2 \underline{\mathbf{E}} \cdot \mathrm{d}\mathbf{l} \,, \quad \underline{I} = \oint_{C_1} \underline{\mathbf{H}} \cdot \mathrm{d}\mathbf{l} \,, \quad \underline{V}_{12} = \frac{\underline{Q}'}{C'} \,, \quad \underline{I} = c\underline{Q}' = \frac{\underline{Q}'}{\sqrt{\varepsilon\mu}} \,, \tag{9.3}$$

where both the integration path between the conductors and the integration contour enclosing a conductor must lie entirely in a transversal plane (defined by a coordinate z), and C' is the capacitance per unit length of the line, Eq.(2.24).

Equations (9.1) tell us that the line voltage and current have the same exponential dependence on the z-coordinate, and hence their ratio is the same (a constant) for every cross section of the line. It is termed the characteristic impedance of the line and is designated by Z_0. Combining Eqs.(9.3) and (6.8), we obtain

$$Z_0 = \frac{\underline{V}_{12}}{\underline{I}} = \frac{\underline{V}_{12}}{c\underline{Q}'} = \frac{1}{cC'} = \frac{\sqrt{\varepsilon\mu}}{C'} = \sqrt{\frac{L'}{C'}} \quad \text{(line characteristic impedance)} \,, \tag{9.4}$$

with L' standing for the inductance per unit length of the line [see Eq.(6.7)].

MATLAB EXERCISE 9.1 **Characteristic impedance of a coaxial cable.** Write a function `chImpedanceCoaxCable()` in MATLAB that calculates the characteristic impedance (Z_0) of a coaxial cable, Fig.2.9(b), filled with a dielectric of permittivity ε and permeability μ. *(chImpedanceCoaxCable.m on IR)*[1]

HINT:
Use Eqs.(9.4) and (2.26) to obtain the following expression for Z_0:

$$Z_0 = \frac{\sqrt{\varepsilon\mu}}{C'} = \frac{1}{2\pi}\sqrt{\frac{\mu}{\varepsilon}} \ln\frac{b}{a} \quad (Z_0 \text{ – coaxial cable}) \,. \tag{9.5}$$

[1] *IR* = Instructor Resources (for the book).

MATLAB EXERCISE 9.2 TEM wave on a lossless coaxial cable with a homogeneous dielectric. Figure 9.2 shows a cross section of an infinitely long lossless coaxial cable carrying a TEM wave of angular frequency ω. With the z-axis of a cylindrical coordinate system adopted along the cable axis, the complex rms voltage in the cross section of the cable defined by $z = 0$ is $\underline{V}_0$. From Eqs.(9.1) and (9.4), the cable voltage and current for an arbitrary coordinate z are

$$\underline{V}_{12}(z) = \underline{V}_0 \, \mathrm{e}^{-\mathrm{j}\beta z}, \quad \underline{I}(z) = \frac{\underline{V}_{12}(z)}{Z_0} \quad (-\infty < z < \infty). \tag{9.6}$$

The distribution of the electric and magnetic fields in a cross section of the cable are as in electrostatics and magnetostatics, respectively, so we can write, based on Eqs.(9.3) and the field symmetries in Fig.9.2,

$$\underline{E}(r, z) = \frac{\underline{V}_{12}(z)}{r \ln(b/a)}, \quad \underline{H}(r, z) = \frac{\underline{I}(z)}{2\pi r} \quad (a < r < b, \ -\infty < z < \infty). \tag{9.7}$$

Referring to Fig.9.2, boundary conditions in Eqs.(6.19) give the following expressions for the surface charge and current densities on the conductor surfaces of the cable:

$$\underline{\rho}_{s1} = \varepsilon \underline{E}(a^+, z), \quad \underline{\rho}_{s2} = -\varepsilon \underline{E}(b^-, z), \quad \underline{J}_{s1} = \underline{H}(a^+, z), \quad \underline{J}_{s2} = -\underline{H}(b^-, z). \tag{9.8}$$

Finally, using Eqs.(6.32), (9.2), (6.33), and (6.31), the time-average Poynting vector (the complex Poynting vector is purely real), $\mathcal{P}$, at an arbitrary location in the dielectric of the cable (Fig.9.2) and the total time-average power flow along the cable, P, come out to be

$$\underline{\mathcal{P}} = \underline{E}\underline{H}^* = \frac{\underline{E}\underline{E}^*}{Z_{\mathrm{TEM}}} = \frac{|\underline{E}|^2}{Z_{\mathrm{TEM}}} = \mathcal{P}(r), \quad P = \int_{r=a}^{b} \mathcal{P}(r) \, \mathrm{d}S = \frac{|\underline{V}_{12}|^2}{Z_0}. \tag{9.9}$$

Implement in MATLAB Eqs.(9.6)–(9.9), and compute the values of all quantities considered in the analysis for the following numerical data: $a = 1$ cm, $b = 3$ cm, $c = 3.5$ cm, $\varepsilon_r = 2.1$, $V_0 = 1$ V,

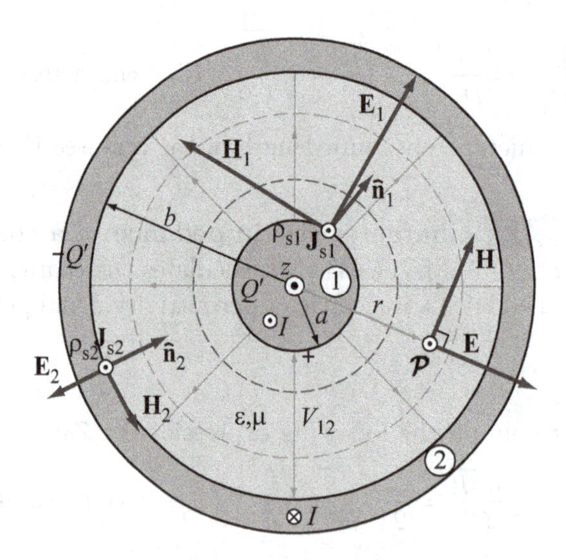

Figure 9.2 Cross section of a lossless coaxial cable with a homogeneous dielectric and TEM wave propagating in the positive z direction; for MATLAB Exercise 9.2.

$f = 1$ GHz, $r = 1.5$ cm, and $z = 20$ cm. Display the results in the Command Window. *(ME9_2.m on IR)* **H**[2]

9.2 Transmission Lines with Small Losses

All real transmission lines have some losses, which, in general, consist of losses in conductors and losses in the dielectric of the line. However, for lines used in engineering practice, these losses, evaluated per unit length of the line, are small. Simply, the conductors and dielectrics in practical transmission lines, if not perfect, are good – by design, such that, having in mind Eqs.(7.18) and (7.17), and denoting the conductivity of the line conductors by σ_c, and that of the line dielectric by σ_d, the following conditions are met:

$$\sigma_\mathrm{c} \gg \omega\varepsilon_0 \quad \text{and} \quad \sigma_\mathrm{d} \ll \omega\varepsilon \quad \text{(transmission line with small losses)} . \qquad (9.10)$$

The losses in a transmission line result in the attenuation of TEM waves along the line, as in Eqs.(7.12) and (7.13), and hence the complex current intensity along the line in Eqs.(9.1), for instance, is now given by

$$\underline{\gamma} = \alpha + \mathrm{j}\beta \quad (\alpha = \alpha_\mathrm{c} + \alpha_\mathrm{d}) \quad \longrightarrow \quad \underline{I}(z) = \underline{I}(0)\,\mathrm{e}^{-\underline{\gamma} z} = \underline{I}(0)\,\mathrm{e}^{-\alpha z}\,\mathrm{e}^{-\mathrm{j}\beta z} , \qquad (9.11)$$

and similarly for the voltage, field intensity vectors, and other z-dependent quantities in the analysis. The attenuation coefficient (namely, the portion of α) corresponding to the losses in the conductors in the structure, α_c, is computed as

$$\alpha_\mathrm{c} = \frac{R'}{2Z_0} , \quad R' = \frac{1}{|\underline{I}|^2} \oint_{C_\mathrm{c}} R_\mathrm{s} |\underline{\mathbf{H}}_\mathrm{tang}|^2 \, \mathrm{d}l , \quad R_\mathrm{s} = \sqrt{\frac{\pi\mu_\mathrm{c} f}{\sigma_\mathrm{c}}} \quad \text{(for line conductors)} \qquad (9.12)$$

(practically always, $\mu_\mathrm{c} = \mu_0$), where R' is the high-frequency resistance per unit length of the transmission line (in Ω/m), C_c denotes the contour of both conductors in the line cross section (Fig.9.1), with dl being an elemental segment along C_c, R_s is the surface resistance of the conductors (with the skin effect pronounced – see Section 7.6), measured in Ω/square, and $\underline{\mathbf{H}}_\mathrm{tang}$ is the tangential component of the complex rms magnetic field intensity vector on the conductor surfaces. On the other side, the attenuation coefficient α_d in Eqs.(9.11), for the losses in the dielectric of the transmission line, amounts to

$$\alpha_\mathrm{d} = \frac{G'}{2Y_0} \quad \left(Y_0 = \frac{1}{Z_0}\right) , \quad G' = \frac{\sigma_\mathrm{d}}{\varepsilon} C' \quad \text{(for line dielectric)} , \qquad (9.13)$$

where Y_0 stands for the characteristic admittance of the line, while G' is the leakage conductance per unit length of the line (unit: S/m).

MATLAB EXERCISE 9.3 **Surface resistance function.** Write a MATLAB function `surfResistance()` that returns the surface resistance of a good conductor (with the skin effect

[2]**H** = recommended to be done also "by hand," i.e., not using MATLAB.

pronounced), R_s, for a given operating frequency, f, and conductivity, σ_c, and relative permeability, μ_{rc}, of the conductor, using the expression in Eqs.(9.12). *(surfResistance.m on IR)*

MATLAB EXERCISE 9.4 **Surface resistance of copper and zinc – versus frequency.** Write a MATLAB program that calculates and plots the surface resistance (R_s) of copper and zinc in the same graph over a frequency range 0.3 – 3 GHz. *(ME9_4.m on IR)*

HINT:
Use function `surfResistance` (from the previous MATLAB exercise) and take the conductivity values from the GUI developed in MATLAB Exercise 3.1.

MATLAB EXERCISE 9.5 **High-frequency resistance p.u.l. of a coaxial cable.** Assume that the coaxial cable in Fig.9.2 has small conductor losses. From Eqs.(9.7), the complex rms magnetic field intensities on surfaces of the inner ($r = a^+$) and outer ($r = b^-$) conductors of the cable are $\underline{H}_1 = \underline{I}/(2\pi a)$ and $\underline{H}_2 = \underline{I}/(2\pi b)$, respectively. The vector $\underline{\mathbf{H}}$ is entirely tangential to both surfaces, so that the integral in Eqs.(9.12) results in the following expression for the high-frequency resistance per unit length of the cable:

$$R' = \frac{1}{|\underline{I}|^2}\left(\oint_{C_{c1}} R_s|\underline{H}_1|^2\,dl + \oint_{C_{c2}} R_s|\underline{H}_2|^2\,dl\right) = \frac{R_s}{2\pi}\left(\frac{1}{a}+\frac{1}{b}\right) \quad (R' \text{ – coaxial cable})\,. \tag{9.14}$$

Write a function `resistanceCoaxCable()` in MATLAB that calculates R'. *(resistanceCoaxCable.m on IR)*

MATLAB EXERCISE 9.6 **Coaxial cable design for minimum attenuation coefficient.** Consider the coaxial cable from the previous MATLAB exercise. For given materials in the structure, copper for the conductors and polyethylene ($\varepsilon_r = 2.25$) for the dielectric, and assuming a fixed outer radius b, $b = 2$ cm, and variable inner radius a of the cable, we wish to design the cable (find a) so that its attenuation coefficient is minimum. Namely, combining Eqs.(9.12), (9.14), (9.5), and (9.2), α for the losses in conductors of the cable can be expressed as

$$\alpha_c = \frac{R'}{2Z_0} = \frac{R_s}{2bZ_{TEM}}\frac{1+x}{\ln x} \quad \left(x = \frac{b}{a}\right) \quad (\alpha_c \text{ – coaxial cable})\,, \tag{9.15}$$

where x stands for the outer to inner conductor radii ratio of the cable. Write a program in MATLAB that calculates and plots the dependence of α_c on x, and finds x that results in the minimum α_c. The program also computes the associated characteristic impedance of the cable. *(ME9_6.m on IR)* **H**

TUTORIAL:
For the surface resistance of the cable conductors, R_s (`Rs`), we call function `surfResistance` (from MATLAB Exercise 8.3), and the wave impedance of the TEM wave along the cable, Z_{TEM} (`Ztem`), is computed using Eqs.(9.2), (2.2), and (7.3), so as $Z_{TEM} = \eta_0/\sqrt{\varepsilon_r} = 120\pi\,\Omega/\sqrt{\varepsilon_r}$,

```
Rs = surfResistance(f,MUR,SIGMA);
Ztem = 120*pi/sqrt(EPSR);
```

Next, the conductor radii ratio b/a is defined as a vector `x`, and the attenuation coefficient α_c is calculated from Eq.(9.15) and is plotted (in blue),

```
x = 1:0.01:30;
alphac = Rs/(2*b*Ztem)*(1+x)./log(x);
plot(x,alphac,'b');
hold on;
```

We find the minimum α_c and the corresponding x using MATLAB function `min` (see MATLAB Exercise 3.12, for instance), and mark it in the graph by a red star (*),

```
[alphaMin,i] = min(alphac);
plot (x(i),alphaMin, 'r*');
hold off;
```

The characteristic impedance of the cable is then obtained from Eqs.(9.5), (2.2), and (7.3), as $Z_0 = 60\ \Omega \ln(b/a)/\sqrt{\varepsilon_r}$,

```
Z0 = 60/sqrt(EPSR)*log(x(i));
```

The results, displayed in the Command Window by means of MATLAB function `fprintf`, are the following:

```
Minimum attenuation coefficient is:  0.00294711 Np/m,
for x = b/a equal to:  3.59.
Z0 is:  51.1261 ohm.
```

The plot of $\alpha_c(x)$ is shown in Fig.9.3. Note that a standard procedure – to find the optimal x, for which α_c is minimum – of equating to zero the derivative of α_c with respect to x yields here a transcendental equation that cannot be solved analytically.

MATLAB EXERCISE 9.7 Coaxial cable design for maximum breakdown rms voltage. In this MATLAB exercise, we wish to optimize the conductor radius a of a coaxial cable (Fig.9.2), for a fixed radius b and a given dielectric strength of the cable dielectric, E_{cr}, such that the cable can withstand the maximum possible applied rms voltage (before its dielectric breaks down). Namely, Eq.(9.7) tells us that the electric field in the cable is the strongest right next to the inner conductor of the cable, for $r = a^+$, meaning that dielectric breakdown occurs when the peak-value (amplitude) of this field reaches the critical field value (dielectric strength), E_{cr}, for the dielectric [also see Eq.(2.34)]. Using the notation from Eq.(9.15), that is, $x = b/a$, the breakdown rms voltage of the cable, $|\underline{V}|_{cr}$, thus comes out to be

$$E_{cr} = \underbrace{|\underline{E}(a^+)|\sqrt{2}}_{\text{Peak-value}} = \frac{|\underline{V}|\sqrt{2}}{a\ln(b/a)} \quad \longrightarrow \quad |\underline{V}(x)|_{cr} = \frac{E_{cr}b}{\sqrt{2}}\frac{\ln x}{x} \,. \tag{9.16}$$

Write a MATLAB code to compute and plot the dependence of $|\underline{V}|_{cr}$ on x, and to find x, using

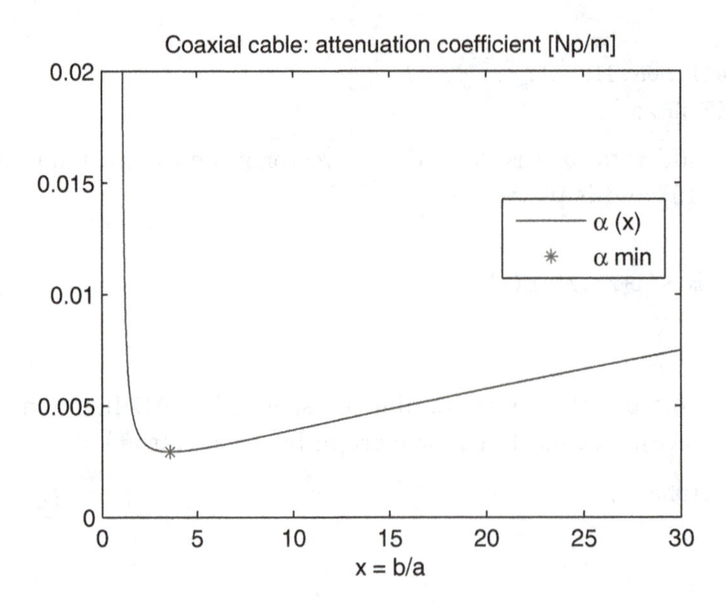

Figure 9.3 Plot of the attenuation coefficient for conductors of a coaxial cable versus the conductor radii ratio $x = b/a$ [value $x = x_{\text{opt}} = 3.59$ giving $\alpha_{\text{c}} = (\alpha_{\text{c}})_{\text{min}}$ is marked in the graph]; for MATLAB Exercise 9.6. *(color figure on CW)*[3]

MATLAB function `max`, such that $|\underline{V}|_{\text{cr}}$ is maximum, as well as this maximum voltage, $(|\underline{V}|_{\text{cr}})_{\text{max}}$ (this is the maximum permissible applied voltage that the cable can carry before a breakdown of its dielectric). Assume that the cable dielectric is polyethylene ($\varepsilon_{\text{r}} = 2.25$, $E_{\text{cr}} = 47$ MV/m) and $b = 5$ cm. *(ME9_7.m on IR)* **H**

HINT:
See the previous MATLAB exercise.

MATLAB EXERCISE 9.8 **Coaxial cable design for maximum permissible power flow.** Repeat the previous MATLAB exercise but for the permissible time-average power flow along the cable (for the safe operation of the cable prior to an eventual dielectric breakdown), $P_{\text{cr}}(x)$, given by a combination of Eqs.(9.9), (9.16), (9.5), and (9.2), namely, by

$$Z_0(x) = \frac{Z_{\text{TEM}}}{2\pi} \ln x \quad \longrightarrow \quad P_{\text{cr}}(x) = \frac{|\underline{V}(x)|_{\text{cr}}^2}{Z_0(x)} = \frac{\pi E_{\text{cr}}^2 b^2}{Z_{\text{TEM}}} \frac{\ln x}{x^2} . \tag{9.17}$$

Based on this, find $(P_{\text{cr}})_{\text{max}}$ and the corresponding x by MATLAB function `max`. *(ME9_8.m on IR)* **H**

---

[3] *CW* = Companion Website (of the book).

MATLAB EXERCISE 9.9 **Three different optimizations of a coaxial cable.** Consider a coaxial cable with a polyethylene ($\varepsilon_r = 2.25$) dielectric and copper ($\sigma_c = 58$ MS/m) conductors at a frequency of $f = 100$ MHz. The outer radius of the cable is $b = 8.6$ mm, the dielectric strength of the dielectric is $E_{cr} = 47$ MV/m, and the losses in the dielectric can be neglected. In MATLAB, compute the attenuation coefficient ($\alpha = \alpha_c$), breakdown rms voltage ($|\underline{V}|_{cr}$), and maximum permissible (breakdown) time-average transferred power (P_{cr}) of the cable for the following values of the inner radius of the cable: (a) $a = b/3.59$ (for which α_c is minimum), (b) $a = b/\mathrm{e}$ (for which $|\underline{V}|_{cr}$ is maximum), and (c) $a = b/\sqrt{\mathrm{e}}$ (for which P_{cr} is maximum). *(ME9_9.m on IR)* **H**

TUTORIAL:

Values of the ratio $x = b/a$ from parts (a), (b), and (c) of the exercise, namely, $x = 3.59$, $x = \mathrm{e}$, and $x = \sqrt{\mathrm{e}}$, respectively, are stored as the three elements of a vector (array) `x`,

```
x = [3.59, exp(1), sqrt(exp(1))];
```

For computing the attenuation coefficient, breakdown voltage, and maximum power flow of the cable, as functions of x, we implement Eqs.(9.15)–(9.17), where the wave impedance of the TEM wave along the cable (Z_{TEM}) and the surface resistance (R_s) of copper are determined as in MATLAB Exercise 9.6. Values of α_c, $|\underline{V}|_{cr}$, and P_{cr} obtained for the three elements of `x` are stored as the corresponding elements of vectors `alphac`, `Vcr`, and `Pcr`; vectors `Ztem` and `Rs` are filled in the same way.

```
Ztem = 120*pi/sqrt(EPSR);
Rs = surfResistance(f,MUR,SIGMA);
alphac = Rs/(2*b*Ztem).*(1+x)./log(x);
Vcr = Ecr*b.*log(x)./(sqrt(2)*x);
Pcr = pi*Ecr^2*b^2/Ztem*log(x)./x.^2;
```

At display of results, vector `label` containing characters `a`, `b`, and `c` is used to label the respective part of the problem, (a), (b), or (c),

```
label = ['a', 'b', 'c'];
for i = 1:length(x)
fprintf('(%s) \n', label(i));
fprintf('alpha = %.4e Np/m.\n', alphac(i));
fprintf('|Vcr| = %.4e V.\n', Vcr(i));
fprintf('Pcr = %.4e W.\n', Pcr(i));
end;
```

and the results (displayed in the Command Window) turn out to be:

```
(a)
alpha = 2.1682e-003 Np/m.
|Vcr| = 1.0176e+005 V.
Pcr = 2.0253e+008 W.
(b)
alpha = 2.2450e-003 Np/m.
|Vcr| = 1.0514e+005 V.
Pcr = 2.7638e+008 W.
```

```
(c)
alpha = 3.1984e-003 Np/m.
|Vcr| = 8.6677e+004 V.
Pcr = 3.7565e+008 W.
```

9.3 Evaluation of Primary and Secondary Circuit Parameters of Transmission Lines

As we shall see in the next chapter, an arbitrary two-conductor transmission line with TEM waves can be analyzed as an electric circuit with distributed parameters, based on a representation of the line by a network of cascaded equal small cells, of length Δz, with lumped elements. These elements are characterized by per-unit-length parameters C', L', R', and G' of the line (studied in this and previous chapters), multiplied by Δz. As C', L', R', and G' are a basis for the circuit analysis of transmission lines (to be presented in the next chapter), they are referred to as primary circuit parameters of a line. The other parameters that will be used in the circuit analysis are the characteristic impedance, Z_0, phase coefficient, β, phase velocity, v_{p}, wavelength, λ_z, and attenuation coefficient, α, of the line. As these parameters can be derived from the primary parameters, they are called secondary circuit parameters of transmission lines. Moreover, once the secondary parameters are known for a given line, they suffice for the analysis (i.e., primary parameters are not needed).

In summary, the capacitance C' in Eqs.(9.3) is determined from a 2-D electrostatic analysis in the cross section of the line, in Fig.9.1. The inductance L' and conductance G' are then obtained from C' using the duality relationships in Eqs.(6.8) and (9.13), respectively, while the resistance R' is evaluated by means of Eqs.(9.12), based on a 2-D magnetostatic analysis in the line cross section (Fig.9.1). The impedance Z_0 is found from Eqs.(9.4), the coefficient β employing the expression in Eqs.(9.1) or as $\beta = \omega\sqrt{L'C'}$ [see Eq.(6.8)], the velocity v_{p} is given by Eq.(7.6), the wavelength along the line equals $\lambda_z = 2\pi/\beta$ [Eq.(7.4)], and the coefficient α is computed from R', G', and Z_0 using Eqs.(9.11)–(9.13).

MATLAB EXERCISE 9.10 **GUI for primary and secondary circuit parameters of transmission lines.** In MATLAB, create a calculator of primary and secondary circuit parameters in the form of a graphical user interface (GUI), with names of structures appearing in a pop-up menu, for low-loss transmission lines with homogeneous dielectrics. For each line, calculate and show the capacitance, inductance, resistance, and conductance per unit length, C', L', R', and G', of the line (primary circuit parameters), and the characteristic impedance, phase coefficient, phase velocity, wavelength, and attenuation coefficient, Z_0, β, v_{p}, λ_z, and α, of the line (secondary circuit parameters). Include the following transmission lines: coaxial cable [C' and R' given by Eqs.(2.26) and (9.14), respectively], thin two-wire line [C' in Eq.(2.27) and $R' = R_{\text{s}}/(\pi a)$], wire-plane transmission line, defined in MATLAB Exercise 6.8 [$C' = 2\pi\varepsilon/\ln(2h/a)$ and $R' = (R_{\text{s1}}/a + R_{\text{s2}}/h)/(2\pi)$, with R_{s1} standing for the skin-effect surface resistance of the wire and R_{s2} for that of the ground plane (distinction between surface resistances of the two conductors of the transmission line is made only in this case)], microstrip line neglecting fringing effects [C' in Eq.(2.29) and $R' = 2R_s/w$], and strip line neglecting fringing [C' in Eq.(2.30) and $R' = R_s/w$]. *[folder ME9_10(GUI) on IR]*

HINT:
See MATLAB Exercise 2.13.

9.4 Transmission Lines with Inhomogeneous Dielectrics

Consider a two-conductor transmission line with an inhomogeneous, lossless ($\sigma_d = 0$), and nonmagnetic ($\mu = \mu_0$) dielectric [an example is the two-wire line with dielectrically coated conductors in Fig.2.13(e)]. We define the effective relative permittivity of the line as

$$\varepsilon_{\rm reff} = \frac{C'}{C_0'} \quad \text{(effective relative permittivity of a transmission line; dimensionless)}\,, \tag{9.18}$$

where C_0' stands for the p.u.l. capacitance of the same line if air-filled [for the example mentioned, the two-wire line in Fig.2.9(c) with $\varepsilon = \varepsilon_0$]. Note that for lines with homogeneous dielectrics ($\varepsilon_r =$ const), $\varepsilon_{\rm reff} = \varepsilon_r$. The phase coefficient [Eqs.(9.1)] of the actual line can now be computed as $\beta = \omega\sqrt{\varepsilon_{\rm reff}}/c_0$ [c_0 is the wave velocity in free space, Eq.(7.7)]. The inductance L' of the line is the same as if the dielectric were air, so that Eq.(6.8) gives $L' = L_0' = \varepsilon_0\mu_0/C_0'$, and the line characteristic impedance is obtained from $Z_0 = \sqrt{L'/C'}$ [Eqs.(9.4)]. Finally, the resistance R' of the line is also the same as for the air-filled line, $R' = R_0'$.

MATLAB EXERCISE 9.11 **Circuit parameters of a line with an inhomogeneous dielectric.** Write a function `circParamInhomogTrLine()` in MATLAB that calculates the primary circuit parameters, C' and L', and secondary circuit parameters, Z_0, β, v_p, and λ_z, of a lossless two-conductor transmission line with an inhomogeneous nonmagnetic dielectric. The input data are: the effective relative permittivity of the line, $\varepsilon_{\rm reff}$, the capacitance per unit length, C_0', of the same line if air-filled, and the operating frequency of the line, f. *(circParamInhomogTrLine.m on IR)*

9.5 Multilayer Printed Circuit Board

Figure 9.4 shows a typical multilayer printed circuit board, which is widely used in digital electronics (e.g., in computers). Observing the interconnects in different layers in the figure, we identify two types of two-conductor transmission lines making up the structure: a microstrip line, Fig.2.9(e), and a strip line, Fig.2.9(f). In MATLAB Exercise 9.10, the primary and secondary circuit parameters of these two lines are computed neglecting the fringing effects. These values are, thus, accurate only if $h \ll w$ in Fig.2.9(e) and (f). Otherwise, there is a considerable fringing field outside the region below the strip in Fig.2.9(e), and the field in this region close to the strip edges is not uniform (edge effects), as illustrated in Fig.9.5, and similarly for the line in Fig.2.9(f). We present here a set of available empirical closed-form formulas for the circuit parameters of microstrip and strip lines.

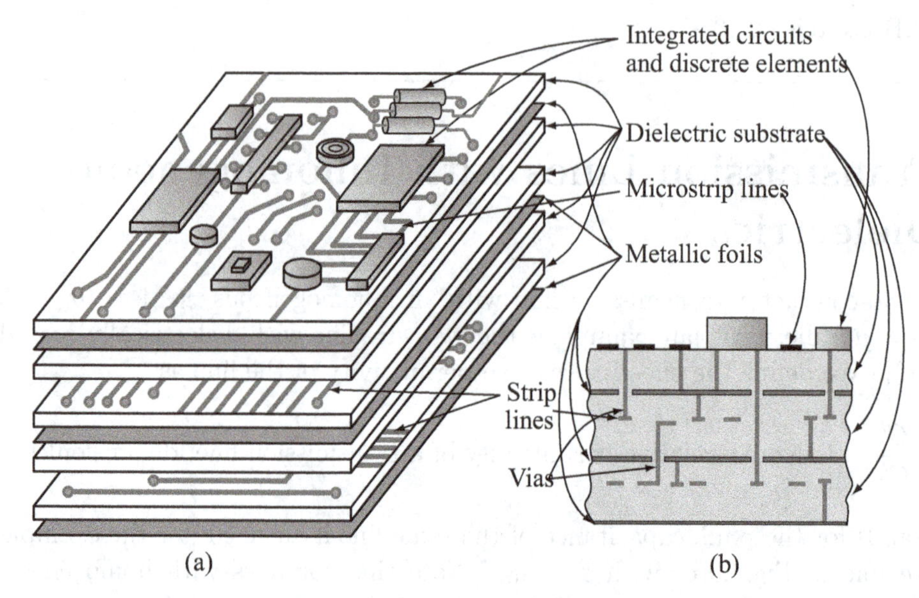

Figure 9.4 Sketch of a typical multilayer printed circuit board: (a) three-dimensional view of the structure and (b) detail of its cross section.

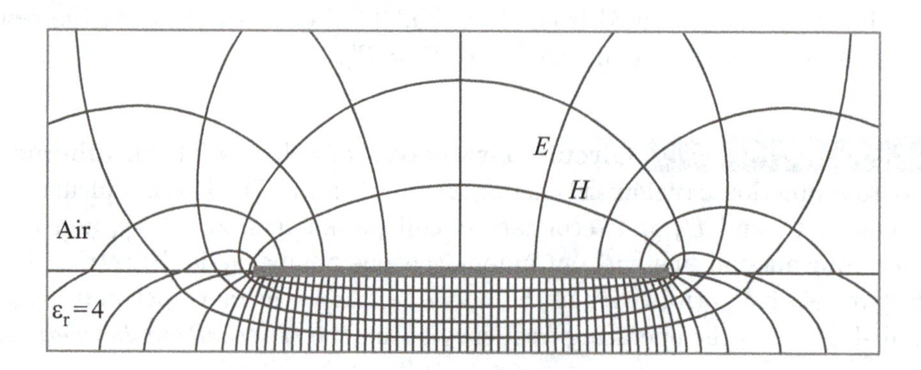

Figure 9.5 Electric (E) and magnetic (H) field lines in a cross section of a microstrip line with the strip width to substrate height ratio $w/h = 5.4$ and substrate relative permittivity $\varepsilon_r = 4$; field pattern plots are obtained by numerical analysis based on a method of moments (see Section 1.8).

Since the electric field in a microstrip line, Fig.9.5, is only partly in the dielectric substrate, of relative permittivity ε_r, and the rest is in air above it, the dielectric of the line is actually inhomogeneous, and the effective relative permittivity of the line, Eq.(9.18), is given by

$$\varepsilon_{\mathrm{reff}} = \frac{\varepsilon_r + 1}{2} + \frac{\varepsilon_r - 1}{2}\left[\left(1 + 12\frac{h}{w}\right)^{-1/2} + p\right] \quad (\varepsilon_{\mathrm{reff}} \text{ – microstrip line}) , \tag{9.19}$$

where $p = 0.04(1 - w/h)^2$ if $w/h < 1$, and $p = 0$ otherwise. The characteristic impedance of the line is then found as

$$Z_0 = \frac{\eta_0}{2\pi\sqrt{\varepsilon_{\rm reff}}} \ln\left(\frac{8h}{w} + \frac{w}{4h}\right) \quad \text{for} \quad \frac{w}{h} \leq 1 ,$$

$$Z_0 = \frac{\eta_0}{\sqrt{\varepsilon_{\rm reff}}}\left[\frac{w}{h} + 1.393 + 0.667\ln\left(\frac{w}{h} + 1.444\right)\right]^{-1} \quad \text{for} \quad \frac{w}{h} > 1$$

$$(Z_0 - \text{microstrip line} - \text{analysis}) , \quad (9.20)$$

with η_0 being the intrinsic impedance of free space, Eq.(7.3). For design (synthesis) purposes, namely, to find w/h for a desired Z_0 of the line and given $\varepsilon_{\rm r}$ of the substrate dielectric, the following formulas are used:

$$\frac{w}{h} = 8\left(\mathrm{e}^{A} - 2\,\mathrm{e}^{-A}\right)^{-1} \quad \text{for} \quad \frac{w}{h} \leq 2 ,$$

$$\frac{w}{h} = \frac{\varepsilon_{\rm r} - 1}{\pi\varepsilon_{\rm r}}\left[\ln(B-1) + 0.39 - \frac{0.61}{\varepsilon_{\rm r}}\right] + \frac{2}{\pi}\left[B - 1 - \ln(2B-1)\right] \quad \text{for} \quad \frac{w}{h} > 2 ,$$

$$A = \pi\sqrt{2(\varepsilon_{\rm r}+1)}\,\frac{Z_0}{\eta_0} + \frac{\varepsilon_{\rm r} - 1}{\varepsilon_{\rm r} + 1}\left(0.23 + \frac{0.11}{\varepsilon_{\rm r}}\right) , \quad B = \frac{\pi}{2\sqrt{\varepsilon_{\rm r}}}\frac{\eta_0}{Z_0} \quad (\text{microstrip line} - \text{synthesis}) . \quad (9.21)$$

The phase coefficient, phase velocity, and wavelength along the line are found as in MATLAB Exercise 9.11, from $\beta = \omega\sqrt{\varepsilon_{\rm reff}}/c_0$, $v_{\rm p} = \omega/\beta$, and $\lambda_z = 2\pi/\beta$. The attenuation coefficient of the line, $\alpha = \alpha_{\rm c} + \alpha_{\rm d}$, is computed as

$$\alpha_{\rm c} = \frac{R_{\rm s}}{wZ_0} , \quad \alpha_{\rm d} = \frac{\pi f \tan\delta_{\rm d}\sqrt{\varepsilon_{\rm reff}}}{c_0}\frac{\varepsilon_{\rm r}}{\varepsilon_{\rm r} - 1}\frac{\varepsilon_{\rm reff} - 1}{\varepsilon_{\rm reff}} \quad (\alpha - \text{microstrip line}) , \quad (9.22)$$

where $\tan\delta_{\rm d} = \sigma_{\rm d}/(\omega\varepsilon)$ is the so-called loss tangent of the substrate.

The dielectric of a strip line, Fig.2.9(f), is homogeneous, and hence $\varepsilon_{\rm reff} = \varepsilon_{\rm r}$. The empirical formulas for the characteristic impedance of the line read

$$Z_0 = \frac{\eta_0}{4\sqrt{\varepsilon_{\rm r}}\,[w/(2h) + 0.441 - s]} , \quad s = \left(0.35 - \frac{w}{2h}\right)^2 \quad \text{for} \quad \frac{w}{h} < 0.7 , \quad s = 0 \quad \text{for} \quad \frac{w}{h} \geq 0.7$$

$$(Z_0 - \text{strip line} - \text{analysis}) . \quad (9.23)$$

To design a line with a certain Z_0 for a given $\varepsilon_{\rm r}$, these equations can be easily solved for w/h (assuming that Z_0 and $\varepsilon_{\rm r}$ are known), and the result is

$$\frac{w}{h} = \frac{\eta_0}{2\sqrt{\varepsilon_{\rm r}}Z_0} - 0.882 \quad \left(\frac{w}{h} \geq 0.7\right) \quad \text{for} \quad \sqrt{\varepsilon_{\rm r}}Z_0 \leq 0.316\eta_0 ,$$

$$\frac{w}{h} = 1.7 - \sqrt{4.164 - \frac{\eta_0}{\sqrt{\varepsilon_{\rm r}}Z_0}} \quad \left(\frac{w}{h} < 0.7\right) \quad \text{for} \quad \sqrt{\varepsilon_{\rm r}}Z_0 > 0.316\eta_0 \quad (\text{strip line} - \text{synthesis}) . \quad (9.24)$$

MATLAB EXERCISE 9.12 **Effective relative permittivity of a microstrip line.** Write a function `epsrEffMicrostrip()` in MATLAB that computes the effective relative permittivity,

$\varepsilon_{\rm reff}$, of a microstrip line [in Fig.2.9(e)] including fringing effects (Fig.9.5), based on Eq.(9.19), for the given substrate relative permittivity, $\varepsilon_{\rm r}$, and strip width to substrate height ratio, w/h. *(epsrEffMicrostrip.m on IR)*

MATLAB EXERCISE 9.13 **Characteristic impedance – microstrip line with fringing – analysis.** Write a function `microstripAnalysis()` in MATLAB that returns the characteristic impedance, Z_0, of the microstrip line from the previous MATLAB exercise. *(microstripAnalysis.m on IR)*

TUTORIAL:

From the input data $\varepsilon_{\rm r}$ (`EPSR`) and w/h (`ratio`), Z_0 (`Z0`) is computed implementing Eqs.(9.20), with the effective relative permittivity, $\varepsilon_{\rm reff}$, of the line being previously found by function `epsrEffMicrostrip` (from the previous MATLAB exercise),

```
function Z0 = microstripAnalysis(EPSR,ratio)
MU0 = 4*pi*10^(-7);
EPS0 = 8.8542*10^(-12);
ETA0 = sqrt(MU0/EPS0);
Ereff = epsrEffMicrostrip(EPSR,ratio);
if (ratio <= 1)
Z0 = ETA0/(2*pi*sqrt(Ereff))*log(8/ratio+ratio/4);
else
Z0 = ETA0/sqrt(Ereff)*(ratio+1.393+0.667*log(ratio+1.444))^(-1);
end;
```

MATLAB EXERCISE 9.14 **Microstrip line – synthesis function.** Write a function `microstripSynthesis()` in MATLAB for design (synthesis) of a microstrip line – to find the strip width to substrate height ratio, w/h, that results in a desired characteristic impedance of the line, Z_0, for a given relative permittivity of the substrate dielectric, $\varepsilon_{\rm r}$. *(microstripSynthesis.m on IR)*

TUTORIAL:

Using Eqs.(9.21), we compute constants A and B, as well as the ratio w/h according to expressions for both cases: $w/h \leq 2$ (result is `ratio1`) and $w/h > 2$ (result is `ratio2`), as we do not know in advance which condition (if any) applies.

```
function [ratio] = microstripSynthesis(Z0,EPSR,display)
MU0 = 4*pi*10^(-7);
EPS0 = 8.8542*10^(-12);
ratio = 0;
ETA0 = sqrt(MU0/EPS0);
A = pi*sqrt(2*(EPSR+1))*Z0/ETA0+(EPSR-1)/(EPSR+1)*(0.23+0.11/EPSR);
B = pi/2/sqrt(EPSR)*ETA0/Z0;
```

```
ratio1 = 8/(exp(A)-2*exp(-A));
ratio2 = (EPSR-1)/pi/EPSR*(log(B-1)+0.39-0.61/EPSR)+2/pi*(B-1-log(2*B-1));
```

Then we check whether the result for w/h obtained from the expression for $w/h \leq 2$ in Eqs.(9.21) actually satisfies this condition ($w/h \leq 2$) or w/h from the other expression satisfies the other inequality ($w/h > 2$), and the solution to the design problem (`ratio`) is the result that satisfies the corresponding condition. However, for some combinations of Z_0 and ε_r at input, both checks fail, i.e., it is impossible to design a microstrip line with these parameters; in such cases, a proper information (no solution) is displayed. Finally, whether or not any result/information is displayed is determined by the input parameter `display`.

```
if ratio1 <= 2
ratio = ratio1;
elseif ratio2 > 2
ratio = ratio2;
else
if display == 1
disp('No solution for given Z0 and EPSR.');
end;
end;
if ratio ~= 0
if display == 1
disp('For given Z0 and EPSR, the solution is');
fprintf('w/h = %g.\n', ratio);
end;
end;
```

MATLAB EXERCISE 9.15 **Attenuation coefficient of a microstrip line.** Write a function `alphaMicrostrip()` to calculate the attenuation coefficient representing the losses in conductors, α_c, and that for the losses in the dielectric, α_d, of a microstrip line, using Eqs.(9.22), as well as the total attenuation coefficient of the line, $\alpha = \alpha_c + \alpha_d$. *(alphaMicrostrip.m on IR)*

MATLAB EXERCISE 9.16 **Characteristic impedance – strip line with fringing – analysis.** Write a function `stripLineAnalysis()` in MATLAB that calculates the characteristic impedance (Z_0) of a strip line, in Fig.2.9(f), including fringing effects, with the use of Eqs.(9.23); the input arguments of the function are ε_r and w/h. *(stripLineAnalysis.m on IR)*

MATLAB EXERCISE 9.17 **Strip line – synthesis function.** Write a function `stripLineSynthesis()` in MATLAB for design of a strip line (whose fringing effects are not negligible) – for finding the geometrical ratio w/h in Fig.2.9(f) for a desired characteristic impedance,

Z_0, of the line and given relative permittivity, ε_r, of the line dielectric. *(stripLineSynthesis.m on IR)*

HINT:
Use Eqs.(9.24), and see MATLAB Exercise 9.14.

MATLAB EXERCISE 9.18 **Microstrip lines with different strip width to height ratios.** Consider a microstrip line with a copper strip and ground plane, dielectric substrate parameters $\varepsilon_r = 4$ and $\tan\delta_d = 10^{-4}$ ($\mu_r = 1$), strip width w, and substrate thickness $h = 2$ mm, and perform the following computations and analysis in MATLAB. Compute the primary and secondary circuit parameters of the line, taking into account the fringing effects, for the following w/h ratios: (a) 0.05, (b) 0.1, (c) 0.5, (d) 1, (e) 2, (f) 10, and (g) 20. (h) Compare the results in cases (d)–(g) with the corresponding values of circuit parameters of the line obtained neglecting the fringing effects (see MATLAB Exercise 9.10). (i) For cases (a)–(d), compare the results to those obtained for a wire-plane transmission line (MATLAB Exercise 9.10) with the conducting strip in Fig.2.9(e) replaced by a thin wire of an equivalent radius equal to $a = w/4$. The results should be tabulated in a text file `microstrip.txt`. *(ME9_18.m on IR)* **H**

HINT:
Compute the effective relative permittivity, $\varepsilon_{\text{reff}}$, characteristic impedance, Z_0, and attenuation coefficients for conductors and dielectric, α_c and α_d, of the microstrip line for the given w/h ratios, cases (a)–(g), with the fringing effects taken into account – using functions `epsrEffMicrostrip`, `microstripAnalysis`, and `alphaMicrostrip`, from MATLAB Exercises 9.12, 9.13, and 9.15, respectively. The phase coefficient, β, phase velocity, v_p, and wavelength, λ_z, along the line can be calculated as in MATLAB Exercise 9.11.

Then, combining Eqs.(9.18), (6.8), and (9.4), the per-unit-length capacitance and inductance of the line can be found as $C' = \sqrt{\varepsilon_{\text{reff}}}/(c_0 Z_0)$ and $L' = \varepsilon_{\text{reff}}\varepsilon_0\mu_0/C'$. From Eqs.(9.12) and (9.13), the high-frequency resistance and conductance p.u.l. of the line come out to be $R' = 2Z_0\alpha_c$ and $G' = 2\alpha_d/Z_0$.

Circuit parameters of the line in cases (d)–(g) neglecting the fringing effects can be obtained using GUI developed in MATLAB Exercise 9.10 or implementing the respective equations from that exercise.

Finally, the same GUI (or the corresponding equations) can be used to evaluate, for cases (a)–(d), circuit parameters of the wire-plane transmission line with $a = w/4$ and the entire half-space above the ground plane filled with the line dielectric.

For MATLAB operations with a text file, see MATLAB Exercise 7.21.

MATLAB EXERCISE 9.19 **Analysis of a microstrip line with and without fringing.** Write a MATLAB program that computes and plots the characteristic impedance of a microstrip line [Fig.2.9(e)] with and without taking fringing effects into account, for $1 \le w/h \le 100$. The program also finds the minimum w/h for which the relative error in computing Z_0 neglecting fringing, with respect to the solution with fringing, is smaller than 5%, displays it in the Command

Window, and marks it on the plot. Input data are: $\varepsilon_r = 4$ and $h = 1$ mm. *(ME9_19.m on IR)*

TUTORIAL:
We first specify the input data and define a vector `ratio` (with w/h values for plotting) and the corresponding vector `w` (with w values),

```
EPSO = 8.8542*10^(-12);
MUO = 4*pi*10^(-7);
ETAO = sqrt(MUO/EPSO);
EPSR = 4;
h = 0.001;
ratio = 1:0.05:100;
w = ratio.*h;
```

In a `for` loop, we then compute Z_0 with fringing effects (`Zfe`) and Z_0 neglecting fringing (`Znf`), for each element of `w`, using function `microstripAnalysis` (from MATLAB Exercise 9.13) and expression $Z_0 = \eta_0 h/(\sqrt{\varepsilon_r}\, w)$ [from Eqs.(9.4), (2.29), and (7.3)], respectively,

```
for i=1:length(w)
Zfe(i) = microstripAnalysis(EPSR,ratio(i));
Znf(i) = ETAO/sqrt(EPSR)*h/w(i);
end;
```

Next, we evaluate the relative error in computing Z_0 neglecting fringing, and the minimum w/h that results in an error smaller than 5% – by means of MATLAB function `min` (see MATLAB Exercise 3.12, for instance),

```
error = abs((Znf-Zfe)./Zfe);
[minE,k]= min(abs(error - 0.05));
ratioMin = ratio(k);
```

The displayed result in the Command Window reads:

```
For w/h greater than 62.05, error is smaller than 5%.
```

The plot of Z_0 against w/h is shown in Fig.9.6.

MATLAB EXERCISE 9.20 **Analysis of a strip line with and without fringing.** Repeat the previous MATLAB exercise but for a strip line [Fig.2.9(f)]. *(ME9_20.m on IR)*

HINT:
Line characteristic impedance (Z_0) with and without fringing effects can be obtained using function `stripLineAnalysis` (from MATLAB Exercise 9.16) and expression $Z_0 = \eta_0 h/(2\sqrt{\varepsilon_r}\, w)$ [from Eqs.(9.4), (2.30), and (7.3)], respectively.

MATLAB EXERCISE 9.21 **Microstrip line design.** In MATLAB, design a microstrip line that has a characteristic impedance of (a) $Z_0 = 75\ \Omega$ and (b) $Z_0 = 50\ \Omega$, respectively,

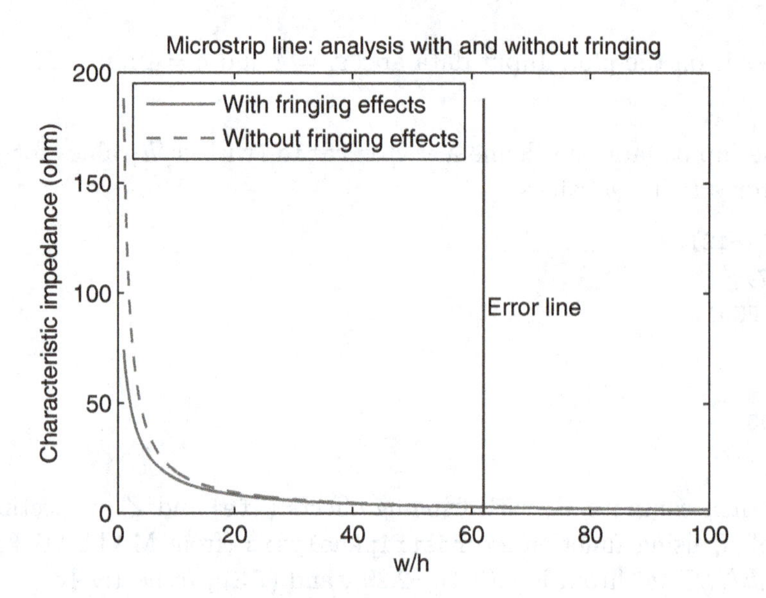

Figure 9.6 Characteristic impedance of a microstrip line with and without taking fringing effects into account, against the strip width to substrate height ratio, w/h (error line marks the minimum ratio w/h for which the relative error in computing Z_0 neglecting fringing, with respect to the results with fringing, is smaller than 5%); for MATLAB Exercise 9.19. *(color figure on CW)*

for a given relative permittivity of the substrate dielectric, $\varepsilon_r = 4$; use synthesis function **microstripSynthesis** (from MATLAB Exercise 9.14). Check the results by means of analysis function **microstripAnalysis** (from MATLAB Exercise 9.13). *(ME9_21.m on IR)* **H**

MATLAB EXERCISE 9.22 **Strip line design.** Using MATLAB, design strip lines with characteristic impedances $Z_0 = 50\ \Omega$ and $Z_0 = 75\ \Omega$, respectively, if the relative permittivity of the line dielectric is $\varepsilon_r = 4$ in both cases. *(ME9_22.m on IR)* **H**

TUTORIAL:

We use synthesis function **stripLineSynthesis** (from MATLAB Exercise 9.17) for each of the two cases, $Z_0 = 50\ \Omega$ and $Z_0 = 75\ \Omega$, respectively ($\varepsilon_r = 4$ in both cases), check the results of synthesis by means of analysis function **stripLineAnalysis** (from MATLAB Exercise 9.16), and display all data in the Command Window,

```
Z01 = 50;
Z02 = 75;
EPSR = 4;
ratio1 = stripLineSynthesis(Z01,EPSR);
ratio2 = stripLineSynthesis(Z02,EPSR);
Z01check = stripLineAnalysis(ratio1,EPSR);
Z02check = stripLineAnalysis(ratio2,EPSR);
fprintf('Strip line synthesis:  For Z0 = %g ohm',Z01);
```

```
fprintf(' and EPSr = %g, w/h = %d.\n',EPSR,ratio1);
fprintf('Checking the results of synthesis:  ');
fprintf('Analysis gives:  Z0 = %g ohm.\n\n',Z01check);
fprintf('Strip line synthesis:  For Z0 = %g ohm',Z02);
fprintf(' and EPSr = %g, w/h = %d.\n',EPSR,ratio2);
fprintf('Checking the results of synthesis:  ');
fprintf('Analysis gives:  Z0 = %g ohm.\n\n',Z02check);
```

Here is the display with results of the synthesis and of the "inverse" analysis (to validate the design):

```
Strip line synthesis:  For Z0 = 50 ohm and EPSr = 4, w/h = 1.001650e+000.
Checking the results of synthesis:  Analysis gives:  Z0 = 50 ohm.

Strip line synthesis:  For Z0 = 75 ohm and EPSr = 4, w/h = 4.145171e-001.
Checking the results of synthesis:  Analysis gives:  Z0 = 75 ohm.
```

MATLAB EXERCISE 9.23 **Design of a microstrip line with same properties as a coaxial cable.** Design a microstrip line that has the same characteristic impedance for the same relative permittivity of the dielectric as a coaxial cable, in Fig.9.2, with $a = 1$ mm, $b = 3.5$ mm, $\varepsilon_r = 2.25$, and $\mu = \mu_0$. *(ME9_23.m on IR)* **H**

HINT:
Design should be carried out in the same way as in MATLAB Exercise 9.21.

MATLAB EXERCISE 9.24 **Design of a strip line with same properties as a two-wire line.** Using MATLAB, design a strip line that has the same Z_0 for the same dielectric as a nonsymmetrical thin two-wire transmission line [the same as in Fig.2.9(c) but with wires of different radii, a and b ($a \neq b$)], whose p.u.l. capacitance is given by $C' = \pi\varepsilon/\ln(d/\sqrt{ab})$, for $a = 6$ mm, $b = 3$ mm, $d = 90$ mm, and $\varepsilon_r = 3$ ($\mu_r = 1$). *(ME9_24.m on IR)* **H**

HINT:
See MATLAB Exercise 9.22.

MATLAB EXERCISE 9.25 **One more strip line design.** Repeat the previous MATLAB exercise but with the following (changed) dimensions of the nonsymmetrical two-wire line: $a = 10$ mm, $b = 8$ mm, and $d = 60$ mm (permittivity of the dielectric is not changed). *(ME9_25.m on IR)* **H**

---

10 CIRCUIT ANALYSIS OF TRANSMISSION LINES

Introduction:

This chapter takes over the primary and secondary circuit parameters of transmission lines computed in the field analysis of lines in the previous chapter, and uses them to solve for the voltage and current along lossless and lossy lines, with various excitations and load terminations. Most importantly, this is a circuit analysis of transmission lines, using only pure circuit-theory concepts to develop the complete frequency-domain and transient analysis of lines as circuits with distributed parameters whose per-unit-length characteristics are already known. The analysis is based on a circuit model of an arbitrary two-conductor transmission line in the form of a ladder network of elementary circuit cells with lumped elements, and on circuit differential equations for this network, termed telegrapher's equations, which can be easily solved for voltages and currents on the network. We also introduce and implement a graphical technique for the circuit analysis and design of transmission lines in the frequency domain based on the so-called Smith chart. Transient analysis of transmission lines will cover arbitrary excitations of lines, including step and pulse signals, and a variety of line terminations, including reactive loads, and both matched and unmatched conditions at either end of the line.

10.1 Telegrapher's Equations and Their Solution

Figure 10.1 (upper part) shows a circuit-theory representation of an arbitrary two-conductor lossy transmission line, where a pair of parallel horizontal thick lines in the schematic diagram, although resembling a two-wire transmission line, symbolizes a structure with conductors of completely arbitrary cross sections (Fig.9.1) and a generally inhomogeneous dielectric. We subdivide the line into short sections, of length Δz, so that, using primary circuit parameters C', L', R', and G' of the line – studied in the previous chapter, each such section can be represented by a circuit cell shown in Fig.10.1. From Kirchhoff's laws for the cells and current-voltage characteristics (element laws) for their elements, we obtain, in the limit of $\Delta z \to 0$, transmission-line equations or telegrapher's equations for the complex rms voltage and current on the line:

$$\frac{\mathrm{d}\underline{V}}{\mathrm{d}z} = -\underbrace{(R' + \mathrm{j}\omega L')}_{\underline{Z}'}\,\underline{I}\,, \quad \frac{\mathrm{d}\underline{I}}{\mathrm{d}z} = -\underbrace{(G' + \mathrm{j}\omega C')}_{\underline{Y}'}\,\underline{V} \quad \text{(telegrapher's equations)}\,, \tag{10.1}$$

whose general solutions are complex exponential functions in z given by

$$\underline{V}(z) = \underbrace{\underline{V}_{\mathrm{i}0}\,\mathrm{e}^{-\underline{\gamma}z}}_{\text{Incident wave}} + \underbrace{\underline{V}_{\mathrm{r}0}\,\mathrm{e}^{\underline{\gamma}z}}_{\text{Reflected wave}}\,, \quad \underline{I}(z) = \underbrace{\frac{\underline{V}_{\mathrm{i}0}}{\underline{Z}_0}\,\mathrm{e}^{-\underline{\gamma}z}}_{\text{Incident wave}} + \underbrace{\left(-\frac{\underline{V}_{\mathrm{r}0}}{\underline{Z}_0}\,\mathrm{e}^{\underline{\gamma}z}\right)}_{\text{Reflected wave}}\,,$$

$$\underline{\gamma} = \sqrt{\underline{Z}'\underline{Y}'} = \alpha + \mathrm{j}\beta\,, \quad \underline{Z}_0 = \sqrt{\frac{\underline{Z}'}{\underline{Y}'}} = |\underline{Z}_0|\,\mathrm{e}^{\mathrm{j}\phi} \quad \text{(solutions for } \underline{V} \text{ and } \underline{I} \text{ on a line)}\,. \tag{10.2}$$

As expected, the total voltage and current waves along the line are, in general, sums of two oppositely directed traveling waves, an incident (forward) wave, propagating in the positive z direction, and a reflected (backward) wave, progressing in the negative z direction [analogously to Eqs.(8.7) and (8.8)]. As in Eqs.(7.12), the expressions for the instantaneous incident voltage and current along the line are

$$v_{\rm i}(z,t) = V_{\rm i0}\sqrt{2}\,{\rm e}^{-\alpha z}\cos(\omega t - \beta z + \theta_{\rm i0})\ , \quad i_{\rm i}(z,t) = \frac{V_{\rm i0}}{|\underline{Z}_0|}\sqrt{2}\,{\rm e}^{-\alpha z}\cos(\omega t - \beta z + \theta_{\rm i0} - \phi) \quad (10.3)$$

($\underline{V}_{\rm i0} = V_{\rm i0}\,{\rm e}^{{\rm j}\theta_{\rm i0}}$), and similarly for $v_{\rm r}$ and $i_{\rm r}$. For transmission lines with small losses [see Eqs.(9.10)], we have $R' \ll \omega L'$ and $G' \ll \omega C'$, and hence the characteristic impedance, $\underline{Z}_0$, and the attenuation and phase coefficients, α and β, of the line can be computed (approximately) using Eqs.(9.4), (9.12), (9.13), (9.1), and (6.8), namely, as

$$\underline{Z}_0 = Z_0 = \sqrt{\frac{L'}{C'}}\ , \quad \alpha = \frac{R'}{2Z_0} + \frac{G'}{2Y_0}\ , \quad \beta = \omega\sqrt{L'C'} \quad \text{(line with small losses)}\ . \quad (10.4)$$

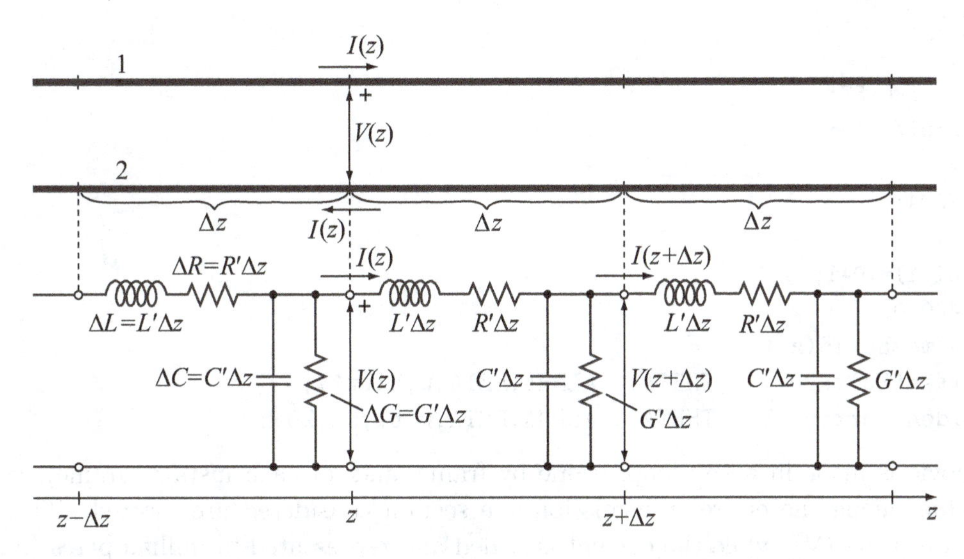

Figure 10.1 Circuit model of a two-conductor lossy transmission line in an ac regime.

MATLAB EXERCISE 10.1 Instantaneous incident voltage and current along a lossy line. Write functions `incidentVoltage()` and `incidentCurrent()` in MATLAB that calculate the instantaneous incident voltage and current, respectively, along a two-conductor lossy transmission line, $v_{\rm i}(z,t)$ and $i_{\rm i}(z,t)$, given by Eqs.(10.3). The input to the functions consists of the rms value and initial phase of the voltage for $z = 0$, $V_{\rm i0}$ and $\theta_{\rm i0}$, operating angular frequency of the line, ω, line attenuation and phase coefficients, α and β, coordinate z, and time t, where, in general, `z` and `t` are arrays or matrices (of values), rather than a single value. Add one more input parameter to function `incidentCurrent`: the complex characteristic impedance of the line, $\underline{Z}_0$. *(incidentVoltage.m and incidentCurrent.m on IR)*[1]

[1] *IR* = Instructor Resources (for the book).

MATLAB EXERCISE 10.2 Travel and attenuation of voltage and current waves – movie. Create a MATLAB movie that visualizes the incident voltage and current waves on a lossy transmission line, in Fig.10.1, according to Eqs.(10.3), for $V_{i0} = 0.1$ V, $\theta_{i0} = 0$, $f = 1$ GHz, $\alpha = 0.55$ Np/m, $\beta = 35$ rad/m, and $\underline{Z}_0 = (36.5 - \mathrm{j}0.57)$ Ω. The movie lasts ten time periods of the waves, $10T$, and spans a range of ten wavelengths along the z-axis, $-10\lambda_z \le z \le 0$ ($\lambda_z = 2\pi/\beta$). *(ME10_2.m on IR)*

TUTORIAL:

For specified input data and defined spatial and temporal meshes, matrices `z1` and `t1`, incident voltage and current, $v_i(z,t)$ and $i_i(z,t)$ (`Vi` and `Ii`), are calculated using functions `incidentVoltage` and `incidentCurrent` (from the previous MATLAB exercise),

```
Vi0 = 0.1;
THETAi0 = 0;
f = 10^9;
ALPHA = 0.55;
BETA = 35;
Z0 = 36.5 - i*0.57;
LAMBDA = 2*pi/BETA;
w = 2*pi*f;
l = 10*LAMBDA;
T = 1/f;
t = (0:0.01:1)*10*T;
z = -l:l/400:0;
[z1,t1] = meshgrid(z,t);
Vi = incidentVoltage(Vi0,THETAi0,ALPHA,BETA,w,t1,z1);
Ii = incidentCurrent(Vi0,THETAi0,ALPHA,BETA,w,t1,z1,Z0);
```

The movie is made in a `for` loop, frame by frame, and, at each instant (frame), voltage and current values along the entire transmission-line section considered are plotted. The voltage is represented in volts (V), while the current is scaled and represented in milliamperes (mA).

```
for k = 1:length(t)
plot(z,Vi(k,:),'b'); hold on;
plot(z,10*Ii(k,:),'r--'); hold off;
xlabel('z[m]');
ylabel('Vi(z,t)[V], 10*Ii(z,t)[A]');
title('Incident signals - on a lossy transmission line');
legend('Vi(z,t)','Ii(z,t)');
ylim(max(max(max(Vi)),max(max(10*Ii)))*1.5*[-1,1]);
M(k) = getframe;
end;
```

Figure 10.2 shows a snapshot (frame) of the movie.

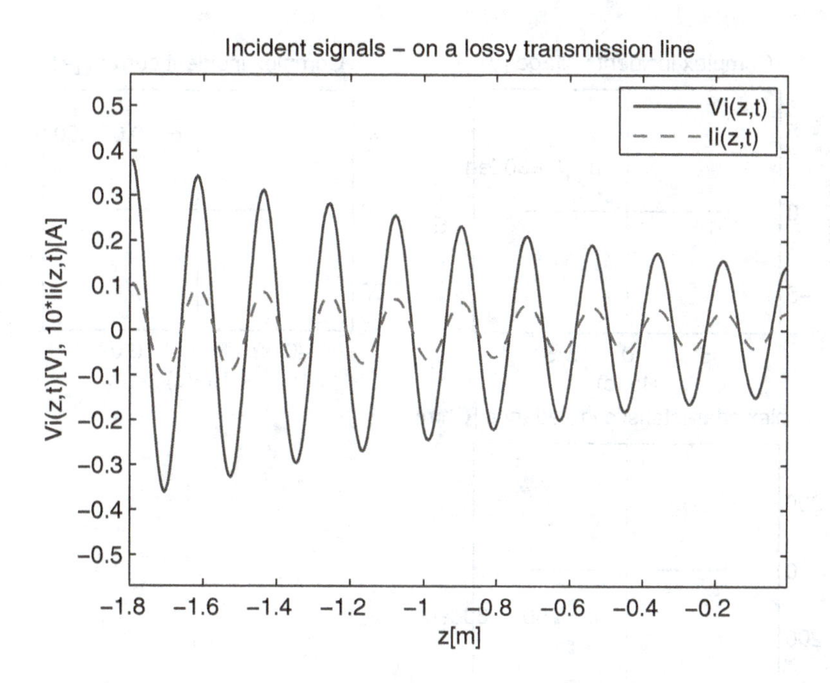

Figure 10.2 Snapshot of a MATLAB movie visualizing the travel and attenuation of incident voltage and current waves along a lossy transmission line (Fig.10.1); for MATLAB Exercise 10.2. *(color figure on CW)*[2]

MATLAB EXERCISE 10.3 **Plotting line voltage and current in the complex plane.** Complex rms voltage and current of an incident wave equal $\underline{V}_{\rm i}(z) = 5\,{\rm e}^{{\rm j}\pi/6}$ V and $\underline{I}_{\rm i}(z) = 20\,{\rm e}^{{\rm j}\pi/3}$ mA, respectively, in a cross section (defined by the coordinate z) of a transmission line (in Fig.10.1). In MATLAB, find the complex characteristic impedance of the line, $\underline{Z}_0$, and plot $\underline{V}_{\rm i}(z)$, $\underline{I}_{\rm i}(z)$, and $\underline{Z}_0$ in three separate graphs in the complex plane, as in Fig.6.14(c). *(ME10_3.m on IR)* **H**[3]

HINT:
From Eqs.(10.2), $\underline{Z}_0$ can be computed as $\underline{Z}_0 = \underline{V}_{\rm i}(z)/\underline{I}_{\rm i}(z)$. Use function `cplxNumPlot`, developed in MATLAB Exercise 6.20, to plot the three complex numbers in the complex plane. The tree graphs are shown in Fig.10.3.

10.2 Reflection Coefficient for Transmission Lines

Let the terminal network at the beginning of a transmission line be a voltage generator of complex rms electromotive force (open-circuit voltage) $\underline{\mathcal{E}}$ and complex internal (series) impedance $\underline{Z}_{\rm g}$, as shown in Fig.10.4. In general, such a generator represents the Thévenin equivalent generator (circuit), with respect to the line input terminals, of an arbitrary input network. In addition, let

[2] *CW* = Companion Website (of the book).
[3] **H** = recommended to be done also "by hand," i.e., not using MATLAB.

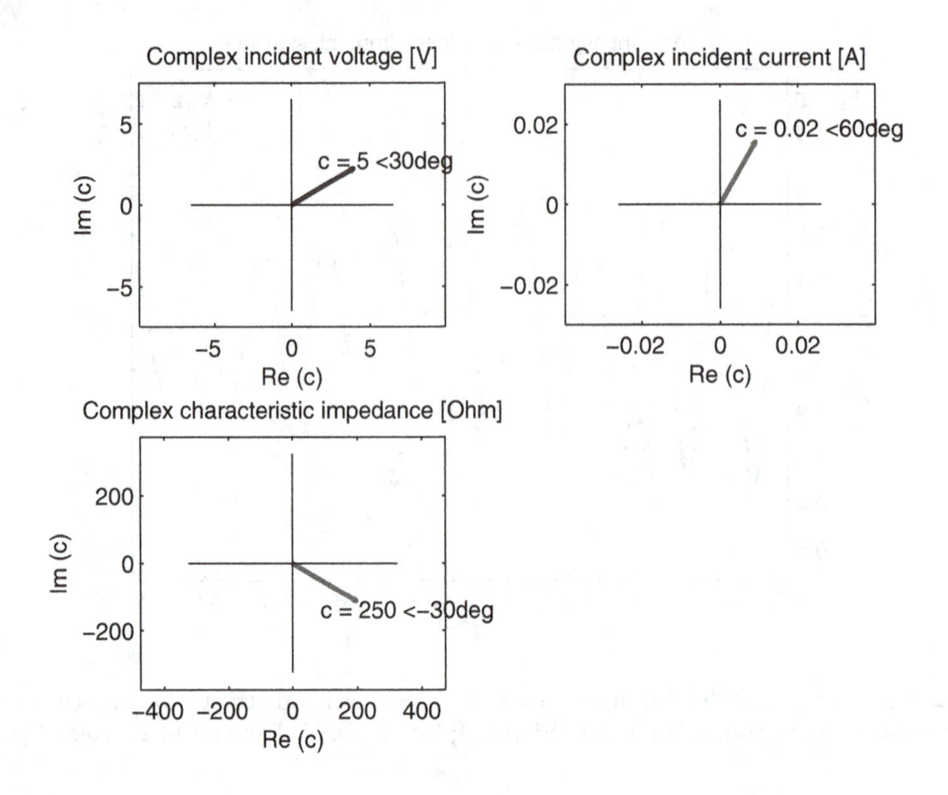

Figure 10.3 MATLAB graphs showing complex voltage and current of an incident wave in a cross section of a transmission line, $\underline{V}_{\mathrm{i}}(z)$ and $\underline{I}_{\mathrm{i}}(z)$, and complex characteristic impedance of the line, $\underline{Z}_0$ – in the complex plane, using function `cplxNumPlot` (from MATLAB Exercise 6.20); for MATLAB Exercise 10.3. *(color figure on CW)*

the other end of the line be terminated in a load of complex impedance $\underline{Z}_{\mathrm{L}}$, which, in general, is an equivalent (input) impedance of an arbitrary passive (with no generators) output network. Finally, we adopt the origin of the z-axis to be at the output terminals of the line (i.e., at the load), so that, denoting the length of the line by l, the location of the line input terminals (generator) is defined by $z = -l$ (Fig.10.4).

Boundary conditions at the load terminals ($z = 0$) in Fig.10.4 for the total voltage and current of the line, given by Eqs.(10.2), result, analogously to Eq.(8.11), in the following solution for the ratio of $\underline{V}_{\mathrm{r0}}$ and $\underline{V}_{\mathrm{i0}}$:

$$\underline{\Gamma}_{\mathrm{L}} = \frac{\underline{V}_{\mathrm{r0}}}{\underline{V}_{\mathrm{i0}}} = \frac{\underline{Z}_{\mathrm{L}} - \underline{Z}_0}{\underline{Z}_{\mathrm{L}} + \underline{Z}_0} \quad \text{(load voltage reflection coefficient)}\,, \tag{10.5}$$

which we term the load voltage reflection coefficient of the line. Note that, from Eqs.(10.2), the load reflection coefficient of the line for currents,

$$\underline{\Gamma}_{\text{for currents}} = \frac{\underline{I}_{\mathrm{r}}(0)}{\underline{I}_{\mathrm{i}}(0)} = -\frac{\underline{V}_{\mathrm{r0}}}{\underline{V}_{\mathrm{i0}}} = -\underline{\Gamma}_{\mathrm{L}}\,, \tag{10.6}$$

comes out to be just opposite to the voltage coefficient. Note also that the complex $\underline{\Gamma}_{\mathrm{L}}$ can be

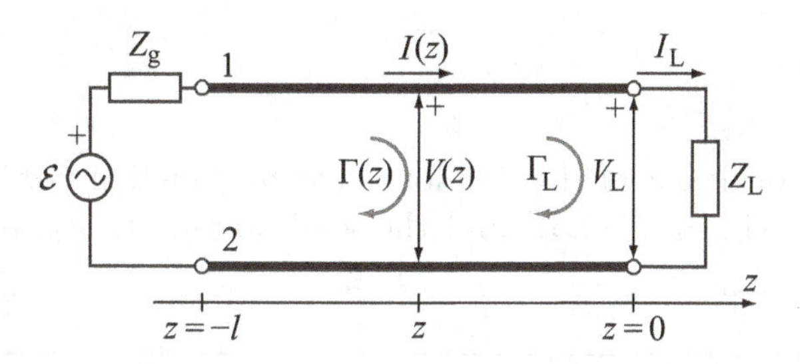

Figure 10.4 Transmission line of Fig.10.1 with a voltage generator (at $z = -l$) and complex impedance load (at $z = 0$) as terminal networks.

written in the exponential form:

$$\underline{\Gamma}_L = |\underline{\Gamma}_L| \, e^{j\psi_L} \quad (0 \leq |\underline{\Gamma}_L| \leq 1 \, ; \; -180^\circ < \psi_L \leq 180^\circ) \, , \tag{10.7}$$

where ψ_L denotes its phase angle. Finally, having in mind Eqs.(10.2), we generalize the concept of the line voltage reflection coefficient at the load, Eq.(10.5), to that at an arbitrary position (defined by the coordinate z) along the line, Fig.10.4,

$$\underline{\Gamma}(z) = \frac{\underline{V}_r(z)}{\underline{V}_i(z)} = \frac{\underline{V}_{r0} \, e^{\underline{\gamma} z}}{\underline{V}_{i0} \, e^{-\underline{\gamma} z}} = |\underline{\Gamma}_L| \, e^{2\alpha z} \, e^{j(2\beta z + \psi_L)} \quad \text{(generalized reflection coefficient)} \, . \tag{10.8}$$

With the use of the coefficient $\underline{\Gamma}(z)$, the total voltage and current along the line, Eqs.(10.2), can now be written as

$$\underline{V}(z) = \underline{V}_i(z) \, [1 + \underline{\Gamma}(z)] \, , \quad \underline{I}(z) = \underline{I}_i(z) \, [1 - \underline{\Gamma}(z)] \, . \tag{10.9}$$

MATLAB EXERCISE 10.4 **Load reflection and transmission coefficients.** Write functions `reflCoeff()` and `reflCoeffCurr()` in MATLAB that calculate the load voltage and current reflection coefficients, respectively, of a transmission line – in Fig.10.4, $\underline{\Gamma}_L$ and $\underline{\Gamma}_{\text{for currents}}$, for the given load impedance, $\underline{Z}_L$, and characteristic impedance of the line, $\underline{Z}_0$. Write also a MATLAB function `transCoeff()` that returns the load voltage transmission coefficient of the line, $\underline{\tau}_L$, given by (Fig.10.4) $\underline{\tau}_L = \underline{V}_L/\underline{V}_{i0} = 2\underline{Z}_L/(\underline{Z}_0 + \underline{Z}_L)$ [note the analogy with Eq.(8.12)]. *(reflCoeff.m, reflCoeffCurr.m, and transCoeff.m on IR)*

TUTORIAL:

For the load voltage reflection coefficient of the line, we implement Eq.(10.5). However, to include all possible values of $\underline{Z}_L$, we separately deal with the case $|\underline{Z}_L| \to \infty$ [open-circuited transmission line – Fig.10.9(b)], in which $\underline{\Gamma}_L = 1$ [see Eq.(10.14)]. So, if function `reflCoeff` is called with a string `'open'` in place of `Zl` ($\underline{Z}_L$), and this is determined by MATLAB function `strcmp`, which compares strings – whether `Zl` equals `'open'`, the returned value for $\underline{\Gamma}_L$ is unity; otherwise, the result is obtained from Eq.(10.5).

```
function Gl = reflCoeff(Zl,Z0)
if strcmp('open',Zl);
Gl = 1;
else
```

```
Gl = (Zl - Z0)/(Zl + Z0) ;
end
```

The load reflection coefficient of the line for currents (function `reflCoeffCurr`) can be computed as $\underline{\Gamma}_{\text{for currents}} = -\underline{\Gamma}_{\text{L}}$ [Eq.(10.6)] and the load voltage transmission coefficient (function `transCoeff`) as $\underline{\tau}_{\text{L}} = 1 + \underline{\Gamma}_{\text{L}}$.

MATLAB EXERCISE 10.5 Total complex voltage and current along a transmission line. Write functions `cplxVoltage()` and `cplxCurrent()` in MATLAB that calculate the total complex rms voltage and current, $\underline{V}(z)$ and $\underline{I}(z)$, respectively, along a lossy transmission line, based on Eqs.(10.9), (10.8), and (10.2). The input to the functions contains the complex rms incident voltage or current for $z = 0$ in Fig.10.4, $\underline{V}_{\text{i0}}$ or $\underline{I}_{\text{i0}}$, the complex propagation coefficient of the line, $\underline{\gamma}$, the load voltage reflection coefficient, $\underline{\Gamma}_{\text{L}}$, and the coordinate z at which the voltage (current) is computed. *(cplxVoltage.m and cplxCurrent.m on IR)*

MATLAB EXERCISE 10.6 Total line voltage as a sum of traveling and standing waves – movie. Being the superposition of two traveling waves with generally different amplitudes (and opposite propagation directions), the transmission-line voltage $\underline{V}(z)$ [or current $\underline{I}(z)$] in Eqs.(10.2) is not a pure standing wave. For a lossless line (in Fig.10.4), $\underline{V}(z)$ can be written as

$$\underline{V}(z) = \underbrace{\underline{\tau}_{\text{L}}\,\underline{V}_{\text{i0}}\,\text{e}^{-\text{j}\beta z}}_{\text{Traveling wave}} + \underbrace{2\text{j}\underline{\Gamma}_{\text{L}}\underline{V}_{\text{i0}}\sin\beta z}_{\text{Standing wave}}\,, \tag{10.10}$$

where $\underline{\Gamma}_{\text{L}}$ and $\underline{\tau}_{\text{L}}$ are, respectively, the load voltage reflection and transmission coefficients of the line. Based on this equation, create a movie in MATLAB to visualize, in space and time, the traveling and standing waves constituting the total voltage along a lossless line with a homogeneous dielectric of permittivity ε and permeability μ, as well as the total voltage. *(ME10_6.m on IR)*

TUTORIAL:

Input to the program consists of the complex rms incident voltage for $z = 0$ (Fig.10.4), `Vi0`, operating frequency, `f`, characteristic impedance of the line, `Z0`, load impedance, `Zl`, and relative permittivity and permeability of the line dielectric, `EPSR` and `MUR`. As the line is lossless and its dielectric is homogeneous [see Eqs.(9.1)], propagation parameters of the incident wave on the line are calculated by function `propParamLossless`, from MATLAB Exercise 7.1. Load reflection and transmission coefficients, `GAMMAl` and `TAUl`, are obtained using functions `reflCoeff` and `transCoeff`, from MATLAB Exercise 10.4. The length of the transmission line is adopted to be $l = 3\lambda_z$ (see Fig.10.5). Spatial and temporal meshes defined for plotting are stored in vectors `z` and `t`, and `z` is used first to compute, based on Eq.(10.10), the traveling- and standing-wave components, `VcmplTW` and `VcmplSW`, of the total complex voltage along the line. Variable `limit` is the axis limit for all three figures, and is chosen accordingly.

```
[w,T,beta,lambda] = propParamLossless(f,EPSR,MUR);
GAMMAl = reflCoeff(Zl,Z0);
```

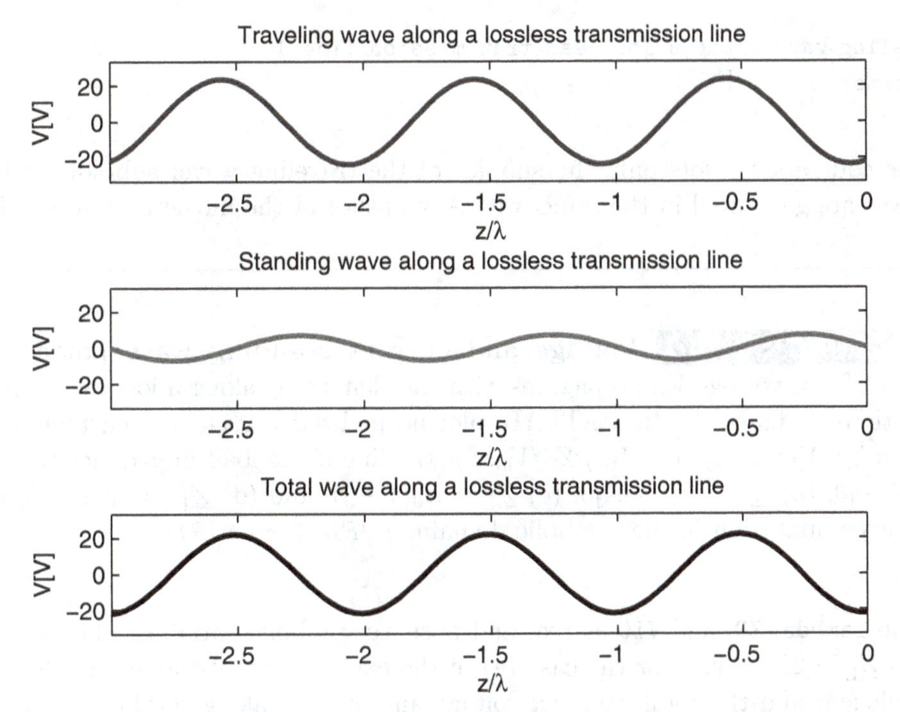

Figure 10.5 Snapshot of a MATLAB movie visualizing the spatial and temporal behavior on a lossless transmission line of the traveling, standing, and total voltage waves, as defined by Eq.(10.10), for $\underline{V}_{\text{i0}} = 10\,\text{e}^{\text{j}0}$ V, $f = 150$ MHz, $Z_0 = 50\ \Omega$, $\underline{Z}_{\text{L}} = (250 + \text{j}0)\ \Omega$, $\varepsilon_{\text{r}} = 4$, and $\mu_{\text{r}} = 1$; for MATLAB Exercise 10.6. *(color figure on CW)*

```
TAUl = transCoeff(Zl,Z0);
L = 3*lambda;
z = -L:0.01*L:0;
t = 0:0.01*T:3*T;
VcmplTW = TAUl*Vi0*exp(-j*beta*z);
VcmplSW = 2*j*GAMMAl*Vi0*sin(beta*z);
limit = 2*max(max(abs(VcmplTW)),max(abs(VcmplSW)));
```

The instantaneous voltages of the traveling, standing, and total waves along the line are calculated for each time instant of the vector `t` and plotted as one frame of a movie, in a `for` loop. Complex voltages `VcmplTW` and `VcmplSW` are converted to their time-domain counterparts, `VtimeTW` and `VtimeSW`, by function `cplx2TimeDomain` from MATLAB Exercise 6.25.

```
for k = 1:length(t);
VtimeTW = cplx2TimeDomain(VcmplTW,w,t(k));
VtimeSW = cplx2TimeDomain(VcmplSW,w,t(k));
Vtot = VtimeTW + VtimeSW;
subplot(3,1,1);
plot(z/lambda,VtimeTW,'b','linewidth',2);
axis([-L 0 -limit limit]);
xlabel('z/\lambda');
ylabel('V[V]');
```

```
title('Traveling wave along a lossless transmission line');
M(k) = getframe;
end;
```

Note that the code above plots only the subplot of the traveling wave; subplots of the standing and total waves are generated in the same way. A snapshot of the movie is shown in Fig.10.5.

MATLAB EXERCISE 10.7 **Voltage and current standing wave patterns.** A time-harmonic wave of rms voltage V_{i0} propagates with wavelength λ_z along a lossless transmission line of characteristic impedance Z_0. In MATLAB, plot normalized voltage and current standing wave patterns, given by $|\underline{V}(z)|/V_{i0}$ and $|\underline{I}(z)|Z_0/V_{i0}$, for the line if the load impedance terminating it is (a) $\underline{Z}_L = 2Z_0 + j0$, (b) $\underline{Z}_L = Z_0/4 + j0$, (c) $\underline{Z}_L = Z_0(1 + j)$, and (d) $\underline{Z}_L = Z_0(1 - j)$, respectively, performing the computation in the symbolic domain. *(ME10_7.m on IR)*

TUTORIAL:

We first define `lambda`, `Z0`, and `Vi0` as real and positive symbolic variables. The load impedance is specified as $\underline{Z}_L = 2Z_0 + j0$ – for the case (a) of the exercise. Symbolic expressions for the load reflection coefficient and the total complex voltage and current along the line are obtained using functions `reflCoeff`, `cplxVoltage`, and `cplxCurrent` (from MATLAB Exercises 10.4 and 10.5), respectively.

```
lambda = sym('lambda','positive');
Z0 = sym('Z0','positive');
Vi0 = sym('Vi0','positive');
Ii0 = Vi0/Z0;
z = -lambda*(0:0.01:1.5);
gamma = i*2*pi/lambda;
Zl = 2*Z0;
Gl = reflCoeff(Zl,Z0);
V = abs(cplxVoltage(Vi0,gamma,Gl,z));
I = abs(cplxCurrent(Ii0,gamma,Gl,z));
```

For plotting, the symbolic results are converted to numerical values, by MATLAB function `double`, and the following lines of code generate a plot for case (a),

```
V1 = double(V/Vi0);
I1 = double(I*Z0/Vi0);
x = double(z/lambda);
figure(1);
plot(x,V1,'b',x,I1,'r--');
legend('|V|/Vi0','|I|*Z0/Vi0');
xlabel('z/\lambda');
ylim([0,2]);
```

This plot is shown in Fig.10.6. Plots for the remaining three cases, (b)–(c), of the exercise are generated by specifying the corresponding load impedances ($\underline{Z}_L$) in the first part of the code.

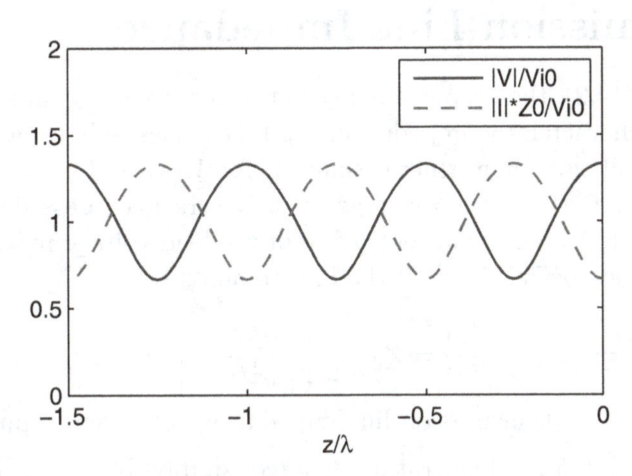

Figure 10.6 MATLAB plots of normalized voltage and current standing wave patterns [$|\underline{V}(z)|/V_{i0}$ and $|\underline{I}(z)|Z_0/V_{i0}$ as functions of z/λ_z] for a lossless transmission line (Fig.10.4) with load impedance $\underline{Z}_L = 2Z_0 + j0$; for MATLAB Exercise 10.7. *(color figure on CW)*

MATLAB EXERCISE 10.8 **Standing wave patterns for a lossy transmission line.** Using MATLAB, plot voltage and current standing wave patterns for a lossy transmission line with attenuation coefficient and phase angle of the complex characteristic impedance ($\underline{Z}_0$) given by (a) $\alpha = 0.3$ Np/λ_z and $\phi = 0$ and (b) $\alpha = 1.2$ Np/λ_z and $\phi = 20°$, respectively, and a purely resistive load with $R_L = |\underline{Z}_0|/4$. *(ME10_8.m on IR)*

HINT:

Figure 10.7 shows the resulting pattern plots for the line with large losses – case (b) of the exercise.

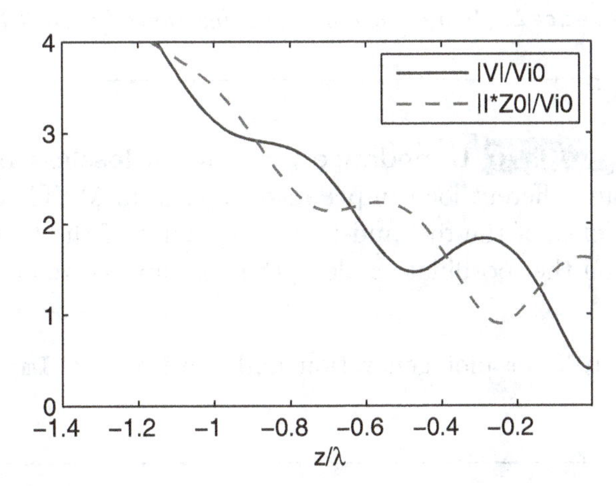

Figure 10.7 MATLAB plots of normalized standing wave patterns for a lossy transmission line with $\alpha = 1.2$ Np/λ_z, $\phi = 20°$ (ϕ is the phase angle of $\underline{Z}_0$), and $R_L = |\underline{Z}_0|/4$; for MATLAB Exercise 10.8. *(color figure on CW)*

10.3 Transmission-Line Impedance

As can be seen in Eqs.(10.2), the proportionality between the voltage and current of a transmission line (Fig.10.4) via its characteristic impedance, $\underline{Z}_0$, takes place only for a single traveling (forward or backward) wave, and not for a general solution for $\underline{V}(z)$ and $\underline{I}(z)$, on the line. Combining Eqs.(10.9), (10.2), and (10.8), the total voltage to current ratio expressed in terms of $\underline{Z}_0$ and either the generalized voltage reflection coefficient, $\underline{\Gamma}(z)$, or the load voltage reflection coefficient, $\underline{\Gamma}_{\rm L}$, and the complex propagation coefficient, $\underline{\gamma}$, of the line amounts to

$$\underline{Z}(z) = \frac{\underline{V}(z)}{\underline{I}(z)} = \underline{Z}_0 \frac{1+\underline{\Gamma}(z)}{1-\underline{\Gamma}(z)} = \underline{Z}_0 \frac{1+\underline{\Gamma}_{\rm L}\,{\rm e}^{2\underline{\gamma}z}}{1-\underline{\Gamma}_{\rm L}\,{\rm e}^{2\underline{\gamma}z}} \quad (-l \le z \le 0)\,, \quad \underline{Z}_{\rm in} = \underline{Z}(-l)$$

(transmission-line impedance, equivalent input impedance) , (10.11)

with $\underline{\Gamma}_{\rm L}$ given by Eq.(10.5). This ratio, denoted simply by $\underline{Z}(z)$, represents the so-called transmission-line impedance, seen at a line cross section defined by the coordinate z looking toward the load. In other words, $\underline{Z}(z)$ equals the complex impedance of an equivalent load that can be used to completely replace the portion of the line beyond this cross section, including the (original) load of impedance $\underline{Z}_{\rm L}$, with respect to the rest of the line (and the generator). For $z = -l$, $\underline{Z}$ at the generator terminals represents the equivalent input impedance, relative to the generator, of the entire line in Fig.10.4, and we mark it as $\underline{Z}_{\rm in}$.

MATLAB EXERCISE 10.9 **Transmission-line impedance and admittance functions.** Write a function `ImpedanceZ()` in MATLAB that calculates the transmission-line impedance, $\underline{Z}(z)$, at a coordinate z in Fig.10.4, based on Eq.(10.11), for the given characteristic impedance ($\underline{Z}_0$), load voltage reflection coefficient ($\underline{\Gamma}_{\rm L}$), and complex propagation coefficient ($\underline{\gamma}$) of the line, and a vector of z-coordinates (`z`). Write also a MATLAB function `AdmittanceZ()` that uses function `ImpedanceZ` and returns the line admittance, $\underline{Y}(z)$, given by $\underline{Y}(z) = 1/\underline{Z}(z)$. The input variables are the same as for `ImpedanceZ`. *(ImpedanceZ.m and AdmittanceZ.m on IR)*

MATLAB EXERCISE 10.10 **Impedance plots for a lossless line.** Consider the lossless transmission line and four different load impedances, $\underline{Z}_{\rm L}$, from MATLAB Exercise 10.7, and plot in MATLAB the dependence of the real and imaginary parts of the transmission-line impedance, $\underline{Z}$, normalized as $\underline{Z}/Z_0$, on the coordinate z along the line, normalized as z/λ_z. *(ME10_10.m on IR)*

HINT:
See MATLAB Exercise 10.7 for plot generation and use function `ImpedanceZ` (from MATLAB Exercise 10.9).

MATLAB EXERCISE 10.11 **Impedance plots for two lossy lines.** Generate MATLAB plots of the behavior of the real and imaginary parts of the transmission-line impedance along the line for the two cases of lossy transmission lines from MATLAB Exercise 10.8. *(ME10_11.m on IR)*

10.4 Complete Solution for Line Voltage and Current

With the concept of the input impedance of a transmission line ($\underline{Z}_{\rm in}$) in hand, and the expression in Eqs.(10.11) for its computation, it is now a very simple matter to express the constant $\underline{V}_{\rm i0}$ in Eqs.(10.2) and (10.5) using the parameters of the voltage generator at the beginning of the line in Fig.10.4, namely, its complex rms emf $\underline{\mathcal{E}}$ and internal impedance $\underline{Z}_{\rm g}$. In the equivalent circuit with the entire line and the load being replaced by $\underline{Z}_{\rm in}$, $\underline{Z}_{\rm in}$ and $\underline{Z}_{\rm g}$ form a voltage divider, that gives the following expression for the voltage $\underline{V}_{\rm g}$ of the generator: $\underline{V}_{\rm g} = \underline{Z}_{\rm in}\underline{\mathcal{E}}/(\underline{Z}_{\rm in} + \underline{Z}_{\rm g})$. From Eqs.(10.9), (10.2), and (10.8), on the other side, this same voltage is $\underline{V}_{\rm g} = \underline{V}(-l) = \underline{V}_{\rm i0}\,{\rm e}^{\underline{\gamma} l}(1 + \underline{\Gamma}_{\rm L}\,{\rm e}^{-2\underline{\gamma} l})$, and hence the solution for $\underline{V}_{\rm i0}$:

$$\underline{V}_{\rm i0} = \frac{\underline{Z}_{\rm in}\underline{\mathcal{E}}\,{\rm e}^{-\underline{\gamma} l}}{(\underline{Z}_{\rm in} + \underline{Z}_{\rm g})(1 + \underline{\Gamma}_{\rm L}\,{\rm e}^{-2\underline{\gamma} l})} \quad \text{(solution for the incident voltage)} . \tag{10.12}$$

With it, and the solution for $\underline{V}_{\rm r0}$ [$\underline{V}_{\rm r0} = \underline{\Gamma}_{\rm L}\underline{V}_{\rm i0}$, Eq.(10.5)], we are then able to express the voltage $\underline{V}(z)$ and current $\underline{I}(z)$ along the line, Eqs.(10.2), in terms of the parameters of the line terminal networks and operating frequency (f) of the structure, in addition to the length (l) and other characteristics of the line itself, i.e., to obtain complete solutions for line voltage and current.

MATLAB EXERCISE 10.12 **Solution for the incident voltage.** Write a function `constantVi0()` in MATLAB that calculates the constant $\underline{V}_{\rm i0}$ in Eqs.(10.2) and (10.5), needed for a complete solution for line voltage and current, using Eq.(10.12). Input parameters of the function are the input impedance of a transmission line ($\underline{Z}_{\rm in}$), the complex rms emf ($\underline{\mathcal{E}}$) and internal impedance ($\underline{Z}_{\rm g}$) of the voltage generator at the beginning of the line (in Fig.10.4), the complex propagation coefficient of the line ($\underline{\gamma}$), the load voltage reflection coefficient ($\underline{\Gamma}_{\rm L}$), and the length of the line (l). *(constantVi0.m on IR)*

MATLAB EXERCISE 10.13 **Complete circuit analysis of a lossless transmission line.** A lossless transmission line of length $l = 4.25$ m and characteristic impedance $Z_0 = 50\ \Omega$ is driven by a time-harmonic voltage generator of frequency $f = 75$ MHz. The emf of the generator has rms value of $\mathcal{E} = 20$ V and zero initial phase; its internal impedance is purely real and equal to $R_{\rm g} = 20\ \Omega$. At the other end, the line is terminated in a load whose complex impedance is $\underline{Z}_{\rm L} = (100 + {\rm j}50)\ \Omega$. The relative permittivity of the line dielectric is $\varepsilon_{\rm r} = 4$ ($\mu_{\rm r} = 1$). Using MATLAB, find complex and instantaneous total voltages and currents along the line, and time-average loss powers in the load and generator, respectively. In addition, create a movie showing instantaneous total voltage and current along the line. Also, plot the complex voltage and current of the line. *(ME10_13.m on IR)* **H**

TUTORIAL:
See MATLAB Exercise 10.6. To find $\underline{\Gamma}_{\rm L}$, $\underline{Z}_{\rm in}$, and $\underline{V}_{\rm i0}$, we use functions `reflCoeff`, `ImpedanceZ`, and `constantVi0`, from MATLAB Exercises 10.4, 10.9, and 10.12, respectively,

```
[w,T,beta,lambda] = propParamLossless(f,EPSR,MUR);
gamma = i*beta;
```

```
GAMMAl = reflCoeff(Zl,Z0);
Zin = ImpedanceZ(Z0,GAMMAl,gamma,-L);
Vi0 = constantVi0(Zin,Vg,Rg,gamma,GAMMAl,L);
Ii0 = Vi0/Z0;
```

The total complex voltage and current along the line are calculated by functions `cplxVoltage` and `cplxCurrent`, from MATLAB Exercise 10.5. They are transferred to time domain using function `cplx2TimeDomain`, from MATLAB Exercise 6.25. Variables `limV` and `limI` are limits needed for plotting the voltage and current along the line (in Fig.10.8). The time-average power dissipated in the load and generator (its internal resistance) are computed as $P_{\mathrm{L}} = \mathrm{Re}\{\underline{V}_{\mathrm{L}}\underline{I}_{\mathrm{L}}^*\} = \mathrm{Re}\{\underline{V}(0)\underline{I}^*(0)\}$ and $P_{R_{\mathrm{g}}} = R_{\mathrm{g}}|\underline{I}_{\mathrm{g}}|^2 = R_{\mathrm{g}}|\underline{I}(-l)|^2$, respectively.

```
z = -L:L/100:0;
V = cplxVoltage(Vi0,gamma,GAMMAl,z);
I = cplxCurrent(Ii0,gamma,GAMMAl,z);
Vtime = cplx2TimeDomain(V,w,t(k));
Itime = cplx2TimeDomain(I,w,t(k));
limV = 2*max(abs(V));
limI = 2*max(abs(I));
Pl = real(V(length(z))*conj(I(length(z))));
Pg = Rg*abs(I(1))^2;
fprintf('\nTime-average loss power in the load is %.2f W',Pl);
fprintf('\nTime-average loss power in the generator is %.2f W',Pg);
```

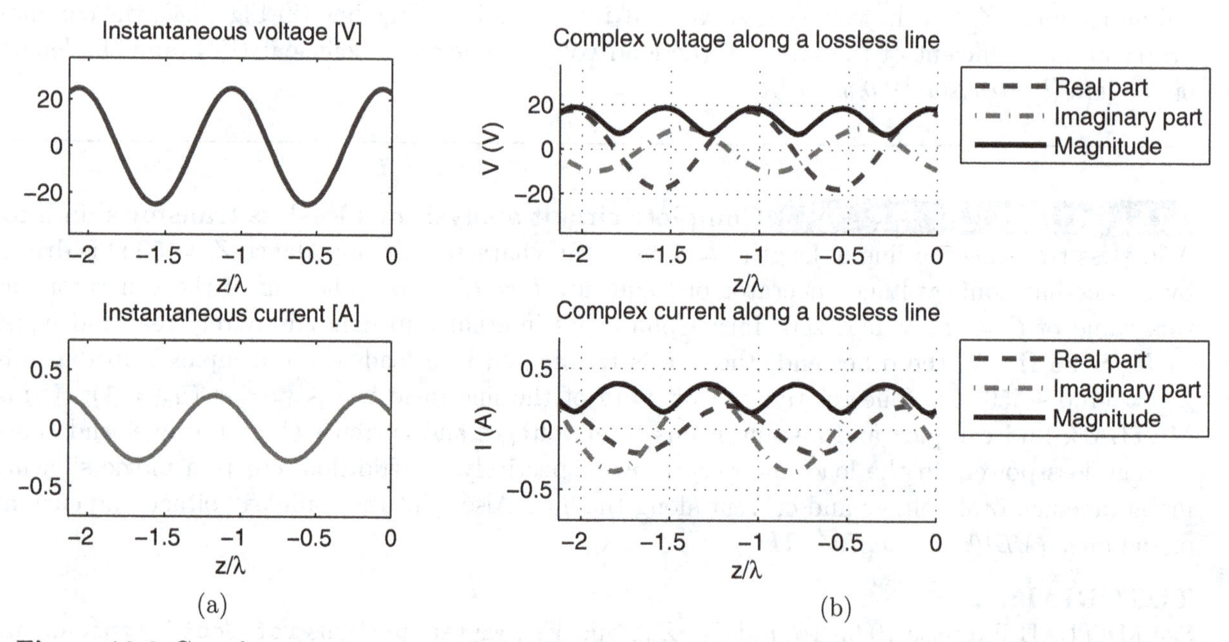

Figure 10.8 Complete circuit analysis of a lossless transmission line (Fig.10.4): (a) snapshot of a MATLAB movie visualizing the computed instantaneous total voltage and current along the line and (b) distribution of the computed complex total voltage and current along the line; for MATLAB Exercise 10.13. *(color figure on CW)*

Finally, a movie showing instantaneous voltage and current along the line is created as in MATLAB Exercise 10.6; Fig.10.8(a) shows a snapshot of the movie. Additionally, plots of real and imaginary parts of complex voltage and current, as well as their magnitudes, are shown in Fig.10.8(b).

MATLAB EXERCISE 10.14 **Complete solution for a low-loss line.** Repeat the previous MATLAB exercise but assuming that the conductors of the transmission line considered have small losses described by the high-frequency resistance per unit length of the line equal to $R' = 1.2\ \Omega/\text{m}$, whereas the line dielectric is the same (lossless). *(ME10_14.m on IR)* **H**

HINT:
Use the expression in Eqs.(10.4) to compute α ($G' = 0$).

10.5 Short-Circuited, Open-Circuited, and Matched Transmission Lines

This section discusses three important special cases of load terminations of a transmission line, in Fig.10.4: a short-circuited, open-circuited, and matched line. If the line is terminated in a short circuit (sc), i.e., if its output terminals (at $z = 0$) are galvanically connected together, as shown in Fig.10.9(a), the load voltage ($\underline{V}_\text{L}$) is zero, and so is the load impedance ($\underline{Z}_\text{L}$). Therefore, the load voltage reflection coefficient ($\underline{\Gamma}_\text{L}$), Eq.(10.5), equals $(0 - Z_0)/(0 + Z_0) = -1$, and Eqs.(10.2), (10.5), (10.11), and (6.26) give the following for the total complex rms voltage ($\underline{V}_\text{sc}$) and current ($\underline{I}_\text{sc}$) on the line and the input impedance ($\underline{Z}_\text{sc}$) of the line (assuming that it is lossless, $\gamma = \text{j}\beta$):

$$\underline{Z}_\text{L} = 0 \quad \longrightarrow \quad \underline{\Gamma}_\text{L} = -1\,, \quad \underline{V}_\text{sc}(z) = -2\text{j}\underline{V}_\text{i0}\sin\beta z\,, \quad \underline{I}_\text{sc}(z) = 2\frac{\underline{V}_\text{i0}}{Z_0}\cos\beta z\,,$$

$$\underline{Z}_\text{sc} = \underline{Z}_\text{in}\big|_{\underline{Z}_\text{L}=0} = \text{j}Z_0\tan\beta l \quad \text{(short-circuited transmission line)}\,. \tag{10.13}$$

On the other side, if the line is terminated in an open circuit (oc), that is, if the output terminals ($z = 0$) in Fig.10.4 are left open, as in Fig.10.9(b), then the load current, $\underline{I}_\text{L}$, is forced to be zero, and hence an infinite load impedance. This, in turn, gives [in Eq.(10.5)] a unity load voltage reflection coefficient of the line [it equals $(\underline{Z}_\text{L} - 0)/(\underline{Z}_\text{L} - 0) = 1$, since Z_0 can be treated as a zero value in comparison with the infinitely large $|\underline{Z}_\text{L}|$], Using the same equations as for the shorted line, we now obtain

$$|\underline{Z}_\text{L}| \to \infty \quad \longrightarrow \quad \underline{\Gamma}_\text{L} = 1\,, \quad \underline{V}_\text{oc}(z) = 2\underline{V}_\text{i0}\cos\beta z\,, \quad \underline{I}_\text{oc}(z) = -2\text{j}\frac{\underline{V}_\text{i0}}{Z_0}\sin\beta z\,,$$

$$\underline{Z}_\text{oc} = \underline{Z}_\text{in}\big|_{|\underline{Z}_\text{L}|\to\infty} = -\text{j}Z_0\cot\beta l \quad \text{(open-circuited transmission line)}\,. \tag{10.14}$$

As the last special case, if the line is terminated in a load whose impedance is equal to the line characteristic impedance (Z_0), Fig.10.9(c), the load voltage reflection coefficient is zero, we

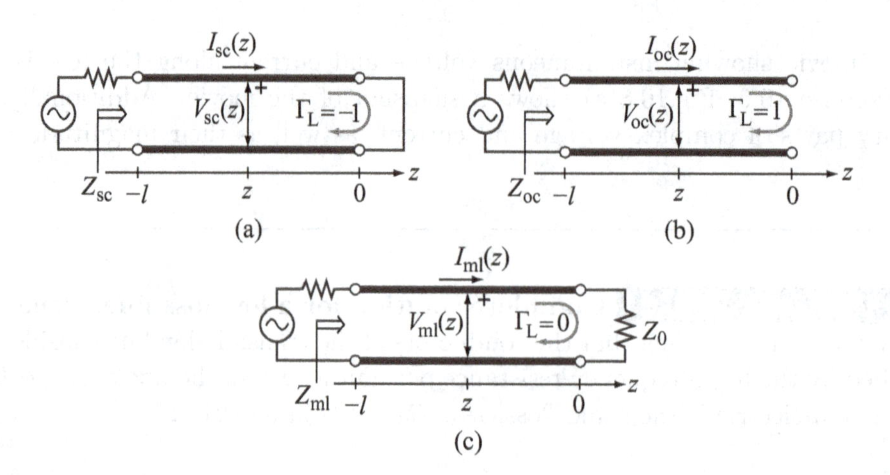

Figure 10.9 Three important special cases of the impedance load termination of a transmission line: (a) short circuit, (b) open circuit, and (c) impedance-matched load.

say that the load is matched (by its impedance) to the line – a matched load (ml), and also refer to the line itself as a matched line. Since there is no reflected wave on the line, we have

$$\underline{Z}_L = Z_0 \quad \longrightarrow \quad \underline{\Gamma}_L = 0\,, \quad \underline{V}_{ml}(z) = \underline{V}_{i0}\,e^{-j\beta z}\,, \quad \underline{I}_{ml}(z) = \frac{\underline{V}_{i0}}{Z_0}\,e^{-j\beta z}\,,$$
$$\underline{Z}_{ml} = \underline{Z}_{in}\big|_{\underline{Z}_L = Z_0} = Z_0 \quad \text{(matched transmission line)}\,. \tag{10.15}$$

MATLAB EXERCISE 10.15 **Input impedance of a shorted line and an open line.** In MATLAB, write functions `inputImpShort()` and `inputImpOpen()` to compute the input impedance of a short-circuited lossless transmission line, Fig.10.9(a), and of an open-circuited line, Fig.10.9(b), using Eqs.(10.13) and (10.14), respectively. At input, the functions take the characteristic impedance, Z_0, phase coefficient, β, and length, l, of the line. *(inputImpShort.m and inputImpOpen.m on IR)*

MATLAB EXERCISE 10.16 **Complete analysis of an open-circuited line.** Repeat MATLAB Exercise 10.13 but for an open circuit as load. *(ME10_16.m on IR)* **H**

MATLAB EXERCISE 10.17 **Standing wave patterns for short, open, and matched loads.** In MATLAB, plot the normalized voltage and current standing wave patterns, $|\underline{V}|/V_{i0}$ and $|\underline{I}|Z_0/V_{i0}$ against z/λ_z (as in MATLAB Exercise 10.7), for a lossless transmission line and three special cases of the impedance load termination in Fig.10.9, using Eqs.(10.13), (10.14), and (10.15). *(ME10_17.m on IR)* **H**

MATLAB EXERCISE 10.18 **Impedance plots for short-circuited lossy lines.** For the two cases of lossy transmission lines from MATLAB Exercise 10.8 and a short-circuit termination [Fig.10.9(a)], generate MATLAB plots of real and imaginary parts of the normalized transmission-line impedance, $\underline{Z}/|\underline{Z}_0|$, against z/λ_z, using function `ImpedanceZ` (from MATLAB Exercise 10.9) or the expression $\underline{Z}_{sc} = \underline{Z}_0 \tanh \underline{\gamma} l$ $(l = -z)$, in place of that in Eqs.(10.13), with 'tanh' standing for the hyperbolic tangent function. *(ME10_18.m on IR)*

10.6 Impedance-Matching Using Short- and Open-Circuited Stubs

We see in Eqs.(10.13) and (10.14) that the input impedance of short- and open-circuited lines is purely imaginary (reactive), $\underline{Z}_{sc} = jX_{sc}$ and $\underline{Z}_{oc} = jX_{oc}$ $(R_{sc} = R_{oc} = 0)$, and conclude that any input reactance X_{sc} or X_{oc} $(-\infty < X_{sc}, X_{sc} < \infty)$ can be realized by simply varying the length l of a short- or open-circuited line, for a given operating frequency (f) of the structure (given λ_z). Owing to these features, short- and open-circuited transmission line segments (called stubs), connected in parallel (shunt stubs) or in series to the existing circuit or device, are extensively used as tuning and compensating elements in impedance-matching applications at higher (microwave) frequencies. Basically, such shunt and series stubs, providing an appropriate reactance X by adjusting the electrical length of the line segment, are used, at an appropriate location on the main transmission line, to compensate (cancel) the imaginary part of the line admittance or impedance, as shown in Fig.10.10(a) and (b), respectively.

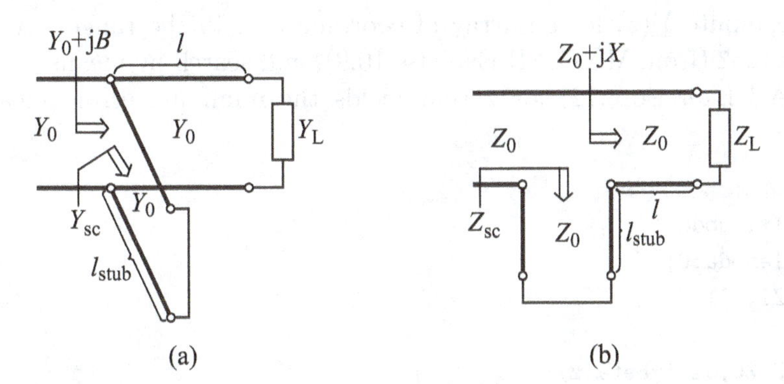

Figure 10.10 (a) Admittance-matching by a shunt short-circuited stub connected at a location on a transmission line where the real part of the complex line admittance equals $Y_0 = 1/Z_0$. (b) Impedance-matching by a series short-circuited stub, at a distance from the load where the real part of the transmission-line impedance equals Z_0.

MATLAB EXERCISE 10.19 **Complete design of a shunt short-circuited stub – numerical solution.** Write a function `shuntShortCircuitedStub()` in MATLAB that carries out a complete design of an admittance-matching transmission-line circuit with a shunt short-circuited stub in Fig.10.10(a), including finding the stub location on the line. Namely, taking at input the characteristic impedance, Z_0, and phase coefficient, β, of a lossless transmission line, and the load impedance, $\underline{Z}_\mathrm{L}$, the function returns both the distance of the stub junction from the load, l, and the length of the stub, l_stub – to match the load to the feeding line. *(shuntShortCircuitedStub.m on IR)*

TUTORIAL:

First, we would like to find a distance l from the load at which the real part of the complex line admittance, $\underline{Y}(-l) = 1/\underline{Z}(-l)$, is equal to the characteristic admittance of the line, $Y_0 = 1/Z_0$. The imaginary part of $\underline{Y}(-l)$ will then be compensated (canceled) by a short-circuited transmission line segment (stub) connected in parallel [Fig.10.10(a)]. In other words, the sum of the input admittances of the two parallel lines at their junction (looking toward their loads) is to match Y_0 (or $Y_0 + \mathrm{j}0$),

$$\underline{Y}(-l) + \underline{Y}_\mathrm{stub} = Y_0 + \mathrm{j}0 \quad \text{(admittance-matching by a shunt stub)}\,, \tag{10.16}$$

where $\underline{Y}_\mathrm{stub}$ is purely imaginary. Since the impedance $\underline{Z}(z)$ and admittance $\underline{Y}(z)$ along a lossless transmission line repeat themselves after each $\lambda_z/2$ (for instance, see MATLAB Exercise 10.10), we search for the solution for l in the range $0 \le l \le \lambda_z/2$, where, according to the equation $\underline{Y}(-l) = Y_0 + \mathrm{j}B$, there are actually two values of l resulting in $\underline{Y}(-l)$ such that its real part is exactly Y_0. These two solutions are complex conjugates of each other, and are given by $\underline{Y}(-l) = Y_0 \pm \mathrm{j}|B|$. To find l, we solve numerically the following equation [the real part of Eq.(10.16)]:

$$\mathrm{Re}\{\underline{Y}(-l)\} = Y_0\,, \quad \text{where} \quad Y_0 = \frac{1}{Z_0}\,. \tag{10.17}$$

Specifically, we compute $\underline{Y}(z)$ for an array of coordinates z in the range $-\lambda_z/2 \le z \le 0$, using function `AdmittanceZ` (from MATLAB Exercise 10.9), and search by means of MATLAB function `min` (see MATLAB Exercise 3.12) for z that yields the minimum error (ideally, zero) between $\mathrm{Re}\{\underline{Y}(-l)\}$ and Y_0.

```
lambda = 2*pi/beta;
dlambda = 0.0001*lambda;
z = -lambda/2:dlambda:0;
Gl = reflCoeff(Zl,Z0);
Y0 = 1/Z0;
Yz = AdmittanceZ(Z0,Gl,i*beta,z);
Yz1 = Yz;
[ErrorMin,k] = min(abs(real(Yz1)-Y0));
Yz1(k)= Yz1(1);
[ErrorMin,p] = min(abs(real(Yz1)-Y0));
```

Comparing the two indices (positions), `p` and `k`, of the minimum values in the array, returned by function `min`, the two solutions for $l = -z$ are arranged so that the first one, `l(1)`, is closer to the load, while the second one, `l(2)`, is farther away,

```
if (p > k)
n = [p k];
```

```
else
n = [k p];
end;
l(1) = -z(n(1));
l(2) = -z(n(2));
Yin(1:2) = [Yz(n(1)) Yz(n(2))];
```

with `Yin(1)` and `Yin(2)` being the corresponding solutions for $\underline{Y}(-l)$. Of course, extensions of `l(1)` and `l(2)` by any integer multiple of $\lambda_z/2$ are also solutions to Eq.(10.17). Note that this completed part of the code (determination of the location on the main line where the stub should be attached) is the most complex part of the stub matching design.

Finally, we solve numerically the imaginary part of the admittance-matching equation, Eq.(10.16), namely,

$$-\mathrm{j}\,\mathrm{Im}\{\underline{Y}(-l)\} = \underline{Y}_{\text{stub}}\,, \quad \text{where} \quad \underline{Y}_{\text{stub}} = \frac{1}{\underline{Z}_{\text{sc}}}\,, \tag{10.18}$$

by searching for a minimum, again by function `min`, of the error between $-\mathrm{j}\,\mathrm{Im}\{\underline{Y}(-l)\}$ and $\underline{Y}_{\text{stub}} = 1/\underline{Z}_{\text{sc}}$ in the range $0 \le l_{\text{stub}} \le \lambda_z/2$, that is, for the same array of coordinates z spanning $-\lambda_z/2 \le z \le 0$ as in solving Eq.(10.17). For $\underline{Z}_{\text{sc}}$ (short-circuited stub), we use function `inputImpShort` (from MATLAB Exercise 10.15), adopting that the stub be cut of the same transmission line (the same Z_0 and β) as the main section. Of course, we have two solutions, `lstub(1)` and `lstub(2)`, corresponding to `l(1)` and `l(2)`, i.e., to `Yin(1)` and `Yin(2)`.

```
ImYin = imag(Yin);
negjImYin = -i*ImYin;
Ystub = 1./inputImpShort(Z0,beta,-z);
[ErrorMin,m(1)] = min(abs(Ystub-negjImYin(1)));
[ErrorMin,m(2)] = min(abs(Ystub-negjImYin(2)));
lstub(1) = -z(m(1));
lstub(2) = -z(m(2));
```

MATLAB EXERCISE 10.20 **Example of a complete shunt-stub circuit design, numerically.** For a lossless transmission line, the magnitude and phase angle of the load reflection coefficient are $|\underline{\Gamma}_{\text{L}}| = 0.38$ and $\psi_{\text{L}} = 138°$, respectively, the wavelength along the line is $\lambda_z = 60$ mm, and the characteristic admittance of the line amounts to $Y_0 = 10$ mS. Design, in MATLAB, an admittance-matching shunt short-circuited stub for this line, namely, find l and l_{stub} in Fig.10.10(a) such that Eq.(10.16) is satisfied. *(ME10_20.m on IR)* **H**

TUTORIAL:

For the given numerical data, we carry out the design numerically, using function `shuntShortCircuitedStub`, developed in the previous MATLAB exercise,

```
lambda = 60*10^(-3);
beta = 2*pi/lambda;
Y0 = 10*10^(-3);
```

```
phil = 138/180*pi;
Gl = 0.38*exp(i*phil);
Z0 = 1/Y0;
Zl = Z0*(Gl+1)/(1-Gl);
[l,lstub] = shuntShortCircuitedStub(Z0,Zl,beta);
fprintf('\nFirst solution:');
fprintf('\nDistance of the stub from the load ');
fprintf('is %4.2f mm',10^3*l(1));
fprintf('\nLength of the shunt short-circuited stub ');
fprintf('is %4.2f mm',10^3*lstub(1));
fprintf('\nSecond solution:');
fprintf('\nDistance of the stub from the load ');
fprintf('is %4.2f mm',10^3*l(2));
fprintf('\nLength of the shunt short-circuited stub ');
fprintf('is %4.2f mm',10^3*lstub(2));
```

The display with results for the two designs (two solutions for l and l_{stub}) in the Command Window reads:

```
First solution:
Distance of the stub from the load is 2.14 mm
Length of the shunt short-circuited stub is 21.57 mm
Second solution:
Distance of the stub from the load is 20.86 mm
Length of the shunt short-circuited stub is 8.43 mm
```

MATLAB EXERCISE 10.21 **Complete design of a shunt open-circuited stub – numerical solution.** Repeat MATLAB Exercise 10.19 but for an open-circuited shunt stub, in the same configuration as the one in Fig.10.10(a) – function `shuntOpenCircuitedStub()`. *(shuntOpenCircuitedStub.m on IR)*

MATLAB EXERCISE 10.22 **Example of admittance-matching by a shunt open-circuited stub.** An antenna whose input impedance at a frequency of $f = 8$ GHz amounts to $\underline{Z}_{\text{L}} = (85 + \text{j}30)\ \Omega$ needs to be admittance-matched to a lossless microstrip transmission line of characteristic impedance $Z_0 = 50\ \Omega$ and phase velocity $v_{\text{p}} = 1.7 \times 10^8$ m/s – using a shunt open-circuited stub, in the configuration as in Fig.10.10(a). If the stub is also a microstrip line, printed on the same substrate and having the same parameters as the main line, find the two locations on the main line closest to the load at which the stub should be attached and the corresponding minimal required lengths of the stub. *(ME10_22.m on IR)* **H**

HINT:
Use function `shuntOpenCircuitedStub()` (from the previous MATLAB exercise).

MATLAB EXERCISE 10.23 **Complete design of a series short-circuited stub – numerical solution.** Repeat MATLAB Exercise 10.19 but for an impedance-matching series short-circuited stub, shown in Fig.10.10(b) [function `seriesShortCircuitedStub()`]. *(seriesShortCircuitedStub.m on IR)*

HINT:
In place of Eq.(10.16), the impedance-matching equation for the circuit in Fig.10.10(b) reads

$$\underline{Z}(-l) + \underline{Z}_{\text{stub}} = Z_0 + \text{j}0 \quad \text{(impedance-matching by a series stub)}, \qquad (10.19)$$

where $\underline{Z}_{\text{stub}} = \underline{Z}_{\text{sc}}$ (short-circuited stub) in this case.

MATLAB EXERCISE 10.24 **Impedance-matching using a series two-wire stub.** A series short-circuited stub is used in an impedance-matching circuit for an antenna with input impedance $\underline{Z}_{\text{L}} = (73 + \text{j}42)\ \Omega$ at a frequency of $f = 800$ MHz, as in Fig.10.10(b). Both the main line and the stub are air two-wire lines of characteristic impedance $Z_0 = 300\ \Omega$. Determine the distance from the load of the location where the stub is inserted, l, and the matching length of the stub, l_{stub}. *(ME10_24.m on IR)* **H**

MATLAB EXERCISE 10.25 **Complete design of a series open-circuited stub – numerical solution.** Repeat MATLAB Exercise 10.23 but with an open-circuited stub [function `seriesOpenCircuitedStub()`]. Then use this function to design a series open-circuited two-wire stub for the two-wire transmission line and the load described in the previous MATLAB exercise. *(seriesOpenCircuitedStub.m and ME10_25.m on IR)* **H**

10.7 The Smith Chart – Construction and Basic Properties

As the last topic in the circuit theory of two-conductor transmission lines in a time-harmonic regime, in this and the following section we present an alternative, graphical, technique for the circuit analysis and design of transmission lines in the frequency domain. The technique is based on the so-called Smith chart. It enables approximate determination, based on graphical manipulations on the chart, of the reflection coefficients, line impedances, voltages, currents, and other quantities of interest for a given transmission-line problem – without actually performing any complex algebra. Of course, it is possible (and advisable) to combine computations in the complex domain using concepts and equations developed in previous sections of this chapter with visualizations and graphical evaluations on the Smith chart.

The Smith chart is, essentially, a polar plot of the generalized voltage reflection coefficient, $\underline{\Gamma}(z)$, given by Eq.(10.8), along a transmission line (in Fig.10.4). Equivalently, representing $\underline{\Gamma}$ via its real and imaginary parts, $\underline{\Gamma} = \Gamma_r + j\Gamma_i$, the chart lies in the complex plane of $\underline{\Gamma}$, i.e., the Γ_r–Γ_i plane, as shown in Fig.10.11. Confining our analysis to the lossless case, the magnitude of $\underline{\Gamma}$ is constant along the line, which corresponds to a circle of radius $|\underline{\Gamma}|$, centered at the coordinate origin in the Γ_r–Γ_i plane. Employing the standing wave ratio of the line, $s = |\underline{V}|_{\max}/|\underline{V}|_{\min} = (1+|\underline{\Gamma}|)/(1-|\underline{\Gamma}|)$ [see Eq.(8.14)], we have

$$|\underline{\Gamma}| = \frac{s-1}{s+1} = \text{const} \quad (-l \le z \le 0) \quad (\text{constant-}|\underline{\Gamma}| \text{ or } s \text{ circle}) , \tag{10.20}$$

and hence the constant-$|\underline{\Gamma}|$ circle in Fig.10.11 is also referred to as the s circle (or SWR circle). Since $0 \le |\underline{\Gamma}| \le 1$, Eq.(10.7), the Smith chart is bounded by the circle defined by $|\underline{\Gamma}| = 1$, called the unit circle. Impedances are displayed on the chart using their normalized (to Z_0) values, $\underline{z}_n$. From Eq.(10.11),

$$\underline{z}_n = \frac{\underline{Z}}{Z_0} = \frac{R}{Z_0} + j\frac{X}{Z_0} = r + jx = \frac{1+\underline{\Gamma}}{1-\underline{\Gamma}} = \frac{1+\Gamma_r+j\Gamma_i}{1-\Gamma_r-j\Gamma_i} , \quad \underline{\Gamma} = \frac{\underline{z}_n - 1}{\underline{z}_n + 1}$$

$$(\text{mapping between complex } \underline{\Gamma} \text{ and } \underline{z}_n) , \tag{10.21}$$

where r and x are the normalized line resistance and reactance, respectively. These simple relationships, defining in the complex plane a mapping of $\underline{\Gamma}$ to $\underline{z}_n$, and vice versa, are the underpinning of the utility of the Smith chart.

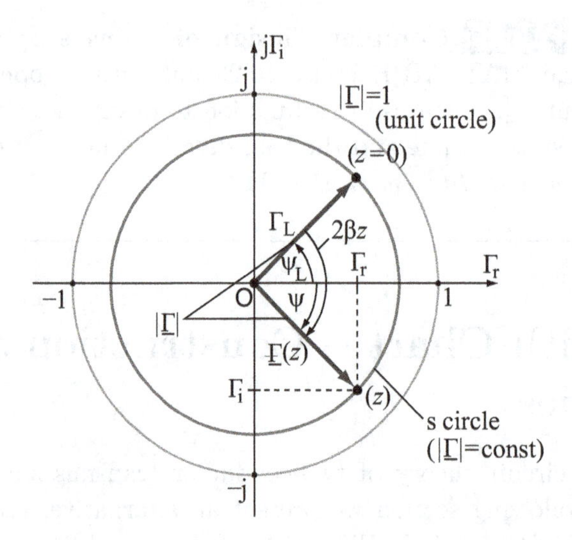

Figure 10.11 Graphical representation in the complex plane of the generalized voltage reflection coefficient $\underline{\Gamma}(z)$, Eq.(10.8), along a lossless transmission line (Fig.10.4), as a basis for construction of the Smith chart.

Equating r to the real part of the last expression for $\underline{z}_n$ in Eqs.(10.21), we obtain an equation of a circle in the Γ_r–Γ_i plane, whose center and radius are

$$r \text{ circle}: \quad \text{center at } (\Gamma_r, \Gamma_i) = \left(\frac{r}{1+r}, 0\right) ; \quad \text{radius} = \frac{1}{1+r} . \tag{10.22}$$

Figure 10.12(a) shows several characteristic representatives of the family of r circles. Similarly, equating the imaginary parts in Eqs.(10.21) leads to an equation describing an x circle in the plane Γ_r–Γ_i,

$$x \text{ circle}: \quad \text{center at } (\Gamma_r, \Gamma_i) = \left(1, \frac{1}{x}\right) ; \quad \text{radius} = \frac{1}{|x|} , \tag{10.23}$$

and several typical x circles are shown in Fig.10.12(b). Of course, only their portions – x arcs – falling within the domain $|\underline{\Gamma}| \leq 1$ are relevant for the analysis.

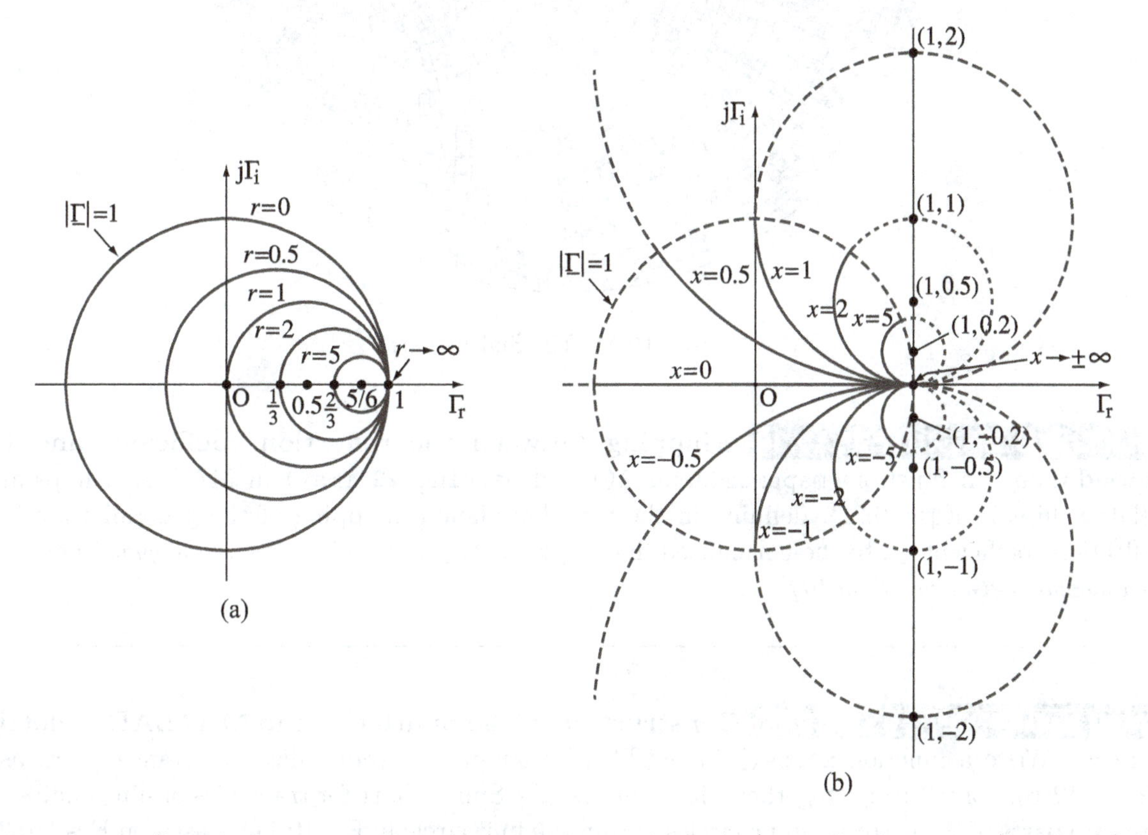

Figure 10.12 Graphical representation of parametric equations of r (normalized-resistance) circles (a) and x (normalized-reactance) arcs (b), Eqs.(10.22) and (10.23), constituting the Smith chart for transmission-line analysis (x arcs are portions of the corresponding circles within the bounds of the unit circle in Fig.10.11).

Superposing the r circles and x arcs, from Fig.10.12(a) and (b), we obtain the Smith chart – shown in Fig.10.13. On the chart, a normalized line impedance given by $\underline{z}_{n0} = r_0 + jx_0$ corresponds to the point of intersection of the $r = r_0$ circle and $x = x_0$ arc. Overall, the Smith chart simultaneously displays values of $\underline{z}_n$ and $\underline{\Gamma}$, according to the relationships in Eqs.(10.21), in a convenient format – for graphical calculations on transmission lines and/or visualization of analysis data.

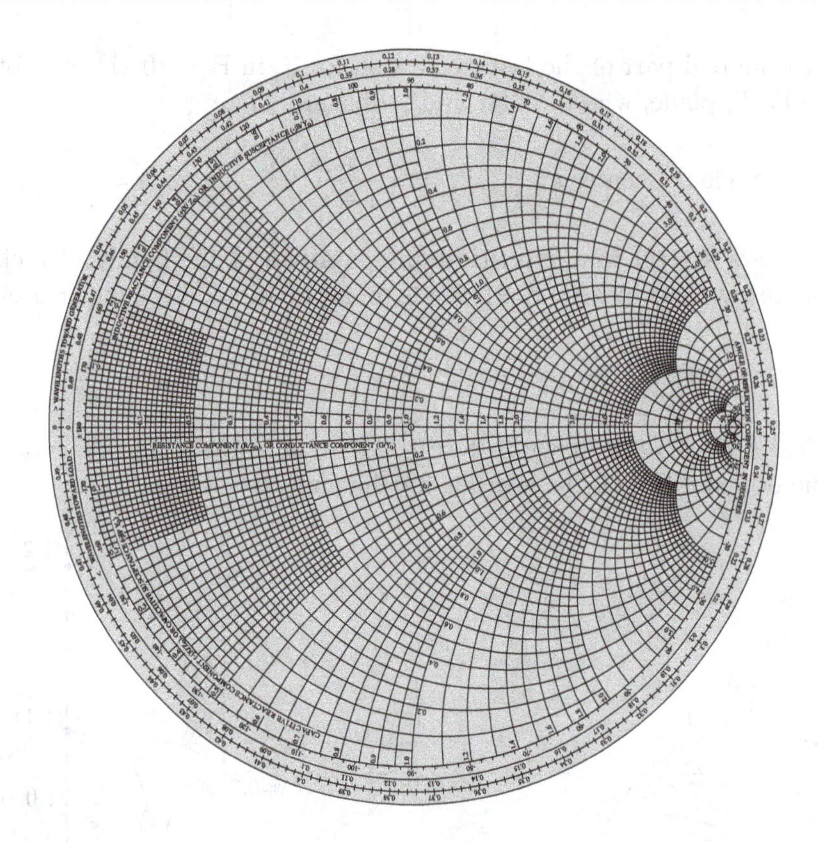

Figure 10.13 The Smith chart.

MATLAB EXERCISE 10.26 **Mapping between the reflection coefficient and line impedance.** In functions `mappingGamma2Z()` and `mappingZ2Gamma()` in MATLAB, implement relationships in Eqs.(10.21), defining in the complex plane a mapping of the generalized voltage reflection coefficient, $\underline{\Gamma}$, to the normalized line impedance, $\underline{z}_\mathrm{n}$, and vice versa. *(mappingGamma2Z.m and mappingZ2Gamma.m on IR)*

MATLAB EXERCISE 10.27 **Construction of the Smith chart in MATLAB – plotting x arcs.** Write a function `xArcs()` in MATLAB that plots x (normalized-reactance) arcs, as in Fig.10.12(b), constituting (together with r circles) the Smith chart for transmission-line analysis (x arcs are parts of the corresponding circles within the unit circle in Fig.10.11), based on Eqs.(10.23). In specific, the input to the function is a positive normalized-reactance x, and the function plots two arcs, for x and $-x$, respectively. *(xArcs.m on IR)*

TUTORIAL:

First, we compute, from Eqs.(10.23), the complex number `a0` ($\underline{a}_0 = 1 + \mathrm{j}/x$) in the Γ_r–Γ_i complex plane in Fig.10.11 or 10.12(b) that corresponds to the center of the x circle (defined by the input value `x`) and the radius `R` ($R = 1/x$; $x > 0$) of the circle. Then, all points on the circle can be obtained as $\underline{a} = \underline{a}_0 + R\mathrm{e}^{\mathrm{j}\theta}$, where $0 \leq \theta \leq 2\pi$ [note the similarity with Eq.(8.13) and Fig.8.9]; the coordinates (real and imaginary parts of respective complex numbers) of all points on the circle are stored in the array **a** with the same length as the array **theta**. In order to plot only the contours within the bounds of the unit circle ($|\underline{\Gamma}| = 1$), in Fig.10.12(b), we multiply **a** by another array, **u**, whose elements are unity if the corresponding element of **a** falls inside the unit circle, while zero otherwise. Finally, we plot the x arc determined by nonzero elements of the resulting (product) array, **gx**. We also plot the complex conjugate of **gx**, which adds the $-x$ arc to the graph.

```
function xArcs(x)
theta = 0:0.001*pi:2*pi;
a0 = 1+i*1/x;
R = 1/x;
a = a0 + R*exp(i*theta);
u = sign(sign(1-abs(a))+1);
gx = a.*u;
gx = nonzeros(gx);
plot(gx);
plot(conj(gx));
```

MATLAB EXERCISE 10.28 **Construction of the Smith chart – plotting r circles.** Repeat the previous MATLAB exercise but for r (normalized-resistance) circles, as in Fig.10.12(a), for the Smith chart, using Eqs.(10.22) [function `rCircles()`]. The input is r (nonnegative), and the output is a plot of the corresponding circle. *(rCircles.m on IR)*

MATLAB EXERCISE 10.29 **MATLAB version of the Smith chart.** Write a function `SmithChart()` in MATLAB to combine a family of r circles and a family of x arcs obtained using functions `rCircles` and `xArcs` from the previous two MATLAB exercises into a MATLAB version of the Smith chart (the actual Smith chart is shown in Fig.10.13). *(SmithChart.m on IR)*

TUTORIAL:

In the code, r circles for a selected set of `r` values are plotted in a `for` loop, and similarly for x arcs. MATLAB function `hold on` is used to keep all the plots of circles and arcs in the graph, instead of just showing the last plot. The $x = 0$ arc, which degenerates into a line, is plotted by MATLAB function `plot`. The default MATLAB axes are removed.

```
function SmithChart()
hold on;
plot([-1 1],[0 0]);
r = [0 0.1 0.3 0.5 1 2 4];
for k = r
rCircles(k);
end;
x = [0.1 0.2 0.5 1 2 4];
for k = x
```

```
xArcs(k);
end;
axis('square');
axis('off');
hold off;
title('Smith chart');
```

Figure 10.14 shows the chart. Of course, a finer grid of contours of constant r and x can be plotted, by selecting more values for arrays `r` and `x` in the code.

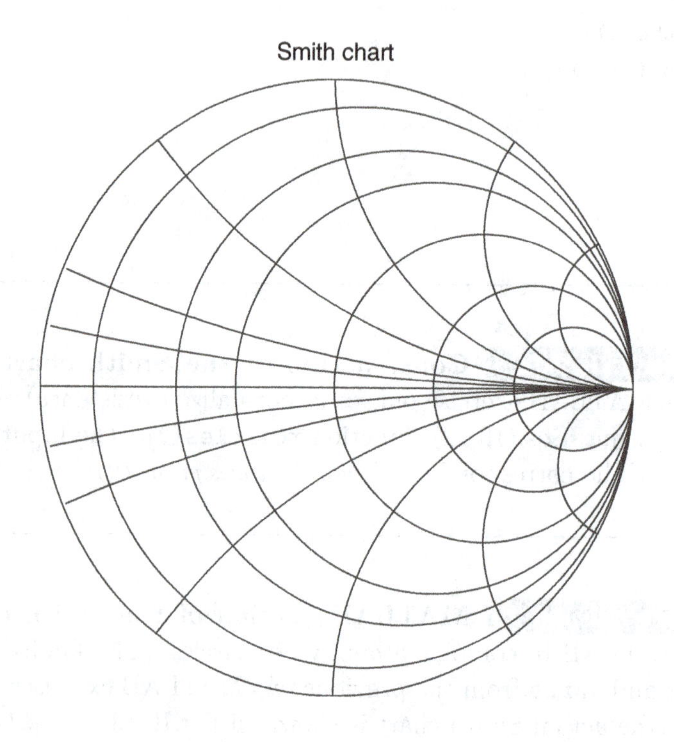

Figure 10.14 MATLAB version of the Smith chart (Fig.10.13); for MATLAB Exercise 10.29. *(color figure on CW)*

MATLAB EXERCISE 10.30 **Adding dots of data on the Smith chart – in movie frames.** Write function `plotOnSCh()` in MATLAB that plots in a movie, frame by frame, an array of points (Γ_r, Γ_i), named `array`, on the Smith chart. Each point is added in a new frame, while all previous plots are kept in the graph. Other input arguments of the function are an array `color` with color symbols for points in `array`, parameter `style` defining the style for marking all points (e.g., *, +, etc.), the ordinal number `k` of the current movie frame, before the plotting (movie) has started, and indices `lim1` and `lim2` of the first and last elements of `array` that are actually plotted. The function also returns the ordinal number of the last movie frame (after the movie is completed) – the new ordinal number, `knew`, for eventual further plotting (adding more frames) in the main program. Note that these styled (and colored) "dots" of data (Γ_r, Γ_i) are superimposed, in the main program, on the Smith chart developed in the previous MATLAB exercise, and this is used for all plots in the transmission-line analysis based on the Smith chart. *(plotOnSCh.m on IR)*

TUTORIAL:
The function is realized in a `for` loop of the movie with limits `lim1` and `lim2`, i.e., from `k` (starting) to `knew`, as follows:

```
function knew = plotOnSCh(array,color,style,k,lim1,lim2)
n = length(color);
for j = lim1:lim2
plot(real(array(j)),imag(array(j)),[color(floor(mod(j,4*n)/4)+1)style],'MarkerSize',5);
pause(0.1);
k = k+1;
pause(0.1);
M(k) = getframe();
end;
knew = k;
```

MATLAB EXERCISE 10.31 **SWR circle in the Smith chart.** Write a function `sCircle()` in MATLAB that, for a given standing wave ratio, s, of a lossless transmission line, returns an array of reflection coefficients $\underline{\Gamma}$ corresponding to points along the entire s circle in the Smith chart (Fig.10.11), as well as the magnitude of the coefficient, $|\underline{\Gamma}|$. These values are needed for plotting s circles on a MATLAB version of the Smith chart. *(sCircle.m on IR)*

10.8 Circuit Analysis of Transmission Lines Using the Smith Chart

To help us efficiently use the Smith chart in analysis, design, and visualization of transmission lines, which is the subject of this section, Fig.10.15 highlights several key features of the chart in Fig.10.13 or 10.14. For instance, the three important special cases of load terminations of a transmission line in Fig.10.9, a short circuit, open circuit, and matched load, are marked in Fig.10.15: points P_{sc}, P_{oc}, and P_{ml}, respectively. The figure also reemphasizes that the upper (lower) half of the chart corresponds to inductive (capacitive) line impedances. From Eq.(10.8), the change in the coordinate z due to a movement along the line (with no losses), in Fig.10.4, results in the following change of the phase angle ψ of the reflection coefficient $\underline{\Gamma}(z)$: $\Delta\psi = 2\beta\Delta z$ $(\beta = 2\pi/\lambda_z)$. Hence, a clockwise rotation on the chart in Fig.10.15 $(\Delta\psi < 0)$ corresponds to a movement toward the generator (G) along the line in Fig.10.4 $(\Delta z < 0)$, while moving toward the load (L) in Fig.10.4 gives the counterclockwise direction of rotation on the chart. In movements in either direction, one complete rotation around the chart corresponds to a half-wave shift along the line,

$$360^\circ \text{ around the Smith chart} \quad \longleftrightarrow \quad \frac{\lambda_z}{2} \text{ along the transmission line} . \tag{10.24}$$

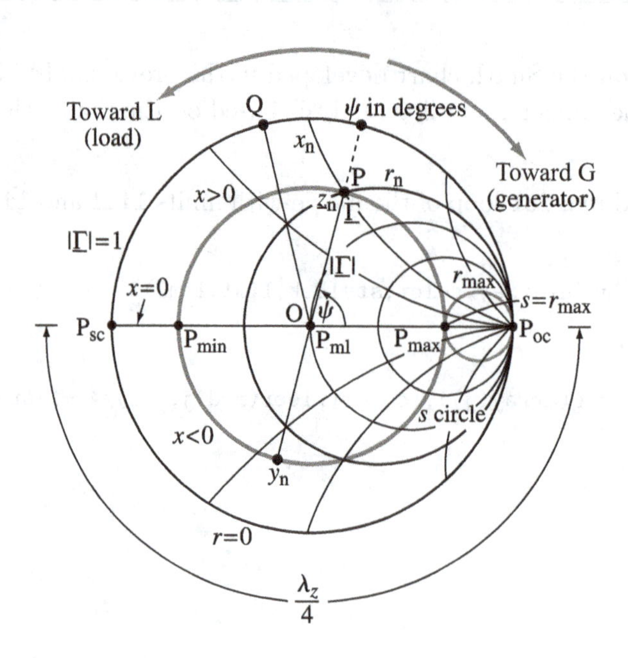

Figure 10.15 Highlighting several key features of the Smith chart in Fig.10.13 and its use.

With the use of Eq.(10.9), we realize that voltage maxima on the line in Fig.10.4 occur at locations where the generalized voltage reflection coefficient is purely real and positive, and this corresponds to the point P_{max} on the Smith chart (Fig.10.15) where the s circle intersects the positive Γ_r-axis (i.e., the line $\overline{OP_{oc}}$),

$$|\underline{V}| = |\underline{V}|_{max} \quad \longleftrightarrow \quad (\Gamma_r, \Gamma_i) = (|\underline{\Gamma}_L|, 0) \quad \text{(point } P_{max} \text{ on the chart)} . \tag{10.25}$$

On the other side, $\underline{\Gamma}$ at locations of voltage minima on the line is purely real and negative, so that the corresponding point P_{min} on the chart is at the intersection of the s circle and the negative Γ_r-axis ($\overline{OP_{sc}}$),

$$|\underline{V}| = |\underline{V}|_{min} \quad \longleftrightarrow \quad (\Gamma_r, \Gamma_i) = (-|\underline{\Gamma}_L|, 0) \quad \text{(point } P_{min}) . \tag{10.26}$$

MATLAB EXERCISE 10.32 **Rotation in the Smith Chart.** Write a function `RotateAroundSCh()` in MATLAB that performs rotation in the Smith Chart. As input, it takes the reflection coefficient $\underline{\Gamma}$ at a starting point on the chart, `gStart`, and the rotation direction, `dir` (`'g'` – toward the generator or `'l'` – toward the load), according to Fig.10.15, the length of a transmission line, `lengthTL`, and the operating wavelength, `lambda`. The function returns an array of coefficients $\underline{\Gamma}$ along the line (in Fig.10.4), which constitute movement around the s circle (constant-$|\underline{\Gamma}|$ circle) in Fig.10.15, `gArray`, and the final value of $\underline{\Gamma}$ (at the end of rotation), `gEnd`. *(RotateAroundSCh.m on IR)*

TUTORIAL:

For the generalized reflection coefficient as a function of the location along a transmission line, we implement Eq.(10.8), namely, $\underline{\Gamma}(l) = \underline{\Gamma}_{start}\, e^{\mp j2\beta l}$ ($\alpha = 0$, $\beta = 2\pi/\lambda_z$), where we use a variable `sign` set in the `switch-case` command to incorporate the rotation direction and a mesh `l` of positions for computing $\underline{\Gamma}$,

```
function [gArray,gEnd] = RotateAroundSCh(gStart,dir,lengthTL,lambda)
switch dir
case'g'
sign = -1;
case'l'
sign = 1;
otherwise
sign = 0;
end;
l = 0:0.01*lambda:lengthTL;
gArray = gStart*exp(sign*i*l/(lambda/2)*2*pi);
gEnd = gArray(length(gArray));
```

MATLAB EXERCISE 10.33 **Smith chart calculations on a transmission line – in a movie.** A lossless transmission line of length $l = 2.8$ m and characteristic impedance $Z_0 = 50\ \Omega$, fed by a time-harmonic generator of frequency $f = 150$ MHz, is terminated at the other end in a load with impedance $\underline{Z}_{\rm L} = (30 + {\rm j}60)\ \Omega$. The parameters of the line dielectric are $\varepsilon_{\rm r} = 4$ and $\mu_{\rm r} = 1$. Find the load reflection coefficient and the input impedance of the line. In specific, create a movie in MATLAB to perform and visualize the calculations – in the Smith chart. *(ME10_33.m on IR)* **H**

TUTORIAL:

See Figs.10.16 and 10.17. We first specify input data and compute the wavelength along the line, as $\lambda_z = c_0/(\sqrt{\varepsilon_{\rm r}\mu_{\rm r}}f)$ [Eqs.(7.5) and (7.6)], and the normalized load impedance, as $\underline{z}_{\rm L} = \underline{Z}_{\rm L}/Z_0$ [Eqs.(10.21)]. Function `mappingZ2Gamma` (from MATLAB Exercise 10.26) is used to find the load reflection coefficient (`gl`), which is displayed in the Command Window.

```
l = 2.8;
Z0 = 50;
f = 150*10^6;
Zl = 30 + i*60;
Er = 4;
ur = 1;
c0 = 299792458;
lambdaz = c0/f/sqrt(Er*ur);
zl = Zl/Z0;
gl = mappingZ2Gamma(zl);
fprintf('\nLoad reflection coefficient is %.2f <%.2f degrees',abs(gl),phase(gl)*180/pi);
```

To make a movie visualizing the Smith chart calculations, we plot a chart by calling function `SmithChart` (from MATLAB Exercise 10.29), and initialize the movie frame counter `k`. MATLAB function `hold on` enables that every new frame is made from the previous one by adding current

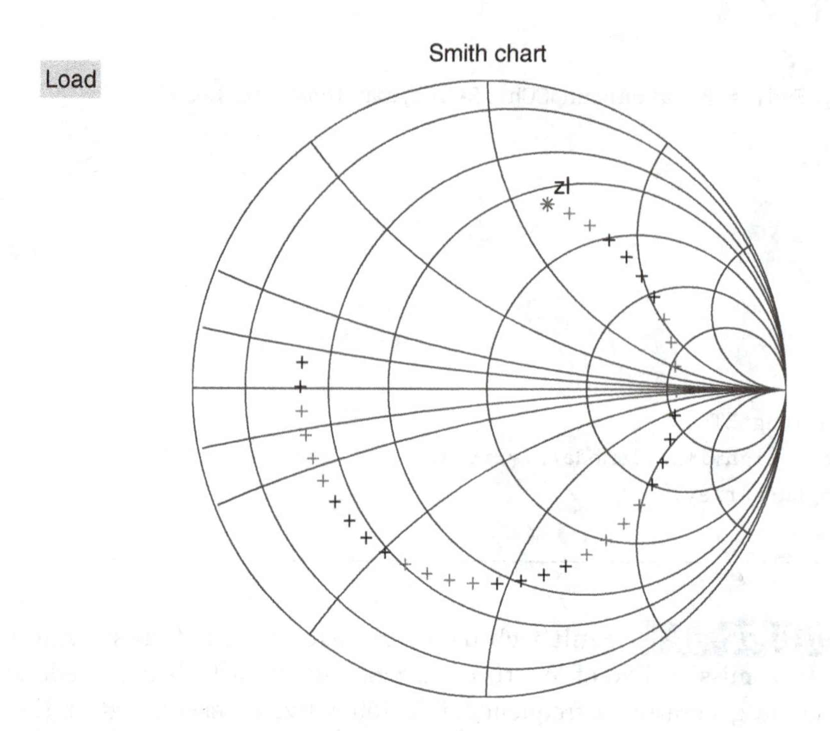

Figure 10.16 Snapshot (frame) of a MATLAB movie visualizing the movement of the point $\underline{z}_n$ ($\underline{\Gamma}$) around the s circle in the Smith chart from the load point (`zl`) toward the generator; for MATLAB Exercise 10.33. *(color figure on CW)*

plots. We start the movie by plotting on the chart the load point (corresponding to the reflection coefficient `gl` and normalized impedance `zl`). Then, an array of points $\underline{\Gamma}$ along the transmission line is calculated and incorporated in the movie using functions `RotateAroundSCh` and `plotOnSCh` (from MATLAB Exercises 10.32 and 10.30). The movie visualizes the movement of the point representing the reflection coefficient $\underline{\Gamma}$ and normalized line impedance $\underline{z}_n$ around the s circle in the Smith chart (equivalent to the movement along the transmission line in Fig.10.4). Since $l > \lambda_z/2$, i.e., we have more than one full rotation on the chart [see Eq.(10.24)], the movie uses different colors to make it easier for the viewer to track the actual moving around the s circle.

```
SmithChart();
k = 1;
hold on;
plot(gl,'b*');
text(real(gl)*1.1,imag(gl)*1.1,'zl');
text(-1.5,1,'Load','BackgroundColor',[0.3 1 1]);
M(k) = getframe(k);
[gA,gin] = RotateAroundSCh(gl,'g',l,lambdaz);
color = ['r','c','k','y','g'];
k = plotOnSCh(gA,color,'+',k,1,length(gA));
plot(gin,'dk');
text(real(gin)*1.1,imag(gin)*1.1,'zin');
```

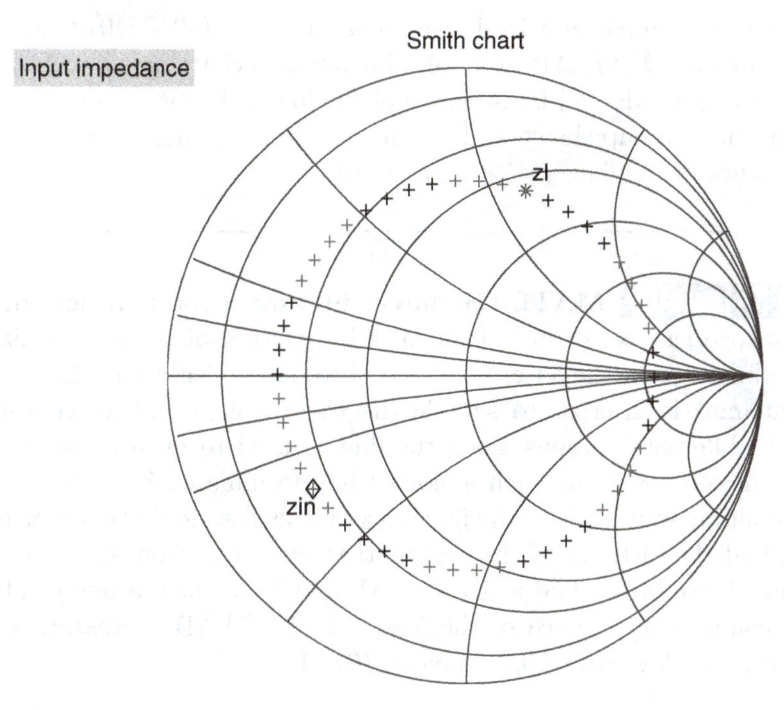

Figure 10.17 Final movie frame showing the generator point (`zin`); for MATLAB Exercise 10.33. *(color figure on CW)*

```
text(-1.5,1,'Input impedance','BackgroundColor',[0.3 1 1]);
hold off;
```

At the end of the code, the normalized input impedance of the line is computed by function `mappingGamma2Z` (from MATLAB Exercise 10.26), and the impedance $\underline{Z}_{\rm in} = \underline{z}_{\rm in} Z_0$ is displayed,

```
zin = mappingGamma2Z(gin);
Zin = Z0*zin;
if (imag(Zin)) < 0
sign = '-';
else
sign = '+';
end;
fprintf('\nInput impedance is %.0f %s i%.0f Ohm',real(Zin),sign,abs(imag(Zin)));
```

The following is the display of the results:

```
Load reflection coefficient is 0.63 <71.57 degrees
Input impedance is 12 - i15 Ohm
```

Figures 10.16 and 10.17 show two snapshots of the movie.

MATLAB EXERCISE 10.34 **Transmission-line analysis using a Smith chart movie.** A lossless transmission line of characteristic impedance $Z_0 = 100\ \Omega$, phase velocity $v_{\rm p} = 2 \times 10^8$ m/s, and length $l = 34$ cm is terminated in a load of impedance $\underline{Z}_L = (30 - j40)\ \Omega$ at a frequency of $f = 1$ GHz. As in the previous MATLAB exercise, play a Smith chart movie in MATLAB for this line, and determine the magnitude and phase angle of the load reflection coefficient, the standing wave ratio of the line, the line impedance and admittance halfway along the line, and the input impedance and admittance of the line. *(ME10_34.m on IR)* **H**

MATLAB EXERCISE 10.35 **MATLAB movie finding a load impedance using the Smith chart.** An instrument consisting of an air-filled section of a rigid coaxial line (cable) with a narrow longitudinal slot in the outer conductor through which a movable (sliding) electric probe (short wire antenna) is inserted to sample the electric field and hence voltage between the line conductors, at different locations along the line, is used to measure load impedances at high frequencies. By measurements on such a slotted line terminated in an unknown load, it is determined that the standing wave ratio of the line is $s = 3$, the distance between successive voltage minima $\Delta l = 40$ cm, and the distance of the first voltage minimum from the load $l_{min} = 12$ cm. The characteristic impedance of the line is $Z_0 = 50\ \Omega$, and losses in the line conductors can be neglected. Find the complex impedance of the load – in MATLAB – creating a movie of the graphical analysis in the Smith chart. *(ME10_35.m on IR)* **H**

HINT:
See MATLAB Exercise 10.33 and Figs.10.18–10.20. Start with a plain Smith chart (use function

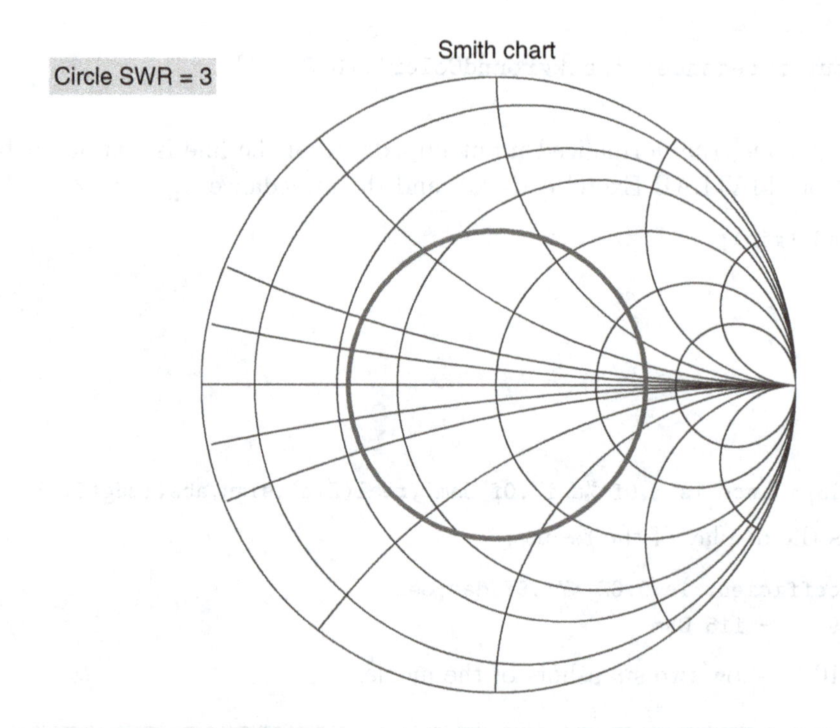

Figure 10.18 MATLAB movie finding an unknown load impedance on a slotted line in the Smith chart: initial snapshot showing a plain Smith chart and the s circle; for MATLAB Exercise 10.35. *(color figure on CW)*

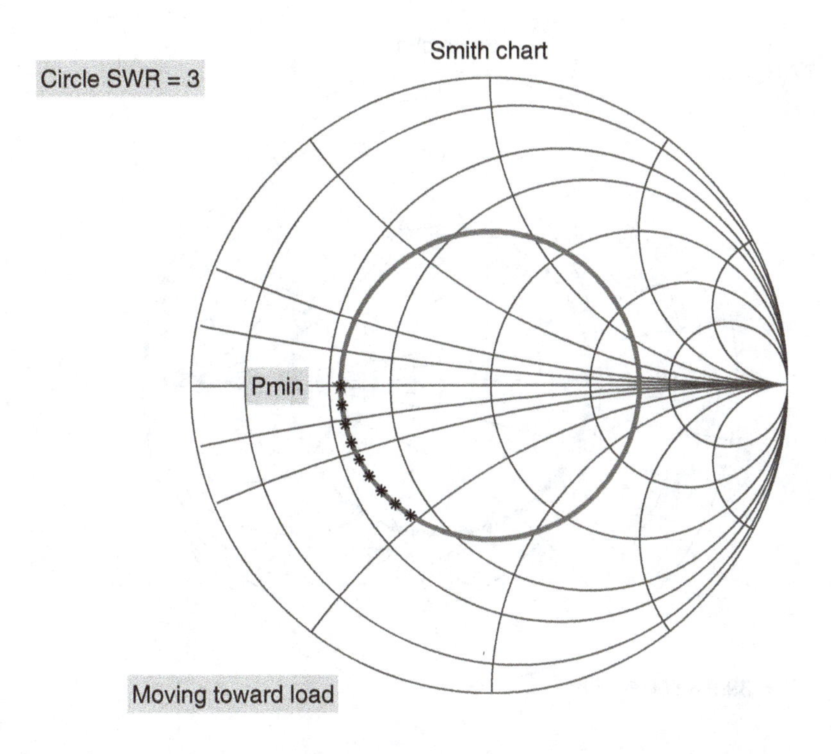

Figure 10.19 MATLAB movie finding an unknown load impedance: snapshot of the movement around the s circle from the point P_{min} toward the load; for MATLAB Exercise 10.35. *(color figure on CW)*

`SmithChart`), and add the s circle [use function `sCircle`, from MATLAB Exercise 10.31, and plot the entire circle in one frame (not point by point)]. Then, start at the point P_{min}, defined by Eq.(10.26), and move, in a movie, around the circle toward the load by a distance l_{min} (calling functions `RotateAroundSCh` and `plotOnSCh`). At the load point (L), take the result for $\underline{\Gamma}_L$ and convert it to $\underline{z}_L$ and $\underline{Z}_L$. Three characteristic frames of the resulting movie are shown in Figs.10.18–10.20.

MATLAB EXERCISE 10.36 **Another MATLAB movie to find an unknown impedance.** Using a MATLAB movie of the Smith chart analysis, find the complex impedance of an unknown load terminating an air-filled slotted line, whose characteristic impedance is $Z_0 = 60\ \Omega$ and losses in conductors are negligible, if measurements on the line give a standing wave ratio of $s = 2$ and the location of the first voltage maximum at $l_{max} = 40$ cm from the load, which is also $\Delta l = 80$ cm to the next maximum. *(ME10_36.m on IR)* **H**

MATLAB EXERCISE 10.37 **Searching for a desired resistance or reactance along a line.** Write a function `findDesiredRorX()` in MATLAB that, as function `RotateAroundSCh` (from MATLAB Exercise 10.32), starts at a point in the Smith chart, where the reflection coefficient

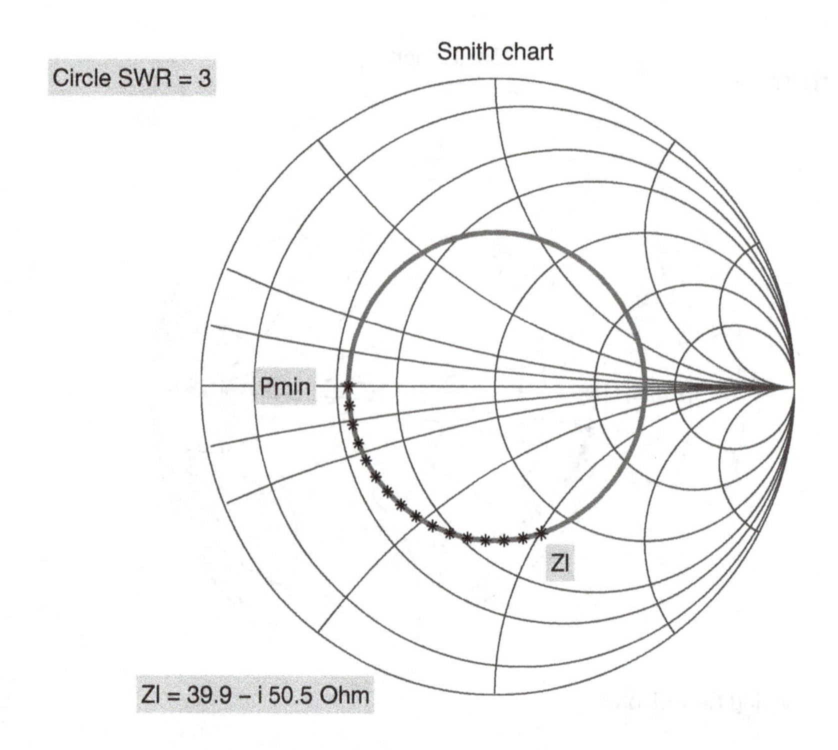

Figure 10.20 MATLAB movie finding an unknown load impedance: final snapshot showing the load point and the obtained load impedance; for MATLAB Exercise 10.35. *(color figure on CW)*

equals **gStart**, and moves around the s circle (along the transmission line) in a given direction, determined by the parameter **dir** – but, as opposed to function **RotateAroundSCh**, ends at a point where either r or x on the line match the given value, **value**. Whether it is r or x is specified by an input argument **term**, which is either **'r'** or **'x'**. So, if, for instance, **term** equals **'r'** and **value** amounts to r_0, the function finds the normalized impedance $r_0 + jx$ on the line. Since the s circle cuts an r circle or an x arc on the chart either twice or once (rarely) or they do not intersect, the last input argument of the function, **a**, tells the function which one of the two solutions should be returned (with one solution or no solution as included special cases); $a = 1$ means return the solution found first (closer to the starting point), while $a = 2$ means return the other solution. The output of the function consists of an array of coefficients $\underline{\Gamma}$ along a $\lambda_z/2$ range, **gArray**, and the array index **N** and the distance along the line from the starting point, **lout**, corresponding to the matched normalized resistance or reactance on the line. *(findDesiredRorX.m on IR)*

TUTORIAL:

First, in case it is not specified as input, the variable **a** takes the default value, $a = 1$ (return the first solution). Then, a variable **sign** is introduced and Eq.(10.8) (with $\alpha = 0$ and $\beta = 2\pi/\lambda_z$) is implemented as in MATLAB Exercise 10.32, with a $\lambda_z/2$ long mesh being defined as in MATLAB Exercise 10.19. The two solutions for the matched normalized resistance or reactance are found in **switch-case** command, based on the input parameter **term**, where function **mappingGamma2Z** (from MATLAB Exercise 10.26) is applied to convert the $\underline{\Gamma}$ array to the array of normalized impedances and the search is performed using MATLAB function **min** (see MATLAB Exercise

10.19), which constitutes a numerical approach to solving this problem. Finally, the solutions are arranged according to their distances from the starting point (closer first) and, depending on the input a, the proper solution is returned.

```
function [gArray,N,lout] = findDesiredRorX(gStart,dir,lambda,term,value,a)
if (nargin == 5)
a=1;
end;
switch dir
case 'g'
sign = -1;
case 'l'
sign = 1;
otherwise
sign = 0;
end;
l = 0:0.005*lambda:lambda/2;
array = gin*exp(sign*i*l/(lambda/2)*2*pi);
gArray = array;
switch term
case 'r'
[ErrorMin,n1] = min(abs(real(mappingGamma2Z(array))- value));
array(n1) = array(1);
[ErrorMin,n2] = min(abs(real(mappingGamma2Z(array))- value));
case 'x'
[ErrorMin,n1] = min(abs(imag(mappingGamma2Z(array))- value));
array(n1) = array(1);
[ErrorMin,n2] = min(abs(imag(mappingGamma2Z(array))- value));
end;
if (n1 < n2)
n = [n1 n2];
else
n = [n2 n1];
end;
N = n(a);
lout = l(N);
```

MATLAB EXERCISE 10.38 Matching the real part of the line impedance – Smith chart movie. Consider a lossless transmission line with characteristic impedance $Z_0 = 50\ \Omega$ and operating wavelength $\lambda_z = 1$ m, terminated in a load whose complex impedance is $\underline{Z}_{\rm L} = (40 + {\rm j}20)\ \Omega$. Write a MATLAB program that calculates numerically the length of the line, l, such that the real part of its input impedance, $\underline{Z}_{\rm in}$, equals Z_0 (normalized, $r = 1$). As there are two solutions for such an input impedance and two solutions for l within a $\lambda_z/2$ range, the user should have an opportunity to choose at input which solution is to be retained, the first (shorter line) or the second (longer line). The program plays a movie showing the design using the Smith chart. It also displays the numerical solution for l and $\underline{Z}_{\text{in}}$. Note that once the real part of the line impedance is thus matched to Z_0, the imaginary part is easily compensated (annulled) by a series stub, as in Fig.10.10(b). *(ME10_38.m on IR)* **H**

TUTORIAL:

See Figs.10.21 and 10.22. We first plot a plain Smith chart (as a background of the entire movie), specify input data, and compute and enter on the chart the load reflection coefficient. In addition, the program offers to the user a choice between the two solutions. Then, the core of the numerical calculation along the transmission line is done by function `findDesiredRorX` (developed in the previous MATLAB exercise). The movie visualizes, frame by frame, the movement around the s circle – until the desired solution is found. Finally, the program displays in the Command Window the obtained results for l and $\underline{Z}_{\text{in}}$.

```
SmithChart();
hold on;
Z0 = 50;
lambdaz = 1;
Zl = 40 + i*20;
zl = Zl/Z0;
gl = mappingZ2Gamma(zl);
```

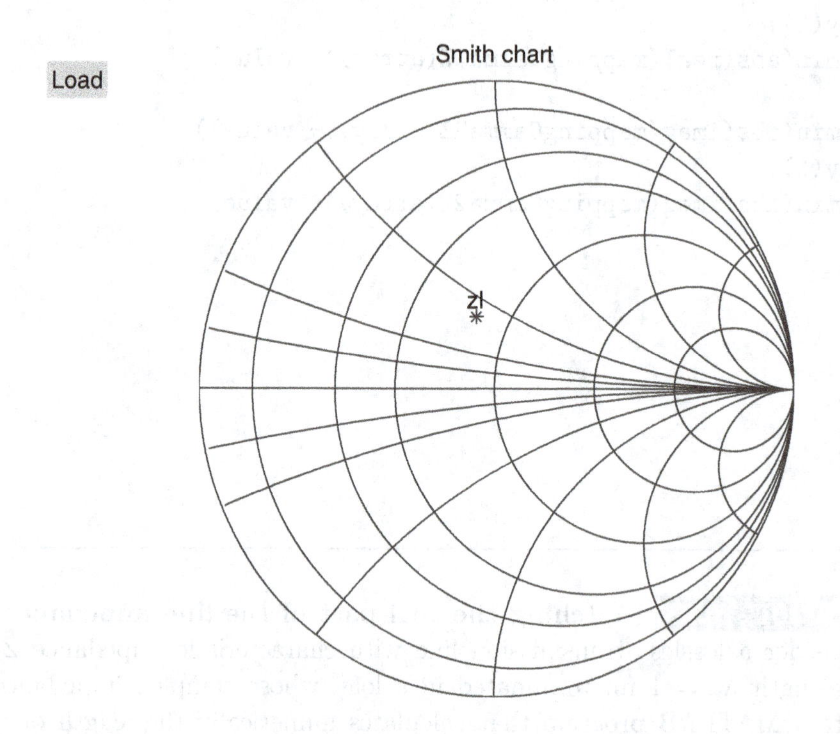

Figure 10.21 MATLAB movie visualizing the numerical design in the Smith chart of a transmission line with input impedance having $r = 1$ (matched real part): initial frame showing the load point (`zl`) on the chart; for MATLAB Exercise 10.38. *(color figure on CW)*

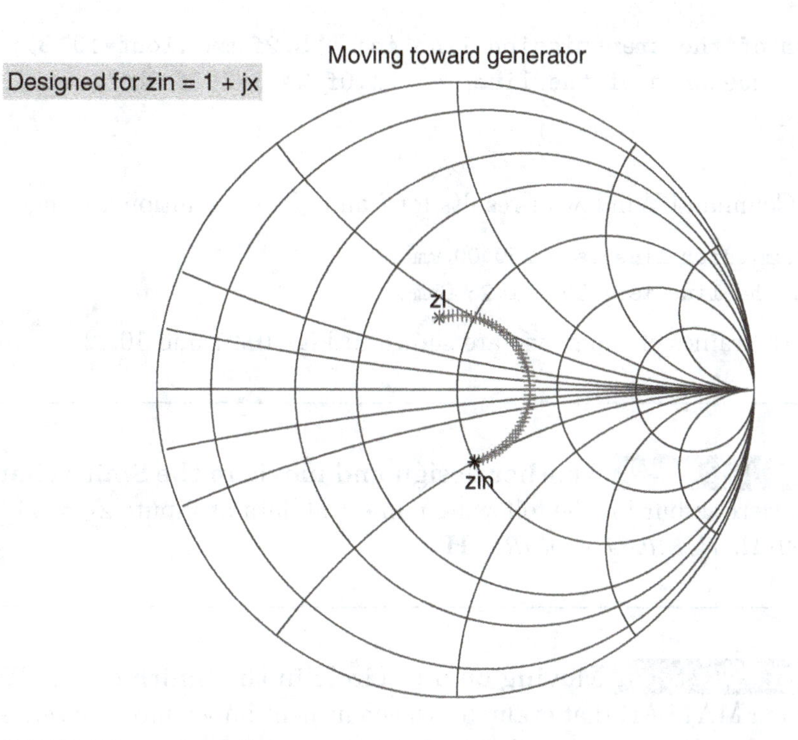

Figure 10.22 MATLAB movie of the numerical design of a line with matched input resistance: final frame showing the generator point (zin) on the chart; for MATLAB Exercise 10.38. *(color figure on CW)*

```
plot(gl,'b*');
text(real(gl)*1.1,imag(gl)*1.1,'zl');
text(-1.5,1,'Load','BackgroundColor',[0.3 1 1]);
a = input('\nChoose first (1) or second (2) solution:  ');
[gArray,N,lout] = findDesiredRorX(gl,'g',lambdaz,'r',1,a);
title('Moving toward generator');
text(real(gl)*1.1,imag(gl)*1.1,'zl');
k = plotOnSCh(gArray,'m','+',1,1,N);
gmatch = gArray(N);
k = k + 1;
plot(gmatch,'*k');
text(real(gmatch)*1.1,imag(gmatch)*1.1,'zin');
text(-1.5,1,'Designed for zin = 1 + jx','BackgroundColor',[0.3 1 1]);
M(k) = getframe();
zin = mappingGamma2Z(gmatch);
Zin = zin*Z0;
if imag(Zin) < 0
sign = '-';
else
sign = '+';
end;
```

```
fprintf('\nLength of the transmission line is:  %5.2f mm',lout*10^3);
fprintf('\nInput impedance of the line is:  %.0f %s i*%.0f Ohm',real(Zin),...
sign,abs(imag(Zin)));
hold off;
```

The display in the Command Window of results for l and $\underline{Z}_{\text{in}}$ for solution 2 (longer line) reads:

```
Length of the transmission line is:  250.00 mm
Input impedance of the line is:  50 - i*25 Ohm
```

and two characteristic frames of the movie are shown in Figs.10.21 and 10.22.

MATLAB EXERCISE 10.39 **Another design and movie in the Smith chart.** Repeat the previous MATLAB exercise but for the following numerical data at input: $Z_0 = 50\ \Omega$, $\lambda_z = 0.1$ m, and $\underline{Z}_{\text{L}} = (100 + \text{j}50)\ \Omega$. *(ME10_39.m on IR)* **H**

MATLAB EXERCISE 10.40 **Moving on an r circle in the Smith chart.** Write a function `movingOnRcircle()` in MATLAB that evaluates the equivalent impedance of a series connection of a transmission line and a short- or open-circuited stub, as in Fig.10.10(b). At input, the function takes the generalized reflection coefficient of the line at the connection location, `gin`, and the reference input reactance of the stub (stub impedance is purely imaginary), `xadd`. At output, the function returns an array of reflection coefficients, `gArray`, as in function `RotateAroundSCh` (MATLAB Exercise 10.32), obtained for a range of stub reactances, from zero to `xadd`, which constitutes movement around the r circle (r = const and x is varied) in the Smith chart [see Figs.10.12(a) and 10.13], as well as the final value of $\underline{\Gamma}$ (for `xadd`), `gend`. *(movingOnRcircle.m on IR)*

TUTORIAL:

As this is a series stub, the function implements the addition of impedances in Eqs.(10.19), as follows:

```
function [gArray,gend] = movingOnRcircle(gin,xadd)
z = mappingGamma2Z(gin);
r = real(z);
x = imag(z);
dx = 0.05*(xadd);
xnew = x:dx:x+xadd;
znew = r + i*xnew;
gArray = mappingZ2Gamma(znew);
gend = gArray(length(gArray));
```

MATLAB EXERCISE 10.41 **Complete matching with a series stub using a Smith chart movie.** Write a MATLAB program that carries out a complete design in the Smith chart of an impedance-matching transmission-line circuit with a series stub, shown in Fig.10.10(b), including finding the stub location on the line. Assume that $Z_0 = 100\ \Omega$, $\lambda_z = 10$ cm, and $\underline{Z}_{\rm L} = (80 - {\rm j}40)\ \Omega$. The user can select one of the two solutions for l and $l_{\rm stub}$, as well as choose whether an open- or short-circuited stub is used. The program plays a movie in the Smith chart for all stages of the design. *(ME10_41.m on IR)* **H**

TUTORIAL:

See Figs.10.23–10.26. The first part of the program is similar to that in MATLAB Exercise 10.38 – it matches the real part of the input impedance of the transmission line (makes its normalized input resistance be $r = 1$) using function **findDesiredRorX** (from MATLAB Exercise 10.37), and plays a movie showing these transformations graphically in the Smith chart.

```
SmithChart()
hold on;
Z0 = 100;
lambdaz = 0.1;
Zl = 80 - i*40;
zl = Zl/Z0;
gl = mappingZ2Gamma(zl);
plot(gl,'b*');
text(real(gl)*1.1,imag(gl)*1.1,'zl');
```

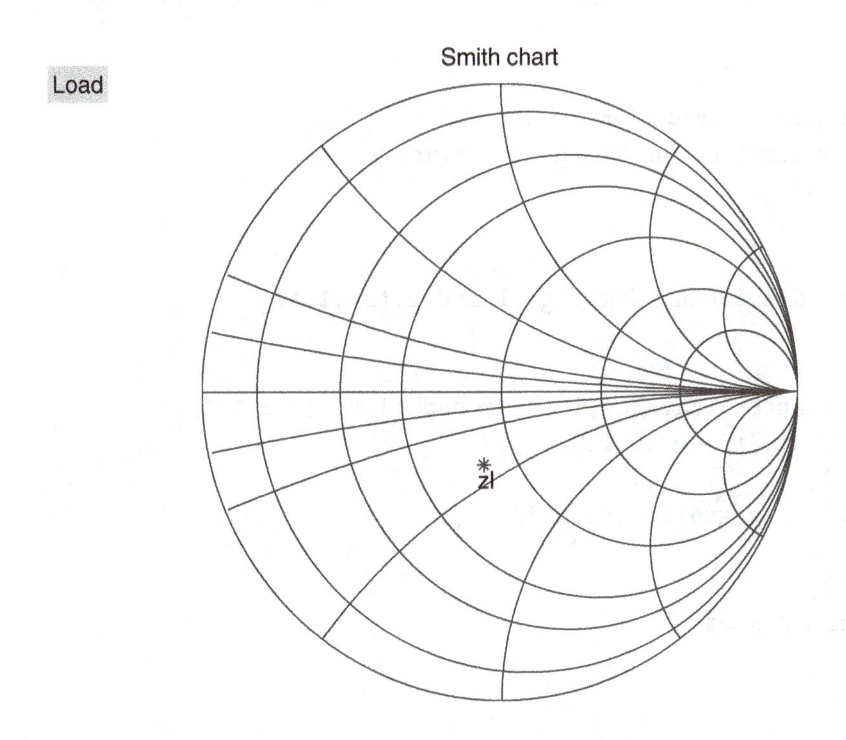

Figure 10.23 MATLAB movie of a complete design in the Smith chart of an impedance-matching transmission-line circuit with a series stub [Fig.10.10(b)]: initial frame showing the load point (`zl`); for MATLAB Exercise 10.41. *(color figure on CW)*

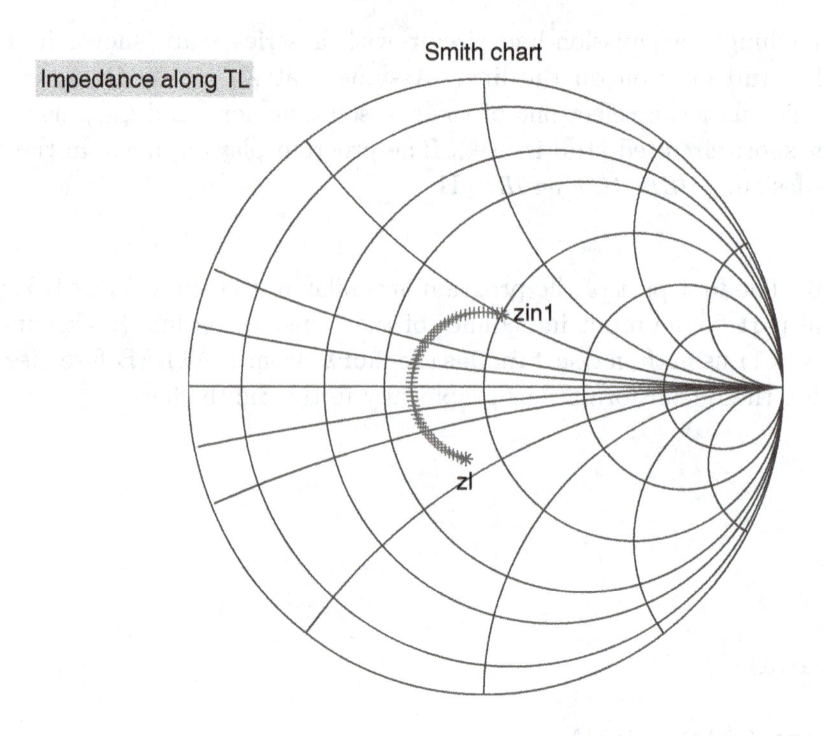

Figure 10.24 MATLAB movie of a complete series-stub design: frame showing movement on the s circle (to match the real part of the line impedance); for MATLAB Exercise 10.41. *(color figure on CW)*

```
text(-1.5,1,'Load','BackgroundColor',[0.3 1 1]);
a = input('\nChoose first (1) or second (2) solution:  ');
if (a > 2) && (a < 1)
a = 1;
end;
[gArray,N,l] = findDesiredRorX(gl,'g',lambdaz,'r',1,a);
gmatch = gArray(N);
lout = l;
text(-1.5,1,'Impedance along TL','BackgroundColor',[0.3 1 1]);
k = plotOnSCh(gArray,'r','+',0,1,N);
plot(gmatch,'r*');
text(real(gmatch),imag(gmatch),'zin1');
k = k+1;
M(k) = getframe();
zeq = mappingGamma2Z(gmatch);
Zeq = zeq*Z0;
if imag(Zeq) < 0
sign = '-';
else
sign = '+';
end;
```

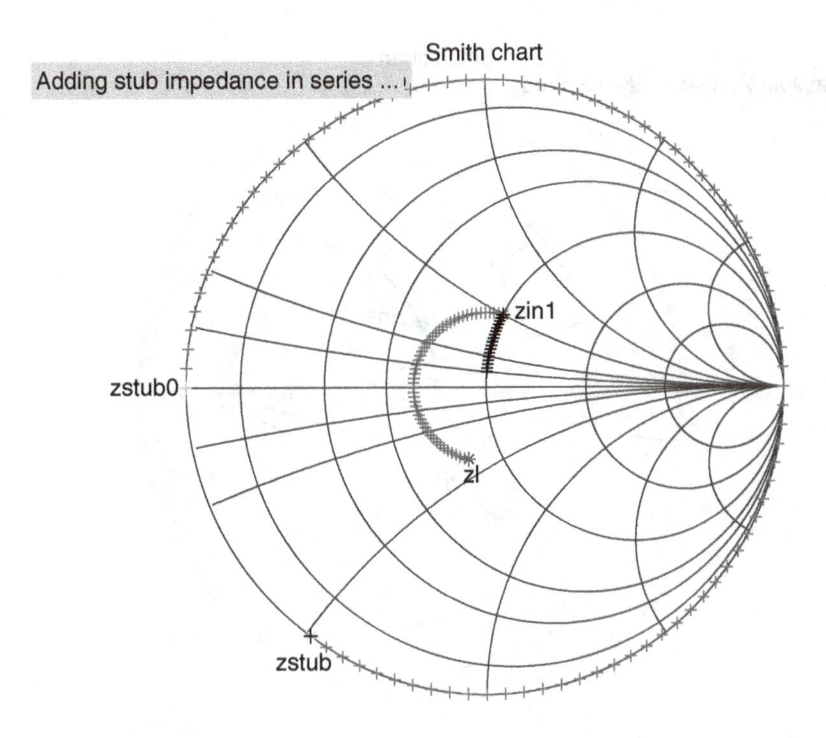

Figure 10.25 MATLAB movie of a complete series-stub design: frame showing movement on the r circle (to annul the imaginary part of the line impedance); for MATLAB Exercise 10.41. *(color figure on CW)*

```
fprintf('\nLength of the transmission line is:  %4.2f mm',lout*1000);
fprintf('\nEquivalent input impedance of the line is:  ');
fprintf('%.0f %s i*%.0f Ohm',real(Zeq),sign,imag(Zeq));
```

The next part calculates the required stub impedance, displays it, and asks the user whether an open- or short-circuited stub is preferred. Based on the entered choice, the stub length is determined, again by means of function **findDesiredRorX**, and the corresponding frames of the movie are realized with the use of function **plotOnSCh** (from MATLAB Exercise 10.30).

```
zstub = -i*imag(zeq);
Zstub = Z0*zstub;
if imag(Zstub) < 0
sign = '-';
else
sign = '+';
end;
fprintf('\nStub impedance is:  %s i*%.0f Ohm', sign,abs(imag(Zstub)));
a = input('\nChoose open stub (1) or shorted stub (2):  ');
switch a
case 1
gs0 = 1 + i*0;
stubName = 'open';
```

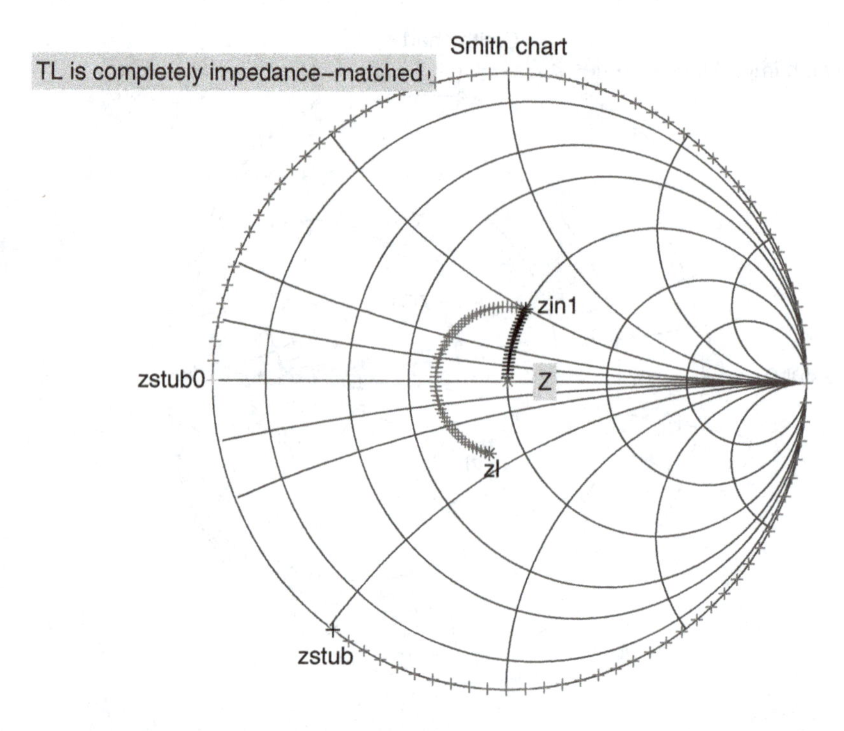

Figure 10.26 MATLAB movie of a complete series-stub design: final frame showing the completely impedance-matched line impedance; for MATLAB Exercise 10.41. *(color figure on CW)*

```
case 2
gs0 = -1 + i*0;
stubName = 'short';
otherwise
fprintf('\n Invalid input!!!' );
end;
text(-1.5,1,['Calculating impedance of ',stubName,' stub'],'BackgroundColor',[0.3 1 1]);
plot(real(gs0),imag(gs0),'+g','MarkerSize',5);
text(1.1*real(gs0),1.1*imag(gs0),'zstub0');
[gsArray,ns,ls] = findDesiredRorX(gs0,'g',lambdaz,'x',imag(zstub));
k = plotOnSCh(gsArray,'m','+',k,2,ns);
gsout = gsArray(ns);
plot(gsout,'c*');
text(1.1*real(gsout),1.1*imag(gsout),'zstub');
fprintf('\nStub length is:  %4.2f mm',1000*ls );
text(-1.5,1,'Adding stub impedance in series ...','BackgroundColor',[0.3 1 1]);
plot(gmatch,'+k');
pause(0.1);
k = k+1;
M(k) = getframe();
plot(gsout,'+k');
```

```
pause(0.1);
k = k+1;
M(k) = getframe();
```

Finally, the stub impedance is added to the line impedance employing function `movingOnRcircle` (from the previous MATLAB exercise), and movie frames showing the associated movement on the r circle in the Smith chart are played.

```
[gArrayAdd,gAdd] = movingOnRcircle(gmatch,imag(zstub));
k = plotOnSCh(gArrayAdd,'k','+',k,1,length(gArrayAdd)-1);
plot(0,0,'*r');
text(0.1,0,'Z','BackgroundColor',[0 1 1]);
text(-1.6,1,'TL is completely impedance-matched','BackgroundColor',[0 1 1]);
hold off;
```

Results for the complete impedance-matching for solution 1 (shorter line) and short-circuited stub are displayed in the Command Window as follows:

```
Choose first (1) or second (2) solution:  1
Length of the transmission line is:  25.00 mm
Equivalent input impedance of the line is:  100 + i*50 Ohm
Stub impedance is:  - i*50 Ohm
Choose open stub (1) or shorted stub (2):  2
Stub length is:  42.50 mm
```

Four characteristic frames of the movie are shown in Figs.10.23–10.26. The user should rerun this program multiple times choosing all different options.

MATLAB EXERCISE 10.42 **Another matching design with a Smith chart movie.** Repeat the previous MATLAB exercise but for $\underline{Z}_{\mathrm{L}} = (40 + \mathrm{j}30)\ \Omega$ (other input data are the same). Run the program for all options (solutions 1 and 2, open and shorted stub). *(ME10_42.m on IR)* **H**

10.9 Transient Analysis of Transmission Lines

Temporary variations of voltages and currents on transmission lines in establishing the steady state of time-harmonic or any other forms of signals in the structure are called transients. A notable example are transients produced by step-like (on or off) abrupt changes of the input voltage or current at the beginning of a transmission line, which corresponds to establishing a time-constant (dc) voltage and current along the line. To start our transient analysis of transmission lines, let us assume that the voltage generator at the beginning of the line in Fig.10.4 is given by its time-varying emf $e(t)$ and internal resistance R_{g}, as well as that $e(t)$ is zero for $t < 0$ and nonzero (with an arbitrary time variation) for $t \geq 0$ (the generator is switched on at $t = 0$). In addition, let us, for the convenience of the study in this and the sections to follow, adopt the origin of the z-axis ($z = 0$) to be at the generator terminals, as shown in Fig.10.27. The load terminals are then at the position defined by $z = l$, with l being the length of the line. Finally, let (for now) the load be purely resistive, of resistance R_L.

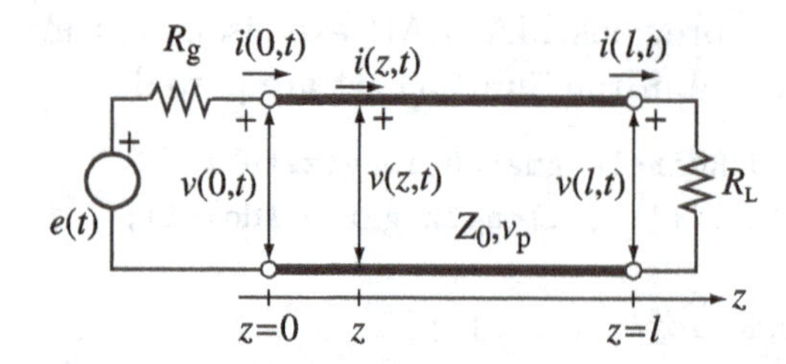

Figure 10.27 Transmission line fed by a time-varying (nonsinusoidal) voltage generator with a purely resistive internal impedance and terminated in a purely resistive load.

Prior to the time $t = 0$, the voltage and current are identically zero at every position along the line. At $t = 0$, the generator launches a voltage v_i, and the accompanying current i_i, to propagate toward the load. The incident wavefront reaches the load at instant $t = T$, where T designates the full one-way transit or delay time along the line,

$$T = \frac{l}{v_p} \quad \text{(one-way delay time of a transmission line)} \tag{10.27}$$

(v_p is the phase velocity along the line). In the general case for Fig.10.27, the load is not matched to the line ($R_L \neq Z_0$), so that a reflected wave, of voltage v_r and current i_r, is generated at the load. The load voltage reflection coefficient (Γ_L) is given by Eq.(10.5) with $\underline{Z}_L$ replaced by R_L. As the reflected wave travels away from the load, in Fig.10.27, its wavefront arrives to the generator at $t = 2T$. If the generator is not matched to the line, that is, if $R_g \neq Z_0$, reflection (namely, re-reflection) of the reflected voltage v_r from the generator occurs. The expression for the voltage reflection coefficient at the generator (Γ_g) is the same as for the load coefficient but with R_g in place of the load resistance, so we have

$$\Gamma_g = \frac{R_g - Z_0}{R_g + Z_0}, \quad \Gamma_L = \frac{R_L - Z_0}{R_L + Z_0} \quad \text{(generator and load reflection coefficients)}. \tag{10.28}$$

In analogy to Eqs.(10.2), the total voltage and current on the line are

$$v(z,t) = v_i(z,t) + v_r(z,t), \quad i(z,t) = i_i(z,t) + i_r(z,t) = \frac{1}{Z_0} v_i(z,t) - \frac{1}{Z_0} v_r(z,t), \tag{10.29}$$

with Z_0 standing for the characteristic impedance of the line.

MATLAB EXERCISE 10.43 **Initial incident voltage and current of a transmission line.** Let the emf in Fig.10.27 be a step time function, defined as

$$e(t) = \mathcal{E}h(t), \quad h(t) = \begin{cases} 0 & \text{for } t < 0 \\ 1 & \text{for } t > 0 \end{cases} \quad \text{(step excitation)}, \tag{10.30}$$

thus representing a voltage generator switched on at an instant $t = 0$, from a zero emf to a time-constant (for $t > 0$) value, $\mathcal{E}$ [the unit step function, $h(t)$, is known as the Heaviside function]. Prior to the return of any backward propagating wave reflected from the load, the only wave on the line is the incident one, traveling in the positive z direction in Fig.10.27. Denoting its voltage and current intensity by v_{i1} and i_{i1}, respectively, the dynamic (for transients) line impedance at the generator terminals, that is, the input dynamic impedance of the line, amounts to

$$(Z_{\rm in})_{\rm dynamic} = \frac{v_{i1}}{i_{i1}} = Z_0 \quad \text{(input dynamic impedance of a transmission line)} . \tag{10.31}$$

In other words, the generator sees, looking into the line, a purely resistive impedance equal to the characteristic impedance of the line, Z_0. Hence, the line can be replaced, with respect to the generator, by an equivalent purely resistive load of resistance Z_0, and from the voltage divider formed by Z_0 and R_g, the incident voltage in this initial period of time (before any reflections have occurred) is given by

$$v_{i1} = \frac{Z_0}{Z_0 + R_g}\mathcal{E} \quad \text{(initial incident voltage on a line)} , \tag{10.32}$$

and the corresponding current is $i_{i1} = v_{i1}/Z_0$. Write functions `initialVoltage()` and `initialCurrent()` in MATLAB that compute v_{i1} and i_{i1}, respectively. The input to the functions consists of the line characteristic impedance, Z_0, and the step emf, $\mathcal{E}$, and internal resistance, R_g, of the voltage generator. *(initialVoltage.m and initialCurrent.m on IR)*

MATLAB EXERCISE 10.44 General code for calculation of transients on a transmission line. Write a function `signalTL()` in MATLAB that calculates a matrix `V` representing a signal (transient voltage or current) along a lossless transmission line, in Fig.10.27, for any excitation. In specific, `V` is a `(N+1)` × `(KN+1)` matrix, where each row carries a signal along the entire transmission line at a specified time, while each column describes the time dependence of the voltage at a specified coordinate. At input, the function takes an array `Es` describing the time dependence of the emf of the voltage generator, $e(t)$, internal resistance of the generator, R_g, load resistance, R_L, characteristic impedance of the line, Z_0, array `z` with a spatial mesh along the line, parameter K defining the total duration of the signal (KT), with T being the one-way time delay of the line, Eq.(10.27), and `type` specifying whether the signal is voltage (`'v'`) or current (`'i'`). Note that, from the definition of the phase velocity, v_p, of the line, the wave travels a distance $dl = v_p\,dt = (l/T)\,dt$ during an elemental time dt, where l is the line length. So, the temporal and spatial meshes are defined as $z = 0 : dz : l$ and $t = 0 : dt : KT$, respectively (`Es` is defined using the step dt). *(signalTL.m on IR)*

TUTORIAL:

The code is based on several relationships for the incident and reflected voltages on a transmission line, $v_i(z,t)$, propagating in the positive z direction, and $v_r(z,t)$, propagating in the opposite direction, in Fig.10.27. Using the generator and load reflection coefficients, in Eqs.(10.28), we can write for locations $z = 0$ and $z = l$ on the line:

$$v_i(0,t) = v_{i1}(0,t) + \Gamma_g v_r(0,t) \quad \text{and} \quad v_r(l,t) = \Gamma_L v_i(l,t) , \tag{10.33}$$

respectively, where v_{i1} is the initial incident voltage of the line. In addition, having in mind that $\mathrm{d}z = v_\mathrm{p}\,\mathrm{d}t$, and the fact that travel of the reflected wave with velocity v_p in the negative z direction can be interpreted as travel with velocity $-v_\mathrm{p}$ in the positive z direction, we have

$$v_\mathrm{i}(z,t) = v_\mathrm{i}(z - \mathrm{d}z, t - \mathrm{d}t) \quad \text{and} \quad v_\mathrm{r}(z,t) = v_\mathrm{r}(z + \mathrm{d}z, t - \mathrm{d}t) \,. \tag{10.34}$$

Finally, the total voltage on the line, $v(z,t)$, is given by the first relationship in Eqs.(10.29). Analogous relationships hold for the current waves, $i_\mathrm{i}(z,t)$, $i_\mathrm{r}(z,t)$, and $i(z,t)$, along the line [see Eqs.(10.29)].

In the first part of the code, we determine the total number of steps $\mathrm{d}z$ in the spatial mesh, N, defined by $l = N\,\mathrm{d}z$, and initialize matrices for incident and reflected waves, `Vinc` and `Vref`, with proper dimensions. Based on the signal type (voltage or current), we compute the generator and load reflection coefficients, `Gg` and `Gl`, using functions `reflCoeff` and `reflCoeffCurr` (from MATLAB Exercise 10.4), as well as the time dependence of the initial incident signal, $v_{i1}(t)$ or $i_{i1}(t)$ – matrix `Vs`, by means of functions `initialVoltage` and `initialCurrent` (from the previous MATLAB exercise).

```
function V = signalTL(Es,Rg,Rl,Z0,z,K,type)
N = length(z)-1;
Vinc = zeros(N+1,K*N+1);
Vref = zeros(N+1,K*N+1);
Vs = zeros(1,K*N+1);
if strcmp(type,'v')
Gg = reflCoeff(Rg,Z0);
Gl = reflCoeff(Rl,Z0);
Vs(1,1:length(Es)) = initialVoltage(Z0,Rg,Es);
else
Gg = reflCoeffCurr(Rg,Z0);
Gl = reflCoeffCurr(Rl,Z0);
Vs(1,1:length(Es)) = initialCurrent(Z0,Rg,Es);
end;
```

The core of the code is a double `for` loop, where the outer loop runs through all time instances starting with $t = 0$, while the inner loop calculates $v_\mathrm{i}(z,t)$ and $v_\mathrm{r}(z,t)$ [or $i_\mathrm{i}(z,t)$ and $i_\mathrm{r}(z,t)$], implementing Eqs.(10.33) and (10.34). At the end, the total signal matrix, `V`, is obtained using Eq.(10.29).

```
for j = 1:K*N+1
for i = 1:N+1
if i == N+1
Vinc(1,j) = Vs(1,j) + Gg*Vref(1,j);
Vref(N+1,j) = Gl*Vinc(N+1,j);
else
if (j > 1)
Vinc(N+2-i,j) = Vinc(N+2-i-1,j-1);
Vref(i,j) = Vref(i+1,j-1);
end;
```

```
end;
end;
end;
V = Vinc + Vref;
```

MATLAB EXERCISE 10.45 **Plotting transient snapshots and waveforms on transmission lines.** Write a function TLplot() in MATLAB to plot transient responses of lossless transmission lines. The plots present the signal (voltage or current) dependence on either spatial (z) or temporal (t) coordinate, i.e., they are either snapshots (scans) of a signal along the transmission line at an instant of time or waveforms during the course of time at a fixed location on the line. The input to the function contains arrays of signal values (**signal**), corresponding z or t coordinates (**coord**), and characters with units (**unit**). The function also appropriately labels signal values and units next to the characteristic signal levels in the graph. Note that the input data for this function can be obtained by extracting entire rows or columns of the signal matrix generated by function **signalTL** (from the previous MATLAB exercise). *(TLplot.m on IR)*

TUTORIAL:
First, the function plots signal values over the coordinate range. Then, it calculates the maximum of the absolute signal values and the maximum of the absolute coordinate values and uses these maximum extents in determining the text (signal value and unit) placements in the graph. In order to avoid writing these textual labels for every data point, we introduce an if branch inside a for loop, and write labels only for new (changed) signal levels.

```
function TLplot(signal,coord,unit)
plot(coord,signal);
maxS = max(abs(signal));
maxC = max(abs(coord));
for k = 2:length(signal)-1
if (k == 2) || (signal(k) ~= signal(k-1))
value = num2str(signal(k),'%.4g');
text(coord(k)+0.05*maxC, signal(k)+0.08*maxS,[value unit]);
end;
end;
```

10.10 Step Response of Transmission Lines with Purely Resistive Terminations

In this and the next section, we employ the codes developed in the previous section to study step and pulse responses of lossless transmission lines terminated in purely resistive loads, which also includes open- and short-circuited lines as special cases; the section to follow will then introduce a graphical tool for the analysis, based on so-called bounce diagrams.

MATLAB EXERCISE 10.46 Complete transient analysis in MATLAB, both line ends unmatched. A lossless coaxial cable of length $l = 30$ cm has a homogeneous dielectric of parameters $\varepsilon_r = 2.25$ and $\mu_r = 1$ (polyethylene). The ratio of the outer to inner conductor radii is such that $\ln b/a = 1.25$. The cable is fed by a voltage generator of step emf $\mathcal{E} = 5$ V applied at $t = 0$ and internal resistance $R_g = 75\ \Omega$. The other end of the cable is terminated in a purely resistive load of resistance $R_L = 25\ \Omega$. In MATLAB, compute and plot the voltage waveforms at both ends of the cable within a time interval $0 \le t \le 9$ ns. In addition, create a MATLAB movie visualizing the wave travel along the cable within an interval $0 \le t \le 3T$. *(ME10_46.m on IR)* **H**

TUTORIAL:

See Figs.10.28 and 10.29. We first specify input data for the analysis. The characteristic impedance (Z_0) of the coaxial cable, phase velocity (v_p) of the waves traveling along it, and one-way time delay (T) of the cable are computed using function `chImpedanceCoaxCable` (from MATLAB Exercise 9.1) and Eqs.(7.6) and (10.27), respectively.

```
l = 30*10^(-2);
a = 1*10^(-3);
b = 3.4934*10^(-3);
EPSR = 2.25;
MUR = 1;
Rg = 75;
Rl = 25;
K = 6;
Z0 = chImpedanceCoaxCable(EPSR,MUR,a,b);
```

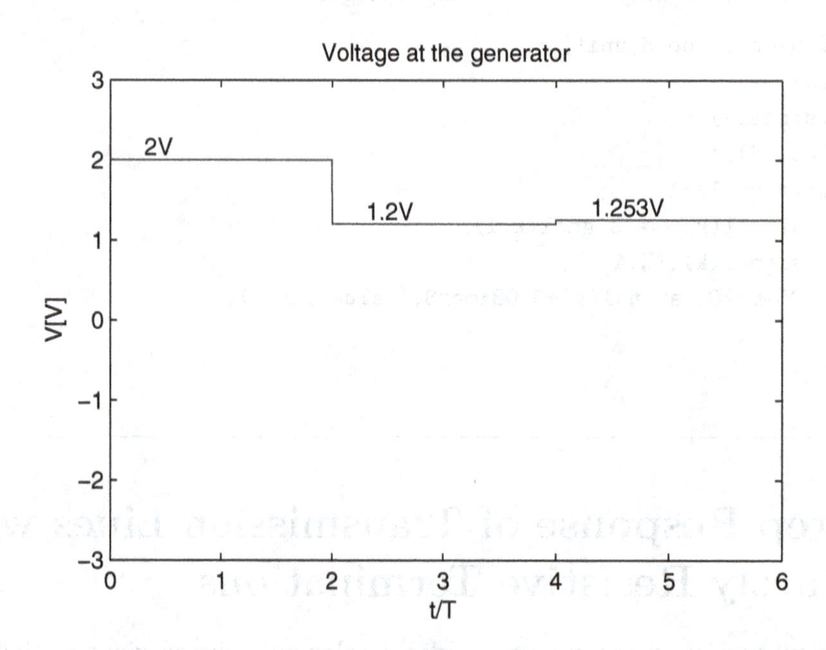

Figure 10.28 MATLAB plot of the voltage waveform at the generator end of a coaxial cable with both load and generator unmatched [$R_L < Z_0$ ($\Gamma_L < 0$) and $R_g > Z_0$ ($\Gamma_g > 0$)]; for MATLAB Exercise 10.46. *(color figure on CW)*

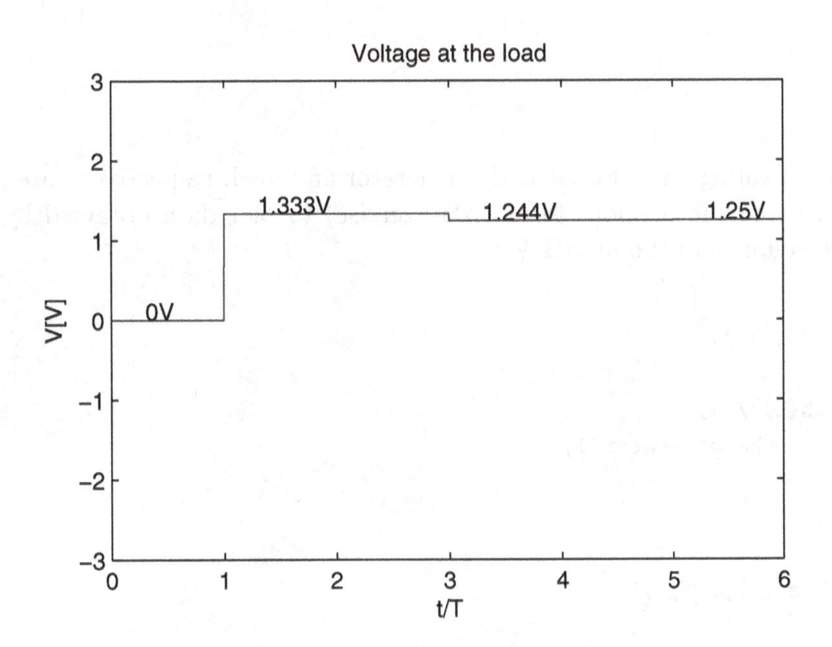

Figure 10.29 MATLAB plot of the voltage waveform at the load end of a coaxial cable with both ends unmatched; for MATLAB Exercise 10.46. *(color figure on CW)*

```
vp = 299792458/sqrt(EPSR*MUR);
T = l/vp;
```

We then define temporal and spatial meshes, conforming to the rule $dl = v_p\,dt$, and call function `signalTL` (from MATLAB Exercise 10.44) to generate the signal matrix `V`,

```
dT = 0.0025*T;
t = 0:dT:K*T;
dl = vp*dT;
z = 0:dl:l;
Es = 5*ones(1,length(t));
V = signalTL(Es,Rg,Rl,Z0,z,K,'v');
```

Within the `for` loop of the movie, we plot, by means of MATLAB function `plot`, row by row of the matrix `V` in each frame,

```
figure(1);
for k = 1:length(t)
plot(z/l,V(:,k));
ylim([-3,3]);
xlim([0,1]);
title('Voltage along the transmission line');
xlabel('z/L');
ylabel('V[V]');
M(k)= getframe();
```

```
if t(k) > 3*T
break;
end;
end;
```

Finally, plots of voltage waveforms at the generator and load, respectively, are generated using function TLplot (from the previous MATLAB exercise) to plot data obtained by extracting the entire respective columns of the matrix V,

```
figure(2);
TLplot(V(1,:),t/T,'V');
ylim([-3,3]);
xlim([0,t(length(t))/T]);
title('Voltage at the generator');
xlabel('t/T');
ylabel('V[V]');
figure(3);
TLplot(V(length(z),:),t/T,'V');
ylim([-3,3]);
xlim([0,t(length(t))/T]);
title('Voltage at the load');
xlabel('t/T');
ylabel('V[V]');
```

Figures 10.28 and 10.29 show these plots.

MATLAB EXERCISE 10.47 **Complete MATLAB analysis, matched generator, open-circuited line.** Repeat the previous MATLAB exercise but assuming that the generator is matched to the line ($R_g = Z_0$) and that the other end of the line is open ($R_L \to \infty$). *(ME10_47.m on IR)* **H**

HINT:
The line of the code calling function `signalTL` (from MATLAB Exercise 10.44) should read: `V = signalTL(Es,Z0,'open',Z0,z,K,'v')` (see MATLAB Exercise 10.4).

MATLAB EXERCISE 10.48 **Matched generator, short-circuited line.** Repeat the previous MATLAB exercise but for the short-circuited transmission line ($R_g = Z_0$ and $R_L = 0$). *(ME10_48.m on IR)* **H**

MATLAB EXERCISE 10.49 **Complete transient analysis, ideal generator, open-circuited line.** Repeat MATLAB Exercise 10.47 but for an ideal step voltage generator feeding the line ($R_g = 0$ and $R_L \to \infty$). In addition, plot the voltage snapshots along the line at times

$t = 2.5T$ and $t = 5.75T$, respectively, T being the one-way time delay of the line. *(ME10_49.m on IR)* **H**

HINT:

The following lines of code show how to extract the entire respective rows of the signal matrix V (obtained by function signalTL – from MATLAB Exercise 10.44) and plot the specified voltage snapshots along the transmission line using function TLplot (from MATLAB Exercise 10.45):

```
figure(4)
[error,k1] = min(abs(t-t1));
TLplot(V(:,k1),z/l,'V');
ylim([-3,3]);
xlim([0,1]);
title('Voltage at t = 2.5T');
xlabel('z/L');
ylabel('V[V]');
figure(5)
[error,k2] = min(abs(t-t2));
TLplot(V(:,k2),z/l,'V');
ylim([-3,3]);
xlim([0,1]);
title('Voltage at = 5.75T');
xlabel('z/L');
ylabel('V[V]');
```

The snapshots are given in Figs.10.30 and 10.31.

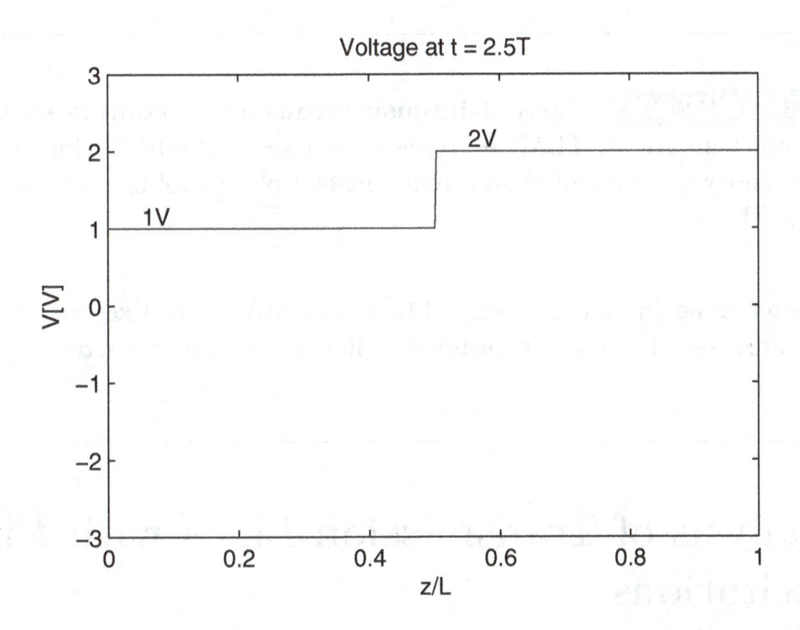

Figure 10.30 MATLAB plot of the voltage snapshot at the time $t = 2.5T$ along an open-circuited transmission line fed by an ideal step voltage generator; for MATLAB Exercise 10.49. *(color figure on CW)*

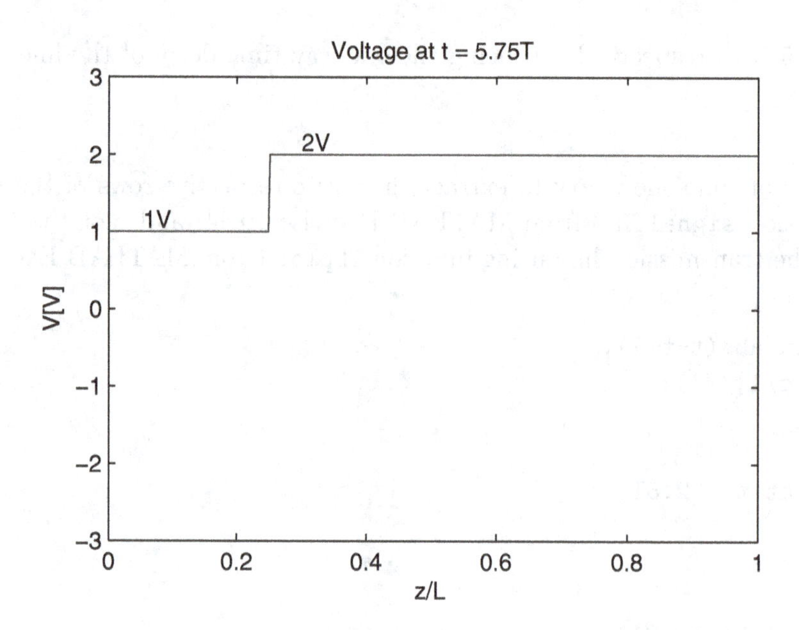

Figure 10.31 MATLAB plot of the voltage snapshot at $t = 5.75T$ along an open-circuited line wih an ideal generator; for MATLAB Exercise 10.49. *(color figure on CW)*

MATLAB EXERCISE 10.50 **Short-circuited line with an ideal generator.** Repeat the previous MATLAB exercise but for a short circuit as load of the line ($R_g = R_L = 0$). *(ME10_50.m on IR)* **H**

MATLAB EXERCISE 10.51 **Current-intensity transients – complete MATLAB analysis.** Repeat the previous five MATLAB exercises (Exercises 10.46–10.50) but for the respective movies, snapshots, and waveforms of the current, instead of the voltage, on transmission lines. *(ME10_51.m on IR)* **H**

HINT:
The input parameter `type` in function `signalTL` (from MATLAB Exercise 10.44), specifying whether the computed signal along a transmission line is voltage or current, should be set to `'i'`.

10.11 Analysis of Transmission Lines with Pulse Excitations

Let us now assume that the emf of the voltage generator in Fig.10.27 is a rectangular pulse function of time, of magnitude $\mathcal{E}$ and duration t_0, triggered at $t = 0$, as shown in Fig.10.32(a). This function is analytically given by

$$e(t) = \mathcal{E}\Pi(t) \,, \quad \Pi(t) = \begin{cases} 1 & \text{for } 0 < t < t_0 \\ 0 & \text{for } t < 0 \text{ and } t > t_0 \end{cases} \quad \text{(pulse excitation)} \,, \tag{10.35}$$

with $\Pi(t)$ standing for the corresponding unit rectangular pulse time function, which is also called the gate or window function. It is obvious from Fig.10.32(b) that $e(t)$ can be viewed as a superposition of two step functions, Eq.(10.30), with opposite polarities and a time shift t_0 between them. Due to the linearity and time invariance of the transmission line in Fig.10.27 [this comes from the linearity of the governing telegrapher's equations and the coefficients in these equations, namely, the per-unit-length parameters of the transmission line, being time-invariant (not changing with time)], including its terminal networks (generator and load), the response of the line to $e(t)$ can be computed combining the individual responses to the two step inputs if applied alone. Namely, marking by $v_{L1}(t)$ the line output response (load voltage in Fig.10.27) to the input $e_1(t) = \mathcal{E}h(t)$ alone, the resultant output response to the combined excitation is obtained as

$$e(t) = \mathcal{E}h(t) - \mathcal{E}h(t - t_0) \quad \longrightarrow \quad v_L(t) = v_{L1}(t) - v_{L1}(t - t_0) \quad \text{(pulse response)} \,, \tag{10.36}$$

i.e., the load voltage is the same superposition of $v_{L1}(t)$ and its flipped-over (multiplied by -1) and delayed (by t_0) version – as for the excitation. Consequently, transient analysis of a transmission line with a rectangular pulse excitation can essentially be reduced to the computation on the same line if excited by a step generator. Analogous transformations from a step response to the pulse one can also be applied for the generator voltage, $v_g(t)$, and for the total voltage and current, $v(z,t)$ and $i(z,t)$, at any position (z) along the line. Of course, pulse responses can as well be found directly – performing the complete transient analysis of the line with the original pulse excitation, $e(t)$, in Eq.(10.35). In this latter fashion, we can obtain transient responses to pulses of non-rectangular shapes, such as triangular pulses in time.

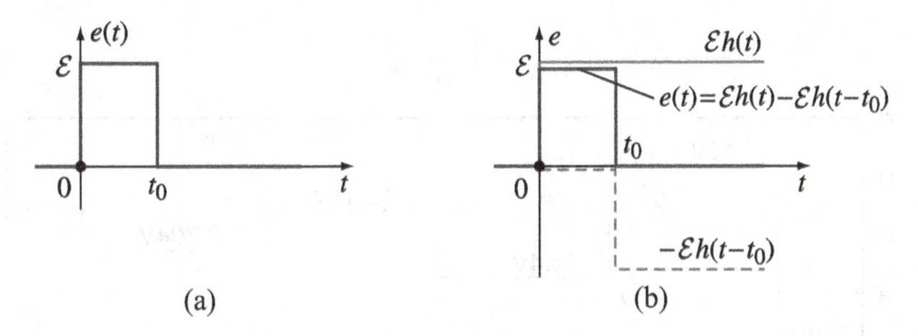

Figure 10.32 Pulse excitation of a lossless transmission line (Fig.10.27): (a) rectangular pulse emf function in time and (b) its representation using two step functions, Eq.(10.30).

MATLAB EXERCISE 10.52 Pulse response of a transmission line – from step analysis. A lossless transmission line of length $l = 45$ cm and characteristic impedance $Z_0 = 50\ \Omega$ is driven by an ideal voltage generator of rectangular pulse emf in Fig.10.32(a) with magnitude $\mathcal{E} = 10$ V and duration $t_0 = 2$ ns. At its other end, the line is terminated in a purely resistive load of resistance $R_L = 200\ \Omega$. The line dielectric has relative permittivity $\varepsilon_r = 4$ and is nonmagnetic. In MATLAB, compute and plot the voltage waveform at the load within a time interval

$0 \leq t \leq 30$ ns. In specific, first compute the output (at the load terminals) step response of the line, then find the flipped-over and delayed by t_0 version of this signal, and finally obtain the output pulse response of the line, according to Eqs.(10.36), as a sum of the two previous waveforms. Plot intermediate signals as well. *(ME10_52.m on IR)* **H**

HINT:

In the following portion of the MATLAB code, the step response of the line at the load, voltage waveform $v_{L1}(t)$ (signal V1), is obtained by extracting the entire respective column of the signal matrix V, previously generated using function **signalTL** (from MATLAB Exercise 10.44), with the step analysis of the line being done similarly to that in MATLAB Exercise 10.46. In addition, the signal $-v_{L1}(t - t_0)$ is formed from V1. Finally, the resultant pulse response (Vtot) is synthesized by superposition.

```
t0 = 2*10^(-9);
ts = 0:dT:t0;
N = length(ts);
V1 = V(length(z),:);
V2 = -[zeros(1,N-1) V1];
Vtot = V1 + V2(1:length(t));
```

The three signals are shown in Figs.10.33–10.35. We see, in Fig.10.35, that the output response of the line to a pulse excitation in Fig.10.32(a) consists of multiple pulses (theoretically, an infinite series of pulses) of decaying magnitudes with alternating polarity, separated in time by $2T - t_0 =$ 4 ns. Essentially, because of unmatched terminations at both ends of the line, multiple reflections from the load and generator result in the reception of additional pulses (in addition to the first one) at the load, which is normally unintended and undesirable. Since $t_0 < 2T$, the adjacent pulses in the sequence do not overlap (or touch each other).

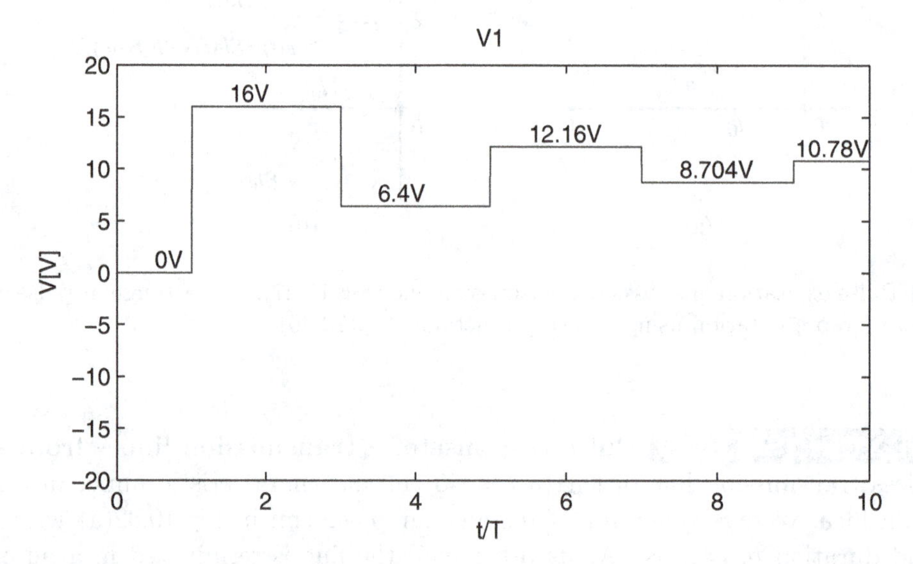

Figure 10.33 MATLAB evaluation of the output pulse response of a lossless transmission line with unmatched load and generator: the line step response; for MATLAB Exercise 10.52. *(color figure on CW)*

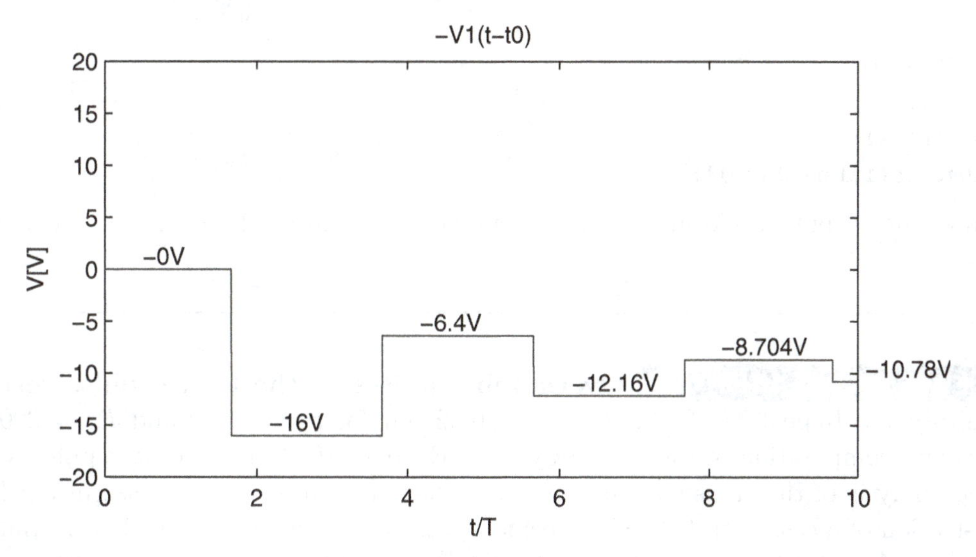

Figure 10.34 Delayed (by the pulse width) negative of the signal in Fig.10.33; for MATLAB Exercise 10.52. *(color figure on CW)*

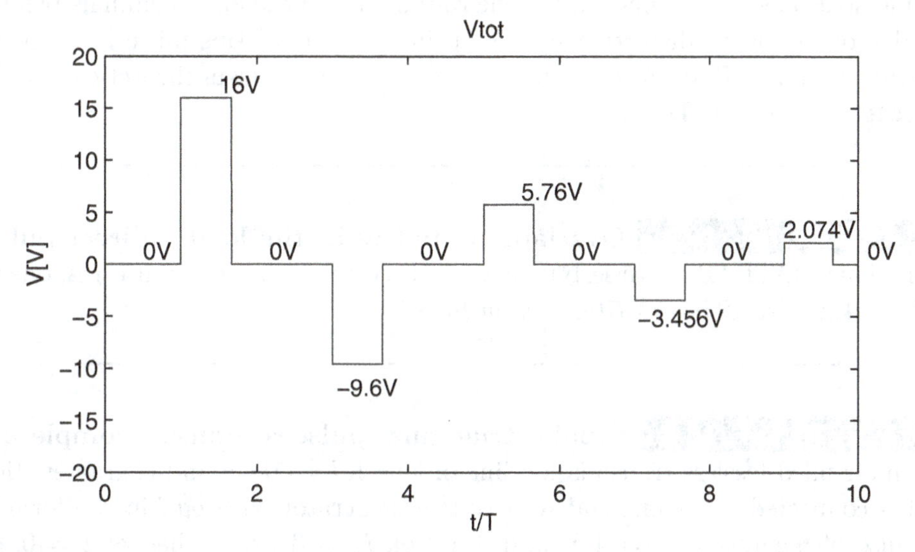

Figure 10.35 Line pulse response, by superposition of signals in Figs.10.33 and 10.34; for MATLAB Exercise 10.52. *(color figure on CW)*

MATLAB EXERCISE 10.53 Direct pulse analysis of a line – implementing a pulse excitation. Redo the previous MATLAB exercise but in an alternative way – by finding the line pulse response directly: perform the complete MATLAB simulation of the line with a pulse excitation, that is, run function `signalTL` (from MATLAB Exercise 10.44) with a pulse emf of the voltage generator (in Fig.10.27), shown in Fig.10.32(a). In addition, make a movie visualizing the voltage variations along the line. *(ME10_53.m on IR)*

HINT:

Based on Eq.(10.35), the pulse emf can be defined in the code as follows:

```
t0 = 2*10^(-9);
ts = 0:dT:t0;
N = length(ts);
Es = 10*ones(1,length(ts));
```

Of course, the output waveform should be identical to that in Fig.10.35 (in the previous MATLAB exercise).

MATLAB EXERCISE 10.54 **Overlapping pulses at the load – pulse response from step analysis.** Repeat MATLAB Exercise 10.52 but for $t_0 = 12$ ns and $R_g = 200\ \Omega$. Do the analysis and computation in the same way as in Exercise 10.52, by finding a pulse response from the step analysis of the transmission line. Note that, as opposed to the case in Fig.10.35, this is an illustration of a case with $t_0 > 2T$ (specifically, $t_0 = 4T$), in which the adjacent pulse responses due to multiple reflections on the line in the output sequence now overlap, resulting in a distorted pulse that is being received by the load. Namely, the load and then generator reflection of the rise edge of the incident pulse comes back, after one round trip, to the load terminals before the arrival of the fall edge of the pulse, distorting it. The principal part of this pulse lasts the same as the generator pulse (from $t = T$ to $t = T + t_0$), but the overall signal lasts theoretically indefinitely (to $t \to \infty$). *(ME10_54.m on IR)* **H**

MATLAB EXERCISE 10.55 **Overlapping pulses at the load – direct pulse analysis.** Redo the previous MATLAB exercise but now by directly finding the pulse response of the line, as in MATLAB Exercise 10.53. *(ME10_55.m on IR)*

MATLAB EXERCISE 10.56 **Bipolar triangular pulse response – complete MATLAB analysis.** An air-filled lossless transmission line of length $l = 60$ cm and characteristic impedance $Z_0 = 100\ \Omega$ is connected at its one end to a voltage generator with emf in the form of a bipolar triangular pulse of magnitude $\mathcal{E} = 4$ V and duration $t_0 = 3$ ns, applied at $t = 0$, as shown in Fig.10.36. The internal resistance of the generator is $R_g = 300\ \Omega$. The load, at the other end of the line, is purely resistive, of resistance $R_L = 400\ \Omega$. Using MATLAB, compute and plot the voltage waveforms at the generator and load terminals, respectively, of the line for $0 \le t \le 10$ ns. Also, play a movie of the transient voltage on the line. *(ME10_56.m on IR)* **H**

HINT:
The emf in Fig.10.36 can be implemented in the code as follows:

```
t0 = 3*10^(-9);
ts = 0:dT:t0;
N = length(ts);
Es = 4/(floor(N/4))*[0:1:floor(N/4),floor(N/4)-1:-1:-floor(N/4),-floor(N/4):1:0];
```

The response of the line to this excitation, and the waveform plots in Figs.10.37 and 10.38, can be

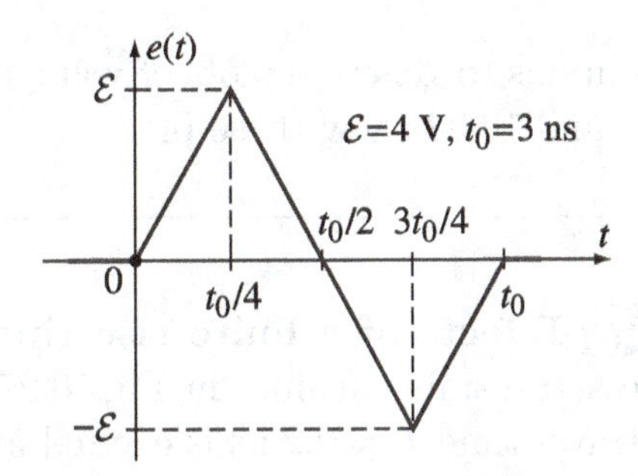

Figure 10.36 Bipolar triangular emf function in time; for MATLAB Exercise 10.56.

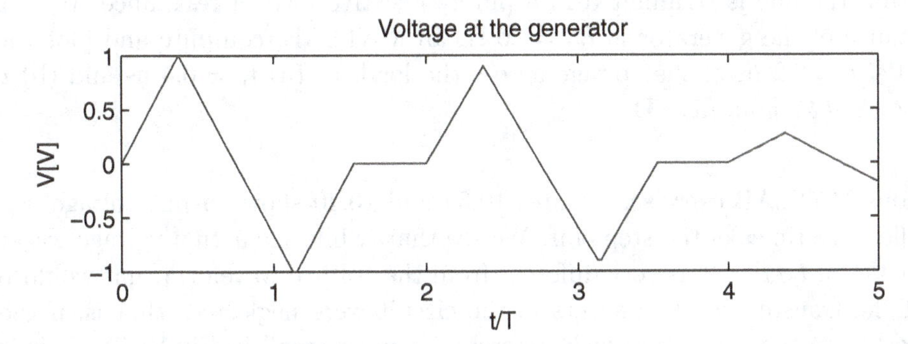

Figure 10.37 MATLAB evaluation of the transient response of a transmission line to a triangular pulse excitation in Fig.10.36: voltage waveform at the generator terminals; for MATLAB Exercise 10.56. *(color figure on CW)*

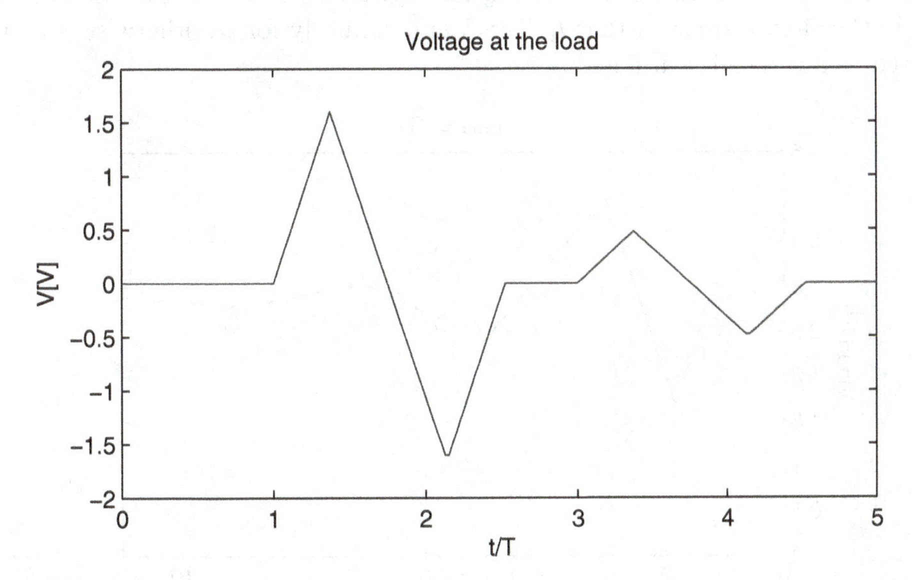

Figure 10.38 MATLAB plot of the load voltage waveform for the line excitation in Fig.10.36; for MATLAB Exercise 10.56. *(color figure on CW)*

found directly, performing the complete MATLAB simulation with the use of function `signalTL` (from MATLAB Exercise 10.44) to generate the signal matrix `V`, and extracting rows of this matrix to make frames of a movie, or columns to generate plots of voltage waveforms at the generator and load, respectively. Figures 10.37 and 10.38 show these plots.

MATLAB EXERCISE 10.57 **Effects of a finite rise time of step signals – complete MATLAB analysis.** A lossless transmission line, in Fig.10.27, with characteristic impedance $Z_0 = 100\ \Omega$ and one-way wave travel time $T = 0.2$ ns is excited at one end by a voltage generator of step emf with magnitude $\mathcal{E} = 3$ V and finite linear-rise time, t_r. This emf is applied at $t = 0$ (it linearly rises during time $0 \leq t \leq t_r$ from zero to $\mathcal{E}$, and is constant, equal to $\mathcal{E}$, afterward). At its other end, the line is terminated in a purely resistive load of resistance $R_L = 400\ \Omega$. The internal resistance of the generator is $R_g = 20\ \Omega$. In MATLAB, compute and plot the waveform during time $0 \leq t \leq 2.2$ ns of the voltage across the load for (a) $t_r = 0.4$ ns and (b) $t_r = 0.8$ ns, respectively. *(ME10_57.m on IR)* **H**

HINT:
See the previous MATLAB exercise. Figures 10.39 and 10.40 show output voltage waveforms for the two specified rise times of the step emf. We see that, while the actual voltage waveform at the load in Fig.10.39, for $t_r/T = 2$, is very different from the voltage waveform that would be observed at the load if the transmission-line effects in the circuit were neglected, that is, if the load were directly connected to the generator, such difference is quite small in Fig.10.40. This implies that for a larger t_r over T ratio, namely, if $t_r/T = 4$, the line appears almost as if nonexistent. In general, a rule of thumb in transient analysis of digital circuits is that the one-way signal delay time T along an interconnect in the circuit can be neglected ($T \approx 0$) if the rise and fall times of pulse signals in the circuit are such that $t_r/T > 5$ and similarly for t_f; otherwise, the interconnect must be treated as a transmission line.

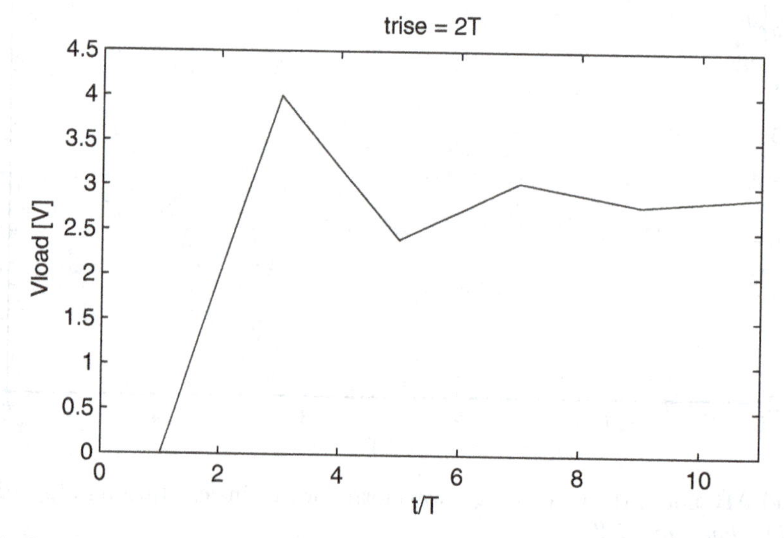

Figure 10.39 Evaluation of effects of a finite rise time of a step voltage excitation of a transmission line: MATLAB plot of the voltage waveform at the load terminals for rise time $t_r = 2T$; for MATLAB Exercise 10.57. *(color figure on CW)*

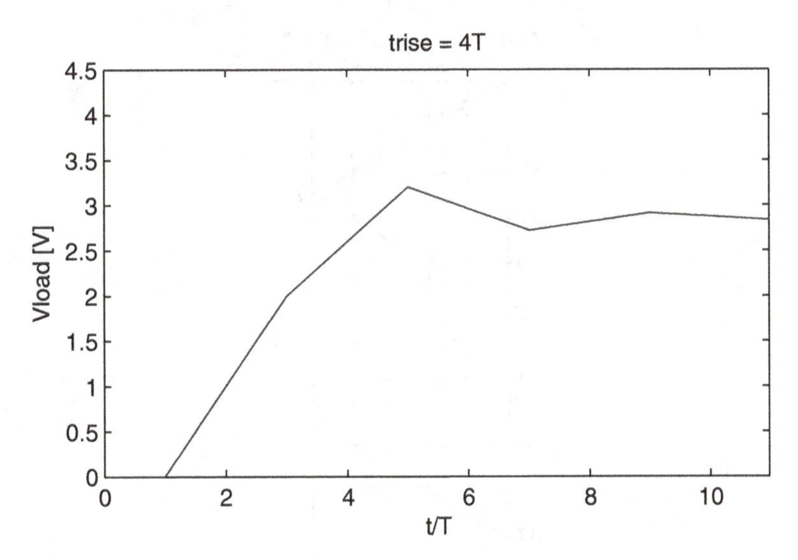

Figure 10.40 The same as in Fig.10.39 but for $t_r = 4T$; for MATLAB Exercise 10.57. *(color figure on CW)*

MATLAB EXERCISE 10.58 **Evaluation of current-intensity responses to pulse excitations.** Repeat MATLAB Exercises 10.52–10.57 but for the current-intensity signals along transmission lines. *(ME10_58.m on IR)* **H**

10.12 Bounce Diagrams

We now present a graphical tool for recording multiple reflection transient processes on lossless transmission lines and computing the total voltage and current intensity at an arbitrary location on the line and any instant of time, called a bounce diagram (also known as a lattice diagram) of the line. This is a space-time (i.e., distance-time) plot of the voltage (or current) state of a transmission line, in Fig.10.27, with the distance (z) from the generator end measured on the horizontal axis and time ($t > 0$) on the vertical axis oriented downward, as shown in Fig.10.41. We assume a step excitation, Eq.(10.30), of the line. The zigzag line in Fig.10.41 indicates the progress of the voltage wave along the line, where each line segment (sloping downward from left to right or right to left) represents a component traveling voltage (forward or backward) and is labeled with the magnitude of the voltage step increment. Starting from the top of the diagram, these voltages are successively obtained by multiplying the magnitude of the preceding component by one of the line reflection coefficients Γ_L and Γ_g, given in Eqs.(10.28), depending on the position (load or generator) where the reflection occurs.

Once the bounce diagram is constructed, the total voltage at a location $z = z_1$ and time $t = t_1$, $v(z_1, t_1)$, can be easily found from the position of the point (z_1, t_1) in Fig.10.41, point P_1, relative to the zigzag voltage line. Namely, $v(z_1, t_1)$ equals the sum of the component voltages of all sloping line segments intersected by the vertical line $z = z_1$ between points $t = 0$ and $t = t_1$, so above P_1,

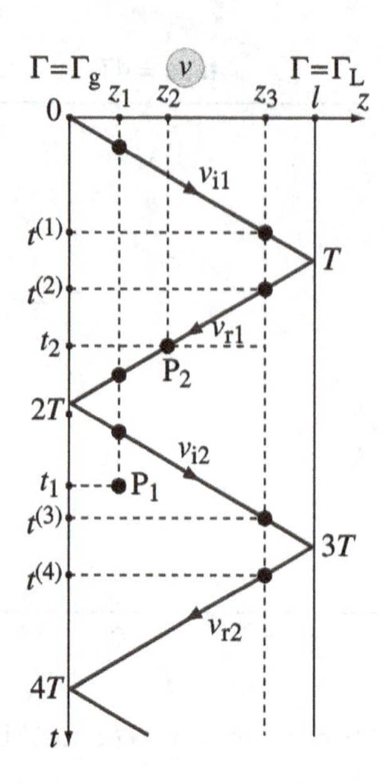

Figure 10.41 Voltage bounce diagram of a lossless transmission line (Fig.10.27) with a step excitation (upon reflections, $v_{r1} = \Gamma_L v_{i1}$, $v_{i2} = \Gamma_g v_{r1}$, $v_{r2} = \Gamma_L v_{i2}$, ...).

on that line (these points are also included in the count if on the zigzag line). Hence, for the choice of z_1 and t_1 in Fig.10.41, we see that $v(z_1, t_1) = v_{i1} + v_{r1} + v_{i2} = (1 + \Gamma_L + \Gamma_L \Gamma_g) v_{i1}$, where the initial incident voltage on the line, v_{i1}, is computed from Eq.(10.32), since the vertical (dashed) line starting at the point marked z_1 on the z-axis intersects the first three sloping line segments along its part above P_1. Similarly, the total voltage scan $v(z, t_2)$ along the transmission line at any given time (t_2), as well as the voltage waveform $v(z_3, t)$ at any fixed location (z_3) on the line, can also be obtained from the diagram, in a straightforward fashion. In the first case, we draw a horizontal (dashed) line from t_2 on the t-axis to the right, and identify its intersection with the zigzag voltage line, point P_2, and the corresponding point z_2 on the z-axis – this is the position on the transmission line where the voltage picture exhibits an abrupt (step-like) change. For example, if the instant t_2 is chosen as in Fig.10.41, so that P_2 belongs to the segment labeled with the component voltage magnitude v_{r1}, we read from the diagram that $v(z, t_2) = v_{i1}$ for $0 \leq z < z_2$ and $v(z, t_2) = v_{i1} + v_{r1}$ for $z_2 \leq z \leq l$. For the second case (fixed location), we look at the intersections of the vertical dashed line $z = z_3$ with the segments of the zigzag line, and the corresponding points on the t-axis, marked as $t^{(1)}$, $t^{(2)}$, ... in Fig.10.41. We then read that $v(z_3, t) = 0$ for $0 < t < t^{(1)}$, $v(z_3, t) = v_{i1}$ for $t^{(1)} < t < t^{(2)}$, $v(z_3, t) = v_{i1} + v_{r1}$ for $t^{(2)} < t < t^{(3)}$, and so on. Finally, the current bounce diagram of a transmission line is constructed and used in the same way as the corresponding voltage diagram, the only difference being in the values of the reflection coefficients – the load and generator reflection coefficients for currents equal $-\Gamma_L$ and $-\Gamma_g$, respectively [see Eq.(10.6)].

MATLAB EXERCISE 10.59 **Numerical simulation of a bounce diagram: bounce-diagram matrix.** Write a function `bounceDiagram()` in MATLAB that provides a numerical simulation of a bounce diagram, in Fig.10.41, in the form of a bounce-diagram matrix. At input, the function takes the initial incident voltage, v_{i1}, or current, i_{i1}, load and generator voltage reflection coefficients, Γ_L and Γ_g, the one-way time delay of the line, T, line length, l, parameter K defining the total duration of the signal (KT), and `type` of the diagram, (`'v'` for the voltage bounce diagram of the line and `'i'` for the current diagram). In addition to the bounce-diagram matrix, `bd`, the function generates as output arrays `z` and `t`, with spatial and temporal meshes, along horizontal and vertical axes in Fig.10.41, respectively, where elemental steps in the two meshes are related to each other as $\mathrm{d}z/l = \mathrm{d}t/T$, as in function `signalTL` (MATLAB Exercise 10.44). The matrix `bd` has dimensions `length(z)`×`length(t)`, and its elements are zero everywhere except at the zigzag signal (voltage or current) line in Fig.10.41, where they equal the value of the component traveling signal (forward or backward), i.e., the magnitude of the signal step increment, at that point, as labeled in Fig.10.41. Hence, this matrix can be visualized as a 2-D dense mesh of nodes covering the $l \times KT$ surface in Fig.10.41, with the nodes along the zigzag signal path carrying the corresponding component signal values and all other nodes being "turned off." *(bounceDiagram.m on IR)*

TUTORIAL:

First, depending on `type`, a variable `sign` is introduced to take into account that the reflection coefficients for currents are the negative of the corresponding voltage coefficients. Then, we define meshes `z` and `t`, and initialize the bounce-diagram matrix, `bd`. Finally, in a `for` loop and `switch-case` format, we compute the component (incident or reflected) voltages or current intensities, by recording multiple reflection transient processes according to Fig.10.41, and fill the matrix `bd` – as per instructions above (component signal values at nodes along the zigzag line in Fig.10.41, zero elsewhere).

```
function [bd,z,t] = bounceDiagram(Vi0,Gl,Gg,T,L,K,type)
switch type
case 'v'
sign = 1;
case 'i'
sign = -1;
end;
z = 0:0.01*L:L;
t = 0:0.01*T:K*T;
N = length(z)-1;
V = Vi0;
bd = zeros(K*N+1,N+1);
for i = 1:K
start = (i-1)*N;
switch (mod(i,2))
case 0
for j = 1:N+1
bd(start+j,N+2-j) = bd(start+j,N+2-j) + V;
end;
V = sign*V*Gg;
case 1
```

```
for j = 1:N+1
bd(start+j,j) = bd(start+j,j) + V;
end;
V = sign*V*Gl;
end;
end;
```

MATLAB EXERCISE 10.60 **Extracting a signal waveform from the bounce diagram.** Write a function `zBD()` in MATLAB that extracts from the bounce-diagram matrix, `bd`, of a transmission line the signal waveform at a fixed location (coordinate z) on the line, for the entire time duration considered. The input to the function are output data of function `bounceDiagram` (from the previous MATLAB exercise), `z`, `t`, and `bd`, as well as the coordinate at which the signal is taken, `zs`. The function generates an array `Vt` whose elements are total signal values at elements (time instants) of array `t`. *(zBD.m on IR)*

TUTORIAL:
Using MATLAB function `min` (see MATLAB Exercise 3.12), we find the element of array `z` nearest in value to the coordinate `zs`, and extract the entire corresponding column of the matrix `bd`, in a cumulative fashion (adding the incident and reflected component signal values as we progress in time),

```
function Vt = zBD(z,t,bd,zs)
[error,kz] = min(abs(z - zs));
bdt(1:length(t)) = bd(1:length(t),kz);
for i = 1:length(bdt)
Vt(i) = sum(bdt(1:i));
end;
```

MATLAB EXERCISE 10.61 **Extracting a signal snapshot from the bounce diagram.** Repeat the previous MATLAB exercise but to extract from the bounce-diagram matrix the signal scan (snapshot) along the transmission line at a given time, `ts` [function `timeBD()`]. The elements of the output array `Vz` are total signal values at elements (coordinates) of array `z`. *(timeBD.m on IR)*

TUTORIAL:
This function finds the element of the array `t` nearest to the time `ts`, and adds together all rows of the matrix `bd` from the first one to the one corresponding to that element – into a single row array, `Vz`, with `length(z)` elements, equal to the values of the total signal at coordinates `z`:

```
function Vz = timeBD(z,t,bd,ts)
[error,kt] = min(abs(t - ts));
Vz(1:length(z)) = sum(bd(1:kt,:),1);
```

MATLAB EXERCISE 10.62 **Voltage bounce diagram and waveforms – complete MATLAB analysis.** For the coaxial cable and its excitation and load described in MATLAB Exercise 10.46, construct and plot in MATLAB the voltage bounce diagram. Based on this diagram, obtain and plot in MATLAB the total voltage scan along the cable at a time instant $t = 5$ ns and total voltage waveform for $0 \le t \le 9$ ns at the cable cross section whose distance from the generator is $z = 7.5$ cm. *(ME10_62.m on IR)* **H**

TUTORIAL:

See Figs.10.42–10.44. We call function `bounceDiagram` (developed in MATLAB Exercise 10.57) to calculate bounce-diagram matrix, `bd`, and associated spatial and temporal meshes, `z` and `t`. A matrix `color` is then introduced – with elements equal to zero everywhere except at the zigzag voltage line in the bounce diagram, where they are unity. A 2-D plot of this matrix using MATLAB function `imagesc` gives a graphical representation of the bounce diagram in MATLAB, as shown in Fig.10.42.

```
T = 1.5*10^(-9);
ts = 5*10^(-9);
zs = 7.5/100;
Gl = -1/3;
Gg = 1/5;
V0 = 2;
K = 6;
```

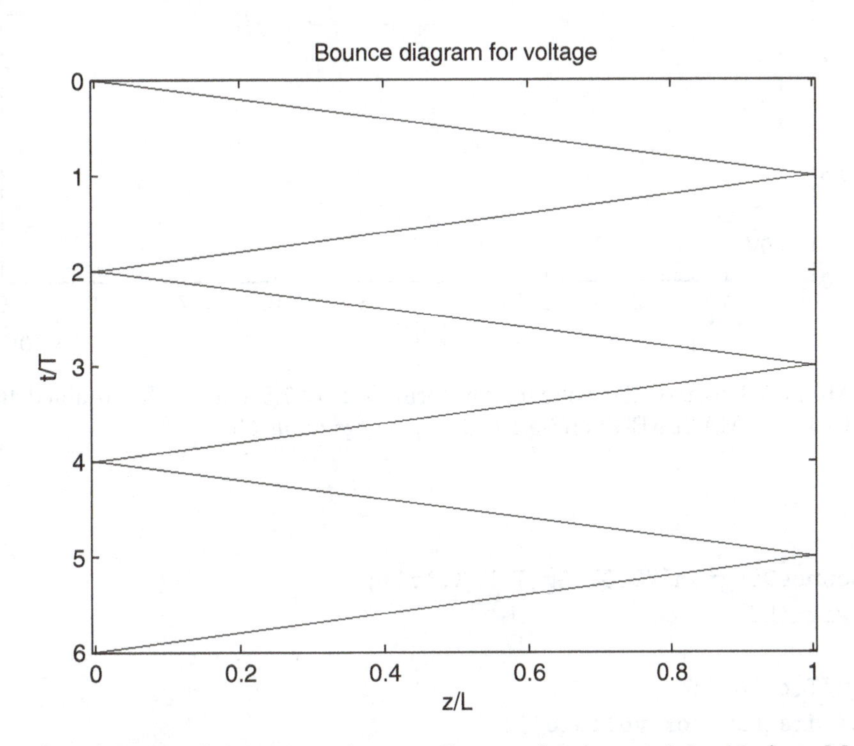

Figure 10.42 MATLAB plot of the voltage bounce diagram for the coaxial cable from MATLAB Exercise 10.46; for MATLAB Exercise 10.62. *(color figure on CW)*

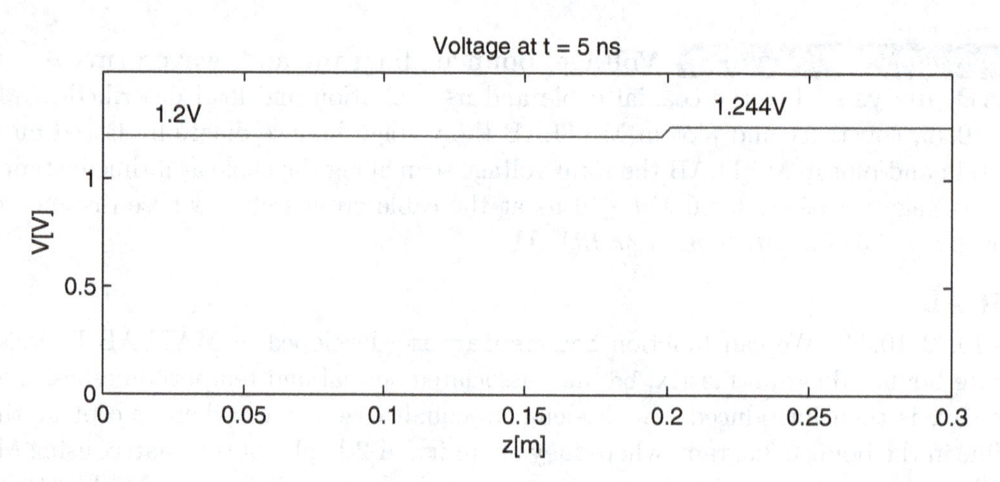

Figure 10.43 MATLAB plot of the voltage snapshot at $t = 5$ ns $= 3\frac{1}{3}\,T$ obtained from the bounce diagram in Fig.10.42; for MATLAB Exercise 10.62. *(color figure on CW)*

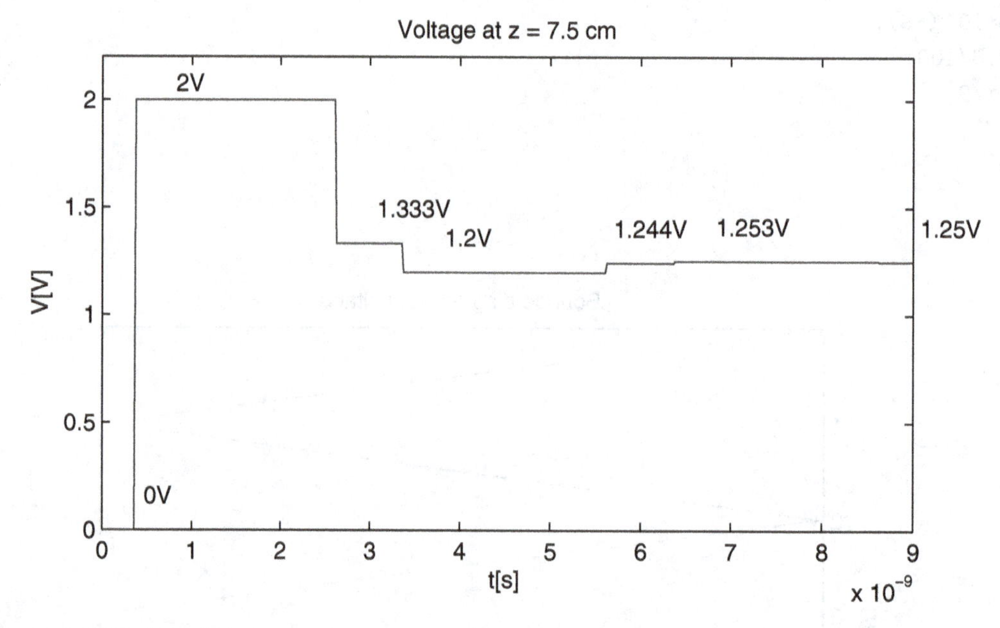

Figure 10.44 MATLAB plot of the voltage waveform at $z = 7.5$ cm $= l/4$ obtained from the bounce diagram in Fig.10.42; for MATLAB Exercise 10.62. *(color figure on CW)*

```
L = 30/100;
[bd,z,t] = bounceDiagram(V0,Gl,Gg,T,L,K,'v');
color = abs(sign(bd));
figure(1);
imagesc(z/L,t/T,color);
title('Bounce diagram for voltage');
colormap('cool');
xlabel('z/L');
```

```
ylabel('t/T');
```

The specified v snapshot (scan) and waveform are obtained with the use of functions `timeBD` and `zBD` (from the previous two MATLAB exercises), and are plotted by means of function `TLplot` (from MATLAB Exercise 10.45),

```
figure(2);
Vz = timeBD(z,t,bd,ts);
TLplot(Vz,z,'V');
title('Voltage at t = 5 ns');
xlabel('z[m]');
ylabel('V[V]');
ylim([0,max(Vz)*1.2]);
xlim([0,L]);
figure(3);
Vt = zBD(z,t,bd,zs);
TLplot(Vt,t,'V');
title('Voltage at z = 7.5 cm');
xlabel('t[s]');
ylabel('V[V]');
xlim([0,t(length(t))]);
ylim([0,1.1*max(Vt)]);
```

These plots are given in Figs.10.43 and 10.44.

MATLAB EXERCISE 10.63 **Current bounce diagram and waveforms – complete MATLAB analysis.** Repeat the previous MATLAB exercise but for the current-intensity signals along the coaxial cable, namely, construct and plot in MATLAB the current bounce diagram, and from it obtain and plot in MATLAB the current scan at $t = 5$ ns and current waveform at $z = 7.5$ cm. *(ME10_63.m on IR)* **H**

MATLAB EXERCISE 10.64 **Another MATLAB solution using bounce diagrams: voltage plots.** A lossless transmission line of characteristic impedance $Z_0 = 100\ \Omega$, phase velocity $v_{\rm p} = 1.75 \times 10^8$ m/s, and length $l = 35$ cm, terminated in a purely resistive load, is driven by a voltage generator of step emf $\mathcal{E} = 4$ V applied at $t = 0$. The internal resistance of the generator is $R_{\rm g} = 60\ \Omega$, while the load resistance is $R_{\rm L} = 200\ \Omega$. Using MATLAB, construct the voltage bounce diagram for the line, and compute and plot the voltage scan along the line at a time instant $t = 9.5$ ns and voltage waveform for $0 \le t \le 12$ ns at a location on the line $z = 21$ cm away from the generator. *(ME10_64.m on IR)* **H**

MATLAB EXERCISE 10.65 **Another MATLAB solution using bounce diagrams: current plots.** Repeat the previous MATLAB exercise but for the current-intensity plots, namely,

the current bounce diagram, scan at $t = 9.5$ ns, and waveform at $z = 21$ cm. *(ME10_65.m on IR)*
H

10.13 Transient Response for Reactive Terminations

Often, transmission-line terminations involve reactive lumped elements, inductors and capacitors. In analysis of high-speed digital circuits, these elements are used to model various lumped inductive and capacitive effects at line terminals that are not already included in the transmission-line model (with distributed inductors and capacitors). In addition, they can model, along with resistors, a complex input impedance of a device (e.g., a transmitting antenna) connected at the load end of the line.

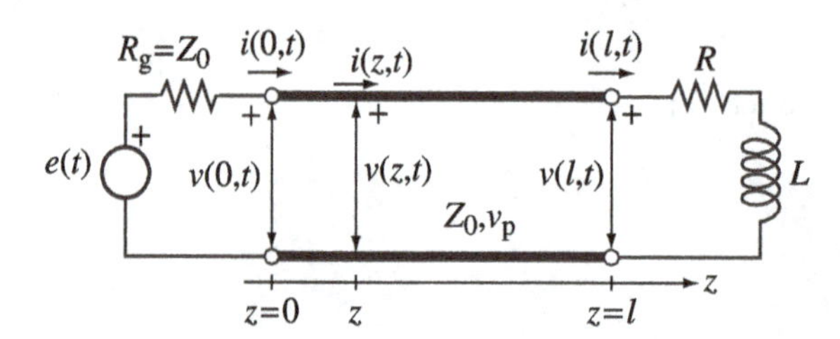

Figure 10.45 Transient analysis of a lossless transmission line with a combined resistive and reactive load.

As an example of deriving equations for the MATLAB transient analysis, consider a lossless transmission line (Fig.10.27) terminated in a load consisting of a resistor of resistance R and an inductor of inductance L connected in series, as shown in Fig.10.45. We assume, for simplicity, that the generator is matched to the line, $R_g = Z_0$ ($\Gamma_g = 0$), so that there are no multiple reflections on the line. Using Eq.(3.11) and the current-voltage characteristic for an inductor, the voltage across the load, $v(l,t)$, can be expressed in terms of the load current, $i(l,t)$, as follows:

$$v(l,t) = Ri(l,t) + L\frac{\mathrm{d}i(l,t)}{\mathrm{d}t} \,. \tag{10.37}$$

On the other side, combining Eqs.(10.29) for $z = l$ and $t > T$ (any time after the arrival of the incident wave), with T being the one-way time delay of the line, Eq.(10.27), in a way to eliminate v_r, we get:

$$v(l,t) = 2v_i(l,t) - Z_0 i(l,t) \quad \text{(Thévenin generator replacing a line, } t > T) \,, \tag{10.38}$$

where $v_i(l,t)$ is the incident voltage at the load. This equation actually tells us that the line in Fig.10.45, with respect to the load (most importantly, this can be any load), behaves like a real (non-ideal) voltage generator – Thévenin equivalent generator, whose emf amounts to $2v_i(l,t)$ and internal resistance to the characteristic impedance of the line, Z_0. With it, Eq.(10.37) becomes

$$\frac{\mathrm{d}i(l,t)}{\mathrm{d}t} = -\frac{Z_0 + R}{L}\, i(l,t) + \frac{2}{L}\, v_i(l,t) \,. \tag{10.39}$$

Solving this differential equation, we obtain the current $i(l,t)$. Equation (10.37) then gives the solution for $v(l,t)$.

The input voltage of the line in Fig.10.45, $v_g(t)$, is given by Eq.(10.32) until the arrival of the reflected wave, at $t = 2T$. In order to compute its temporal form after that time, we first determine the voltage of the reflected wave, v_r, on the line. Since this is the only one reflected voltage (for $T < t < \infty$), it equals the difference of the total and incident voltages at a given location (z) along the line. At the load terminals ($z = l$), from the first equation in Eqs.(10.29),

$$v_r(l,t) = v(l,t) - v_i(l,t) \ . \tag{10.40}$$

For $t > 2T$, the reflected voltage at the generator terminals ($z = 0$), $v_r(0,t)$, has the same form as that at the load end, just with an additional delay T relative to the signal $v_r(l,t)$, so $v_r(0,t) = v_r(l, t-T)$. Finally, the voltage of the generator is obtained as

$$v_g(t) = v_i(0,t) + v_r(0,t) = v_i(0,t) + v_r(l,t-T) \quad (t > 2T) \ . \tag{10.41}$$

MATLAB EXERCISE 10.66 Time derivative function for ordinary differential equation solver. Write a function `ODE()` in MATLAB that at input takes an instant of time, t, and the associated value of a signal, S, and returns the time derivative of S, $\mathrm{d}S/\mathrm{d}t$, obtained as $aS + bV$, where a and b are constants defined in the main program, and V is the incident voltage at a reactive load of a transmission line, equal to a rectangular pulse of magnitude V_0 staring at T and ending at $T + t_0$. Variables a, b, V_0, T, and t_0 are declared as global, so that their values are global (the same) for the function and the main program. This function will be used by MATLAB ordinary differential equation (ODE) solver `ode23tb()`, whose input arguments contain a reference to the function calculating the derivative of the unknown signal, namely, `ODE` in our case (`@ODE`), to solve a differential equation given by

$$\frac{\mathrm{d}S}{\mathrm{d}t} = aS + bV \ , \tag{10.42}$$

as needed in transient analysis of transmission lines with reactive (or combined resistive and reactive) loads [e.g., see Eq.(10.39)]. *(ODE.m on IR)*

TUTORIAL:

First, we declare global variables (the same should be done in the main program as well). Then, given that t (`time`) is just one time instant (as required by solver `ode23tb`), we calculate `V` at that instant only. Finally, we compute $\mathrm{d}S/\mathrm{d}t$ as $aS + bV$, according to Eq.(10.42).

```
function dSdt = ODE(time,S)
global V0 T t0 a b
if (time < T) || (time >= T+t0)
V = 0;
else
V = V0;
end;
dSdt = a*S + b*V;
```

MATLAB EXERCISE 10.67 Line with an RL load – complete analysis in MATLAB. A lossless transmission line, for which the one-way wave travel time is $T = 2$ ns, is fed by a voltage generator of rectangular pulse emf with magnitude $\mathcal{E} = 5$ V and width $t_0 = 1$ ns, applied at $t = 0$. The characteristic impedance of the line and internal resistance of the generator are $Z_0 = R_g = 50\ \Omega$, and the line load consists of a resistor of resistance $R = 30\ \Omega$ and an inductor of inductance $L = 80$ nH connected in series. In MATLAB, compute and plot the waveform of the voltage across the load. *(ME10_67.m on IR)* **H**

TUTORIAL:

The voltage and current of the load (in Fig.10.45), $v(l,t)$ and $i(l,t)$, are governed by Eqs.(10.37) and (10.39), where we note that differential equations in Eqs.(10.39) and (10.42) are of the same form, with

$$a = -\frac{Z_0 + R}{L} \quad \text{and} \quad b = \frac{2}{L}\,. \tag{10.43}$$

In the code, we first declare global variables (as in function `ODE`, from the previous MATLAB exercise), specify input data, find the incident voltage by means of function `initialVoltage` (from MATLAB Exercise 10.43), and define a time array `t` for a total duration of $3T$. Then, we compute the constants a and b, using Eq.(10.43), needed for function `ODE`, which, in turn, is called (`@ODE`) by MATLAB ODE solver `ode23tb`, to solve Eq.(10.39) for $i(l,t)$, with a zero initial condition (the last input parameter in `ode23tb`). Finally, $v(l,t)$ is computed from $i(l,t)$ implementing Eq.(10.37), and plotted (for $0 \leq t \leq 3T$) – as shown in Fig.10.46.

```
global V0 a b T t0
Vg = 5;
L = 80*10^(-9);
Z0 = 50;
Rg = 50;
Rl = 30;
t0 = 1*10^(-9);
V0 = initialVoltage(Z0,Rg,Vg);
T = 2*10^-(9);
t = 0:0.0025*t0:3*T;
a = -(Rl+Z0)/L;
b = 2/L;
[time,i] = ode23tb(@ODE,t,0);
Vl = L*diff(i)./diff(time);
Vr = i*Rl;
Vload = Vr + [Vl;0];
clear i;
figure(1);
plot(time/T,Vload);
ylim(1.1*Vg*[-1 1]);
xlabel('t/T');
ylabel('VL');
title('Pulse voltage response at the load');
```

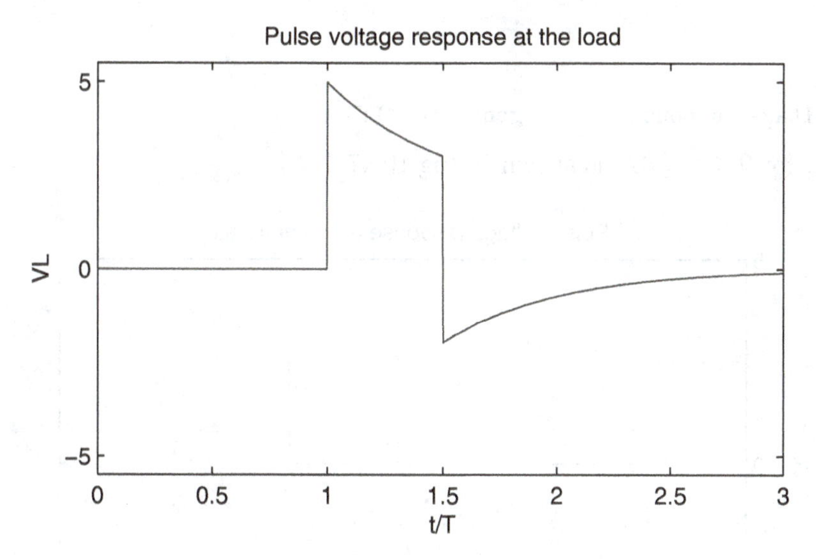

Figure 10.46 MATLAB plot of the pulse voltage response of a transmission line terminated in a series connection of a resistor and an inductor (Fig.10.45) observed at the load; for MATLAB Exercise 10.67. *(color figure on CW)*

MATLAB EXERCISE 10.68 **Generator voltage computation for a line with an *RL* load.** Compute and plot the voltage waveform at the generator end of the transmission line terminated in a series connection of a resistor and an inductor analyzed in the previous MATLAB exercise. *(ME10_68.m on IR)* **H**

TUTORIAL:

In a continuation of the code developed in the previous exercise, the incident voltage (v_i) at the generator terminals in Fig.10.45 is defined in a similar fashion to that in function `ODE` (MATLAB Exercise 10.66), and the voltage pulse response of the line at these terminals, $v_g(t) = v(0,t)$, is obtained from the previously determined load voltage, $v_L(t) = v(l,t)$, and the incident voltage using Eqs.(10.40) and (10.41), as follows:

```
dt = time(2)-time(1);
ts = 0:dt:t0;
tT = 0:dt:T;
N = length(tT);
Vinc = zeros(length(time),1);
Vinc(1:length(ts)-1) = V0*ones(length(ts)-1,1);
Vincl = [zeros(N-1,1);Vinc];
Vrefl = Vload - Vincl(1:length(time));
Vrefgen = [zeros(N-1,1);Vrefl];
Vgen = Vinc + Vrefgen(1:length(time),1);
figure(2);
plot(time/T,Vgen);
ylim(Vg*[-1 1]);
```

```
xlabel('t/T');
ylabel('VG');
title('Pulse voltage response at the generator');
```

The plot of $v_g(t)$, for $0 \leq t \leq 3T$, is shown in Fig.10.47.

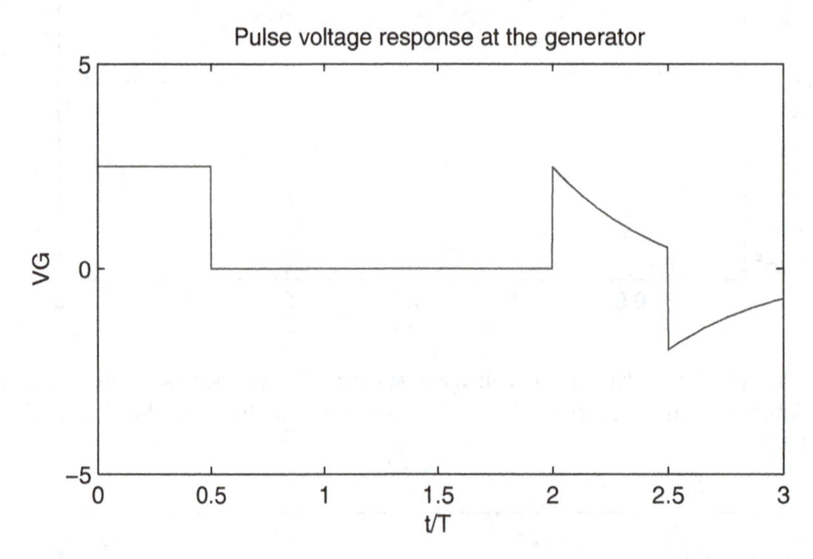

Figure 10.47 MATLAB plot of the pulse voltage response of the transmission line from the previous MATLAB exercise observed at the generator; for MATLAB Exercise 10.68. *(color figure on CW)*

MATLAB EXERCISE 10.69 Line with an *RC* load – complete analysis in MATLAB. Repeat MATLAB Exercise 10.67 but for a capacitor of capacitance $C = 20$ pF in place of the inductor in the line load (transmission line terminated in a resistor and a capacitor connected in series, load voltage computation). *(ME10_69.m on IR)* **H**

HINT:

In place of Eq.(10.39), derive a differential equation with the voltage of the capacitor as unknown quantity, and solve it as in MATLAB Exercise 10.67.

MATLAB EXERCISE 10.70 Generator voltage computation for a line with an *RC* load. Repeat MATLAB Exercise 10.68 but for the line load consisting of a resistor and a capacitor in series from the previous MATLAB exercise. *(ME10_70.m on IR)* **H**

---

11 WAVEGUIDES AND CAVITY RESONATORS

Introduction:

At frequencies in the microwave region, waveguides in the form of metallic tubes are used for energy and information transfer in electromagnetic devices and systems. Essentially, electromagnetic waves travel along such tubes by means of multiple reflections from the metallic walls, through the dielectric filling the tube (most frequently, air), so the waves are guided by the tube conductor. In general, the principal advantage of metallic waveguides, which have one conductor, over transmission lines (e.g., coaxial cables), with two (or more) conductors, at frequencies above several GHz is considerably smaller attenuation along the structure and its larger power transmission capacity. In addition to metallic waveguides for energy/information transmission, waveguide sections closed at both ends, thus forming rectangular metallic cavities, represent microwave resonators – also with widespread applications. Although arbitrary cross sections of metallic tubes and cavities are theoretically possible, our focus here will be on rectangular metallic waveguides and cavity resonators, which are involved most frequently in practical microwave devices and systems.

11.1 Analysis of Rectangular Waveguides Based on Multiple Reflections of Plane Waves

Consider an infinitely long uniform rectangular metallic waveguide with cross-sectional interior dimensions a and b, filled by a homogeneous dielectric of permittivity ε and permeability μ, as shown in Fig.11.1. We assume that the waveguide is lossless, i.e., that its walls are made of a perfect electric conductor (PEC), and that the dielectric is also perfect. We would like to find a solution for a time-harmonic electromagnetic wave, of frequency f (and angular frequency $\omega = 2\pi f$), that propagates inside the waveguide, along the z-axis. One such solution is a normally (or TE – transverse electric) polarized uniform plane wave obliquely incident on a PEC boundary in Fig.8.10(a) (also see MATLAB Exercise 8.17). The wave propagates in the positive z direction by bouncing back and forth, at an incident angle θ_i, between the walls at $x = 0$ and at $x = a$ in Fig.11.1 [note that the coordinate axes in Fig.11.1 are set up differently from Fig.8.10(a)], where the boundary condition stipulating that the tangential component of the total electric field vector be zero at the second plane [similarly to Eqs.(8.6)] gives

$$\sin\beta_\mathrm{n}a = 0\ , \quad \beta_\mathrm{n} = \frac{\omega\cos\theta_\mathrm{i}}{c} \quad \longrightarrow \quad \theta_\mathrm{i} = \arccos\left(m\,\frac{c}{2af}\right) \quad (\arccos\equiv\cos^{-1}) \quad (m = 1, 2, \ldots)\ , \tag{11.1}$$

with β_n standing for the phase coefficient of the wave travel in the direction normal to the PEC boundary in Fig.8.10(a) and c for the intrinsic phase velocity of the waveguide dielectric, that is, the velocity of electromagnetic waves in the medium of parameters ε and μ, Eq.(7.6). Using Eqs.(11.1), the longitudinal phase coefficient (in the z direction), β_z, of the waveguide in Fig.11.1, which is the principal phase coefficient, β, for the structure, comes out to be

$$\beta = \beta_z = \frac{\omega \sin\theta_\mathrm{i}}{c} = \frac{\omega}{c}\sqrt{1-\cos^2\theta_\mathrm{i}} = \frac{\omega}{c}\sqrt{1-\frac{f_\mathrm{c}^2}{f^2}}\,, \quad f_\mathrm{c} = (f_\mathrm{c})_{m0} = m\,\frac{c}{2a} \quad (n=0)$$

$$(\beta \text{ – waveguide phase coefficient})\,, \qquad (11.2)$$

where the frequency f_c is called the cutoff frequency of the waveguide – for a particular, TE_{m0}, mode. Namely, each integer value of m determines a possible field solution in the waveguide, and these distinct waves that can exist in a waveguide are referred to as modes. As we shall see in the next section, waveguide modes with both m and n being arbitrary nonnegative integers are also possible (if properly excited) in a waveguide (in Fig.11.1). The frequency f_c has the same role as the plasma frequency (f_p), in Eq.(7.21). Analogously to a plasma medium, the waveguide in Fig.11.1 behaves like a high-pass filter, letting only waves whose frequency is higher than the cutoff frequency, $f > f_\mathrm{c}$, propagate through it.

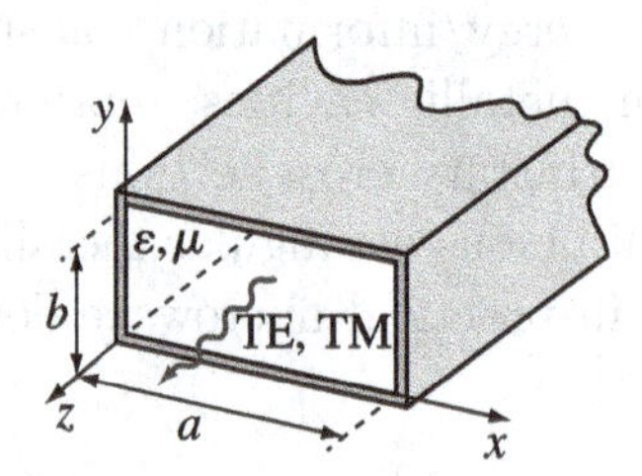

Figure 11.1 Rectangular waveguide with a TE or TM wave.

As it is customary to always denote the transverse dimensions of a rectangular waveguide (Fig.11.1) such that $a \geq b$, the lowest waveguide mode is TE_{10}, for $m = 1$ and $n = 0$. There is an exclusive frequency range, that between $(f_\mathrm{c})_{10}$ and the cutoff frequency of the next higher order mode, in which only one mode, the TE_{10} mode, can propagate, and hence its name – the dominant mode. From Fig.8.10(a) and Eqs.(11.2), its field components are given by

$$\mathrm{TE}_{10}\!: \;\; \underline{E}_y = -\mathrm{j}\omega\mu\,\frac{a}{\pi}\,\underline{H}_0 \sin\left(\frac{\pi}{a}x\right)\mathrm{e}^{-\mathrm{j}\beta z}\,, \quad \underline{H}_x = \mathrm{j}\beta\,\frac{a}{\pi}\,\underline{H}_0 \sin\left(\frac{\pi}{a}x\right)\mathrm{e}^{-\mathrm{j}\beta z}\,,$$

$$\underline{H}_z = \underline{H}_0 \cos\left(\frac{\pi}{a}x\right)\mathrm{e}^{-\mathrm{j}\beta z}\,, \quad \underline{H}_0 = -\frac{2\pi\underline{E}_{\mathrm{i}0}}{\omega\mu a}\,, \quad (f_\mathrm{c})_{10} = \frac{c}{2a} \quad \text{(dominant mode)} \qquad (11.3)$$

($\underline{E}_x\,, \underline{E}_z\,, \underline{H}_y = 0$). Of course, this is a transverse electric (TE) wave, since $\underline{E}_z = 0$ and $\underline{H}_z \neq 0$.

Note, however, that the presented analysis of rectangular metallic waveguides based on multiple reflections of uniform plane waves should not be considered as a general method for field computation in arbitrary guiding structures.

MATLAB EXERCISE 11.1 **3-D drawing of a rectangular waveguide.** Write a function `waveguideGeometry()` in MATLAB that draws the geometry of a rectangular metallic waveguide of transverse dimensions a and b, as in Fig.11.1. The input arguments of the function are a and b, as well as the length of the structure to be shown, c. All dimensions are entered in centimeters. If the user specifies only two arguments, the function returns a 2-D plot [cross section of the waveguide, of dimensions (a, b), or any of the two longitudinal sections, of dimensions (b, c) and (a, c), respectively]; otherwise it returns a 3-D plot. *(waveguideGeometry.m on IR)*[1]

TUTORIAL:

We check the number of input arguments by MATLAB function **nargin** (as in MATLAB Exercise 1.36, for instance). We introduce three vectors of length two, **x**, **y**, and **z** (two vectors in the case of a cross-sectional or another 2-D drawing), and fill them with x, y, and z coordinates of the vertices of the waveguide structure (of length c). Waveguide edges are drawn using MATLAB function **line**, where the input arguments of the function (**x**, **y**, and **z** vectors) are reordered – in accordance to the notation in Fig.11.1. Namely, the default x-coordinate in **line** (first argument) serves as the z-coordinate in Fig.11.1, while coordinates y and z in **line** represent x and y, respectively, in Fig.11.1. In addition, a zero vector, **zeros(1,N)**, and unity vector, **ones(1,N)**, of the same length, N, as **x**, **y**, and **z** [N = **length(x)**] are used in cases when a coordinate of both points defining an edge equals zero or a guide dimension (a, b, or c).

```
function waveguideGeometry(a,b,c)
if nargin == 2;
x = [0, a];
y = [0, b];
N = length(x);
line(x,zeros(1,N));
line(x,b*ones(1,N));
line(zeros(1,N),y);
line(a*ones(1,N),y);
xlabel('x [cm]'); ylabel('y [cm]');
axis equal;
else
x = [0, a];
y = [0, b];
z = [0, c];
N = length(x);
line(zeros(1,N),x,b*ones(1,N));
line(c*ones(1,N),x,b*ones(1,N));
line(zeros(1,N),x,zeros(1,N));
line(c*ones(1,N),x,zeros(1,N));
line(zeros(1,N),a*ones(1,N),y);
line(c*ones(1,N),a*ones(1,N),y);
line(zeros(1,N),zeros(1,N),y);
line(c*ones(1,N),zeros(1,N),y);
line(z,a*ones(1,N),zeros(1,N));
line(z,zeros(1,N),zeros(1,N));
line(z,a*ones(1,N),b*ones(1,N));
line(z,zeros(1,N),b*ones(1,N));
view([-30 30]);
```

[1] *IR* = Instructor Resources (for the book).

```
axis equal;
xlabel('z [cm]'); ylabel('x [cm]'); zlabel('y [cm]');
end;
```

For example, typing `waveguideGeometry(4,2,8)` in the Command Window results in the plot shown in Fig.11.2. Note that by changing azimuthal and elevation angles in MATLAB function `view` (in the code), we can get different views of the structure.

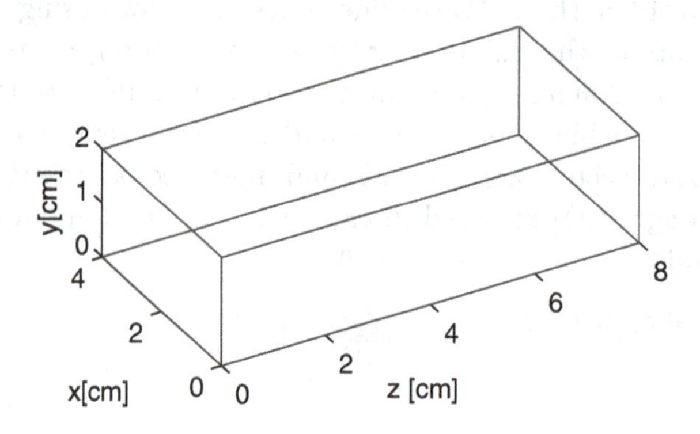

Figure 11.2 MATLAB 3-D drawing of the geometry of a rectangular metallic waveguide of transverse dimensions $a = 4$ cm and $b = 2$ cm (Fig.11.1) and length $c = 8$ cm; for MATLAB Exercise 11.1. *(color figure on CW)*[2]

MATLAB EXERCISE 11.2 **Ray paths of TE modes in a rectangular waveguide – 3-D movie.** For a rectangular metallic waveguide in Fig.11.1, $a = 6$ cm, the dielectric is air, and the operating frequency of the structure is $f = 10$ GHz. In MATLAB, create a 3-D movie that visualizes the ray paths corresponding to the first three TE_{m0} modes ($m = 1, 2, 3$) in the guide. *(ME11_2.m on IR)*

TUTORIAL:

We refer to the waveguide geometry and Cartesian coordinate system in Fig.11.2 (previous MATLAB exercise). In the first portion of the code, we specify the input data, namely, waveguide transverse dimensions, a and b (`a` and `b`), where we adopt $b = a/2 = 3$ cm, the length of the guide, c (`c`), adopted to be $c = 1$ m, mode index, m (`m`), operating frequency, f (`f`), and parameters of the waveguide dielectric (air).

```
a = 6*10^(-2);
b = 3*10^(-2);
c = 1;
m = 1;
f = 10^10;
EPS0 = 8.8542*10^(-12);
MU0 = 4*pi*10^(-7);
```

The incident angle on waveguide walls, θ_i, is found from Eqs.(11.1). We also compute the

[2] *CW* = Companion Website (of the book).

longitudinal projection (along the z-axis in Fig.11.2) of one segment of the zigzag ray path [see Fig.8.10(a)], $d = a \tan \theta_\text{i}$.

```
c0 = 1/sqrt(EPS0*MU0);
THETAi = acos(m*c0/2/a/f);
d = a*tan(THETAi);
```

Coordinates z ($0 \le z \le c$) of the ray path line are stored in the vector **z** (z=0:d/20:c). To calculate coordinates x ($0 \le x \le a$) of the line, we introduce a local vector **zLocal** with local z-coordinates in the interval $0 \le z \le 2d$ (one "rise" and "fall" of the ray in the zigzag path); "mapping" of **z** onto **zLocal** is done using MATLAB function **floor**, which rounds the elements of the vector to the nearest integers less than or equal to the element. Coordinates x are computed within two branches in the code: ray "fall" (zLocal > d) and "rise" (zLocal < d).

```
z = 0:d/20:c;
zLocal = z-floor(z./2/d).*2*d;
x = zeros(1,length(z));
for i = 1:length(z)
if zLocal(i) > d
x(i) = a - (zLocal(i)-d)*cot(THETAi);
else
x(i) = zLocal(i)*cot(THETAi);
end;
end;
```

Within the **for** loop of the movie [number of frames is length(z)], we plot the ray path in 3-D by means of MATLAB function **plot3**, where the y-coordinate is fixed, $y = b/2$, in the plot (for the order of arguments for the function, see the explanation in the tutorial to the previous MATLAB exercise), and draw the waveguide geometry calling function **waveguideGeometry** (from the previous exercise). For a better view of the structure and the ray bounces, we scale up the horizontal (a) and vertical (b) dimensions of the guide, as they are significantly smaller than the length (c).

```
scale = 10;
figure(1)
for i = 1:length(z)
plot3(z(1:i), x(1:i)*scale, b/2*ones(i)*scale);
hold on;
waveguideGeometry(a*scale, b*scale, c);
hold off;
xlabel('z [m]'); ylabel('x*scale [m]'); zlabel('y*scale [m]');
M(i) = getframe;
end;
```

A snapshot (frame) of the movie for the TE_{10} waveguide mode is shown in Fig.11.3.

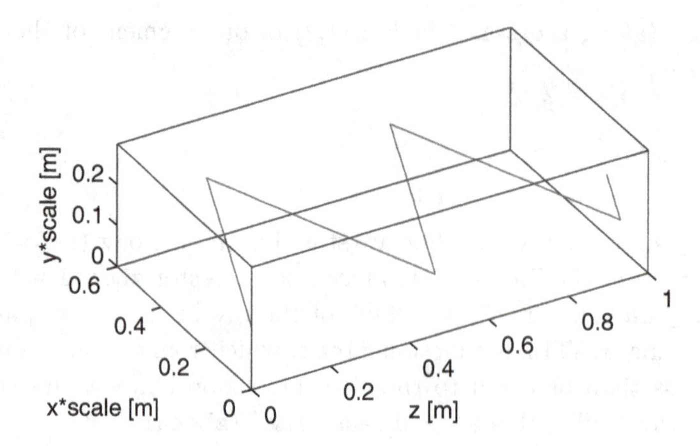

Figure 11.3 Snapshot of a 3-D MATLAB movie visualizing the travel and multiple reflections of a plane wave giving rise to a TE_{10} mode ($m = 1$) in a rectangular metallic waveguide (Fig.11.1); for MATLAB Exercise 11.2. *(color figure on CW)*

MATLAB EXERCISE 11.3 **Ray paths of TE waveguide modes – 2-D movie.** Create a 2-D version (2-D movie), in the xz-cut ($y = b/2$), of the movie (code) developed in the previous MATLAB exercise. For presenting the waveguide geometry, use the 2-D drawing option in function `waveguideGeometry` (from MATLAB Exercise 11.1). *(ME11_3.m on IR)*

MATLAB EXERCISE 11.4 **Phase coefficient of a rectangular waveguide.** Write a function `phaseCoeffWaveguide()` in MATLAB that calculates the phase coefficient β of a rectangular waveguide, Fig.11.1, given by Eq.(11.2), with the following input arguments: the intrinsic phase velocity of the waveguide dielectric, c [Eq.(7.6)], operating frequency of the waveguide, f, and cutoff frequency of the waveguide for a particular mode, f_{c}. In the case $f \le f_{\mathrm{c}}$, a "no propagation" label is returned instead of β. *(phaseCoeffWaveguide.m on IR)*

MATLAB EXERCISE 11.5 **Plots of the electric field of the dominant waveguide mode.** Consider a rectangular waveguide (Fig.11.1) with transverse dimensions $a = 16.51$ cm and $b = 8.225$ cm, and air dielectric, operating at a frequency of $f = 1.5$ GHz. In MATLAB, plot the real and imaginary parts of the electric field intensity of the dominant (TE_{10}) mode, $\underline{E}_y$, in planes defined by $z = 0$ and $x = a/2$, respectively, assuming that the constant $\underline{E}_{\mathrm{i}0}$ in Eqs.(11.3) equals $\underline{E}_{\mathrm{i}0} = 1\,\mathrm{e}^{\mathrm{j}0}$ V/m. In the latter case, the plot should extend along a section of the waveguide $l = 40$ cm long. Redo the plots also for the waveguide filled with a dielectric of relative permittivity $\varepsilon_{\mathrm{r}} = 2.25$ ($\mu_{\mathrm{r}} = 1$). *(ME11_5.m on IR)* **H**[3]

TUTORIAL:

First, we specify the input data (note that for the waveguide filled with the dielectric having

[3]**H** = recommended to be done also "by hand," i.e., not using MATLAB.

$\varepsilon_r = 2.25$, we change the corresponding code line accordingly):

```
EPS0 = 8.8542*10^(-12);
MU0 = 4*pi*10^(-7);
EPSR = 1;
MUR = 1;
MU = MU0*MUR;
EPS = EPS0*EPSR;
a = 16.51*10^(-2);
b = 8.225*10^(-2);
l = 40*10^(-2);
f = 1.5*10^9;
Ei0 = 1;
```

In addition, we define coordinates x, y, and z for field computation and plotting (see Figs.11.4 and 11.5). Because $\underline{E}_y$ does not depend on y, we set $y = 0$, and introduce coordinate vectors **x** ($0 \leq x \leq a$) and **z** ($0 \leq z \leq l$). We also define the two planes for plotting, $z = 0$ (`Z = 0`) and $x = a/2$ (`X = a/2`), respectively.

```
x = 0:a/100:a;
y = 0;
z = 0:l/100:l;
Z = 0;
X = a/2;
```

Next, the intrinsic phase velocity of the waveguide dielectric and cutoff frequency of the dom-

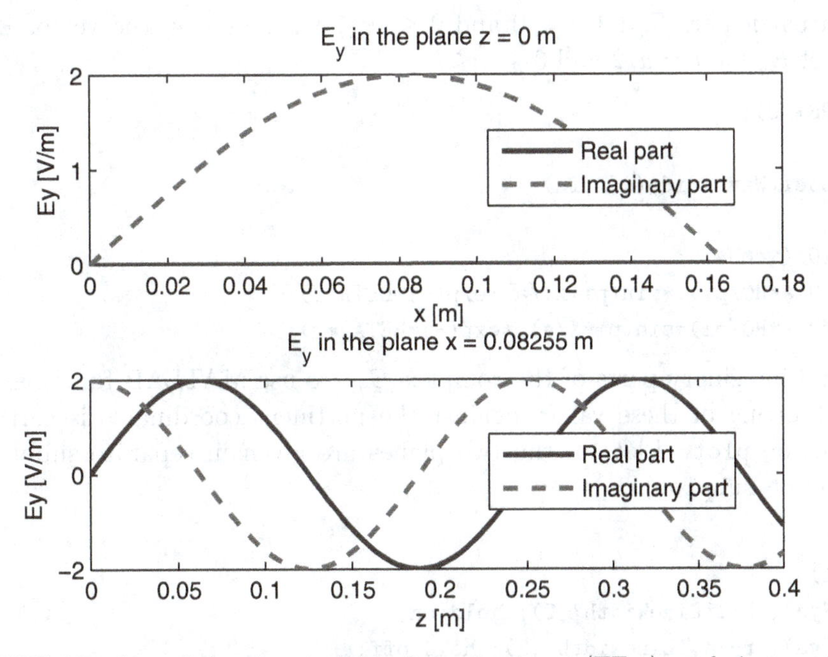

Figure 11.4 MATLAB plots of the electric field of the dominant (TE_{10}) mode in two characteristic cuts (planes) in an air-filled waveguide ($\varepsilon_r = 1$); for MATLAB Exercise 11.5. *(color figure on CW)*

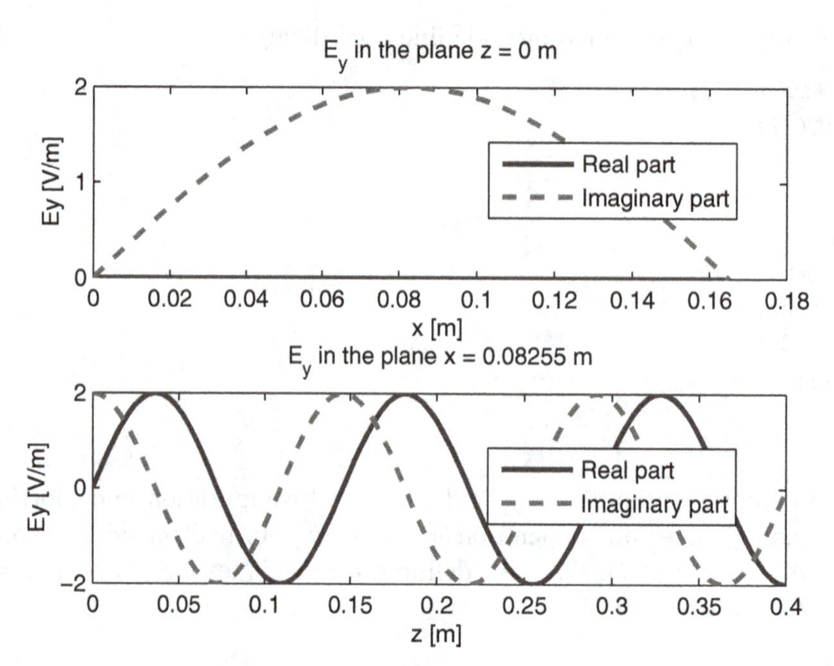

Figure 11.5 MATLAB plots of the electric field of the dominant mode in two planes – in a waveguide filled with a dielectric with $\varepsilon_\mathrm{r} = 2.25$; for MATLAB Exercise 11.5. *(color figure on CW)*

inant mode are found from Eqs.(7.6) and (11.3), respectively, while the phase coefficient of the waveguide is obtained calling function `phaseCoeffWaveguide` (from the previous MATLAB exercise). For field $\underline{E}_y$, we implement the first and the fourth (for the constant $\underline{H}_0$) expressions in Eqs.(11.3). In particular, $\underline{E}_y$ for $z = 0$ and $0 \leq x \leq a$ is stored in the vector `Eya`, whereas `Eyb` carries values of $\underline{E}_y$ for $x = a/2$ and $0 \leq z \leq l$.

```
c = 1/sqrt(EPS*MU);
fc = c/(2*a);
BETA = phaseCoeffWaveguide(c,f,fc);
w = 2*pi*f;
H0 = -2*pi*Ei0/(w*MU*a);
Eya = (-i*w*MU*a*H0/pi).*sin(pi*x./a)*exp(-i*BETA*Z);
Eyb = (-i*w*MU*a*H0/pi)*sin(pi*X/a).*exp(-i*BETA.*z);
```

To find real and imaginary parts of the complex $\underline{E}_y$, we use MATLAB functions `real` and `imag`, respectively. Plotting of these values against the pertinent coordinates is carried out applying MATLAB function `plot`. Plots in the two planes are given in separate subplots (by means of MATLAB function `subplot`).

```
figure(1)
subplot(2,1,1)
plot(x,real(Eya),'b','LineWidth',2); hold on;
plot(x,imag(Eya),'r--','LineWidth',2); hold off;
xlabel('x [m]'); ylabel('Ey [V/m]');
legend('Real part','Imaginary part','Location','Best');
```

```
title(['Ey in the plane z = ',num2str(Z),' m']);
subplot(2,1,2)
plot(z,real(Eyb),'b','LineWidth',2); hold on;
plot(z,imag(Eyb),'r--','LineWidth',2); hold off;
xlabel('z [m]'); ylabel('Ey [V/m]');
legend('Real part','Imaginary part','Location','Best');
title(['Ey in the plane x = ',num2str(X),' m']);
```

The field plots for $\varepsilon_\mathrm{r} = 1$ and $\varepsilon_\mathrm{r} = 2.25$ are shown in Figs.11.4 and 11.5, respectively.

MATLAB EXERCISE 11.6 **Plots of the z-component of H field of the dominant mode.** Repeat the previous MATLAB exercise but for the TE_{10} field component $\underline{H}_z$ in planes defined by $z = l$ and $x = 0$, respectively. *(ME11_6.m on IR)* **H**

11.2 Arbitrary TE and TM Modes in a Rectangular Waveguide

From Maxwell's equations for a perfect dielectric of parameters ε and μ (waveguide dielectric), Eqs.(6.15), and boundary conditions at the waveguide (PEC) walls, Eqs.(6.19), the field components of a TE_{mn} mode in a rectangular waveguide, Fig.11.1, are found to be

$$\underline{E}_x = \frac{\mathrm{j}\omega\mu}{k^2}\frac{n\pi}{b}\underline{H}_0\cos\left(\frac{m\pi}{a}x\right)\sin\left(\frac{n\pi}{b}y\right)\mathrm{e}^{-\mathrm{j}\beta z}, \quad \underline{E}_y = -\frac{\mathrm{j}\omega\mu}{k^2}\frac{m\pi}{a}\underline{H}_0\sin\left(\frac{m\pi}{a}x\right)\cos\left(\frac{n\pi}{b}y\right)$$
$$\times\,\mathrm{e}^{-\mathrm{j}\beta z}, \quad \underline{E}_z = 0, \quad \underline{H}_x = \frac{\mathrm{j}\beta}{k^2}\frac{m\pi}{a}\underline{H}_0\sin\left(\frac{m\pi}{a}x\right)\cos\left(\frac{n\pi}{b}y\right)\mathrm{e}^{-\mathrm{j}\beta z},$$
$$\underline{H}_y = \frac{\mathrm{j}\beta}{k^2}\frac{n\pi}{b}\underline{H}_0\cos\left(\frac{m\pi}{a}x\right)\sin\left(\frac{n\pi}{b}y\right)\mathrm{e}^{-\mathrm{j}\beta z}, \quad \underline{H}_z = \underline{H}_0\cos\left(\frac{m\pi}{a}x\right)\cos\left(\frac{n\pi}{b}y\right)\mathrm{e}^{-\mathrm{j}\beta z},$$
$$k^2 = \left(\frac{m\pi}{a}\right)^2 + \left(\frac{n\pi}{b}\right)^2 \quad (\mathrm{TE}_{mn}\text{ mode; } m, n = 0, 1, 2, \ldots), \tag{11.4}$$

with the restriction that only one of the mode indices can be zero. The electromagnetic field of a TM_{mn} mode is given by

$$\underline{E}_x = -\frac{\mathrm{j}\beta}{k^2}\frac{m\pi}{a}\underline{E}_0\cos\left(\frac{m\pi}{a}x\right)\sin\left(\frac{n\pi}{b}y\right)\mathrm{e}^{-\mathrm{j}\beta z}, \quad \underline{E}_y = -\frac{\mathrm{j}\beta}{k^2}\frac{n\pi}{b}\underline{E}_0\sin\left(\frac{m\pi}{a}x\right)\cos\left(\frac{n\pi}{b}y\right)$$
$$\times\,\mathrm{e}^{-\mathrm{j}\beta z}, \quad \underline{E}_z = \underline{E}_0\sin\left(\frac{m\pi}{a}x\right)\sin\left(\frac{n\pi}{b}y\right)\mathrm{e}^{-\mathrm{j}\beta z}, \quad \underline{H}_x = \frac{\mathrm{j}\omega\varepsilon}{k^2}\frac{n\pi}{b}\underline{E}_0\sin\left(\frac{m\pi}{a}x\right)\cos\left(\frac{n\pi}{b}y\right)$$
$$\times\,\mathrm{e}^{-\mathrm{j}\beta z}, \quad \underline{H}_y = -\frac{\mathrm{j}\omega\varepsilon}{k^2}\frac{m\pi}{a}\underline{E}_0\cos\left(\frac{m\pi}{a}x\right)\sin\left(\frac{n\pi}{b}y\right)\mathrm{e}^{-\mathrm{j}\beta z}, \quad \underline{H}_z = 0$$
$$(\mathrm{TM}_{mn}\text{ mode; } m, n = 1, 2, \ldots), \tag{11.5}$$

where the parameter k (or k^2) is the same as in Eqs.(11.4). Note that this is a TM (transverse magnetic) wave because $\underline{H}_z = 0$ and $\underline{E}_z \neq 0$. Note also that the lowest TM mode is TM_{11}. The expression for the waveguide phase coefficient, β, is that in Eqs.(11.2) in both TE and TM cases, with the cutoff frequency, of the TE_{mn} or TM_{mn} mode, computed as

$$f_\mathrm{c} = (f_\mathrm{c})_{mn} = \frac{c}{2}\sqrt{\left(\frac{m}{a}\right)^2 + \left(\frac{n}{b}\right)^2} \quad \text{(cutoff frequency, TE}_{mn} \text{ or TM}_{mn} \text{ mode)} . \tag{11.6}$$

MATLAB EXERCISE 11.7 **Modal parameter k for a waveguide.** Write a function `kParameter()` in MATLAB that returns the k parameter in Eqs.(11.4), for an arbitrary propagating mode in a rectangular waveguide (Fig.11.1). The input to the function consists of waveguide transverse dimensions, a and b, and mode indices, m and n. *(kParameter.m on IR)*

MATLAB EXERCISE 11.8 **Magnitude of the x-component of E field for TE modes.** Write a function `ExTE()` in MATLAB that computes the magnitude of the x-component of the electric field vector, $|\underline{E}_x|$, of an arbitrary TE_{mn} wave mode, in a rectangular waveguide, Fig.11.1. *(ExTE.m on IR)*

TUTORIAL:

The field component $\underline{E}_x$ is given in Eqs.(11.4). The input arguments of the function are operating angular frequency, ω (`w`), permeability of the guide dielectric, μ (`MU`), k parameter of the waveguide (`k`), complex constant $\underline{H}_0$ (`H0`), dimensions of the guide cross section, a (`a`) and b (`b`), mode indices, m (`m`) and n (`n`), and coordinates x and y, defined as vectors `x` and `y`, respectively. Having in mind that $|\mathrm{e}^{-\mathrm{j}\beta z}| = 1$ and $|\mathrm{j}| = 1$, $|\underline{E}_x|$ is computed as follows:

```
function [Ex] = ExTE(w,MU,k,H0,a,b,m,n,x,y)
Const = w*MU*n*pi*H0/(k^2*b);
Ex = abs(Const.*cos(m*pi*x/a).*sin(n*pi*y/b));
return;
```

MATLAB EXERCISE 11.9 **Other TE field components, all field components of TM modes.** Repeat the previous MATLAB exercise but for the electric-field y-component and for each of the three components of the magnetic field vector of a TE_{mn} mode in the waveguide – write functions `EyTE()`, `HxTE()`, `HyTE()`, and `HzTE()` that return $|\underline{E}_y|$, $|\underline{H}_x|$, $|\underline{H}_y|$, and $|\underline{H}_z|$, respectively, based on Eqs.(11.4). Also, write MATLAB functions `ExTM()`, `EyTM()`, `EzTM()`, `HxTM()`, and `HyTM()` that compute magnitudes of nonzero components of the electric and magnetic field vectors of an arbitrary TM_{mn} waveguide mode, $|\underline{E}_x|$, $|\underline{E}_y|$, $|\underline{E}_z|$, $|\underline{H}_x|$, and $|\underline{H}_y|$, using Eqs.(11.5). *(EyTE.m, HxTE.m, HyTE.m, HzTE.m, ExTM.m, EyTM.m, EzTM.m, HxTM.m, and HyTM.m on IR)*

MATLAB EXERCISE 11.10 **Cutoff frequency of an arbitrary TE or TM mode.** Write a function `cutoffFreq()` in MATLAB that calculates the cutoff frequency of an arbitrary TE or TM mode in a rectangular waveguide (Fig.11.1), $f_c = (f_c)_{mn}$, based on Eq.(11.6), with the following input arguments: the intrinsic phase velocity of the waveguide dielectric, c, waveguide transverse dimensions, a and b, and mode indices, m and n. *(cutoffFreq.m on IR)*

MATLAB EXERCISE 11.11 **Determining all possible propagating modes in a waveguide.** Consider a rectangular waveguide with transverse dimensions a and b, filled with a dielectric of relative permittivity ε_r and permeability μ_r. Write a MATLAB program that determines what wave modes can propagate along this waveguide at a frequency f (data entry should be from the keyboard). *(ME11_11.m on IR)*

TUTORIAL:

Data entry from the keyboard is realized using MATLAB command `input`,

```
EPS0 = 8.8542*10^(-12);
MU0 = 4*pi*10^(-7);
a = input('Enter width of the waveguide, a:  ');
b = input('Enter height of the waveguide, b:  ');
EPSR = input('Enter relative permittivity of the dielectric:  ');
MUR = input('Enter relative permeability of the dielectric:  ');
f = input('Enter operating frequency of the waveguide:  ');
c = 1/sqrt(EPS0*MU0*EPSR*MUR);
```

Then, we set $n = 0$ and increase m [m and n are mode indices in Eq.(11.6)] starting with $m = 1$ to find the maximum m, m_{max}, for which (and $n = 0$) the propagating condition $f > f_c$ is satisfied. This value (m_{max}) will be used as the upper bound in a two-fold search for all possible indices (m, n) of propagating modes, with both m and n being varied concurrently. To compute the cutoff frequency, f_c (`fc`), we use function `cutoffFreq` (from the previous MATLAB exercise).

```
m = 1;
n = 0;
fc = cutoffFreq(c,a,b,m,n);
mmax = 0;
while fc < f
mmax = m;
m = m + 1;
fc = cutoffFreq(c,a,b,m,n);
end
```

In a similar way, we find n_{max} that satisfies the condition $f > f_c$ with $m = 0$,

```
m = 0;
n = 1;
fc = cutoffFreq(c,a,b,m,n);
nmax = 0;
```

```
while fc < f
nmax = n;
n = n + 1;
fc = cutoffFreq(c,a,b,m,n);
end
```

Finally, within two nested `for` loops varied in m and n and bounded by $m_{\max}$ and $n_{\max}$, respectively, we make a final list of possible propagating modes (m, n), and determine their total number (`counter`). The results are stored in a matrix `modes`, whose first two columns are filled with indices m and n of all propagating modes, while the third column contains the associated cutoff frequencies of the modes.

```
counter = 1;
for i = 0:mmax
for j = 0:nmax
fc = cutoffFreq(c,a,b,i,j);
if fc < f
modes(counter,1:3) = [i,j,fc];
counter = counter + 1;
end
end
end
```

We display the matrix `modes` in the Command Window (by MATLAB function `disp`) in a tabular format, with its elements (numerical data) previously converted into strings by MATLAB function `num2str`,

```
disp('');
disp('Cutoff frequencies of propagating modes');
disp('m n fc (Hz)');
for i = 2:counter-1
disp([num2str(modes(i,1:2)), ' ',num2str(modes(i,3))]);
end
```

MATLAB EXERCISE 11.12 **First several modal cutoff frequencies in a standard waveguide.** Consider a standard waveguide (rectangular waveguide with a 2 : 1 aspect ratio, $a = 2b$) with transverse dimensions $a = 8$ cm and $b = 4$ cm, and air dielectric. In MATLAB, determine what wave modes can propagate along this waveguide at a frequency of $f = 4.5$ GHz. *(ME11_11.m on IR)* **H**

TUTORIAL:

We run the program developed in the previous MATLAB exercise. The displayed results are:

```
Cutoff frequencies of propagating modes
m n fc (Hz)
0 1 3747403146.9955
```

```
1 0 1873701573.4978
1 1 4189724087.8893
2 0 3747403146.9955
```

Note that in a frequency range defined as

$$(f_c)_{10} = \frac{c}{2a} < f \leq \frac{c}{a} = (f_c)_{01} = (f_c)_{20} \quad \text{(dominant frequency range)}\,, \tag{11.7}$$

only the dominant mode (TE_{10}) can propagate along a standard waveguide, and that is why this range is referred to as the dominant range.

MATLAB EXERCISE 11.13 **Modal cutoff frequencies in WR-975 and WR-340 waveguides.** For an air-filled WR-975 commercial rectangular waveguide, with transverse dimensions $a = 24.766$ cm and $b = 12.383$ cm, determine using MATLAB (a) cutoff frequencies of the first three TE modes and first three TM modes and (b) all possible propagating modes at a frequency of $f = 2$ GHz. (c) Repeat (a) and (b) for a WR-340 waveguide, with $a = 8.636$ cm and $b = 4.318$ cm (and air dielectric). *(ME11_13.m on IR)* **H**

MATLAB EXERCISE 11.14 **GUI with three types of field visualization, arbitrary mode.** Create a graphical user interface (GUI) in MATLAB that visualizes the spatial distribution of any electric or magnetic field component of an arbitrary mode in a rectangular waveguide, by providing, in a pop-up menu, three types of plots: a 3-D plot using MATLAB function `surf` and 2-D plots applying MATLAB functions `pcolor` and `contour`, respectively. *[folder ME11_14(GUI) on IR]*

TUTORIAL:

See MATLAB Exercise 2.1 – for the GUI development. Choose the following components from the interface palette: Axes, Pop-up Menu, Panel, Push Button, and Static Text. Group Static Text, Pop-up Menu, and Push Button inside the Panel field (space) on one side and locate the Axes field on the other side of the interface. Type "Select plotting function" in the Static Text. In the Pop-up Menu, activate Property Inspector (right click on the Pop-up menu, find String section), leave the first row blank, and type the names of plotting functions: `surf`, `pcolor`, and `contour`, respectively. Then save the GUI as `WaveguideField`.

To activate and control the Pop-up Menu in `WaveguideField.m`, we implement the `case-switch` command in the section `popupmenu1_Callback(hObject, eventdata, handles)`. When the user chooses the blank row from the Pop-up Menu (`case 1`), the Axes field is reset (cleared) (`cla reset`) and the Push Button is in off regime [`set(handles.pushbutton1,'Enable','off')`]. In the remaining three cases, the Push Button is in on regime [`set(handles.pushbutton1, 'Enable','on')`]. Also, here we introduce a global variable `i` that indicates the current choice of the plot type (`i = 1` for `surf`, `i = 2` for `pcolor`, and `i = 3` for `contour`).

```
function popupmenu1_Callback(hObject, eventdata, handles)
global i;
```

```
switch get(handles.popupmenu1,'Value')
case 1
set(handles.pushbutton1,'Enable','off');
cla reset;
case 2
set(handles.pushbutton1,'Enable','on');
i = 1;
case 3
set(handles.pushbutton1,'Enable','on');
i = 2;
case 4
set(handles.pushbutton1,'Enable','on');
i = 3;
end;
```

Finally, plots of a given field component of a given mode are done in function `pushbutton1_Callback(hObject, eventdata, handles)` of the m file. In this function, we first define all input parameters and a mesh of points (`x` and `y` matrices) using MATLAB function `meshgrid`, and then compute the field component (named `Field`) applying one of the functions written in MATLAB Exercises 11.8 and 11.9. These lines of code are not shown here, but are given in the next MATLAB exercise for a specific case, the TE_{10} waveguide mode. Depending on the choice of the plot type (variable `i`), the plot is realized within the `if-else` branching. For implementations of MATLAB functions `surf`, `pcolor`, and `contour`, see – for instance – MATLAB Exercises 1.39, 6.17, and 1.29, respectively.

```
function pushbutton1_Callback(hObject, eventdata, handles)
global i;
if i == 1;
surf(x,y,Field); rotate3d on;
else if i == 2;
pcolor(x,y,Field); rotate3d off;
else
contour(x,y,Field,'linewidth',2); rotate3d off;
end;
end;
xlabel('x [m]'); ylabel('y [m]'); colorbar;
```

MATLAB EXERCISE 11.15 **GUI with field plots for the dominant mode.** For the waveguide from MATLAB Exercise 11.12, use GUI WaveguideField, developed in the previous MATLAB exercise, to plot all nonzero field components of the dominant mode, TE_{10}, assuming that $\underline{H}_0 = 0.1\,e^{j0}$ A/m. *[folder ME11_15(GUI) on IR]*

TUTORIAL:

See Fig.11.6. First, we open the file `WaveguideField.m` and specify the input data in function

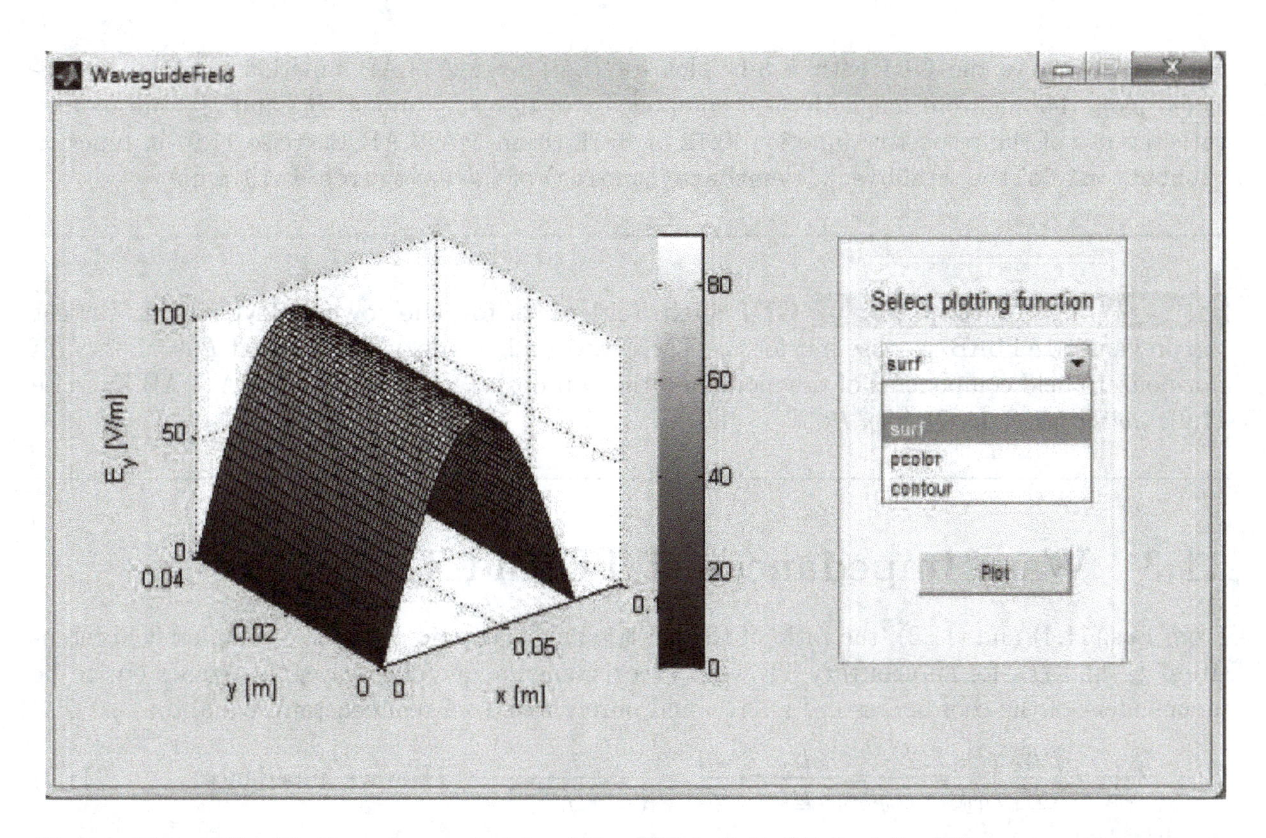

Figure 11.6 GUI with three types of plots of the electric field ($|\underline{E}_y|$) of the TE_{10} waveguide mode; for MATLAB Exercise 11.15. *(color figure on CW)*

`pushbutton1_Callback(hObject,eventdata,handles)`, as follows:

```
MU0 = 4*pi*10^(-7);
MUR = 1;
MU = MU0*MUR;
a = 8*10^(-2);
b = 4*10^(-2);
f = 4.5*10^9;
w = 2*pi*f;
H0 = 0.1;
m = 1;
n = 0;
```

Then, we compute the k parameter (`k`) of the waveguide, applying function `kParameter` (from MATLAB Exercise 11.7), and define a grid of points for plotting (matrices `x` and `y`), using function `meshgrid`. The magnitude of the field component $\underline{E}_y$ is evaluated by means of function `EyTE` (from MATLAB Exercise 11.9).

```
k = kParameter(a,b,m,n);
[x,y] = meshgrid(0:a/50:a,0:b/50:b);
Field = EyTE(w,MU,k,H0,a,b,m,n,x,y);
```

Figure 11.6 shows the GUI with a 3-D plot (of $|\underline{E}_y|$) by MATLAB function **surf** as the selected plot. The nonzero magnetic field components of the TE_{10} mode, $\underline{H}_x$ and $\underline{H}_z$, are plotted with the use of the respective function **HxTE** or **HzTE** (from MATLAB Exercise 11.9) in function **pushbutton1_Callback(hObject,eventdata,handles)** of the **WaveguideField.m** file.

MATLAB EXERCISE 11.16 **GUI with field plots for the lowest TM mode.** Repeat the previous MATLAB exercise but for the TM_{11} waveguide mode, assuming that $\underline{E}_0 = 1\,\mathrm{e}^{\mathrm{j}0}$ V/m (for nonzero field components of the mode, use the appropriate functions from MATLAB Exercise 11.9). *[folder ME11_16(GUI) on IR]*

11.3 Wave Impedances of TE and TM Waves

From Eqs.(11.4) and (11.2), the ratio of the electric and magnetic transverse complex field intensities, $\underline{E}_\mathrm{t}$ and $\underline{H}_\mathrm{t}$, for an arbitrary TE wave in a rectangular metallic waveguide comes out to be independent of the coordinates in Fig.11.1, and purely real, so a real constant, equal to

$$Z_\mathrm{TE} = \left(\frac{\underline{E}_\mathrm{t}}{\underline{H}_\mathrm{t}}\right)_\mathrm{TE} = \frac{\underline{E}_x}{\underline{H}_y} = -\frac{\underline{E}_y}{\underline{H}_x} = \frac{\omega\mu}{\beta} = \frac{\eta}{\sqrt{1 - f_\mathrm{c}^2/f^2}} \quad \text{(TE wave impedance)}\,, \tag{11.8}$$

since $\mu c = \eta$, where $c = 1/\sqrt{\varepsilon\mu}$ and $\eta = \sqrt{\mu/\varepsilon}$ are the intrinsic phase velocity and impedance, respectively, of the waveguide dielectric. The ratio $\underline{E}_\mathrm{t}/\underline{H}_\mathrm{t}$, in turn, defines the wave impedance of a TE wave, analogously to the TEM case (for a transmission line), in Eqs.(9.2). Of course, the cutoff frequency of the waveguide, f_c, for a mode (m,n) is given in Eq.(11.6), and Z_TE, for a TE_{mn} mode, depends on mode indices m and n. Similarly, Eqs.(11.5) and (11.2) give the following expression for the wave impedance of an arbitrary TM wave (TM_{mn} mode) in Fig.11.1:

$$Z_\mathrm{TM} = \left(\frac{\underline{E}_\mathrm{t}}{\underline{H}_\mathrm{t}}\right)_\mathrm{TM} = \frac{\beta}{\omega\varepsilon} = \eta\sqrt{1 - \frac{f_\mathrm{c}^2}{f^2}} \quad \text{(TM wave impedance)}\,. \tag{11.9}$$

Note that, unlike the TEM case, both TE and TM impedances are functions of the frequency (f) of the propagating wave.

MATLAB EXERCISE 11.17 **TE and TM wave impedances for a rectangular waveguide.** Write functions **TEWaveImpedance()** and **TMWaveImpedance()** in MATLAB that return the wave impedance of a TE wave (Z_TE) and that of a TM wave (Z_TM), respectively, in a rectangular waveguide (Fig.11.1), given by Eqs.(11.8) and (11.9), with the following input parameters: the intrinsic impedance of the waveguide dielectric (η), waveguide cutoff frequency (f_c), and operating frequency of the wave (f) (in general, **f** is a vector of frequencies). *(TEWaveImpedance.m and TMWaveImpedance.m on IR)*

MATLAB EXERCISE 11.18 **Wave impedance plot for the dominant mode.** Write a MATLAB code that calls function `TEWaveImpedance` (from the previous MATLAB Exercise) and plots the wave impedance for a given TE_{mn} mode [given (m, n)] propagating along a waveguide of transverse dimensions a and b, in a frequency range $f_c \leq f \leq kf_c$, where $f_c = (f_c)_{mn}$ is the cutoff frequency of the mode and k is an integer specified by the user. In specific, plot the wave impedance of the dominant mode propagating along an air-filled standard waveguide ($a = 2b$) with $a = 1.58$ cm (take $k = 2$). *(ME11_18.m on IR)*

HINT:
The resulting plot is shown in Fig.11.7.

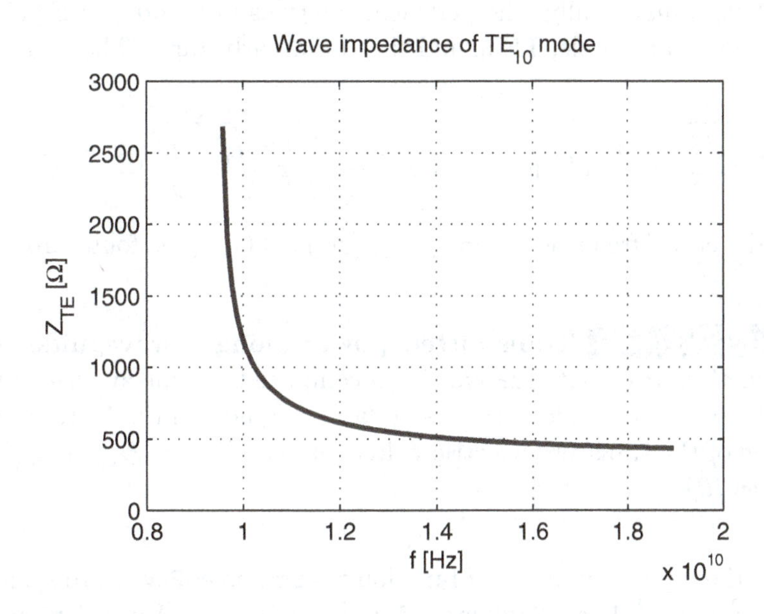

Figure 11.7 Wave impedance of the TE_{10} mode in a standard waveguide; for MATLAB Exercise 11.18. *(color figure on CW)*

MATLAB EXERCISE 11.19 **Wave impedance plot for the lowest TM mode.** Repeat the previous MATLAB exercise but for the wave impedance of a given TM_{mn} mode (make use of function `TMWaveImpedance`, from MATLAB Exercise 11.17). Generate a plot for the TM_{11} mode in the air-filled standard waveguide ($a = 1.58$ cm and $k = 2$). *(ME11_19.m on IR)*

11.4 Power Flow Along a Waveguide

We now evaluate the power flow associated with a traveling TE or TM wave along a rectangular waveguide, in Fig.11.1. Combining Eqs.(6.31)–(6.33), (11.8), and (11.9), the time-average transmitted power along the guide equals the real part of the complex power carried by the wave along the z-axis, i.e., the flux of the complex Poynting vector of the wave through a cross section of the guide dielectric, S_d, and can be computed as follows:

$$P = P_\mathrm{ave} = \mathrm{Re}\left\{\int_{S_\mathrm{d}} \underline{\mathcal{P}} \cdot \mathrm{d}\mathbf{S}\right\} = \mathrm{Re}\left\{\int_{S_\mathrm{d}} (\underline{\mathbf{E}} \times \underline{\mathbf{H}}^*) \cdot \mathrm{d}S\,\hat{\mathbf{z}}\right\} = \frac{1}{Z}\int_{x=0}^{a}\int_{y=0}^{b}\left(|\underline{E}_x|^2 + |\underline{E}_y|^2\right)\underbrace{\mathrm{d}x\,\mathrm{d}y}_{\mathrm{d}S}$$

$$\text{(power flow along a waveguide)}, \quad (11.10)$$

with the x and y electric-field components, $\underline{E}_x$ and $\underline{E}_y$, of a TE or TM wave, that is, an arbitrary TE_{mn} or TM_{mn} mode, being given in Eqs.(11.4) and (11.5), and Z standing for the corresponding (TE or TM) wave impedance in Eq.(11.8) or (11.9).

As examples, solving (analytically) the pertinent integrals in x and y in Eq.(11.10), the transmitted power by the dominant mode, TE_{10}, and that for an arbitrary TM_{mn} wave mode come out to be

$$P = \eta\,\frac{ab}{2}\,\frac{f^2}{f_\mathrm{c}^2}\sqrt{1-\frac{f_\mathrm{c}^2}{f^2}}\,|\underline{H}_0|^2 \;\; (\mathrm{TE}_{10}) \quad \text{and} \quad P = \frac{ab}{4\eta}\,\frac{f^2}{f_\mathrm{c}^2}\sqrt{1-\frac{f_\mathrm{c}^2}{f^2}}\,|\underline{E}_0|^2 \;\; (\mathrm{TM}_{mn}), \qquad (11.11)$$

respectively, where the cutoff frequency $f_\mathrm{c} = (f_\mathrm{c})_{mn}$ [or $f_\mathrm{c} = (f_\mathrm{c})_{10}$] is found from Eq.(11.6).

MATLAB EXERCISE 11.20 **Transmitted power along a waveguide, symbolic integration.** Write a function `transmittedPower()` to compute the time-average transmitted power (P) of an arbitrary TE or TM wave along a rectangular waveguide, in Fig.11.1, by symbolic integration of the magnitude of the transverse electric field squared, $|\underline{E}_\mathrm{t}|^2 = |\underline{E}_x|^2 + |\underline{E}_y|^2$, in Eq.(11.10). *(transmittedPower.m on IR)*

TUTORIAL:

Based on Eq.(11.10), the input arguments of function `transmittedPower` are symbolic expressions for the integrand $|\underline{E}_x|^2 + |\underline{E}_y|^2$, `FieldMagSquared`, wave impedance [Eqs.(11.8) and (11.9)], `Z`, and dimensions of the guide cross section, `a` and `b`, provided from the main MATLAB code.

At the beginning of the function, we introduce spatial coordinates x and y (`x` and `y`) as symbolic variables (by means of MATLAB function `syms`) of integration. Symbolic integration is performed applying MATLAB function `int` twice: first, to solve the integral in x, and then to integrate the result of that integration again – in y, in Eq.(11.10).

```
function [power] = transmittedPower(FieldMagSquared,Z,a,b)
syms x y;
I = int(FieldMagSquared,x,0,a);
power = 1/Z*int(I,y,0,b);
return;
```

MATLAB EXERCISE 11.21 **Symbolic computation of power, arbitrary TE mode.** Write a MATLAB code to obtain a symbolic expression for the time-average power carried by an arbitrary TE_{mn} wave mode along a rectangular waveguide, using function `transmittedPower` (from the previous MATLAB exercise). The input data to the code are the mode indices, m and

n, entered from the keyboard. Test the code for the dominant (TE_{10}) waveguide mode; compare the result with that in Eqs.(11.11). *(ME11_21.m on IR)*

TUTORIAL:

The first step is data entry from the keyboard,

```
disp('Select TEmn mode:');
m = input('Enter index m:  ');
n = input('Enter index n:  ');
```

Next, we specify variables for symbolic computation: x and y coordinates in Fig.11.1 (`x` and `y`), waveguide dimensions a and b (`a` and `b`), complex constant $\underline{H}_0$ present in field expressions (`H0`), dielectric parameters ε and μ, (`EPS` and `MU`), intrinsic impedance η (`ETA`), and operating and cutoff frequencies, f and f_c (`f` and `fc`), respectively.

```
syms x y a b H0 EPS MU ETA f fc
```

The rest of the code is written in an `if-else` pattern to rule out an "unallowed" situation $m = n = 0$, for which the following error message is displayed in the Command Window: `'Error - mode TE00 does not exist!'`. Otherwise, if either m or n are different from zero [expressed in MATLAB notation as `m ~= 0 || n ~= 0`], symbolic expressions for the k parameter (`k`) and wave impedance of the given TE mode are obtained applying functions `kParameter` and `TEWaveImpedance` from MATLAB Exercises 11.7 and 11.17, respectively, and the integrand $|\underline{E}_x|^2 + |\underline{E}_y|^2$ in Eq.(11.10) is computed, in symbolic form, from Eqs.(11.4). The transmitted power along the waveguide is then evaluated, symbolically, invoking function `transmittedPower` – to carry out symbolic integration in Eq.(11.10).

```
if m ~= 0 || n ~= 0
c = 1/sqrt(EPS*MU);
w = 2*pi*f;
ZTE = TEWaveImpedance(ETA,fc,f);
k = kParameter(a,b,m,n);
Const1 = w*MU*n*pi*abs(H0)/(k^2*b);
ExMagSquared = (Const1*cos(m*pi*x/a)*sin(n*pi*y/b))^2;
Const2 = w*MU*m*pi*abs(H0)/(k^2*a);
EyMagSquared = (Const2*sin(m*pi*x/a)*cos(n*pi*y/b))^2;
EtSquared = ExMagSquared + EyMagSquared;
P = transmittedPower(EtSquared,ZTE,a,b);
```

Symbolic expressions for `P` and `fc`, where the latter one is found using function `cutoffFreq` (from MATLAB Exercise 11.10), are displayed in the Command Window by MATLAB function `pretty`,

```
disp('P = ');
pretty(P);
fprintf('\n');
fc = cutoffFreq(c,a,b,m,n);
disp('where fc is:  ');
pretty(fc);
```

Finally, we close the **if-else** statement displaying the error message (if $m = n = 0$),

```
else
disp('Error - mode TE00 does not exist!');
end;
```

MATLAB EXERCISE 11.22 **Symbolic computation of power, arbitrary TM mode.** Repeat the previous MATLAB exercise but for an arbitrary TM_{mn} waveguide mode. Test the code for the lowest mode (TM_{11}); compare the result with the second expression in Eqs.(11.11) for $f_\mathrm{c} = (f_\mathrm{c})_{11}$. *(ME11_22.m on IR)*

MATLAB EXERCISE 11.23 **Power transfer by TE_{02}, TE_{11}, and TM_{21} wave modes.** (a) Compute the (real) power flow through an arbitrary cross section of a rectangular metallic waveguide (Fig.11.1) associated with TE_{02} and TE_{11} waves, respectively, traveling along the structure. Perform the computation using symbolic programming in MATLAB, namely, running the code developed in MATLAB Exercise 11.21. Compare the results with the analytical solution to the problem, for the two modes. (b) Repeat (a) but for a TM_{21} wave propagating along the waveguide: run the code written in the previous MATLAB Exercise for $m = 2$ and $n = 1$. Compare the obtained symbolic expression to that in Eqs.(11.11) (second equation) for $f_\mathrm{c} = (f_\mathrm{c})_{21}$. **H**

11.5 Waveguides With Small Losses

To take into account conductor and dielectric losses in a rectangular waveguide (Fig.11.1), with a TE or TM wave, we assume that these losses are small, i.e., that the conditions in Eqs.(9.10) are satisfied. In place of Eqs.(9.12), the attenuation coefficient for the waveguide conductor is given by

$$\alpha_\mathrm{c} = \frac{P'_\mathrm{c}}{2P}\,, \quad P'_\mathrm{c} = \oint_C R_\mathrm{s}|\underline{\mathbf{H}}_\mathrm{tang}|^2\,\mathrm{d}l \quad \text{(losses in waveguide walls)}\,, \tag{11.12}$$

where P'_c is the time-average power of Joule's losses in the conductor (i.e., in four waveguide walls) per unit length of the structure and P is the time-average power transmitted along the guide, Eq.(11.10). Carrying out the indicated integration along the interior contour C of the conductor, we obtain the following expression for α_c of the dominant mode [the expression for P is that in Eqs.(11.11)]:

$$\alpha_\mathrm{c} = \frac{R_\mathrm{s}}{\eta a}\frac{a/b + 2f_\mathrm{c}^2/f^2}{\sqrt{1 - f_\mathrm{c}^2/f^2}} \quad \left(R_\mathrm{s} = \sqrt{\frac{\pi\mu_\mathrm{c} f}{\sigma_\mathrm{c}}}\right) \quad (\alpha_\mathrm{c} - \mathrm{TE}_{10}\ \text{mode})\,. \tag{11.13}$$

On the other side, the attenuation coefficient for the waveguide dielectric is, as a generalization of Eqs.(9.13), computed as

$$\alpha_{\rm d} = \frac{P'_{\rm d}}{2P} , \quad P'_{\rm d} = \int_{S_{\rm d}} \sigma_{\rm d} |\underline{\mathbf{E}}|^2 \, {\rm d}S \quad \longrightarrow \quad \alpha_{\rm d} = \frac{\omega \tan \delta_{\rm d}}{2c\sqrt{1 - f_{\rm c}^2/f^2}} \quad \text{(any mode)} , \tag{11.14}$$

with $S_{\rm d}$ denoting a cross section of the dielectric, as in Eq.(11.10), $\sigma_{\rm d}$ and $\tan \delta_{\rm d}$ being, respectively, the conductivity and loss tangent of the dielectric, and the resulting expression for $\alpha_{\rm d}$ holding true for any TE or TM mode in the waveguide (with cutoff frequency $f_{\rm c}$).

MATLAB EXERCISE 11.24 **Usable frequency range of a standard waveguide.** From the frequency dependence of the attenuation coefficient $\alpha_{\rm c}$ for the dominant mode, in Eq.(11.13), it is obvious that the lower part of the dominant frequency range (ensuring a single-mode operation) of a standard waveguide ($a = 2b$) in Eq.(11.7) is unusable, due to a very high signal attenuation [for frequencies just above the cutoff ($f = f_{\rm c}^+$), $\alpha_{\rm c} \to \infty$]. On the other side, while considering frequencies close to the upper limit of the dominant range, it is always preferable to have some safety margin (frequency separation) with respect to the next higher order mode. Consequently, a good rule of thumb is to design a waveguide such that the operating frequency (or frequencies) of the dominant mode be within the following range:

$$1.25 \frac{c}{2a} = 0.625 \frac{c}{a} < f < 0.95 \frac{c}{a} \quad \text{(usable frequency range)} , \tag{11.15}$$

which is often called the usable frequency range of the waveguide. Write a function `usableFreqRange()` in MATLAB that returns the lower and upper frequency limits of the dominant frequency range and of the usable frequency range of a standard waveguide, as defined by Eqs.(11.7) and (11.15), respectively; it returns one of the two ranges at the time. The input to the function consists of the relative permittivity $\varepsilon_{\rm r}$ of the guide dielectric, which is assumed to be nonmagnetic ($\mu_{\rm r} = 1$), the larger transverse dimension of the waveguide, a, and a string determining the type of the frequency range (`'dominant'` or `'usable'`). *(usableFreqRange.m on IR)*

MATLAB EXERCISE 11.25 **Attenuation coefficient for the waveguide conductor, dominant mode.** Write a function `alphaCondTE10()` in MATLAB that computes the attenuation coefficient for the waveguide conductor, $\alpha_{\rm c}$, for the dominant (TE_{10}) mode in a rectangular waveguide (Fig.11.1), based on Eq.(11.13). The input arguments of the function are: waveguide transverse dimensions, a and b, relative permittivity and permeability of the guide dielectric, $\varepsilon_{\rm r}$ and $\mu_{\rm r}$, conductivity and relative permeability of the guide conductor, σ and $\mu_{\rm rc}$, and operating frequency of the waveguide, f (in general, `f` is a vector of frequencies). *(alphaCondTE10.m on IR)*

MATLAB EXERCISE 11.26 **Attenuation coefficient for the waveguide dielectric, any mode.** Write a function `alphaDiel()` in MATLAB that computes the attenuation coefficient for the waveguide dielectric, $\alpha_{\rm d}$, for any TE or TM mode in a rectangular waveguide (Fig.11.1), given by Eq.(11.14), for the following input data: dimensions a and b, dielectric parameters $\varepsilon_{\rm r}$, $\mu_{\rm r}$, and

$\tan\delta_d$ (loss tangent), mode indices, m and n, and operating frequency, f (vector). *(alphaDiel.m on IR)*

MATLAB EXERCISE 11.27 Plots of attenuation coefficients in the usable frequency range. Write a MATLAB code to plot the attenuation coefficients for the waveguide conductor and dielectric, α_c and α_d, respectively, versus frequency for the dominant mode in a standard ($a = 2$ cm, $b = 1$ cm) copper ($\sigma_c = 58$ MS/m, $\mu_c = \mu_0$) rectangular waveguide (Fig.11.1), filled with an imperfect dielectric ($\varepsilon_r = 4$, $\tan\delta_d = 10^{-4}$). The plots show α_c and α_d expressed in dB/m, in the usable frequency range of the waveguide. *(ME11_27.m on IR)*

HINT:
Use functions `usableFreqRange`, `alphaCondTE10`, and `alphaDiel` (from the previous three MATLAB exercises), as well as `Np2dB` (conversion from Np/m to dB/m) from MATLAB Exercise 7.14. The resulting plots are shown in Fig.11.8.

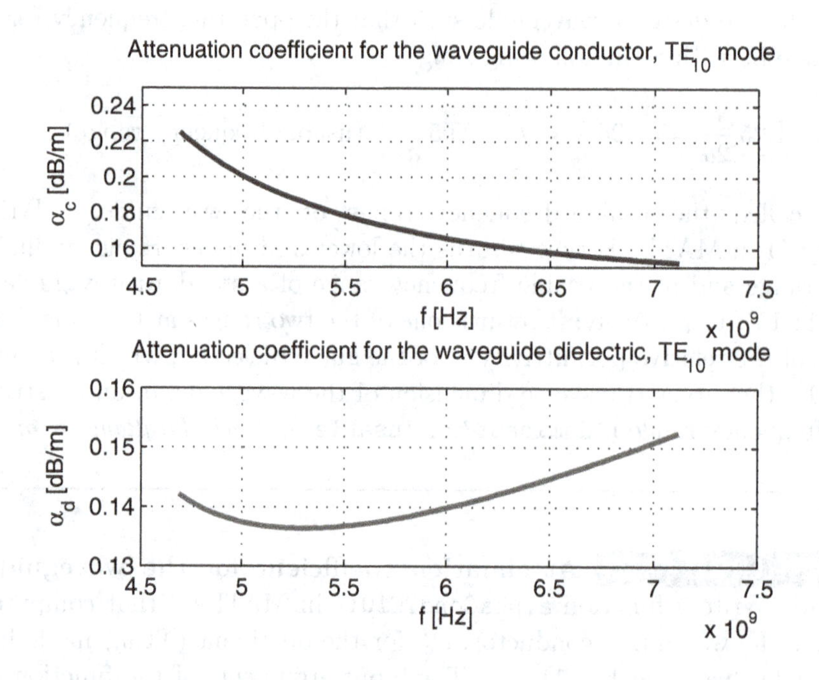

Figure 11.8 Attenuation coefficients α_c and α_d versus frequency for the TE_{10} mode in the usable frequency range of a standard waveguide; for MATLAB Exercise 11.27. *(color figure on CW)*

11.6 Waveguide Dispersion and Group Velocity

Since the phase coefficient β in Eq.(11.2), of the rectangular waveguide in Fig.11.1, is a nonlinear function of the angular frequency, ω, of a propagating TE or TM wave, the phase velocity of the wave, v_p, is frequency dependent,

$$v_\mathrm{p} = \frac{\omega}{\beta} = \frac{c}{\sqrt{1 - f_\mathrm{c}^2/f^2}} \quad \text{(waveguide phase velocity)} \tag{11.16}$$

($c = 1/\sqrt{\varepsilon\mu}$), and the waveguide represents a dispersive propagation medium. Because of the waveguide dispersion, the velocity of travel of electromagnetic energy and information carried by an electromagnetic (TE or TM) wave through the waveguide, called the group velocity (also energy velocity or signal velocity) and denoted as v_g, is different from v_p, and is computed, from Eq.(11.2), as

$$v_\mathrm{g} = \frac{1}{\mathrm{d}\beta/\mathrm{d}\omega} = c\sqrt{1 - \frac{f_\mathrm{c}^2}{f^2}} \quad \text{(waveguide group velocity)} \ . \tag{11.17}$$

MATLAB EXERCISE 11.28 **Phase and group velocities in a rectangular waveguide.** Write functions `waveguidePhaseVelocity()` and `waveguide- GroupVelocity()` in MATLAB to compute the phase velocity (v_p) and group velocity (v_g), respectively, of a propagating wave in a rectangular waveguide (Fig.11.1), using Eqs.(11.16) and (11.17). The input to the functions consists of the intrinsic phase velocity of the waveguide dielectric (c), operating frequency of the waveguide (f), in general, as a vector of values, and cutoff frequency of the waveguide for a particular mode (f_c). *(waveguidePhaseVelocity.m and waveguideGroupVelocity.m on IR)*

MATLAB EXERCISE 11.29 **Plots of phase and group velocities vs. frequency, several modes.** Write a MATLAB program that calculates and plots the phase and group velocities in a rectangular waveguide, Fig.11.1, versus frequency, in the range $f_\mathrm{c} \leq f \leq 10f_\mathrm{c}$, for TE_{10}, TE_{01}, and TE_{11} modes, respectively – all in the same graph, where f_c is the cutoff frequency of the mode. Generate the graph for a standard rectangular air-filled waveguide with $a = 2b = 6$ cm. *(ME11_29.m on IR)*

HINT:
Use functions `waveguidePhaseVelocity`, `waveguideGroupVelocity`, and `cutoff- Freq` (from MATLAB Exercises 11.28 and 11.10, respectively). The resulting graph is shown in Fig.11.9.

MATLAB EXERCISE 11.30 **Plots of phase and group velocities vs. frequency, several dielectrics.** In MATLAB, plot the phase and group velocities in a standard waveguide with $a = 2b = 6$ cm for the dominant wave mode (TE_{10}) in the frequency range $f_\mathrm{c} \leq f \leq 10f_\mathrm{c}$, if the guide is filled with air ($\varepsilon_\mathrm{r} = 1$), teflon ($\varepsilon_\mathrm{r} = 2.1$), polystyrene ($\varepsilon_\mathrm{r} = 2.56$), and quartz ($\varepsilon_\mathrm{r} = 5$), respectively – all in the same graph, where f_c is the cutoff frequency for the given dielectric. *(ME11_30.m on IR)*

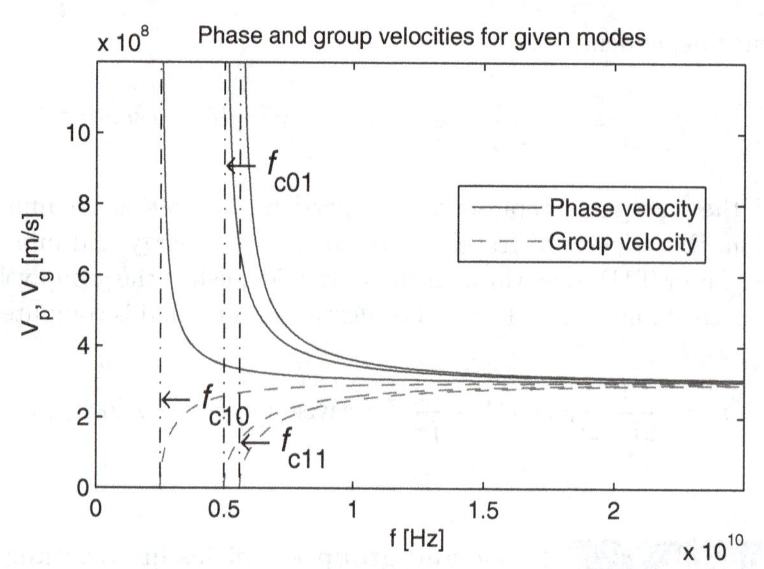

Figure 11.9 Frequency variation of phase and group velocities for three modes in a standard air-filled waveguide; for MATLAB Exercise 11.29. *(color figure on CW)*

11.7 Rectangular Cavity Resonators

Next, we study electromagnetic resonators made from rectangular metallic waveguides with TE or TM waves. We recall that the Fabry–Perot resonator (MATLAB Exercise 8.6) is obtained, essentially, by placing two parallel metallic (PEC) planes, that are a multiple of half-wavelengths apart, in the field of a uniform plane electromagnetic wave, perpendicularly to the direction of wave propagation, so that the resultant standing plane wave exists trapped between the two planes. Similarly, a section of a waveguide closed at both ends with new transversal conducting walls, thus forming a rectangular metallic box (cavity), as shown in Fig.11.10, represents a three-dimensional resonant wave structure (at certain resonant frequencies), called a rectangular cavity resonator. Waveguide cavity resonators are typically used at frequencies higher than 1 GHz, and are important elements in a wide range of microwave applications, including oscillator circuits, filters, tuned amplifiers, frequency-meters (wavemeters), high-field generators, and microwave ovens. Laser cavities are also cavity resonators.

Combining the condition $\beta d = p\pi$ $(p = 1, 2, \ldots)$, as in Eq.(8.6), with the expressions for the waveguide phase coefficient, β, in Eq.(11.2), and the cutoff frequency of an arbitrary TE_{mn} or TM_{mn} mode, $f_\mathrm{c} = (f_\mathrm{c})_{mn}$, in Eq.(11.6), we obtain the following expression for the resonant frequency (f_res) of a mode (m, n, p), i.e., TE_{mnp} or TM_{mnp}, in a cavity of dimensions a, b, and d and dielectric parameters ε and μ (Fig.11.10):

$$f = f_\mathrm{res} = (f_\mathrm{res})_{mnp} = \frac{c}{2}\sqrt{\left(\frac{m}{a}\right)^2 + \left(\frac{n}{b}\right)^2 + \left(\frac{p}{d}\right)^2} \quad (m, n, p = 0, 1, 2, \ldots)$$

$$\text{(cavity resonance, } \mathrm{TE}_{mnp} \text{ or } \mathrm{TM}_{mnp} \text{ mode)}. \quad (11.18)$$

If the cavity dimensions are not all the same and coordinate axes (x, y, and z) in Fig.11.10 are chosen such that $a > b$ and $d > b$, out of all solutions the TE_{101} mode ($m = p = 1$, $n = 0$) has the

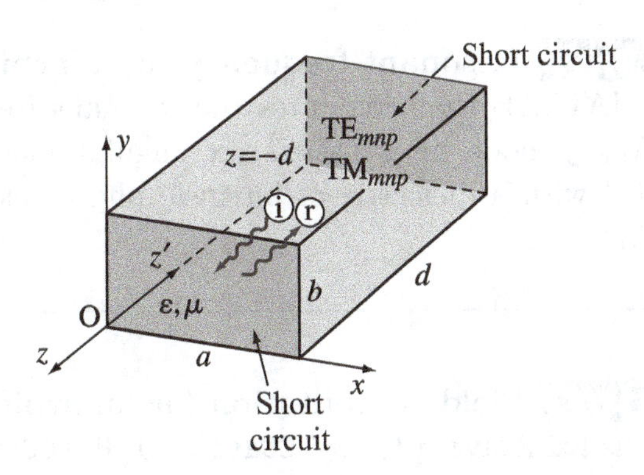

Figure 11.10 Rectangular cavity resonator, obtained by short-circuiting the rectangular metallic waveguide in Fig.11.1 in two transversal planes.

lowest frequency, given by

$$(f_{\rm res})_{101} = \frac{c}{2}\sqrt{\frac{1}{a^2}+\frac{1}{d^2}} \quad (\mathrm{TE}_{101},\ \text{dominant cavity mode})\,, \tag{11.19}$$

and is hence termed the dominant cavity mode.

Considering a TE$_{mn}$ wave described by Eqs.(11.4) as the incident wave that reflects at PEC boundaries $z = 0$ and $z = -d$ in Fig.11.10, and computing the resultant (incident plus reflected) wave, as in Eqs.(8.4), we get the expressions for the electric and magnetic fields of an arbitrary TE$_{mnp}$ wave mode in the cavity:

$$\underline{E}_x = \frac{2\omega\mu}{k^2}\frac{n\pi}{b}\underline{H}_0\cos\left(\frac{m\pi}{a}x\right)\sin\left(\frac{n\pi}{b}y\right)\sin\left(\frac{p\pi}{d}z\right), \quad \underline{E}_y = -\frac{2\omega\mu}{k^2}\frac{m\pi}{a}\underline{H}_0\sin\left(\frac{m\pi}{a}x\right)$$
$$\times\cos\left(\frac{n\pi}{b}y\right)\sin\left(\frac{p\pi}{d}z\right), \quad \underline{E}_z = 0\,, \quad \underline{H}_x = \frac{2\mathrm{j}}{k^2}\frac{m\pi}{a}\frac{p\pi}{d}\underline{H}_0\sin\left(\frac{m\pi}{a}x\right)\cos\left(\frac{n\pi}{b}y\right)\cos\left(\frac{p\pi}{d}z\right),$$
$$\underline{H}_y = \frac{2\mathrm{j}}{k^2}\frac{n\pi}{b}\frac{p\pi}{d}\underline{H}_0\cos\left(\frac{m\pi}{a}x\right)\sin\left(\frac{n\pi}{b}y\right)\cos\left(\frac{p\pi}{d}z\right), \quad \underline{H}_z = -2\mathrm{j}\underline{H}_0\cos\left(\frac{m\pi}{a}x\right)\cos\left(\frac{n\pi}{b}y\right)$$
$$\times\sin\left(\frac{p\pi}{d}z\right) \quad (\mathrm{TE}_{mnp}\ \text{mode};\ m,n,p = 0,1,2,\ldots)\,, \tag{11.20}$$

where the parameter k (k^2) is the same as in Eqs.(11.4), and $\omega = 2\pi f$, with $f = (f_{\rm res})_{mnp}$ being the resonant frequency of the TE$_{mnp}$ mode in the cavity, Eq.(11.18). Similarly, the field components of an arbitrary TM$_{mnp}$ resonance mode in the cavity come out to be

$$\underline{E}_x = -\frac{2}{k^2}\frac{m\pi}{a}\frac{p\pi}{d}\underline{E}_0\cos\left(\frac{m\pi}{a}x\right)\sin\left(\frac{n\pi}{b}y\right)\sin\left(\frac{p\pi}{d}z\right), \quad \underline{E}_y = -\frac{2}{k^2}\frac{n\pi}{b}\frac{p\pi}{d}\underline{E}_0\sin\left(\frac{m\pi}{a}x\right)$$
$$\times\cos\left(\frac{n\pi}{b}y\right)\sin\left(\frac{p\pi}{d}z\right), \quad \underline{E}_z = 2\underline{E}_0\sin\left(\frac{m\pi}{a}x\right)\sin\left(\frac{n\pi}{b}y\right)\cos\left(\frac{p\pi}{d}z\right),$$
$$\underline{H}_x = \frac{2\mathrm{j}\omega\varepsilon}{k^2}\frac{n\pi}{b}\underline{E}_0\sin\left(\frac{m\pi}{a}x\right)\cos\left(\frac{n\pi}{b}y\right)\cos\left(\frac{p\pi}{d}z\right), \quad \underline{H}_y = -\frac{2\mathrm{j}\omega\varepsilon}{k^2}\frac{m\pi}{a}\underline{E}_0\cos\left(\frac{m\pi}{a}x\right)$$
$$\times\sin\left(\frac{n\pi}{b}y\right)\cos\left(\frac{p\pi}{d}z\right), \quad \underline{H}_z = 0 \quad (\mathrm{TM}_{mnp}\ \text{mode};\ m,n,p = 1,2,\ldots)\,. \tag{11.21}$$

MATLAB EXERCISE 11.31 **Resonant frequency of an arbitrary cavity mode.** Write a function `resoFreq()` in MATLAB that computes the resonant frequency, $f_{\rm res} = (f_{\rm res})_{mnp}$, of an arbitrary TE_{mnp} or TM_{mnp} mode in a lossless rectangular metallic cavity resonator, with dimensions a, b, and d, filled with a dielectric of intrinsic phase velocity c, Fig.11.10, based on Eq.(11.18). *(resoFreq.m on IR)*

MATLAB EXERCISE 11.32 **Field computation for an arbitrary TE mode in a resonant cavity.** Write functions `ExTECavity()`, `EyTECavity()`, `HxTECavity()`, `HyTECavity()`, and `HzTECavity()` in MATLAB that calculate, respectively, the magnitudes of x- and y-components of the electric field vector, $|\underline{E}_x|$ and $|\underline{E}_y|$, and magnitudes of x-, y-, and z-components of the magnetic field vector, $|\underline{H}_x|$, $|\underline{H}_y|$, and $|\underline{H}_z|$, of an arbitrary TE_{mnp} wave mode in the resonant cavity described in the previous MATLAB exercise. *(ExTECavity.m, EyTECavity.m, HxTECavity.m, HyTECavity.m, and HzTECavity.m on IR)*

HINT:
Use Eqs.(11.20) and see MATLAB Exercise 11.8.

MATLAB EXERCISE 11.33 **GUI with field plots for the dominant cavity mode.** Apply (modify) GUI WaveguideField from MATLAB Exercise 11.14 to plot all nonzero field components of the dominant cavity mode (TE_{101}) in an air-filled rectangular metallic cavity resonator (Fig.11.10) of dimensions $a = 16$ cm, $b = 8$ cm, and $d = 18$ cm, assuming that $\underline{H}_0 = 1\,\mathrm{e}^{\mathrm{j}0}$ A/m. *[folder ME11_33(GUI) on IR]*

HINT:
See MATLAB Exercise 11.15. Use functions `resoFreq`, `kParameter`, `EyTECavity`, `HxTECavity`, and `HzTECavity` (from MATLAB Exercises 11.31, 11.7, and 11.32, respectively) with $m = 1$, $n = 0$, and $p = 1$. A 3-D plot of the electric field distribution (y-component of the electric field vector) in the cavity by MATLAB function `surf` as the selected plot type is shown in Fig.11.11.

11.8 Electromagnetic Energy Stored in a Cavity Resonator

During the course of time, the stored electromagnetic energy in a rectangular cavity resonator, Fig.11.10, like in any resonant electromagnetic structure (e.g., a simple resonant LC circuit), periodically oscillates between the electric and magnetic fields, as they alternate between maximum and zero values. In the case of the Fabry–Perot resonator (MATLAB Exercise 8.6), such energy fluctuation can be observed by playing the movie of instantaneous electric and magnetic energy densities, $w_{\rm e}$ and $w_{\rm m}$, of the standing plane wave developed in MATLAB Exercise 8.7. It is, essentially, a consequence of a 90° phase shift between instantaneous electric and magnetic fields (difference in "j" in complex field expressions) at every point of the resonant structure. Namely,

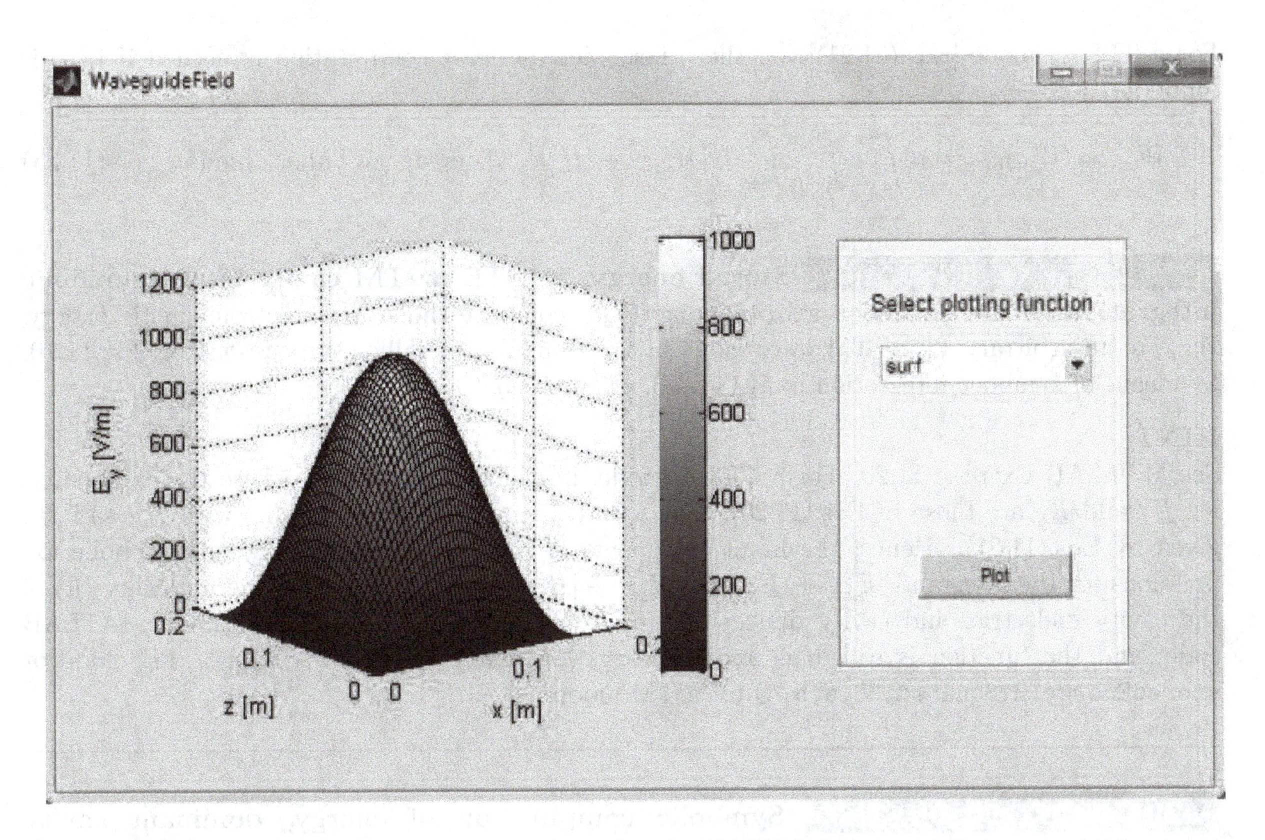

Figure 11.11 Using GUI with three types of field plots, by MATLAB functions `surf`, `pcolor`, and `contour`, respectively, developed in MATLAB Exercise 11.14, to visualize the electric field ($|\underline{E}_y|$) of the TE_{101} wave mode in a cavity resonator; for MATLAB Exercise 11.33. *(color figure on CW)*

for the resonant cavity, the complex electric and magnetic field vectors of an arbitrary (TE_{mnp} or TM_{mnp}) cavity mode, given by Eqs.(11.20) and (11.21), can be written as $\underline{\mathbf{E}} = \underline{H}_0\mathbf{A}_1$ and $\underline{\mathbf{H}} = \mathrm{j}\underline{H}_0\mathbf{A}_2$, where $\mathbf{A}_1$ and $\mathbf{A}_2$ are purely real vectors, and the complex constant $\underline{H}_0$ is substituted by $\underline{E}_0$ for TM waves. So, there are instants of time (t_1) at which the electric field is maximum, while the magnetic field is zero, at every point (x, y, z) in the cavity; at some other times ($t_2 = t_1 \pm T/4$, $T = 1/f$), the situation is just opposite. In other words, the instantaneous electromagnetic energy of the cavity, $W_{\text{em}}(t)$, which is constant in time (assuming no Joule's losses in the resonator), is all electric at instants t_1, and all magnetic for $t = t_2$. At intermediate times, the energy is partly electric and partly magnetic, as it moves from the electric to the magnetic field, and vice versa.

For a TE_{mnp} mode ($\underline{E}_z = 0$), W_{em} is most simply obtained at instants of time (t_1) when it is all electric, $W_{\text{em}} = (W_{\text{e}})_{\text{max}}$ ($W_{\text{m}} = 0$), by integrating the peak electric energy density throughout the volume of the cavity dielectric, v_{d}, that is, the entire cavity interior (Fig.11.10). In addition, the electric field vector in the cavity [Eqs.(11.20)] is linearly polarized, so that the peak-value (E_{max}) equals $\sqrt{2}$ times rms (E_{rms}) of its instantaneous magnitude, and we have

$$W_{\text{em}} = (W_{\text{e}})_{\text{max}} = \int_{v_{\text{d}}} (w_{\text{e}})_{\text{max}}\,\mathrm{d}v = \int_{v_{\text{d}}} \frac{1}{2}\,\varepsilon E_{\text{max}}^2\,\mathrm{d}v = \varepsilon \int_{x=0}^{a}\int_{y=0}^{b}\int_{z=-d}^{0} \left(|\underline{E}_x|^2 + |\underline{E}_y|^2\right) \underbrace{\mathrm{d}x\,\mathrm{d}y\,\mathrm{d}z}_{\mathrm{d}v}$$

$$\text{(stored energy, TE}_{mnp}\text{ mode)}\,. \qquad (11.22)$$

For a TM_{mnp} mode [Eqs.(11.21)], on the other side, the least computation is needed if W_{em} is found as

$$W_{\mathrm{em}} = (W_{\mathrm{m}})_{\mathrm{max}} = \mu \int_{x=0}^{a} \int_{y=0}^{b} \int_{z=-d}^{0} \left(|\underline{H}_x|^2 + |\underline{H}_y|^2 \right) \mathrm{d}x\, \mathrm{d}y\, \mathrm{d}z \quad (\mathrm{TM}_{mnp} \text{ mode}) . \qquad (11.23)$$

MATLAB EXERCISE 11.34 Stored energy, any TE or TM cavity mode, symbolic integration. Write a function `storedEnergy()` to compute the stored electromagnetic energy, W_{em}, of an arbitrary TE or TM wave mode in a rectangular metallic cavity resonator, Fig.11.10, by means of symbolic integration in MATLAB. *(storedEnergy.m on IR)*

HINT:
See MATLAB Exercise 11.20. For a TE_{mnp} mode, implement Eq.(11.22), where the expressions for $\underline{E}_x$ and $\underline{E}_y$ are those in Eqs.(11.20). For a TM_{mnp} mode, use Eq.(11.23), with $\underline{H}_x$ and $\underline{H}_y$ given by Eqs.(11.21). Hence, the input arguments of function `storedEnergy` are symbolic expressions for the integrand $|\underline{E}_x|^2 + |\underline{E}_y|^2$ or $|\underline{H}_x|^2 + |\underline{H}_y|^2$, permittivity ($\varepsilon$) or permeability ($\mu$) of the cavity dielectric, and cavity dimensions (a, b, and d), provided from the main MATLAB code, and the function is called as `storedEnergy(EtSquared,EPS,a,b,d)` for a TE mode or `storedEnergy(HtSquared,MU,a,b,d)` for a TM mode.

MATLAB EXERCISE 11.35 Symbolic computation of energy, dominant cavity mode. Write a MATLAB code to obtain a symbolic expression for the electromagnetic energy of the TE_{101} mode stored in a rectangular cavity resonator. *(ME11_35.m on IR)* **H**

TUTORIAL:
See MATLAB Exercise 11.21. The integrand $|\underline{E}_y|^2$ ($\underline{E}_x = 0$ for the TE_{101} mode) in Eq.(11.22) is computed, in symbolic form, from Eqs.(11.20) with $m = p = 1$ and $n = 0$,

```
m = 1;
n = 0;
p = 1;
syms x y z a b d H0 f EPS MU
w = 2*pi*f;
k = kParameter(a,b,m,n);
Const1 = 2*w*MU*n*pi*abs(H0)/(k^2*b);
ExMagSquared = (Const1*cos(m*pi*x/a)*sin(n*pi*y/b)*sin(p*pi*z/d))^2;
Const2 = 2*w*MU*m*pi*abs(H0)/(k^2*a);
EyMagSquared = (Const2*sin(m*pi*x/a)*cos(n*pi*y/b)*sin(p*pi*z/d))^2;
EtSquared = ExMagSquared + EyMagSquared;
```

The stored energy is then evaluated, symbolically, applying function `storedEnergy` (from the previous MATLAB exercise) – to perform symbolic integration in Eq.(11.22), as follows:

```
Wem = storedEnergy(EtSquared,EPS,a,b,d);
disp('Wem = ');
pretty(Wem);
```

```
fprintf('\n');
c = 1/sqrt(EPS*MU);
f = resoFreq(c,a,b,d,m,n,p);
disp('where f = (fres)101:  ');
pretty(f);
```

The resulting symbolic expression, as displayed in the Command Window, is shown in Fig.11.12. We see that the displayed associated expression for the resonant frequency of the dominant cavity mode, $f = (f_{\mathrm{res}})_{101}$, is that in Eq.(11.19).

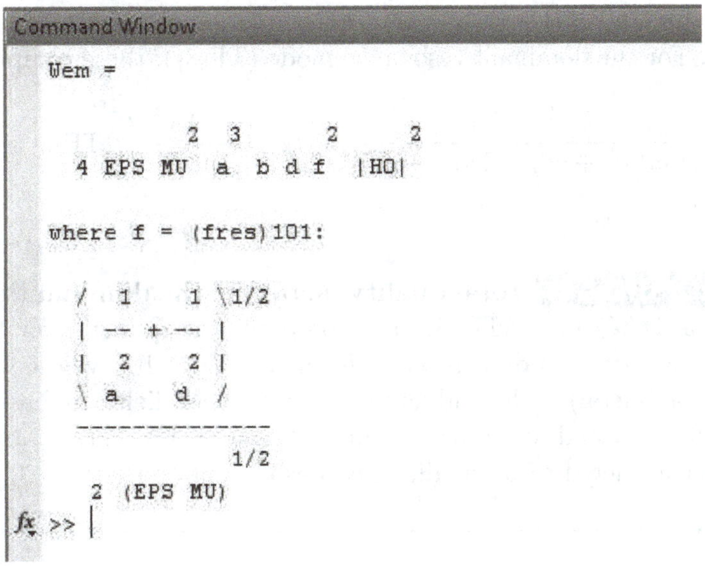

```
Command Window
Wem =

          2  3      2     2
 4 EPS MU  a  b d f  |H0|

where f = (fres)101:

 / 1    1 \1/2
 | -- + -- |
 |  2    2 |
 \ a    d /
 ---------------
             1/2
   2 (EPS MU)
fx >>
```

Figure 11.12 Display in the Command Window in MATLAB of the obtained symbolic expression for the electromagnetic energy of the TE_{101} cavity mode in the resonator in Fig.11.10; for MATLAB Exercise 11.35. *(color figure on CW)*

MATLAB EXERCISE 11.36 **Stored energy of a TM_{111} wave in a rectangular cavity.** Using symbolic programming in MATLAB, obtain the expression for the stored electromagnetic energy of a TM_{111} wave in a rectangular metallic cavity resonator, Fig.11.10. *(ME11_36.m on IR)*
H

HINT:
See the previous MATLAB exercise.

11.9 Quality Factor of Rectangular Cavities with Small Losses

Neglecting the losses in a cavity resonator, its electromagnetic energy, $W_{\mathrm{em}}(t)$, once established remains the same indefinitely (to $t \to \infty$). In a real (lossy) resonator, on the other hand, $W_{\mathrm{em}}(t)$

decreases exponentially with time, as $e^{-2t/\tau}$, where the time constant τ is proportional to the so-called quality factor, Q, of the structure, so the higher the Q the slower the damping of the resonator ($Q \to \infty$ for an ideal resonator). For cavity resonators with small losses, Q can be obtained as follows:

$$Q = \frac{Q_c Q_d}{Q_c + Q_d} \, , \quad Q_c = \omega_{res} \frac{W_{em}}{P_c} \, , \quad Q_d = \omega_{res} \frac{W_{em}}{P_d} \quad \text{(quality factor)} \tag{11.24}$$

($\omega_{res} = 2\pi f_{res}$), where the time-average power of Joule's losses in the cavity conductor (metallic walls), P_c, and that for the imperfect dielectric in Fig.11.10, P_d, are computed similarly to the evaluation of conductor and dielectric per-unit-length losses in a waveguide in Eqs.(11.12) and (11.14), respectively. For the dominant resonance mode (TE_{101}), these computations give

$$Q_c = \frac{\pi\eta}{2R_s(f_{res})} \frac{b(a^2+d^2)^{3/2}}{ad(a^2+d^2)+2b(a^3+d^3)} \, , \quad Q_d = \frac{1}{\tan\delta_d} \quad (TE_{101} \text{ mode}) \, . \tag{11.25}$$

MATLAB EXERCISE 11.37 **Total quality factor for the dominant cavity mode.** Write a function `QFactorCavity()` in MATLAB that computes the Q factor for the TE_{101} resonance mode of a rectangular cavity, with dimensions a, b, and d (Fig.11.10), whose walls are made from a nonmagnetic conductor (metal) with conductivity σ and whose dielectric has relative permittivity ε_r, relative permeability μ_r, and loss tangent $\tan\delta_d$, based on Eqs.(11.24) and (11.25). Losses in the resonator can be considered to be small. *(QFactorCavity.m on IR)*

---

12 ANTENNAS AND WIRELESS COMMUNICATION SYSTEMS

Introduction:

Although any conductor with a time-varying (e.g., time-harmonic) current (Chapter 6) radiates electromagnetic energy into the surrounding space, some conductor configurations are specially designed to maximize electromagnetic radiation, in desired directions at given frequencies. Such systems of conductors, which sometimes also include dielectric parts, are called antennas. In other words, antennas are electromagnetic devices designed and built to provide a means of efficient transmitting or receiving of radio waves. More precisely, they provide transition from a guided electromagnetic wave, in a transmission line (Chapters 9 and 10) or waveguide (Chapter 11) feeding the antenna, to a radiated unbounded electromagnetic wave (in free space or other ambient medium) in the transmitting (radiating) mode of operation, and vice versa for an antenna operating in the receiving mode. In many discussions, we shall study not only antennas but wireless communication systems with antennas at the two ends. In this, we shall use concepts and equations describing the propagation of uniform plane electromagnetic waves, from Chapters 7 and 8.

12.1 Electromagnetic Field due to a Hertzian Dipole

Consider the simplest antenna, a so-called Hertzian dipole, which is an electrically short ($l \ll \lambda$) straight metallic wire segment with a time-varying current that does not change along the wire, as shown in Fig.12.1. Assuming a time-harmonic regime of the dipole, let it be fed at its center by a lumped generator, of frequency f, and let its complex rms current intensity be $\underline{I}$. As the current is nonzero at the wire ends, it must be terminated by charges $\underline{Q}$ and $-\underline{Q}$ that accumulate on a pair of small metallic spheres (Fig.12.1), or conductors of other shapes, attached to these ends. The complex magnetic vector potential at the point P in Fig.12.1 has the form of a spherical electromagnetic wave emanating from the dipole center, and it equals

$$\underline{\mathbf{A}} = \frac{\mu \underline{I}\, l\, \mathrm{e}^{-\mathrm{j}\beta r}}{4\pi r}\,\hat{\mathbf{z}} \quad (\beta = \omega\sqrt{\varepsilon\mu}) \quad \text{(magnetic potential of a Hertzian dipole)}\,, \tag{12.1}$$

where β is the phase coefficient (wavenumber) for the ambient medium and given operating frequency, Eq.(7.4). Using Eqs.(6.34), (4.24), and (5.2), we then obtain, from $\underline{\mathbf{A}}$, the following expressions, in the spherical coordinate system in Fig.12.1, for the complex electric and magnetic field intensity vectors of the antenna:

$$\underline{\mathbf{E}} = -\frac{\eta\beta^2 \underline{I}\, l\, \mathrm{e}^{-\mathrm{j}\beta r}}{4\pi}\left\{\left[\frac{1}{(\mathrm{j}\beta r)^2} + \frac{1}{(\mathrm{j}\beta r)^3}\right] 2\cos\theta\,\hat{\mathbf{r}} + \left[\frac{1}{\mathrm{j}\beta r} + \frac{1}{(\mathrm{j}\beta r)^2} + \frac{1}{(\mathrm{j}\beta r)^3}\right]\sin\theta\,\hat{\boldsymbol{\theta}}\right\},$$

$$\underline{\mathbf{H}} = -\frac{\beta^2 \underline{I}\, l\, \mathrm{e}^{-\mathrm{j}\beta r}\sin\theta}{4\pi}\left[\frac{1}{\mathrm{j}\beta r} + \frac{1}{(\mathrm{j}\beta r)^2}\right]\hat{\boldsymbol{\phi}} \quad \text{(field vectors of a dipole)}\,, \tag{12.2}$$

where $\eta = \sqrt{\mu/\varepsilon}$ is the intrinsic impedance of the medium, Eq.(7.2).

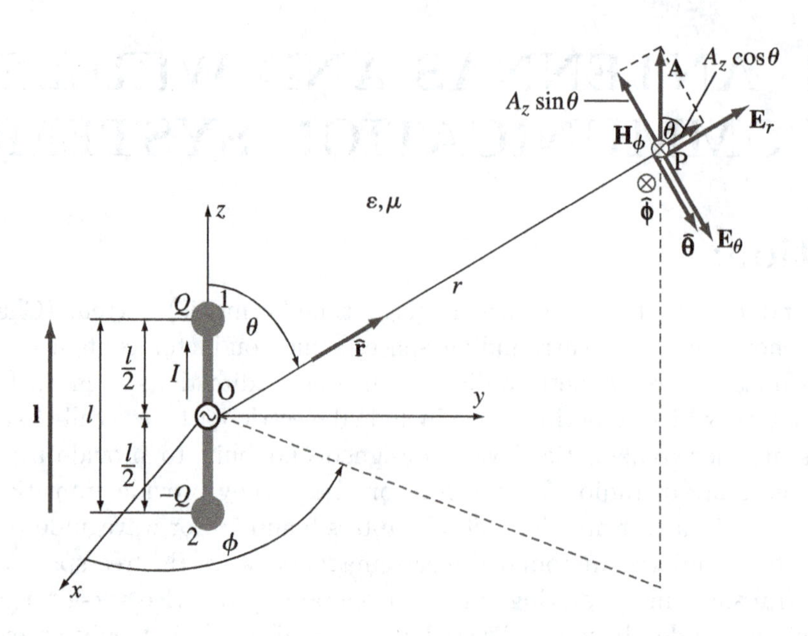

Figure 12.1 Hertzian dipole.

Finally, combining Eqs.(6.32) and (12.2), the complex Poynting vector at the point P in Fig.12.1 is

$$\underline{\boldsymbol{\mathcal{P}}} = \eta \left(\frac{\beta^2 I l}{4\pi}\right)^2 \left\{\left[\frac{\mathrm{j}}{(\beta r)^3} + \frac{\mathrm{j}}{(\beta r)^5}\right] \sin 2\theta\, \hat{\boldsymbol{\theta}} + \left[\frac{1}{(\beta r)^2} - \frac{\mathrm{j}}{(\beta r)^5}\right] \sin^2\theta\, \hat{\mathbf{r}}\right\} , \tag{12.3}$$

where $\underline{I}\underline{I}^* = |\underline{I}|^2 = I^2$, so that the time average of the instantaneous Poynting vector due to the antenna, Eq.(6.33), comes out to be

$$\boldsymbol{\mathcal{P}}_{\text{ave}} = \text{Re}\{\underline{\boldsymbol{\mathcal{P}}}\} = \frac{\eta\beta^2 I^2 l^2 \sin^2\theta}{16\pi^2 r^2}\, \hat{\mathbf{r}} \quad \text{(time-average Poynting vector)} . \tag{12.4}$$

MATLAB EXERCISE 12.1 **Field vectors due to a Hertzian dipole.** Write a function `hertzianDipoleFields()` in MATLAB that computes r-, θ-, and ϕ-components (in the spherical coordinate system) of the complex electric and magnetic field vectors, $\underline{\mathbf{E}}$ and $\underline{\mathbf{H}}$, due to a Hertzian dipole in Fig.12.1, based on Eqs.(12.2). The input to the function consists of the complex rms current intensity ($\underline{I}$), length (l), and operating frequency (f) of the dipole, the relative permittivity and permeability of the ambient medium, ε_r and μ_r, and the spherical coordinates (r, θ, ϕ) of the field (observation) point. *(hertzianDipoleFields.m on IR)*[1]

MATLAB EXERCISE 12.2 **Poynting vector due to a Hertzian dipole.** Repeat the previous MATLAB exercise but to compute r-, θ-, and ϕ-components of the complex and time-average Poynting vectors, $\underline{\boldsymbol{\mathcal{P}}}$ and $\boldsymbol{\mathcal{P}}_{\text{ave}}$, due to a Hertzian dipole in Fig.12.1 [function

[1] *IR* = Instructor Resources (for the book).

`hertzianDipolePoynting()`], using Eqs.(12.3) and (12.4), respectively. *(hertzianDipolePoynting.m on IR)*

12.2 Far Field

This section introduces an important special case of the electromagnetic field due to a Hertzian dipole (Fig.12.1): the far field, for observation locations that are electrically far away from the antenna. In specific, in the far zone the distance r of the field point P in Fig.12.1 from the origin is much larger than the operating wavelength λ of the dipole (for the ambient medium), given in Eqs.(7.5) and (7.4), βr is much larger than unity, and we can write

$$r \gg \lambda \quad \longrightarrow \quad \beta r \gg 1 \quad \longrightarrow \quad \frac{1}{\beta r} \gg \frac{1}{(\beta r)^2} \gg \frac{1}{(\beta r)^3} \quad \text{(far zone)} . \tag{12.5}$$

In practice, a useful rule of thumb quantifying the far-field condition is: $r > 10\lambda$. Therefore, the dominant terms in both field expressions in Eqs.(12.2) are those with the smallest inverse powers of r (or βr), that is, the $1/r$ terms. These expressions can thus be replaced by much simpler approximate ones as follows:

$$\underline{\mathbf{E}} \approx \frac{\mathrm{j}\eta\beta\underline{I}\,l\,\mathrm{e}^{-\mathrm{j}\beta r}\sin\theta}{4\pi r}\,\hat{\boldsymbol{\theta}}\,, \quad \underline{\mathbf{H}} \approx \frac{\mathrm{j}\beta\underline{I}\,l\,\mathrm{e}^{-\mathrm{j}\beta r}\sin\theta}{4\pi r}\,\hat{\boldsymbol{\phi}} \quad \text{(far fields, Hertzian dipole)} . \tag{12.6}$$

Since $\mathbf{E}$ and $\mathbf{H}$ in the far zone are in phase, the associated complex Poynting vector is purely real,

$$\underline{\mathcal{P}} = \underline{E}_\theta \underline{H}_\phi^* \,\hat{\mathbf{r}} = \frac{\eta\beta^2 I^2 l^2 \sin^2\theta}{16\pi^2 r^2}\,\hat{\mathbf{r}} = \mathcal{P}_{\text{ave}} \quad \text{(far-zone Poynting vector)} , \tag{12.7}$$

and $\underline{\mathcal{P}}$ equals the time-average Poynting vector, $\mathcal{P}_{\text{ave}}$, in Eq.(12.4).

MATLAB EXERCISE 12.3 Error in far electric field computation for a Hertzian dipole. Write a MATLAB program that compares the magnitude of the complex electric field vector due to a Hertzian dipole radiating in free space computed by the far-zone expression in Eqs.(12.6), E_{far}, to that obtained from the general expression in Eqs.(12.2), E. In specific, the relative error δ in the far-field computation is evaluated for $r = 30\lambda$ (λ is the operating free-space wavelength of the dipole) as a function of the zenith angle, θ, for $0 \le \theta \le 180°$ in Fig.12.1, and is defined as $\delta = |E_{\text{far}} - E|/E_{\max}$, where $E_{\max}$ is the maximum of all computed values of E. Find the maximum δ and the corresponding θ. Assume that the operating frequency of the dipole is $f = 1$ GHz and that the dipole length equals $l = \lambda/100$. *(ME12_3.m on IR)*

TUTORIAL:

First, we specify input data and necessary parameters for the analysis,

```
f = 10^9;
w = 2*pi*f;
MU0 = 4*pi*10^(-7);
```

```
ETA = 376.73;
BETA = w*MU0/ETA;
c = 299792458;
LAMBDA = c/f;
I = 1;
L = LAMBDA/100;
r = 30*LAMBDA;
```

Then, we define a spatial mesh in terms of the angle θ (array `theta`), and calculate, at all nodes, E (`absE`) and E_{far} (`absEfar`) based on Eqs.(12.2) and (12.6), respectively, as well as the maximum field, E_{max} (`absEmax`), relative error δ (`error`), and maximum error (`errormax`), where the two maxima are found using MATLAB function `max` (see MATLAB Exercise 8.11, for instance),

```
theta = 0:pi/100:pi;
Coeff = -ETA*BETA^2*I*L*exp(-i*BETA*r)/4/pi;
Er = Coeff*(1/(i*BETA*r)^2 + 1/(i*BETA*r)^3)*2*cos(theta);
Etheta = Coeff*(1/(i*BETA*r) + 1/(i*BETA*r)^2 + 1/(i*BETA*r)^3)*sin(theta);
absE = sqrt(abs(Er).^2 + abs(Etheta).^2);
absEfar = abs(i*ETA*BETA*I*L*exp(-i*BETA*r)*sin(theta)/(4*pi*r));
[absEmax,j] = max(absE);
error = abs(absEfar - absE)./absEmax *100;
[errormax,k] = max(error);
```

Note that `absE` can alternatively be obtained with the use of function `hertzianDipoleFields` written in MATLAB Exercise 12.1. Finally, we display the results for `errormax` and the corresponding angle θ in the Command Window, and plot the dependence of `error` on `theta`,

```
fprintf('Maximum relative error is:  %.4f %%\n',errormax);
fprintf('and it occurs for theta = %.2f rad\n',theta(k));
figure(1)
plot(theta,error);
title('Error in far electric field computation, Hertzian dipole');
xlabel('theta [rad]');
ylabel('Relative error (%)');
```

The maximum error in this case turns out to be 1.061% for $\theta = 0$.

MATLAB EXERCISE 12.4 **Error in far magnetic field computation.** Repeat the previous MATLAB exercise but for the magnitude of the complex magnetic field vector due to a Hertzian dipole, computed by the far-zone expression in Eqs.(12.6) and the general expression in Eqs.(12.2), respectively. *(ME12_4.m on IR)*

HINT:

Figure 12.2 shows the results for the error δ as a function of θ.

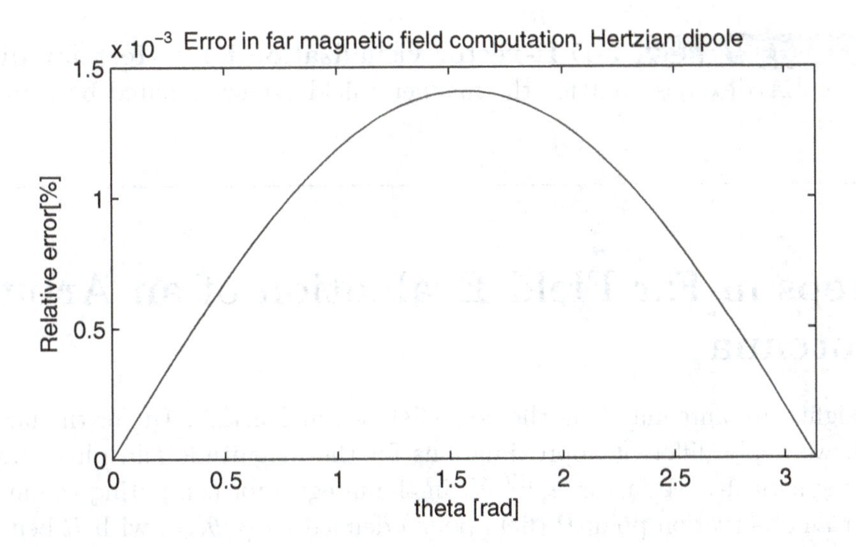

Figure 12.2 Relative error in far magnetic field computation for a Hertzian dipole: comparison of $|\underline{\mathbf{H}}|$ values obtained by the far-zone expression in Eqs.(12.6) and the general field expression in Eqs.(12.2) for $r = 30\lambda$ and $0 \le \theta \le 180°$ in Fig.12.1; for MATLAB Exercise 12.4. *(color figure on CW)*[2]

MATLAB EXERCISE 12.5 **Error in far-zone Poynting vector computation.** Repeat MATLAB Exercise 12.3 but for the magnitude of the complex Poynting vector due to a Hertzian dipole, obtained from the far-zone expression in Eq.(12.7) and the general expression in Eq.(12.3), respectively. Based on the results of this MATLAB exercise and the previous two, does the maximum error in the far-zone computation of $|\underline{\mathbf{E}}|$, $|\underline{\mathbf{H}}|$, and $|\underline{\mathcal{P}}|$ occur for the same θ? *(ME12_5.m on IR)*

MATLAB EXERCISE 12.6 **3-D E-vector visualization for a Hertzian dipole using `quiver3`.** For a Hertzian dipole radiating in free space, plot the distribution of its electric field, in vector form, over a sphere of radius $r = 20\lambda$ in Fig.12.1. *(ME12_6.m on IR)*

HINT:
Use the far-field expression in Eqs.(12.6) to compute the vector $\underline{\mathbf{E}}$ due to the dipole and MATLAB function `quiver3` (see MATLAB Exercise 1.3) to plot $\underline{\mathbf{E}}$. To convert the coordinates of nodes of the spherical mesh (of radius r), e.g., nodes defined by $r = 20\lambda$, $\theta = [0 : \pi/20 : \pi]$, and $\phi = [-\pi : \pi/20 : \pi]$, at which $\underline{\mathbf{E}}$ is computed and plotted, from spherical to Cartesian coordinates (needed for `quiver3`), employ function `sph2Car` (MATLAB Exercise 1.24). Use the rotation button in MATLAB to observe the vector field distribution [both direction (polarization) and magnitude of $\underline{\mathbf{E}}$] of the dipole at different views in the plot.

---

[2] CW = Companion Website (of the book).

MATLAB EXERCISE 12.7 **3-D H-vector visualization for a Hertzian dipole.** Repeat the previous MATLAB exercise but for the magnetic field vector radiated by a Hertzian dipole. *(ME12_7.m on IR)*

12.3 Steps in Far Field Evaluation of an Arbitrary Antenna

Consider a straight wire antenna along the z-axis, shown in Fig.12.3. Under the far-field assumption, Eq.(12.5), we apply different approximations for the magnitude and phase (as indicated in Fig.12.3) of the spherical-wave factor $\mathrm{e}^{-\mathrm{j}\beta R}/R$ in the integral for computing the magnetic vector potential, $\underline{\mathbf{A}}$, at an observation point P (field point) defined by (r, θ, ϕ), with R being the variable source-to-field distance for an arbitrary point P′ at the wire axis (source point). This integral (in terms of the coordinate z along the wire antenna) thus becomes

$$\underline{\mathbf{A}} = \frac{\mu}{4\pi} \int_l \frac{\underline{I}(z)\, \mathrm{d}l\, \mathrm{e}^{-\mathrm{j}\beta R}}{R} \approx \frac{\mu\, \mathrm{e}^{-\mathrm{j}\beta r}}{4\pi r}\, \hat{\mathbf{z}} \int_l \underline{I}(z)\, \mathrm{e}^{\mathrm{j}\beta z \cos\theta}\, \mathrm{d}z \quad \text{(radiation integral)} . \tag{12.8}$$

It is called the radiation integral, and its solution, for a given current distribution $\underline{I}(z)$, is the basis for analysis of the wire antenna in Fig.12.3. Similar integrals are in place for arbitrary (curvilinear) wire antennas, and for surface and volume antennas.

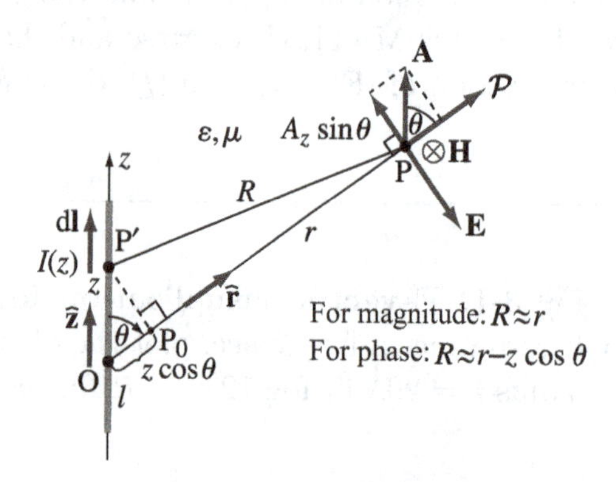

Figure 12.3 Straight wire antenna with an arbitrary current distribution: evaluation of the magnetic vector potential and electric and magnetic field vectors in the far zone.

From Fig.12.1 and Eqs.(12.6) and (12.1), we can write for the far electric and magnetic field components of a z-directed Hertzian dipole: $\underline{E}_\theta = -\mathrm{j}\omega \underline{A}_\theta = \mathrm{j}\omega \underline{A}_z \sin\theta$ and $\underline{H}_\phi = \underline{E}_\theta/\eta$. Since the wire antenna in Fig.12.3 can be represented as a chain of many z-directed Hertzian dipoles, the same relationships hold true for that antenna as well. For an arbitrary transmitting antenna, the vectors $\underline{\mathbf{E}}$ and $\underline{\mathbf{H}}$ have both θ- and ϕ-components (but no radial component) in the far zone, so these relationships should be extended to:

$$\underline{E}_\theta = -\mathrm{j}\omega \underline{A}_\theta\,, \quad \underline{E}_\phi = -\mathrm{j}\omega \underline{A}_\phi\,, \quad \underline{H}_\phi = \frac{\underline{E}_\theta}{\eta}\,, \quad \underline{H}_\theta = -\frac{\underline{E}_\phi}{\eta} \quad \text{(arbitrary antenna)} . \tag{12.9}$$

Once $\underline{\mathbf{A}}$ is obtained from Eq.(12.8), or a similar radiation integral, the remaining steps in the radiation analysis of the antenna are straightforward, and the same for any antenna type and geometry: $\underline{\mathbf{E}}$ and $\underline{\mathbf{H}}$ in the far zone are easily found from Eqs.(12.9).

MATLAB EXERCISE 12.8 Symbolic radiation integral for an arbitrary straight wire antenna. Applying symbolic programming in MATLAB, write a function `radiationIntegral()` that evaluates the radiation integral in Eq.(12.8), to obtain, in symbolic form, the far-zone magnetic vector potential, $\underline{\mathbf{A}}$, due to a given current $\underline{I}(z)$, $z_1 \leq z \leq z_2$, along a straight wire antenna (of an arbitrary length) in Fig.12.3. The ambient medium is free space, the operating wavelength of the antenna is λ, and the observation point P in Fig.12.3 is defined by its Cartesian coordinates (x_0, y_0, z_0). *(radiationIntegral.m on IR)*

MATLAB EXERCISE 12.9 Symbolic magnetic potential of a Hertzian dipole. Using function `radiationIntegral` developed in the previous MATLAB exercise, compute the magnetic vector potential due to a Hertzian dipole at far-zone points along the x-axis in Fig.12.1. Compare the results to those obtained by Eq.(12.1) for $\lambda = 1$ m, $I = 1$ A, $l = 5$ cm, and $100 \text{ m} \leq x \leq 200$ m. *(ME12_9.m on IR)* **H**[3]

HINT:

See MATLAB Exercise 1.11. The results are shown in Fig.12.4.

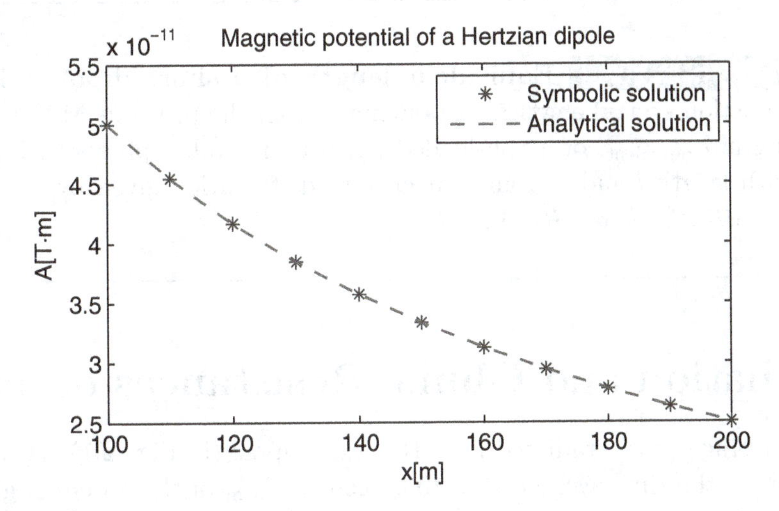

Figure 12.4 Magnetic vector potential ($\underline{A}_z$) due to a Hertzian dipole at points along the x-axis in Fig.12.1: comparison of results obtained by symbolic integration in Eq.(12.8) using function `radiationIntegral` from the previous MATLAB exercise to the analytical expression in Eq.(12.1); for MATLAB Exercise 12.9. *(color figure on CW)*

---

[3]**H** = recommended to be done also "by hand," i.e., not using MATLAB.

MATLAB EXERCISE 12.10 **Equivalent length of an arbitrary short wire antenna.** Consider an arbitrary electrically short straight wire antenna along the z-axis, as in Fig.12.3, radiating in free space. The antenna is centrally fed with current $\underline{I}_0$ and its current distribution is described by a given function $\underline{I}(z)$, $-l/2 \leq z \leq l/2$, where $\underline{I}(0) = \underline{I}_0$. Since $l \ll \lambda = 2\pi/\beta$, the radiation integral in Eq.(12.8) reduces to

$$\underline{\mathbf{A}} = \frac{\mu_0 \, \mathrm{e}^{-\mathrm{j}\beta r}}{4\pi r} \, \hat{\mathbf{z}} \underbrace{\int_l \underline{I}(z) \, \mathrm{d}z}_{\underline{I}_0 \, l_{\mathrm{equivalent}}} \quad \left(\mathrm{e}^{\mathrm{j}\beta z \cos\theta} \approx 1 \right) \quad \text{(short wire antenna)} \, . \tag{12.10}$$

Hence, we can define the equivalent length of the antenna as

$$l_{\mathrm{equivalent}} = \frac{1}{\underline{I}_0} \int_{-l/2}^{l/2} \underline{I}(z) \, \mathrm{d}z \quad \text{(equivalent length of a short antenna)} \, , \tag{12.11}$$

namely, as the length of an equivalent Hertzian dipole, Fig.12.1, with a current-length product equal to $\underline{I}_0 \, l_{\mathrm{equivalent}}$, which can be interpreted as characterizing a dipole with the same current (uniform along the wire) as the feed current of the original antenna, but of length equal to $l_{\mathrm{equivalent}}$. Based on Eq.(12.10), the magnetic potential of the two antennas (the original short antenna and the equivalent Hertzian dipole) at any point in the far zone is the same. In other words, an arbitrary short straight wire antenna can, as the magnetic potential in the far zone is concerned, be replaced by a Hertzian dipole with current $\underline{I}_0$ and length $l_{\mathrm{equivalent}}$ in Eq.(12.11). Write a function `equivLengthShortAntenna()` in MATLAB that symbolically finds $l_{\mathrm{equivalent}}$. *(equivLengthShortAntenna.m on IR)*

MATLAB EXERCISE 12.11 **Equivalent length of a short dipole with a triangular current.** Using function `equivLengthShortAntenna` (from the previous MATLAB exercise), find the equivalent length, $l_{\mathrm{equivalent}}$, of a non-loaded (with no metallic spheres in Fig.12.1) short wire dipole antenna with length l and a triangular current distribution given by $\underline{I}(z) = \underline{I}_0(1 - 2|z|/l)$ $(-l/2 \leq z \leq l/2)$. *(ME12_11.m on IR)* **H**

12.4 Radiation and Ohmic Resistances of an Antenna

To find the time-average power radiated by a Hertzian dipole (in Fig.12.1), P_{rad}, we compute the flux [see Eq.(6.31)] of the time-average Poynting vector, $\mathcal{P}_{\mathrm{ave}}$, of the antenna, given in Eq.(12.4), through a spherical surface, of radius r, centered at the coordinate origin [for integration, we use an elementary surface $\mathrm{d}S$ in the form of a thin ring of radius $r \sin\theta$ and width $r \, \mathrm{d}\theta$, as in Fig.1.9 and Eq.(1.20)]. For an arbitrary antenna, P_{rad} is proportional to the magnitude of the feed current at the antenna input terminals, $\underline{I}_0$, squared. The constant of proportionality is a resistance, called the radiation resistance of the antenna and denoted as R_{rad}. For the Hertzian dipole, with $\underline{I} = \underline{I}_0$, we thus have

$$R_{\mathrm{rad}} = \frac{P_{\mathrm{rad}}}{I_0^2} = \frac{2\pi\eta}{3} \left(\frac{l}{\lambda} \right)^2 \quad \text{(radiation resistance, Hertzian dipole)} \, , \tag{12.12}$$

where $I_0 = |\underline{I}_0|$ and η is the intrinsic impedance of the ambient medium, Eq.(7.2).

In addition, all real antennas exhibit some ohmic (Joule's) losses in the lossy materials constituting the antenna body, because of the conduction current flow through the materials, and the time-average ohmic power of the antenna, P_{ohmic}, can be written as $P_{\text{ohmic}} = R_{\text{ohmic}} I_0^2$, where R_{ohmic} is the total high-frequency (with the skin effect pronounced) ohmic resistance of the antenna. On the other side, P_{ohmic} in the antenna metallic parts can be computed similarly to the integration in Eqs.(11.12). For a straight wire antenna extending along the z-axis, in Fig.12.3, the magnetic field on the surface of the wire, of radius a, equals $\underline{H} = \underline{I}(z)/2\pi a$ [from the generalized Ampère's law in integral form, Eq.(5.1)], so that R_{ohmic} can be found from the following integral:

$$R_{\text{ohmic}} = \frac{P_{\text{ohmic}}}{I_0^2} = \frac{1}{I_0^2} \int_l \frac{R_{\text{s}}}{2\pi a} |\underline{I}(z)|^2 \, \mathrm{d}z \quad \text{(antenna ohmic resistance)} , \qquad (12.13)$$

with R_{s} being the surface resistance of the antenna conductors, Eqs.(9.12), and $\underline{I}(z)$ the current intensity along the antenna. For a Hertzian dipole, the current is uniform along the wire, which leads to

$$\underline{I}(z) = \underline{I}_0 \quad \longrightarrow \quad R_{\text{ohmic}} = \frac{R_{\text{s}} l}{2\pi a} \quad \text{(ohmic resistance of a Hertzian dipole)} . \qquad (12.14)$$

Given that, in general, P_{rad} and P_{ohmic} represent, respectively, the desired and undesired parts of the input power, $P_{\text{in}} = P_{\text{rad}} + P_{\text{ohmic}}$, delivered by the generator (see Fig.12.1) to the antenna, we can define the radiation efficiency of the antenna, customarily measured in percent, through the following ratio [based on Eqs.(12.12) and (12.13)]:

$$\eta_{\text{rad}} = \frac{P_{\text{rad}}}{P_{\text{in}}} = \frac{R_{\text{rad}}}{R_{\text{rad}} + R_{\text{ohmic}}} \quad (0 \le \eta_{\text{rad}} \le 1) \quad \text{(radiation efficiency)} . \qquad (12.15)$$

For an ideal (lossless) antenna, $\eta_{\text{rad}} = 100\%$.

MATLAB EXERCISE 12.12 **Radiation and ohmic resistances of a Hertzian dipole.** Write a MATLAB program that calculates the radiation and ohmic resistances and radiation efficiency of a Hertzian dipole, of given dimensions (length and wire radius) and material properties (conductivity and relative permeability), in a given frequency range (of course, the dipole must be electrically short in the entire range). Input data are entered from the keyboard, and the results are presented graphically. *(ME12_12.m on IR)*

TUTORIAL:

The first part of the code deals with the data entry (note that all values are scaled based on the used multipliers of fundamental units):

```
f1 = input('Enter the starting frequency (in MHz):  ');
f2 = input('Enter the ending frequency (in MHz):  ');
df = 0.01*(f2-f1);
f = (f1:df:f2)*10^6;
c = 299792458;
LAMBDA = c./f;
fprintf('\n The shortest wavelength in the frequency range')
```

```
fprintf(' is %.4f mm\n',10^3*LAMBDA(length(LAMBDA)));
L = input('Enter the length of the Hertzian dipole (in m):  ');
a = input('Enter the wire radius of the dipole (in mm):  ');
a = a/10^3;
ETA = 376.73;
SIGMA = input('Enter the conductivity of the dipole (in MS/m):  ');
SIGMA = SIGMA*10^6;
MU0 = 4*pi*10^(-7);
MUR = input('Enter the relative permeability of the dipole:  ');
```

The radiation and ohmic resistances of the dipole, R_{rad} (`Rrad`) and R_{ohmic} (`Rohmic`), are computed using Eqs.(12.12) and (12.14) and function `surfResistance` (from MATLAB Exercise 8.3), and the dipole radiation efficiency (in percent), η_{rad} (`radEff`), is obtained based on Eq.(12.15),

```
Rrad = 2*pi*ETA/3*(L./LAMBDA).^2;
Rs = surfResistance(f,MUR,SIGMA);
Rohmic = Rs.*L/2/pi/a;
radEff = Rrad./(Rrad + Rohmic)*100;
```

Finally, we plot the results, in the given frequency range,

```
figure(1)
subplot(2,1,1);
plot(f,Rrad,'r'); hold on;
plot(f, Rohmic,'b');
xlabel('f [Hz]');
ylabel('R [\Omega]');
title('Radiation and ohmic resistances of a Hertzian dipole');
legend('Radiation resistance','Ohmic resistance','Location','Best');
subplot(2,1,2);
plot(f,radEff,'g');hold off;
xlabel('f [Hz]');
ylabel('[%]');
title('Radiation efficiency of a Hertzian dipole');
legend('Radiation efficiency','Location','Best');
```

MATLAB EXERCISE 12.13 Radiation and ohmic resistances of a nonloaded short dipole. Repeat the previous MATLAB exercise but for the non-loaded short wire dipole antenna with a triangular current distribution from MATLAB Exercise 12.11. *(ME12_13.m on IR)* **H**

HINT:

Using the equivalent length, $l_{\text{equivalent}}$, of the antenna found in MATLAB Exercise 12.11, obtain the radiation resistance of the antenna as that of the equivalent Hertzian dipole – with current $\underline{I}_0$ and length $l_{\text{equivalent}}$ – by means of Eq.(12.12) with $l = l_{\text{equivalent}}$. To find the ohmic resistance of the antenna, perform symbolic integration to solve the integral in Eq.(12.13) with $\underline{I}(z) = \underline{I}_0(1-2\,|z|/l)$

$(-l/2 \leq z \leq l/2)$. For the antenna radiation efficiency, implement Eq.(12.15), as in the previous MATLAB Exercise.

MATLAB EXERCISE 12.14 **Radiation efficiency of a short dipole with cosine current.** An electrically short wire antenna of length $l = 1$ m and radius $a = 3$ mm, placed in free space at the coordinate origin along the z-axis of a spherical coordinate system, has a current of intensity $\underline{I}(z) = \underline{I}_0 \cos(\pi z/l)$, for $-l/2 \leq z \leq l/2$, and frequency $f = 15$ MHz. The antenna is made out of steel ($\sigma = 2$ MS/m and $\mu_r = 2000$). In MATLAB, find the radiation resistance, the high-frequency ohmic resistance, and the radiation efficiency of the antenna. *(ME12_14.m on IR)* **H**

HINT:
Once the equivalent length ($l_{\text{equivalent}}$) of the antenna is determined using function `equivLengthShortAntenna` (from MATLAB Exercise 12.10), R_{rad}, R_{ohmic}, and η_{rad} for the antenna are obtained as in the previous MATLAB exercise.

12.5 Antenna Radiation Patterns, Directivity, and Gain

The far electric field intensity vector, $\underline{\mathbf{E}}$, of an arbitrary antenna is proportional to the feed current at the antenna input terminals, $\underline{I}_0$. Having also in mind the spherical-wave dependence on r in Eq.(12.8), it is convenient to write $\underline{\mathbf{E}}$, given by Eqs.(12.9), in the following form:

$$\underline{\mathbf{E}}(r,\theta,\phi) = \frac{\mathrm{j}\eta}{2\pi}\,\underline{I}_0\,\frac{\mathrm{e}^{-\mathrm{j}\beta r}}{r}\,\underline{\mathbf{F}}(\theta,\phi) \quad (\underline{\mathbf{F}} - \text{characteristic radiation function})\,, \tag{12.16}$$

where $\underline{\mathbf{F}}(\theta,\phi)$ is termed the characteristic radiation function of the antenna (a dimensionless quantity); it represents the part of the field expression that is characteristic for individual antennas, i.e., that differs from antenna to antenna, while the remaining terms in the expression are the same for all antennas. Independent of r, and thus only a function of the direction of antenna radiation, defined by angles θ and ϕ, $\underline{\mathbf{F}}(\theta,\phi)$ determines the directional properties of the antenna. As an example, comparing Eqs.(12.6) and (12.16) we realize that $\underline{\mathbf{F}}$ of a Hertzian dipole (Fig.12.1) is (the radiation of the dipole is azimuthally symmetrical – does not depend on ϕ)

$$\underline{\mathbf{F}}(\theta) = \frac{\beta l}{2}\sin\theta\,\hat{\boldsymbol{\theta}} \quad (\text{radiation function of a Hertzian dipole})\,. \tag{12.17}$$

Different aspects of the characteristic radiation function, in Eq.(12.16), presented graphically, give different radiation patterns of the antenna under consideration. Most frequently, we plot the normalized field pattern of the antenna, $f(\theta,\phi)$, defined as

$$f(\theta,\phi) = \frac{|\underline{\mathbf{F}}(\theta,\phi)|}{|\underline{\mathbf{F}}(\theta,\phi)|_{\text{max}}} \quad \{[f(\theta,\phi)]_{\text{max}} = 1\} \quad (\text{normalized field pattern})\,, \tag{12.18}$$

and a typical pattern of a directional antenna – as a three-dimensional (3-D) polar plot – is shown in Fig.12.5.

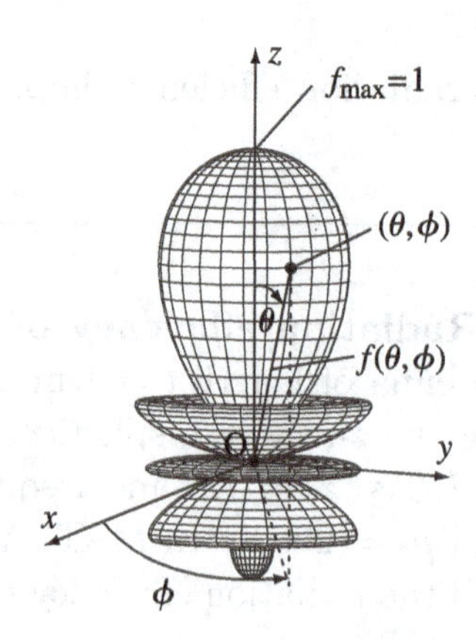

Figure 12.5 Typical normalized field pattern, Eq.(12.18), of a directional antenna.

To further describe and quantify directional properties of transmitting antennas, the directivity of an antenna in a given direction, $D(\theta, \phi)$, is defined via the far-zone time-average Poynting vector and the total time-average radiated power of the antenna [see Eqs.(12.4) and (12.12)] as

$$D(\theta, \phi) = \frac{4\pi r^2 \mathcal{P}_{\text{ave}}(r, \theta, \phi)}{P_{\text{rad}}} = \frac{\eta \, |\underline{\mathbf{F}}(\theta, \phi)|^2}{\pi R_{\text{rad}}} \quad \text{(antenna directivity)} , \tag{12.19}$$

where R_{rad} is the radiation resistance of the antenna. Similarly, the gain of an antenna is computed as

$$G(\theta, \phi) = \frac{4\pi r^2 \mathcal{P}_{\text{ave}}(r, \theta, \phi)}{P_{\text{rad}} + P_{\text{ohmic}}} = \frac{\eta \, |\underline{\mathbf{F}}(\theta, \phi)|^2}{\pi (R_{\text{rad}} + R_{\text{ohmic}})} \quad \text{(antenna gain)} , \tag{12.20}$$

with R_{ohmic} being the ohmic resistance of the antenna, Eq.(12.13). Frequently, D is used without specifying the direction of radiation, in which case the maximum directivity is implied, so $D \equiv D_{\text{max}} = [D(\theta, \phi)]_{\text{max}}$, and analogously $G \equiv G_{\text{max}}$.

MATLAB EXERCISE 12.15 **3-D polar radiation pattern plot, arbitrary radiation function.** Write a function `polar3D()` in MATLAB that generates a 3-D polar plot of a given normalized field pattern, $f(\theta, \phi)$, Eq.(12.18), for a transmitting antenna, like the one in Fig.12.5. The function takes at input three equally-sized matrices: **`theta`** ($0 \le \theta \le \pi$), **`phi`** ($0 \le \phi \le 2\pi$), and **`f`**, where f is the value of the normalized field pattern in the radiation direction defined by (θ, ϕ). *(polar3D.m on IR)*

TUTORIAL:

We first find the magnitude (absolute value) of f and normalize $|f|$ to its maximum value, as in Eq.(12.18), to remedy a possible situation that the radiation function f provided at the input is complex (not given as magnitude) and/or not normalized, i.e., that the maximum of $|f|$ is not unity. The pattern is plotted using MATLAB function **`surf`** (for instance, see MATLAB Exercises 1.39 and 3.4), with the matrix **`color`** representing the optional input to the function that defines the color of the plot – in our case, the same at all points. Since – in the pattern plot in Fig.12.5 – the distance from the coordinate origin (O) to a point on the pattern surface along the direction (θ, ϕ) represents the value $f(\theta, \phi)$, the radial coordinate of the point equals $r = f$. Furthermore, as **surf** requires Cartesian coordinates at input, we convert $(r = f, \theta, \phi)$ to (x, y, z) by means of function `sph2Car` (from MATLAB Exercise 1.24). However, this function needs to first be modified to enable equally sized matrices as input parameters **r**, **theta**, and **phi**, rather than single values. Such modified function `sph2Car` should be used in all similar coordinate transformations in this chapter.

```
function polar3D(theta,phi,f)
f = abs(f);
f = f/max(max(f));
[x,y,z] = sph2Car(f,theta,phi);
[N,M] = size(x);
color = ones(N,M);
surf(x,y,z,color);
```

Finally, in addition to the radiation pattern plot, we draw the x, y, and z axes, as vectors – using MATLAB function `quiver3` (see MATLAB Exercise 1.3), and label them,

```
hold on;
quiver3(-1.2,0,0,2.4,0,0,0,'Color','k');
quiver3(0,-1.2,0,0,2.4,0,0,'Color','k');
quiver3(0,0,-1.2,0,0,2.4,0,'Color','k');
hold off;
text(1.1,0.05,0.05,'x');
text(0.05,1.1,0.05,'y');
text(0.05,0.05,1.1,'z');
xlim([-1.2 1.2]);
ylim([-1.2 1.2]);
zlim([-1.2 1.2]);
axis equal;
axis off;
```

MATLAB EXERCISE 12.16 **Radiation pattern cuts in three characteristic planes.** In practice, 3-D radiation patterns are usually presented as a series of 2-D patterns, representing characteristic cuts (containing the coordinate origin) through the 3-D diagram, and most frequently these cuts are made in one or more of the planes xy ($z = 0$), xz ($y = 0$), and yz ($x = 0$). This MATLAB exercise does exactly that: it develops a function `cutPattern()` to cut the 3-D polar radiation pattern of an arbitrary normalized radiation function, $f(\theta, \phi)$, from the previous MATLAB exercise in a specified plane, xy, xz, or yz. The input to the function contains the same three matrices **theta**, **phi**, and **f** as in function **polar3D** (previous MATLAB exercise), plus the name of the cutting plane. *(cutPattern.m on IR)*

TUTORIAL:

See the code developed in the previous MATLAB exercise. However, instead of considering the radiation function over the entire spherical spatial mesh, we extract and use here only those values lying in the specific (specified) plane. The position of a point in this plane is defined by the radial coordinate r ($r = f$) and angular coordinate α in a local polar 2-D coordinate system. By MATLAB command `switch-case`, the code switches between three separate cases, as determined by the name of the cutting plane, `cutName`, at input: cases `'xy'`, `'xz'`, and `'yz'`, respectively; only the section of the code corresponding to the considered case is executed. As the xy-plane is described in the spherical coordinate system by $\theta = \pi/2$ and the whole range of ϕ ($0 \leq \phi \leq 2\pi$), the cutting plane xy is determined by $\theta = \pi/2$, $\alpha = \phi$, and $r = f$, and analogously for the other two cases. In specific, the xz-plane is described effectively by two half-planes (to cover all radiation directions in the plane), defined by $\phi = 0$ and $\phi = \pi$, respectively, while the half-planes $\phi = \pi/2$ and $\phi = 3\pi/2$ constitute the yz-plane (and all of its radiation directions), with each of the half-planes being described by the entire range of coordinate θ ($0 \leq \theta \leq \pi$) in both cases; polar coordinates α and r are then defined accordingly. The cut of the 3-D radiation pattern in the given plane, i.e., the 2-D polar normalized field pattern, is plotted using MATLAB function `polar`, in terms of α (`alpha`) and r (`r`), where we also make the choice of color for the drawing as an optional input to the function (we choose different colors for the three cuts).

```
function cutPattern(theta,phi,f,cutName)
f = abs(f);
f = f/max(max(f));
[M,N] = size(theta);
switch cutName
case 'xy'
r(1:M) = f(1:M,(N+1)/2);
alpha(1:M) = phi(1:M,(N+1)/2);
polar(alpha,r,'b');
xlabel('x');
ylabel('y');
title('XY cut');
case 'xz'
r(1:2*N) = [f(1,1:N),f((M+1)/2,N:-1:1)];
alpha(1:2*N) = [theta(1,1:N),2*pi-theta((M+1)/2,N:-1:1)];
polar(alpha,r,'g');
xlabel('z');
ylabel('x');
title('XZ cut');
case 'yz'
r(1:2*N) = [f((M-1)/4 + 1,1:N),f((M-1)*3/4 + 1,N:-1:1)];
alpha(1:2*N) = [theta((M-1)/4 + 1,1:N),2*pi-theta((M-1)*3/4 + 1,N:-1:1)];
polar(alpha,r,'r');
xlabel('z');
ylabel('y');
title('YZ cut');
end;
```

MATLAB EXERCISE 12.17 **3-D and 2-D radiation patterns of a Hertzian dipole.** In MATLAB, plot a 3-D polar radiation pattern of a Hertzian dipole in Fig.12.1, as well as its cuts in planes xy, yz, and xz, respectively. *(ME12_17.m on IR)* **H**

HINT:
From Eqs.(12.18) and (12.17), f of a Hertzian dipole is $f(\theta) = \sin\theta$, where the maximum radiation occurs in the equatorial plane (xy-plane), for $\theta = \theta_{\text{max}} = 90°$, in Fig.12.1. Use functions `polar3D` and `cutPattern` (from the previous two MATLAB exercises). The resulting MATLAB graphs are shown in Figs.12.6 and 12.7.

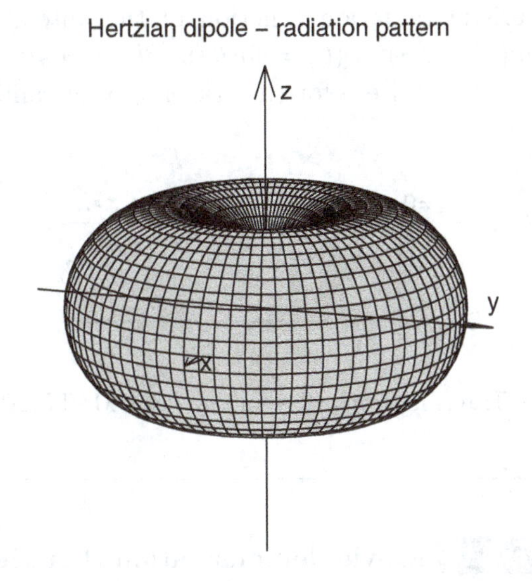

Figure 12.6 MATLAB plot of the 3-D normalized field polar radiation pattern of a Hertzian dipole (Fig.12.1); for MATLAB Exercise 12.17. *(color figure on CW)*

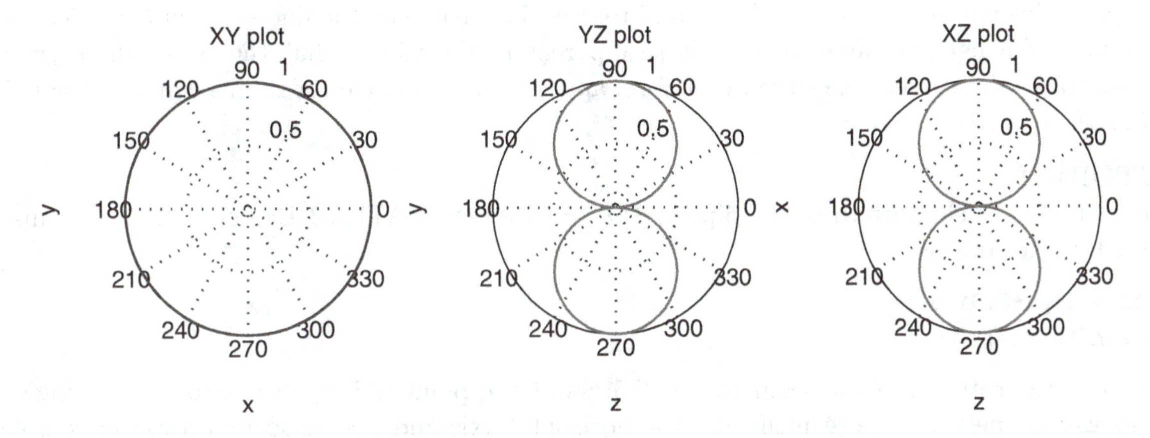

Figure 12.7 Cuts in three characteristic planes of the radiation pattern in Fig.12.6; for MATLAB Exercise 12.17. *(color figure on CW)*

MATLAB EXERCISE 12.18 **Radiation patterns of a traveling-wave antenna.** Figure 12.8 shows an end-fed traveling-wave wire antenna. It consists of a horizontal PEC wire that is l long, driven at $z = 0$ by a current of complex rms intensity $\underline{I}_0$ and angular frequency ω, and terminated at $z = l$ by a purely resistive load, of resistance R_L, adopted such that the current distribution along the antenna is a wave traveling in the positive z direction with a velocity equal to the speed of light, c_0 [Eq.(7.7)]. The current is given by $\underline{I}(z) = \underline{I}_0 \, e^{-j\beta z}$ $(0 \leq z \leq l)$, where the phase coefficient of the wave is $\beta = \omega/c_0$, Eq.(7.4). Neglecting the influence of the ground plane, and thus assuming that the antenna operates in free space, as well as the radiation of currents in vertical wire pieces at the two antenna ends, Eqs.(12.8), (12.9), and (12.16) result in the following expression for the characteristic radiation function of the antenna, for the usual setup of the spherical coordinate system (Fig.12.8): $\underline{\mathbf{F}}(\theta) = \sin\theta \sin[\beta l(1-\cos\theta)/2] \, e^{-j\beta(1-\cos\theta)l/2} \, \hat{\boldsymbol{\theta}}/(1-\cos\theta)$. Repeat the previous MATLAB exercise – for the traveling-wave antenna. *(ME12_18.m on IR)*

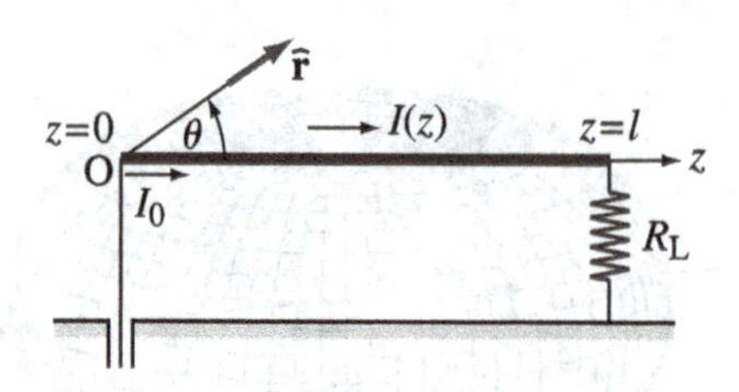

Figure 12.8 Traveling-wave wire antenna; for MATLAB Exercise 12.18.

MATLAB EXERCISE 12.19 **Movie demonstration that Hertzian dipole pattern cuts are circles.** In MATLAB, play a movie that graphically demonstrates that the two closed curves constituting a normalized field polar radiation pattern in a plane $\phi = \text{const}$ of a Hertzian dipole (Fig.12.1), namely, any one of the last two pattern cuts in Fig.12.7, are circles. In specific, the code plots this pattern cut for the dipole and rotates, in the movie, the line segment $\xi = 0.5$ about the center of a pattern curve, so that it is apparent to the viewer that the tip of the segment exactly traces the curve. The curve, therefore, appears to be a circle (of radius equal to $\xi = 0.5$). *(ME12_19.m on IR)*

TUTORIAL:

The normalized field pattern of the dipole, $f(\theta) = \sin\theta$ (see MATLAB Exercise 12.17), is implemented in the code as

```
theta = 0:pi*0.01:pi;
f = sin(theta);
```

To enable the rotation of the segment $\xi = 0.5$ about the point $(0.5, 0)$, we introduce an angle α (array `alpha`) between the segment and the horizontal axis and the associated array of complex numbers `r`, whose real and imaginary parts determine the coordinates of the rotating point (tip of the segment). Within the `for` loop of the movie, the pattern cut of the dipole is plotted (in red) by MATLAB function `polar`, in terms of θ and $r = f$, and then $2\pi - \theta$ and $r = f$ (to cover all radiation directions in a plane $\phi = \text{const}$, i.e., both circles in one of the last two cuts in Fig.12.7), and MATLAB function `line` is used to draw the rotating segment, in each movie frame.

```
alpha = 0:0.01*pi:2*pi;
r = 0.5 + 0.5*exp(i*alpha);
for j = 1:length(alpha)
polar(theta,f,'r');
hold on;
polar(2*pi-theta,f,'r');
plot(0,0.5,'*r');
text(-0.1,0.5,'O');
line([0,imag(r(j))],[0.5,real(r(j))]);
text(0.5*imag(r(j)),0.7*real(r(j)),'r = 0.5')
view (-90,90);
hold off;
axis equal;
M(j) = getframe;
end;
```

Note that MATLAB function `view`, namely, `view(-90,90)`, is applied to rotate the whole graph (movie frames) from the default layout (for function `polar`) to the one with the dipole axis in the vertical position. A snapshot (frame) of the movie is shown in Fig.12.9.

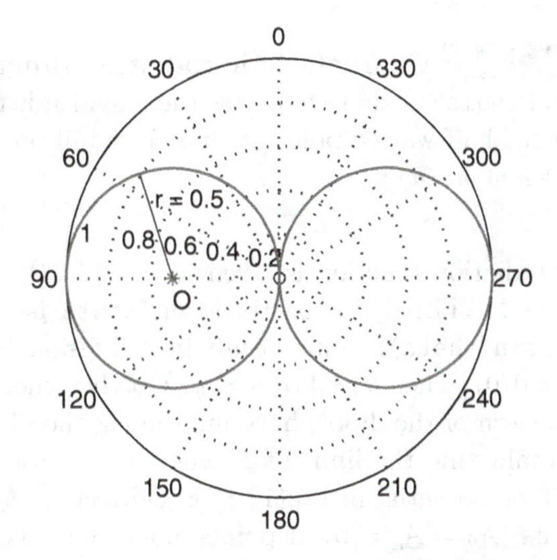

Figure 12.9 Snapshot of a MATLAB movie graphically demonstrating that ϕ = const pattern cuts of a Hertzian dipole (last two cuts in Fig.12.7) are circles; for MATLAB Exercise 12.19. *(color figure on CW)*

12.6 Wire Dipole Antennas of Arbitrary Lengths

We now consider a symmetrical (centrally fed) straight wire dipole antenna with an arbitrary length, l, and free ends (no capacitive loads at wire ends), and assume a sinusoidal (sine-wave) current distribution along the antenna (Fig.12.3),

$$\underline{I}(z) = \underline{I}_{\rm m} \sin \beta(h - |z|) \quad (-h \le z \le h) \quad \text{(sine-wave dipole current)}\,, \tag{12.21}$$

where β is the phase coefficient of the antenna radiation and $h = l/2$ is the length of each of the arms of the dipole. Note that the current at the antenna input terminals, $z = 0$, is $\underline{I}_0 = \underline{I}(0) = \underline{I}_{\rm m} \sin \beta h$. Using Eqs.(12.8), (12.21), (12.9), and (12.16), the characteristic radiation function of the dipole is found to be

$$\underline{\mathbf{F}}(\theta) = \frac{\cos(\beta h \cos\theta) - \cos\beta h}{\sin\beta h \sin\theta}\,\hat{\boldsymbol{\theta}} \quad \text{(radiation function, arbitrary dipole)}\,. \tag{12.22}$$

The most important dipole is by far that for $l = \lambda/2$. This simple wire antenna, known as a half-wave dipole, is, in fact, one of the most widely used of all antenna types. The dipole arm length being $h = \lambda/4$, we have $\beta h = 2\pi h/\lambda = \pi/2$, with which the characteristic radiation function in Eq.(12.22) becomes

$$\underline{\mathbf{F}}(\theta) = \frac{\cos\left(\frac{\pi}{2}\cos\theta\right)}{\sin\theta}\,\hat{\boldsymbol{\theta}} \quad \text{(radiation function, half-wave dipole)}\,. \tag{12.23}$$

It is maximum, $F_{\rm max} = 1$, in the equatorial plane ($\theta = 90^\circ$).

MATLAB EXERCISE 12.20 Characteristic radiation function of a half-wave dipole. Write a function `HalfWaveDipoleF()` that computes the magnitude of the characteristic radiation function, $\underline{\mathbf{F}}(\theta)$, of a z-directed half-wave dipole antenna, for a full range of angles θ (matrix `theta`), $0 \le \theta \le \pi$. *(HalfWaveDipoleF.m on IR)*

TUTORIAL:

The dipole characteristic radiation function is given in Eq.(12.23). However, a direct implementation of $F(\theta) = |\underline{\mathbf{F}}(\theta)|$ in MATLAB, in matrix form (`theta` is a matrix of values) – as `F = cos(pi/2*cos(theta))./sin(theta)`, would result in a division by zero in MATLAB, and an undefined value of F ($F = 0/0$), when $\theta = 0$ or $\theta = \pi$. On the other hand, the radiation function and the associated field pattern of the dipole have nulls along the dipole axis (for $\theta = 0$ or $\theta = \pi$), which can be proved by evaluating the limit of F when $\theta \to 0$ (or π), and also comes from the fact that the magnetic vector potential in Eq.(12.8) is z-directed, $\underline{\mathbf{A}} = \underline{A}\,\hat{\mathbf{z}}$. Namely, $\underline{\mathbf{A}}$ does not have transverse components ($\underline{A}_\theta = \underline{A}_\phi = 0$) at points along the z-axis, so that the components of the far electric field vector in Eqs.(12.9) are zero at these points as well (note that this holds true for a dipole of any length, so irrespective of the current distribution of the antenna). To remedy this situation, we introduce a matrix `u` equal to `u1.*u2`, where `u1` is unity whenever $\theta \neq 0$ and is zero for $\theta = 0$, while `u2` analogously switches between unity for $\theta \neq \pi$ and zero for $\theta = \pi$, and this is implemented using the absolute value of the function signum (MATLAB function `sign`), as follows:

```
function [result] = HalfWaveDipoleF(theta)
u1 = abs(sign(theta));
u2 = abs(sign(theta - pi));
u = u1.*u2;
```

```
result = u.*cos(pi/2*cos(theta))./(1 - u + sin(theta));
```

Obviously, the result equals $|\underline{\mathbf{F}}(\theta)|$ in Eq.(12.23) whenever $\theta \neq 0$ and $\theta \neq \pi$; otherwise, the result is zero.

MATLAB EXERCISE 12.21 **Radiation patterns of a half-wave dipole.** Repeat MATLAB Exercise 12.17 but for a half-wave dipole antenna. *(ME12_21.m on IR)* **H**

HINT:
Use function `HalfWaveDipoleF` (from the previous MATLAB exercise) to compute the characteristic radiation function of the antenna. Figure 12.10 shows the resulting radiation pattern cuts in planes xy, yz, and xz, respectively.

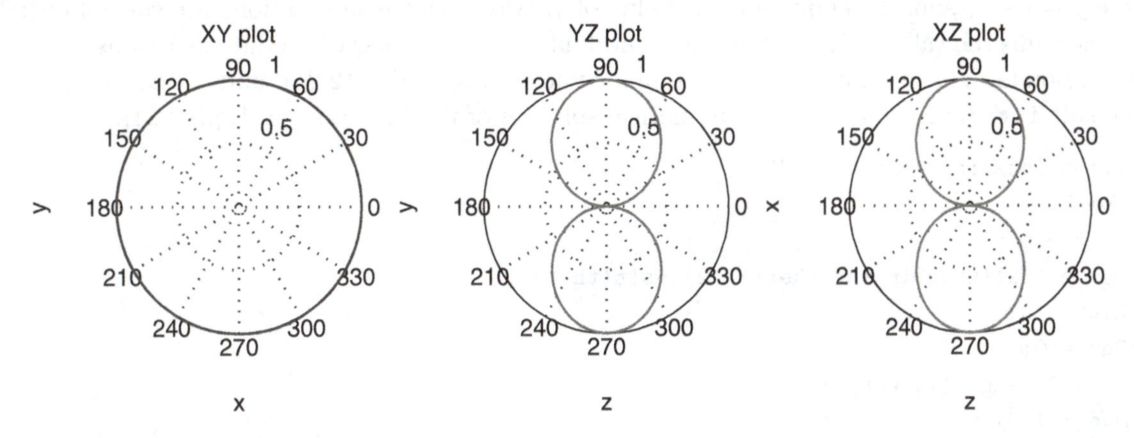

Figure 12.10 Radiation pattern cuts in three characteristic planes of a z-directed half-wave dipole antenna; for MATLAB Exercise 12.21. *(color figure on CW)*

MATLAB EXERCISE 12.22 **Numerical integration by Simpson's rule.** Finding the flux (through a spherical surface, of radius $r \gg \lambda$, centered at the coordinate origin) of the time-average Poynting vector of a wire dipole antenna (of any length) radiating in free space – in analogy to the computation in Eq.(12.12) and with the use of Eqs.(12.16) and (7.3) – the radiation resistance of the dipole can be obtained via the radiated power (P_{rad}) as

$$R_{\text{rad}} = \frac{P_{\text{rad}}}{I_0^2} = (60\ \Omega)\, I\,, \quad I = \int_{\theta=0}^{\pi} |\underline{\mathbf{F}}(\theta)|^2 \sin\theta\, \mathrm{d}\theta \quad (R_{\text{rad}} - \text{wire dipole})\,, \tag{12.24}$$

where $\underline{\mathbf{F}}(\theta)$ is the dipole characteristic radiation function, Eq.(12.22). Except for an electrically short dipole ($l \ll \lambda$), the integral I, in θ, cannot be evaluated analytically in a closed form, without very extensive mathematical manipulations, but can be computed numerically in a very straightforward fashion. Write a function `Simpson()` in MATLAB that numerically evaluates I for a $\lambda/2$ dipole antenna, namely, for $\underline{\mathbf{F}}(\theta)$ in Eq.(12.23), using Simpson's rule for numerical integration. This rule subdivides the integration interval $[a, b]$ into an even number, N, of equal segments, of length h, and is based on approximating the function that is integrated (integrand), $g(x)$, by a parabolic (quadratic) arc on each segment; it is given by

$$\int_a^b g(x)\,dx = \frac{h}{3}\left[g(x_0) + 2\sum_{j=1}^{N/2-1} g(x_{2j}) + 4\sum_{j=1}^{N/2} g(x_{2j-1}) + g(x_N)\right] , \qquad (12.25)$$

where $x_0, x_1, \ldots, x_N$ are points defining the boundaries of segments. *(Simpson.m on IR)*

TUTORIAL:
Based on the input `N`, the function finds `h` and calculates the integrand, `integF`, at boundaries of segments (points $x_0, x_1, \ldots, x_N$), using function `HalfWaveDipoleF` (from MATLAB Exercise 12.20) to find the characteristic radiation function of the antenna (half-wave dipole). The counter (index) $j = 0, 1, \ldots, N$ in Eq.(12.25) translates into the MATLAB `for` loop counter $i = 1, 2, \ldots, N + 1$, where $j = i - 1$, and, depending on the value of j, two separate summations are carried out: the first one sums the values of the integrand at odd integration points, while the other sums the values at even points. The obtained sums are combined according to Eq.(12.25), and the function returns the result `integral`, representing a numerical solution of the integral I in Eq.(12.24).

```
function integral = Simpson(N)
h = pi/N;
theta = 0:h:pi;
integF = (HalfWaveDipoleF(theta).^2).*sin(theta);
sumEven = 0;
sumOdd = 0;
for i = 2:length(integF)-1
if mod(i-1,2) == 0
sumEven = sumEven + integF(i);
else
sumOdd = sumOdd + integF(i);
end
end
integral = (h/3)*(integF(1) + 2*sumEven + 4*sumOdd + integF(length(integF)));
```

MATLAB EXERCISE 12.23 **Radiation resistance of a half-wave dipole using Simpson's rule.** Using function `Simpson` (from the previous MATLAB exercise) and Eq.(12.24), compute the radiation resistance, $R_{\rm rad}$, of a half-wave dipole in free space. In specific, run `Simpson` for $N = 10, 20, 100$, and 300, and evaluate the convergence of the solution. Determine the minimum N giving the accurate resistance. *(ME12_23.m on IR)*

MATLAB EXERCISE 12.24 **Directivity of a half-wave dipole.** Write a program in MATLAB that computes the directivity of a half-wave dipole for an arbitrary angle θ $(0 \le \theta \le \pi)$, $D(\theta)$ – using Eq.(12.19), function `HalfWaveDipoleF` from MATLAB Exercise 12.20 [for the characteristic radiation function of the dipole, $F(\theta)$], and the code from the previous MATLAB exercise (for the radiation resistance of the dipole, R_{rad}). This program also finds the angle θ_{max} for which D is maximum, and that maximum D, as $D_{max} = [D(\theta)]_{max}$ (using MATLAB function `max`). *(ME12_24.m on IR)*

MATLAB EXERCISE 12.25 **Radiation function and pattern plots of an arbitrary wire dipole.** Write a program in MATLAB that calculates and plots a 3-D normalized field polar radiation pattern of an arbitrary wire dipole antenna (of arbitrary length) placed along the z-axis, as well as its cuts in planes xy, yz, and xz, respectively. The user is able to input the electrical length of the dipole, $k = l/\lambda$, λ being the operating wavelength of the antenna. *(ME12_25.m on IR)*

HINT:
The characteristic radiation function, $\underline{\mathbf{F}}(\theta)$, of an arbitrary dipole antenna is given in Eq.(12.22); see MATLAB Exercise 12.20 for its implementation in MATLAB. The 3-D radiation pattern and the three characteristic cuts should be plotted as in MATLAB Exercise 12.17 (using functions `polar3D` and `cutPattern`). Assume $\lambda = 1$ m (or any other numerical value). Run the program for k equal to 1/2, 1, 5/4, and 3/2, respectively.

MATLAB EXERCISE 12.26 **Dependence of radiation pattern on antenna length – movie.** Write a function `patternVsLengthMovie()` in MATLAB that plays a movie visualizing the dependence of the characteristic radiation function of a wire dipole antenna on its electrical length, $k = l/\lambda$. The input to the function contains an array `k` of given values of k, an array `theta` with values of θ, $0 \le \theta \le \pi$, and a `length(k)` × `length(theta)` matrix `f` with the corresponding values of the normalized radiation function $f(k,\theta)$ of the dipole, computed using the code developed in the previous MATLAB exercise. In the movie, $f(k,\theta)$ is plotted as a polar diagram (in terms of θ) for k as a parameter that is varied in time. Therefore, the movie shows the way the radiation pattern changes due to the change of the electrical length of the antenna. *(patternVsLengthMovie.m on IR)*

TUTORIAL:
We plot the radiation function $f(k,\theta)$ using MATLAB function `polar`, in terms of θ and $r = f$, and then $2\pi - \theta$ and $r = f$, in the same way as in MATLAB Exercise 12.19, and apply MATLAB function `view` to rotate the default plot arrangement so that the antenna axis becomes vertical in the graph (also as in MATLAB Exercise 12.19),

```
function M = patternVsLengthMovie(k,theta,f)
for i = 1:length(k)
polar(theta,f(i,:));
hold on;
polar(2*pi-theta,f(i,:));
hold off;
view(-90,90);
```

```
title('Radiation function');
xlabel('z-axis');
```

The rest of the movie for loop visualizes the antenna (wire) as it changes in time,

```
hold on;
line([-1 -1],[-1,-0.65],'Color','k');
plot(-1,-1,'k+');
plot(-1,-0.65,'k+');
plot(-1,-0.65 - 0.35*k(i)/k(length(k)),'r*');
text(-1,-1.1,'k');
text(-1.1,-0.9,num2str(k(length(k))));
text(-1.1,-0.55,num2str(k(1)));
hold off;
M(i) = getframe;
end;
```

MATLAB EXERCISE 12.27 **Playing the pattern movie for different dipole lengths.** Write a program in MATLAB that prepares the input data and calls function `patternVsLengthMovie` (from the previous MATLAB exercise) to play the movie for antennas of lengths (l) ranging from 0.3λ to 4.5λ, assuming that the ambient medium is free space and the operating frequency of the antennas amounts to $f = 600$ MHz. *(ME12_27.m on IR)*

TUTORIAL:

We compute the characteristic radiation function of the antenna, $\underline{\mathbf{F}}(\theta)$, implementing Eq.(12.22); see MATLAB Exercise 12.20. It is then normalized, to obtain the function $f(\theta)$, i.e., $f(k, \theta)$, for the input to `patternVsLengthMovie`, according to Eq.(12.18). Note that this computation is the same as in MATLAB Exercise 12.25.

```
f = 600*10^6;
c0 = 299792458;
lambda = c0/f;
beta = 2*pi/lambda;
k = 0.3:0.01:4.5;
theta = 0:pi/100:pi;
F = zeros(length(k),length(theta));
f = zeros(length(k),length(theta));
for i = 1:length(k);
betah = beta*k(i)*lambda/2;
u1 = abs(sign(theta));
u2 = abs(sign(theta - pi));
u = u1.*u2;
F(i,:)  = abs(u.*(cos(betah*cos(theta))-cos(betah))./(sin(betah).*(1-u+sin(theta))));
Fmax(i) = max(F(i,:));
f(i,:)  = F(i,:)/Fmax(i);
```

```
end;
M = patternVsLengthMovie(k,theta,f);
```

Figure 12.11 shows the final frame of the movie – corresponding to a dipole of length $l = 4.5\lambda$.

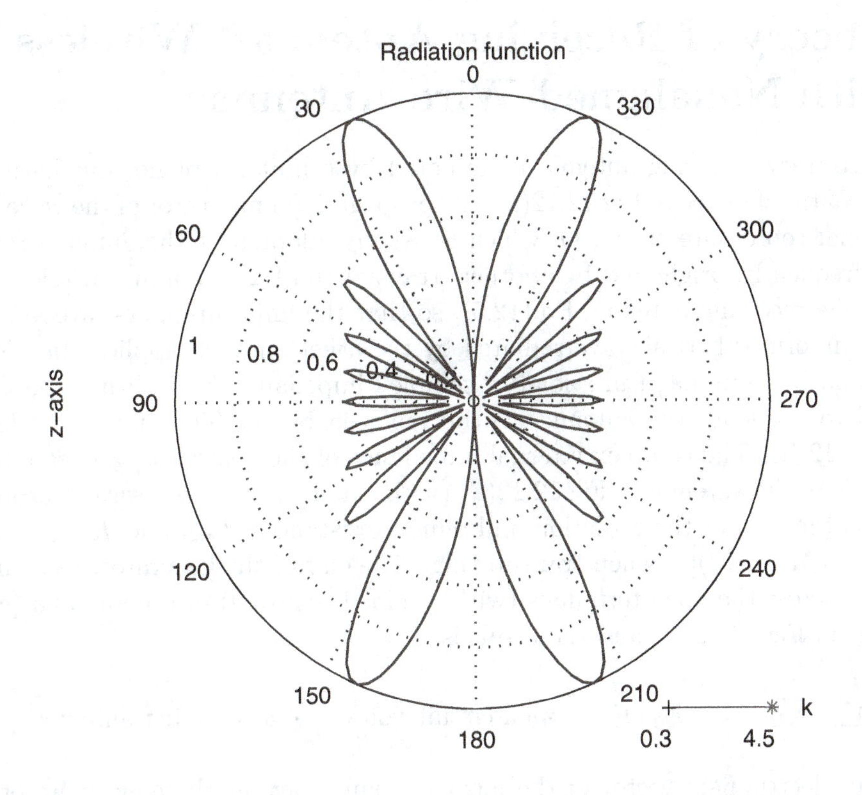

Figure 12.11 MATLAB movie visualizing radiation patterns of wire dipole antennas with different electrical lengths, $k = l/\lambda$, in a range $0.3 \le k \le 4.5$: final snapshot – for a dipole with $k = 4.5$; for MATLAB Exercise 12.27. *(color figure on CW)*

MATLAB EXERCISE 12.28 **Pattern movie for traveling-wave antennas of different lengths.** Repeat the previous MATLAB exercise but for a traveling-wave wire antenna from MATLAB Exercise 12.18. *(ME12_28.m on IR)*

MATLAB EXERCISE 12.29 **Radiation patterns of a quarter-wave monopole.** Consider a quarter-wave vertical wire monopole antenna, which is the upper half of a half-wave dipole ($l = \lambda/2$) attached to a horizontal ($z = 0$) PEC ground plane (the length of the monopole is $h = \lambda/4$) and fed against it. The current intensity along the wire, $\underline{I}(z)$, is the same as the upper half (for $0 \le z \le h$) of the current distribution of the equivalent dipole antenna. By image theory, the electromagnetic field radiated by the monopole at an arbitrary point in air above the ground plane is the same as the field at the corresponding point in the upper half-space in the system with the dipole radiating in free space. The field in the ground plane, for $z \leq 0$, is zero. Repeat MATLAB Exercise 12.21 – for the quarter-wave monopole antenna. *(ME12_29.m on IR)* **H**

12.7 Theory of Receiving Antennas. Wireless Links with Nonaligned Wire Antennas

Consider an arbitrary receiving antenna illuminated by a uniform plane time-harmonic wave, of wavelength λ. With reference to Fig.12.12(a), the propagation unit vector of the wave, $\hat{\mathbf{n}}$, is directed toward the global coordinate origin, O, which is usually adopted at the antenna terminals. This wave is, most frequently, originated by another (transmitting) antenna in a wireless link, which is far away from the receiving antenna, Eq.(12.5), so that the uniform-plane-wave approximation of the actual nonuniform spherical wave radiated by the other antenna applies. In addition, let the receiving antenna be terminated in a load of complex impedance $\underline{Z}_{\mathrm{L}}$. With respect to its output terminals, and to the load, the antenna can be replaced by the Thévenin equivalent generator, shown in Fig.12.12(b). The complex internal impedance of the generator, $\underline{Z}_{\mathrm{T}}$, equals the complex input impedance of the antenna in Fig.12.12(a) (with the incident plane wave "turned off"), so its real part equals the sum of the radiation and ohmic resistances, R_{rad} and R_{ohmic}, of the antenna [see Eqs.(12.12) and (12.13)] – when transmitting. The emf of the generator, $\underline{\mathcal{E}}_{\mathrm{T}}$, can be found as the voltage $\underline{V}_{\mathrm{oc}}$ across the open terminals (with the load removed) of the antenna (excited by the incoming wave) in Fig.12.12(a), and the result is

$$\underline{V}_{\mathrm{oc}} = \underline{\mathcal{E}}_{\mathrm{T}} = \frac{\lambda}{\pi}\,\underline{\mathbf{E}}_0 \cdot \underline{\mathbf{F}} \quad \text{(open-circuit voltage of a receiving antenna)}\,, \tag{12.26}$$

where $\underline{\mathbf{E}}_0$ is the electric field vector of the incident plane wave at the coordinate origin and $\underline{\mathbf{F}}$ is the characteristic radiation function [Eq.(12.16)] that the antenna would have if transmitting in the direction of the wave incidence, i.e., direction defined by $\hat{\mathbf{r}} = -\hat{\mathbf{n}}$, as indicated in Fig.12.12(a). So, the field pattern of a receiving antenna, showing how well it captures the incident signal in different directions in 3-D space, is identical to the radiation pattern of the antenna when in the transmitting mode, Fig.12.5, that is, the transmit and receive patterns of an arbitrary antenna are identical. Finally, the current of the antenna load, $\underline{I}_{\mathrm{L}}$, in Fig.12.12(a) and the power delivered to the load are determined from the equivalent circuit in Fig.12.12(b).

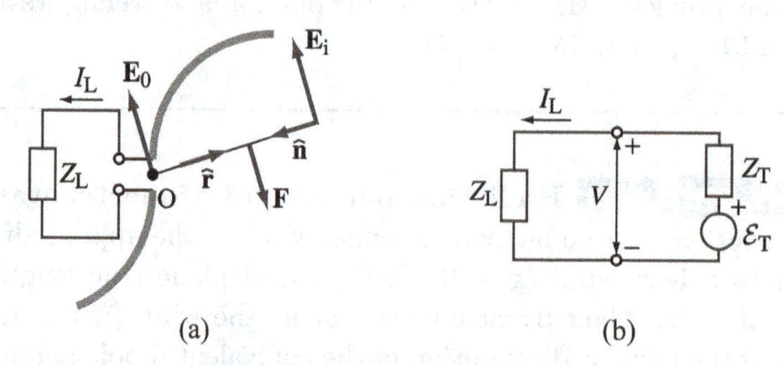

Figure 12.12 (a) Receiving antenna in the field of a uniform plane time-harmonic wave and (b) its Thévenin equivalent representation.

In a communication link, the dot product in Eq.(12.26) determines the polarization match (or mismatch) between the two antennas (transmitting and receiving) at the two ends of the link. For example, in the wireless link with two nonaligned half-wave wire dipole antennas shown in Fig.12.13, the transmitting antenna is placed in the xz-plane at an angle γ_1 with respect to the z-axis, while the receiving antenna lies in the $x'y'$-plane, where it makes an angle γ_2 with the x'-axis. The distance between the transmit and receive ends is $r \gg \lambda$. The open-circuit voltage of the receiving dipole, $\underline{V}_{\rm oc}$, is given by Eq.(12.26) with $\underline{\mathbf{E}}_0 = \underline{\mathbf{E}}_{\rm t}$ being the far electric field intensity vector of the transmitting antenna computed at point O′ in Fig.12.13, and $\underline{\mathbf{F}} = \underline{\mathbf{F}}_{\rm r}$ the characteristic radiation function of the receiving antenna that it would have if transmitting in the direction toward point O. In a local spherical coordinate system with the $z_{\rm t}$-axis along the transmitting dipole, in Fig.12.13, $\underline{\mathbf{E}}_{\rm t}$ has only a $\theta_{\rm t}$-component, where $\theta_{\rm t}$ coincides with γ_1, and the dipole characteristic radiation function, Eq.(12.23), is determined by this local $\theta_{\rm t}$ angle. On the other side, if we attach a local spherical coordinate system whose $z_{\rm r}$-axis is along the receiving dipole, the local $\theta_{\rm r}$ angle between the dipole and the direction toward the transmit end determines both the magnitude and polarization of $\underline{\mathbf{F}}_{\rm r}$ ($\underline{\mathbf{F}}_{\rm r}$ is $\theta_{\rm r}$-directed), where, in Fig.12.13, $\theta_{\rm r} = 90°$.

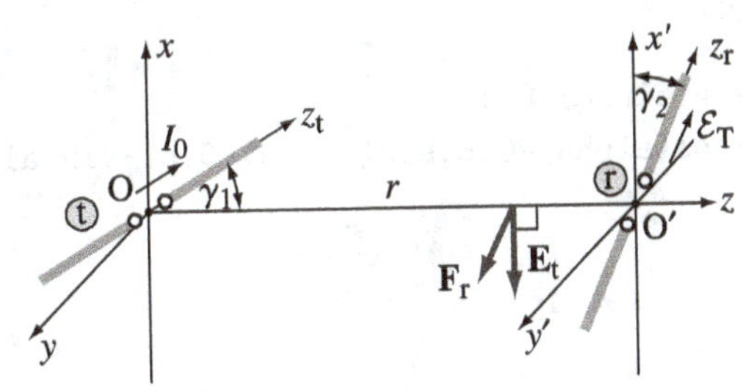

Figure 12.13 Wireless link with two nonaligned half-wave dipole antennas.

MATLAB EXERCISE 12.30 Geometrical preprocessing for arbitrarily oriented T-R antennas. Consider a free-space wireless system (link) with two wire dipole antennas (transmitting and receiving antennas) with arbitrary positions and orientations defined in a global Cartesian coordinate system, like the link in Fig.12.13, but with possibilities that each of the antennas can lie in any of the coordinate planes and make any angle with coordinate axes defining that plane. Write a function `antennaXYZ()` in MATLAB that, in a local spherical coordinate system with the local z-axis along any one of the two antennas in the link, determines the angle θ and the associated unit vector $\hat{\boldsymbol{\theta}}$ [needed for computing the characteristic radiation function of the antenna, $\underline{\mathbf{F}} = \underline{F}(\theta)\,\hat{\boldsymbol{\theta}}$], as well as the Cartesian coordinates of the ends of the antenna (needed for drawing the antenna). Assume that the antenna length is unity (this can later be scaled based on the actual length of the antenna). The input to the function contains the information on the plane, xy, xz, or yz (i.e., the plane name `'xy'`, `'xz'`, or `'yz'`), in which the antenna lies, the angle α that antenna makes with one of the coordinate axes, namely, the x-axis for an antenna in the xy- or xz-plane, and z-axis for an antenna in the yz-plane, and the information on whether the antenna is a transmitting (`'t'`) or receiving (`'r'`) one. *(antennaXYZ.m on IR)*

TUTORIAL:

The following lines of code find θ (`theta`), $\hat{\boldsymbol{\theta}}$ (`unitvec`), and coordinates x, y, and z of antenna ends (`coord`) – separately, using the `switch-case` command, for three planes containing the antenna (cases 'xz', 'xy', and 'yz'), and within each case separately for the transmitting and receiving antennas (cases 't' and 'r'). For instance, if the antenna lies in the xz-plane in Fig.12.13, $\hat{\theta} = -\hat{x}$ ([-1,0,0]), and $\theta = \pi/2 - \alpha$ (α is the angle measured from the x-axis to the antenna axis) if the antenna is on the transmitting side, while $\theta = \pi/2 + \alpha$ for the receiving antenna, and so on.

```
function [theta,unitvec,coord] = antennaXYZ(plane,tr,alpha)
switch plane
case 'xz'
unitvec = [-1,0,0];
theta = pi/2 - alpha;
if tr == 'r'
theta = pi/2 + alpha;
end;
coord = [cos(alpha)*0.5,-cos(alpha)*0.5;0,0;sin(alpha)*0.5,-sin(alpha)*0.5];
case 'xy'
theta = pi/2;
unitvec = [-cos(alpha),-sin(alpha),0];
coord = [cos(alpha)*0.5,-cos(alpha)*0.5;sin(alpha)*0.5,-sin(alpha)*0.5;0,0];
case 'yz'
unitvec = [0,-1,0];
theta = alpha;
if tr == 'r'
theta = pi - alpha;
end;
coord = [0,0;sin(alpha)*0.5,-sin(alpha)*0.5;cos(alpha)*0.5,-cos(alpha)*0.5];
end;
```

MATLAB EXERCISE 12.31 **Visualization of a wireless system with two antennas.** For the wireless system considered in the previous MATLAB exercise, write a function antennaDraw() in MATLAB that draws the system. At input, the function takes unit vectors $\hat{\theta}$ of the far electric field vector of the transmitting antenna ($\underline{\mathbf{E}}_t$) and of the characteristic radiation function of the receiving antenna ($\underline{\mathbf{F}}_r$) (see Fig.12.13, for example), vectors unitt and unitr, respectively, the operating wavelength of the system, λ (lambda), and coordinates x, y, and z of ends of the transmitting antenna (coordT) and of the receiving antenna (coordR). The function draws the two antennas, all coordinate axes, and vectors unitt and unitr. Of course, the distance between the antennas, r, should be scaled down for the purpose of plotting. *(antennaDraw.m on IR)*

TUTORIAL:

We first draw the axes x, y, z, x', and y', as in Fig.12.13, namely, coordinate systems xyz and $x'y'z$, using MATLAB functions line to plot axis lines and text to label them,

```
function antennaDraw(unitt,unitr,lambda,coordT,coordR)
figure(1);
```

```
hold on;
% coordinate system xyz
line([0 0],[0 2*lambda],[0 0],'Color','k');
text(0,2*lambda,lambda/20,'z');
line(0.5*[-lambda lambda],[0 0],[0 0],'Color','k');
text(0.5*lambda,lambda/20,lambda/20,'y');
line([0 0],[0 0],0.5*[-lambda lambda],'Color','k');
text(lambda/20,lambda/20,0.5*lambda,'x');
% coordinate system x'y'z
line([0 0],[1.9*lambda 1.9*lambda],0.5*[-lambda lambda],'Color','k');
text(0.5*lambda,2*lambda,lambda/20,'y''');
line(0.5*[-lambda lambda],[1.9*lambda 1.9*lambda],[0 0],'Color','k');
text(lambda/20,2*lambda,0.5*lambda,'x''');
```

Then, we draw the antennas, by function `line`, based on coordinates `coordT` and `coordR`, as well as vectors `unitt` and `unitr`, with the use of MATLAB function `quiver3` (see MATLAB Exercise 1.3), which visualizes the mutual position of vectors $\underline{\mathbf{E}}_{\mathrm{t}}$ and $\underline{\mathbf{F}}_{\mathrm{r}}$,

```
% antennaT
plot3(0,0,0,'ro');
text(lambda/5,0,lambda/5,'transmitter');
line([coordT(2,:)],[coordT(3,:)],[coordT(1,:)],'Color','b','LineWidth',2);
% antennaR
plot3(0,1.9*lambda,0,'ro');
text(lambda/5,1.9*lambda,lambda/5,'receiver');
line([coordR(2,:)],[coordR(3,:)]+1.9*lambda,[coordR(1,:)],'Color','g','LineWidth',2);
% Et
quiver3(0,1.5*lambda,0,unitt(2),unitt(3),unitt(1),0.5*lambda,'Color','b','LineWidth',2);
text(1.1*0.5*lambda*unitt(2),1.1*(0.5*lambda*unitt(3)+1.5*lambda),1.1*0.5*lambda*...
unitt(1),'Et');
% Fr
quiver3(0,1.51*lambda,0,unitr(2),unitr(3),unitr(1),0.5*lambda,'Color','g','LineWidth',2);
text(0.9*0.5*lambda*unitr(2),0.9*(0.5*lambda*unitr(3)+1.5*lambda),0.9*0.5*lambda*...
unitr(1),'Fr');
text(lambda/20,lambda,lambda/20,'r');
hold off;
axis equal;
axis off;
```

MATLAB EXERCISE 12.32 **Wireless link with nonaligned antennas – complete analysis in MATLAB.** Consider the wireless link in free space in Fig.12.13 and assume that $r = 200$ m, $\gamma_1 = 45°$, and $\gamma_2 = 60°$, as well as that the input power of the transmitting antenna, operating at a frequency of $f = 300$ MHz, is $P_{\text{in}} = 10$ W. Losses in each of the dipoles are negligible, and their radiation resistance amounts to $R_{\text{rad}} = 73\ \Omega$. Using MATLAB, find the magnitude of the open-circuit voltage of the receiving antenna. Present graphically the system in MATLAB, and verify the figure against Fig.12.13. *(ME12_32.m on IR)* **H**

TUTORIAL:

First, we specify the input numerical data and calculate the basic parameters of the system [note that, from Eq.(12.12), $I_0 = \sqrt{P_{\text{in}}/R_{\text{rad}}}$, since $R_{\text{ohmic}} = 0$ and $P_{\text{in}} = P_{\text{rad}}$],

```
Pin = 10;
Rrad = 73;
f = 300*10^6;
r = 200;
GAMA1 = 45/180*pi;
GAMA2 = 60/180*pi;
c = 299792458;
LAMBDA = c/f;
BETA = 2*pi/LAMBDA;
ETA = 376.73;
I0 = sqrt(Pin/Rrad);
```

We then call function antennaXYZ (developed in MATLAB Exercise 12.30) to find θ, $\hat{\boldsymbol{\theta}}$, and coordinates x, y, and z of wire ends for the transmitting antenna (THETAt,UNITt,coordT) and for the receiving one (THETAr,UNITr,coordR). Note that, since the angle α at input of antennaXYZ is measured from the x-axis (to the antenna axis) for antennas lying in both xz- and xy-planes, we have that $\alpha = \pi/2 - \gamma_1$ on the transmitting side and $\alpha = -\gamma_2$ on the receiving side of the system in Fig.12.13. Based on the output of antennaXYZ, we obtain, with the use of Eq.(12.16) and function HalfWaveDipoleF (from MATLAB Exercise 12.20), the far electric field vector of the transmitting antenna (half-wave dipole), $\underline{\mathbf{E}}_{\text{t}}$ (vector Et), and the characteristic radiation function of the receiving antenna ($\lambda/2$ dipole), $\underline{\mathbf{F}}_{\text{r}}$ (vector Fr). Magnitudes of these vectors, calculated by means of function vectorMag (from MATLAB Exercise 1.1), are displayed in the Command Window.

```
[THETAt,UNITt,coordT] = antennaXYZ('xz','t',pi/2-GAMA1);
Et = i*ETA/2/pi*I0*exp(-i*BETA*r)/r*HalfWaveDipoleF(THETAt)*UNITt;
fprintf('|Et| = %f mV/m\ n',vectorMag(Et)*10^3);
[THETAr,UNITr,coordR] = antennaXYZ('xy','r',-GAMA2);
Fr = HalfWaveDipoleF(THETAr)*UNITr;
fprintf('|Fr| = %f\ n',vectorMag(Fr));
```

Next, we implement Eq.(12.26) to find the magnitude of the open-circuit voltage of the receiving antenna (received voltage), $|\underline{V}_{\text{oc}}|$, and display the result,

```
Voc = LAMBDA/pi*dot(Et,Fr);
fprintf('|Voc| = %f mV\ n',abs(Voc)*10^3);
```

Finally, we call function antennaDraw (from the previous MATLAB exercise) to graphically present the whole system, and this is shown in Fig.12.14. The user can use the rotate button in MATLAB to view the system geometry and pertinent vectors at different angles, which is extremely important and helpful for visualizing and understanding the actual orientations and mutual positions of various elements (antennas and vectors) of the system.

```
antennaDraw(UNITt,UNITr,LAMBDA,coordT*LAMBDA/2,coordR*LAMBDA/2);
view(45,10);
```

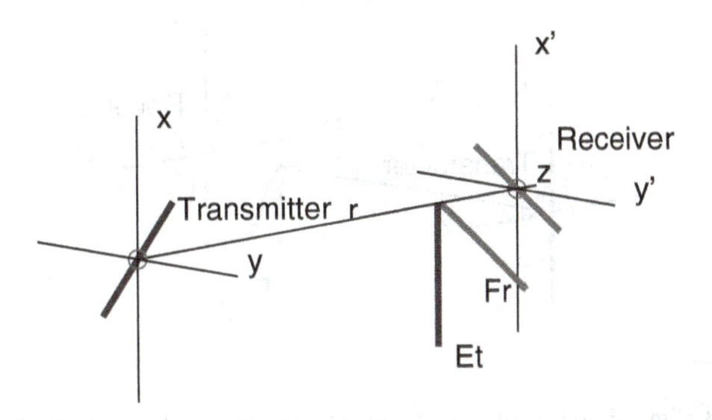

Figure 12.14 MATLAB visualization of the wireless system in Fig.12.13 – using function `antennaDraw` (see the previous MATLAB exercise) (user can rotate the figure in MATLAB to view the system geometry and vectors at different angles); for MATLAB Exercise 12.32. *(color figure on CW)*

MATLAB EXERCISE 12.33 **Switching places of T-R antennas – complete analysis in MATLAB.** Assume that in the wireless link with two nonaligned half-wave dipole antennas shown in Fig.12.13 and described in the previous MATLAB exercise, the transmitting and receiving antennas switch places. Namely, the dipole on the right-hand side (lying in the $x'y'$-plane) is now fed by the input power $P_{\text{in}} = 10$ W and is transmitting at the frequency $f = 300$ MHz, while the terminals of the dipole on the left-hand side (in the xz-plane) are left open. Under these circumstances, find the magnitude of the open-circuit voltage received by the latter antenna – performing all the steps of the analysis, in MATLAB, like in the previous MATLAB exercise. *(ME12_33.m on IR)* **H**

MATLAB EXERCISE 12.34 **Reception of a CP wave by a wire dipole – complete analysis in MATLAB.** At the transmit end of a wireless link, two half-wave wire dipole antennas positioned perpendicularly with respect to each other, along the x- and y-axis of a Cartesian coordinate system, are fed with currents of complex rms intensities $\underline{I}_{01} = 3$ A and $\underline{I}_{02} = \text{j}3$ A, respectively, and the same frequency $f = 500$ MHz (dipole currents are in time-phase quadrature, that is, 90° out of phase relative to one another). At the receive end, $r = 30$ m away from the crossed dipoles, along the z-axis, another half-wave dipole lies in the plane $x'y'$, with all the coordinate axes as in Fig.12.13, and makes an angle γ with the x'-axis. Compute the total (due to both transmitting dipoles) received rms voltage, $|\underline{V}_{\text{oc}}|$, across the open terminals of the receiving dipole. Perform the analysis completely in MATLAB (like in MATLAB Exercise 12.32). Change the angle γ and show that it does not influence $|\underline{V}_{\text{oc}}|$. *(ME12_34.m on IR)* **H**

HINT:
The resulting graphical representation of the system, using function `antennaDraw` (from MATLAB Exercise 12.31), is shown in Fig.12.15.

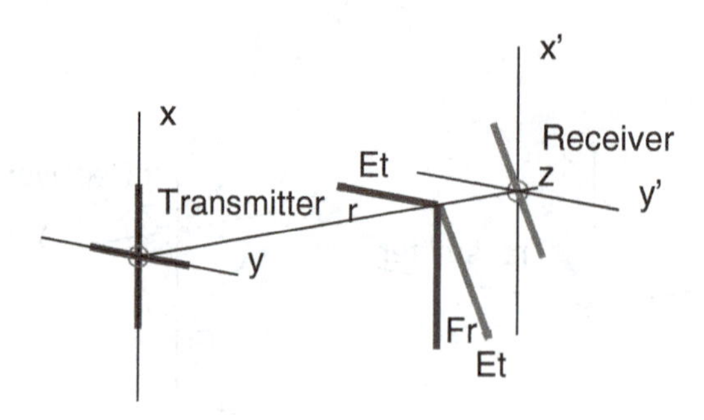

Figure 12.15 MATLAB visualization of the wireless communication system with two crossed dipoles radiating a circularly polarized electromagnetic wave [see Eqs.(7.24) and Fig.7.10] and a receiving dipole in a transversal plane of the wave; for MATLAB Exercise 12.34. *(color figure on CW)*

MATLAB EXERCISE 12.35 **Reception of an EP wave – complete analysis in MATLAB.** Repeat the previous MATLAB exercise but for $\underline{I}_{02} = \mathrm{j}9$ A ($\underline{I}_{01} = 3$ A), which results in the two crossed dipoles now radiating an elliptically polarized total electromagnetic wave [see Eqs.(7.25) and Fig.7.11]. What γ makes the received voltage maximum, and what is the maximum voltage? *(ME12_35.m on IR)* **H**

12.8 Friis Transmission Formula for a Wireless Link

In practice, we normally aim at ensuring antenna matching and orientation conditions that would maximize the power transfer in a wireless system, and in this section we restrict our attention to such an ideal case. Let us consider a general wireless communication link consisting of two antennas at a far distance r, Eq.(12.5), in free space, as shown in Fig.12.16. For an ideal case, we assume an impedance match of the load to the receiving antenna [$\underline{Z}_\mathrm{L} = \underline{Z}_\mathrm{T}^*$ in Fig.12.12(b)], polarization match between the antennas [$|\underline{V}_\mathrm{oc}| = \lambda|\underline{\mathbf{E}}_0||\underline{\mathbf{F}}|/\pi$ in Eq.(12.26)], and that both antennas are aligned and pointed toward each other for maximum gains [Eq.(12.20)], which equal G_t and G_r for the transmitting and receiving antennas, respectively.[4] The attenuation in decibels [Eq.(7.14)] between the transmit and receive ends in the link, expressed through the ratio of the time-average input power that the transmitting antenna accepts at its terminals, P_in, to the time-average power received by the load, P_r (Fig.12.16), can be computed as follows:

$$A_\mathrm{dB} = 10\log\frac{P_\mathrm{in}}{P_\mathrm{r}} = 20\log\frac{4\pi r}{\lambda} - 10\log G_\mathrm{t} - 10\log G_\mathrm{r} \quad \text{(Friis transmission formula)}, \qquad (12.27)$$

which is known as the Friis transmission formula. The last two terms in the final expression are dB gains of the two antennas, and the first one, $(A_\mathrm{dB})_\text{free space} = 20\log(4\pi r/\lambda)$, we refer to as the attenuation in free space, as it is completely independent of the particular antennas in the system and is fixed for a given electrical separation between the antennas, r/λ.

[4]For the wireless system in Fig.12.13 (analyzed in MATLAB Exercise 12.32), for example, the ideal power transfer conditions are established with $\gamma_1 = 90^\circ$, $\gamma_2 = 0$, and $\underline{Z}_\mathrm{L} = R_\mathrm{rad} = 73\ \Omega$.

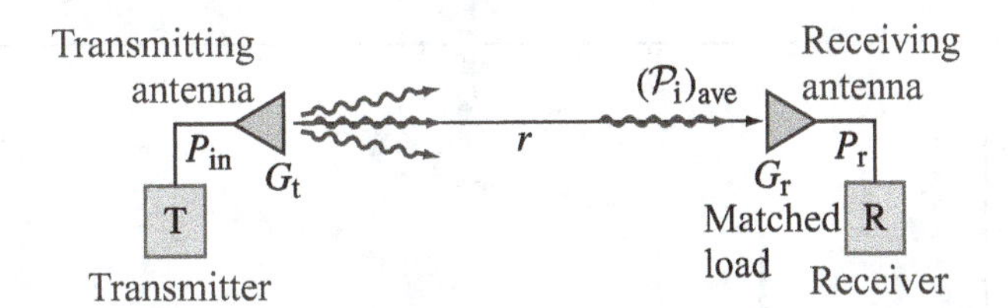

Figure 12.16 Wireless communication link with ideal antenna matching and orientation conditions (load impedance match to the receiving antenna, polarization match of antennas, and orientation of both antennas for maximum gains).

MATLAB EXERCISE 12.36 Attenuation in a wireless link, Friis transmission formula. Write a function `attenuationFriis()` in MATLAB that calculates the attenuation in a wireless link (Fig.12.16) based on the Friis transmission formula, Eq.(12.27). In specific, the function returns the dB attenuation in free space, $(A_{dB})_{free\ space}$, if the number of input arguments of the function is two (the operating wavelength in the system, λ, and the distance between the antennas, r), or the total dB attenuation in the link, A_{dB}, if the number of input arguments is four (λ, r, and maximum gains G_t and G_r of the transmitting and receiving antennas, respectively, given in natural numbers). *(attenuationFriis.m on IR)*

TUTORIAL:

We check the number of input arguments using MATLAB function `nargin` (see MATLAB Exercise 1.36),

```
function [A] = attenuationFriis(lambda,r,Gt,Gr)
if nargin == 2;
A = 20*log10(4*pi*r./lambda);
else
A = 20*log10(4*pi*r./lambda) - 10*log10(Gt) - 10*log10(Gr);
end;
```

MATLAB EXERCISE 12.37 Wireless link with ideal antenna matching and orientation conditions. In a free-space communication system in Fig.12.16, the frequency is $f = 800$ MHz and the antennas are the same, with gains $G_t = G_r = 100$. Compute and plot the dB attenuation of the system against the distance between the antennas, 10 km $\leq r \leq$ 100 km. Also plot the corresponding dB attenuation in free space. *(ME12_37.m on IR)*

HINT:

Use function `attenuationFriis` (from the previous MATLAB exercise). The resulting plots are shown in Fig.12.17.

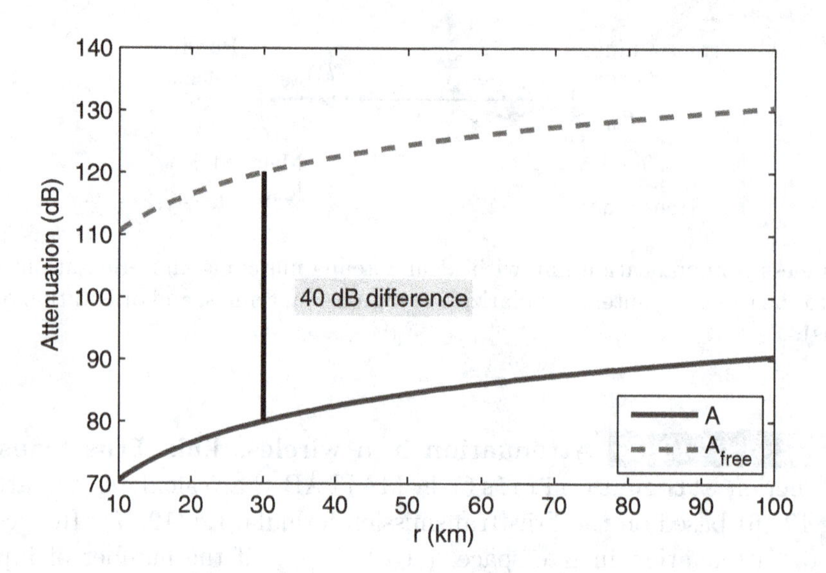

Figure 12.17 MATLAB plots of the dB attenuation of an antenna communication system (Fig.12.16), Eq.(12.27), and the corresponding dB attenuation in free space, $(A_{\rm dB})_{\rm free\ space}$, versus the distance between antennas; for MATLAB Exercise 12.37. *(color figure on CW)*

12.9 Antenna Arrays

Antenna arrays are spatial arrangements of identical antennas (array elements), equally oriented in space (e.g., wire array elements are parallel to each other or collinear), and excited independently, with feed currents of generally different magnitudes and phases, but of the same frequency. An array of a large number of electrically relatively small or medium-sized element antennas can be used to obtain a similar performance to that of a single electrically large antenna. In addition, arrays provide great flexibility and new degrees of freedom in synthesizing radiation patterns of desired shapes (for example, by varying the phases of feed currents of array elements). Most frequently, centers of array elements lie along a straight line – linear arrays.

Consider a linear array of N point sources (having uniform radiation in all directions), with complex rms current intensities $\underline{I}_k = I_k \,\mathrm{e}^{\mathrm{j}\alpha_k}$ $(I_k = |\underline{I}_k|)$, $k = 1, 2, \ldots, N$, placed along the z-axis and radiating in an ambient medium for which the phase coefficient is β. The locations of sources in the array are defined by their coordinates z_k $(k = 1, 2, \ldots, N)$, which can be arbitrary. This array is simply an antenna with discrete spatial current distribution, and we can evaluate its radiation (far-zone) magnetic vector potential, $\underline{\mathbf{A}}$, using the discrete form of the radiation integral in Eq.(12.8), with the integral along the length l of the antenna in Fig.12.3 now becoming a sum over array constituents,

$$\underline{F}_{\rm a} = \mathrm{AF} = \sum_{k=1}^{N} \underline{I}_k \,\mathrm{e}^{\mathrm{j}\beta z_k \cos\theta} = \sum_{k=1}^{N} I_k \,\mathrm{e}^{\mathrm{j}(\beta z_k \cos\theta + \alpha_k)} \quad \text{(array factor)} \,. \tag{12.28}$$

Based on Eq.(12.8), the total $\underline{\mathbf{A}}$ due to an antenna array can be represented as $\underline{\mathbf{A}} = \underline{\mathbf{A}}_{\rm element}\, \underline{F}_{\rm a}$, namely, as a product of the vector potential $\underline{\mathbf{A}}_{\rm element}$ that a single array element antenna (point source in this case) with current $\underline{I}_0 = 1\,\mathrm{e}^{\mathrm{j}0}$ A would radiate if placed at the coordinate origin (reference point) and a complex scalar function $\underline{F}_a$, in Eq.(12.28). This function provides a complete far-field characterization of the array itself (regardless of the characteristics of its elements), and is called accordingly the array factor (AF). Most importantly, $\underline{\mathbf{A}}_{\text{element}}$ can be replaced by the far-zone potential due to an arbitrary antenna as the reference antenna (array element).

The far electric field vector, $\underline{\mathbf{E}}$, due to the antenna array is obtained from $\underline{\mathbf{A}}$, like for any transmitting antenna, using Eqs.(12.9). Since $\underline{F}_a$ is a scalar, taking the transverse components of $\underline{\mathbf{A}}$ in Eqs.(12.9) applies only to $\underline{\mathbf{A}}_{\text{element}}$, and hence the total field of the array is the field of the reference element, $\underline{\mathbf{E}}_{\text{element}}$, times the array factor. Having in mind Eq.(12.16), the same can be written for the characteristic radiation function of the antenna array,

$$\underline{\mathbf{F}} = \underline{\mathbf{F}}_{\text{element}}\, \underline{F}_a \quad \text{(pattern multiplication for antenna arrays)}\,, \tag{12.29}$$

and this is known as the pattern multiplication theorem for antenna arrays. Simply, multiplying the element radiation function (pattern) and array factor we obtain the overall array pattern.

MATLAB EXERCISE 12.38 **Array factor.** Write a function `ArrayF()` in MATLAB that computes the array factor of an array with an arbitrary number, N (`N`), of point sources lying along the z-axis and radiating with the phase coefficient β (`beta`). The points are equally spaced with respect to each other, and the interelement spacing is d (`d`). Feed currents of sources in the array have equal phase shifts between adjacent points, equal to δ (`delta`). The AF is computed for a given array or matrix of θ angles (array or matrix `theta`). An additional optional input to the function is an array of magnitudes of feed currents (`I`); if this input is missing, the currents are of the same magnitude, adopted to be unity (default assumption for the function). *(ArrayF.m on IR)*

TUTORIAL:

Placing the coordinate origin (O) at the center of the array, as shown in Fig.12.18, the z-coordinates of element points are given by $z_1 = -(N-1)d/2$, $z_2 = z_1 + d$, ..., $z_N = z_1 + (N-1)d = (N-1)d/2$. Similarly, element phases amount to $\alpha_1 = -(N-1)\delta/2$, $\alpha_2 = \alpha_1 + \delta$, ..., $\alpha_N = \alpha_1 + (N-1)\delta = (N-1)\delta/2$. With these values for z_k and α_k ($k = 1, 2, \ldots, N$) in Eq.(12.28), the array factor (`AF`) is evaluated for all angles in the array (matrix) `theta` as follows (note that if `I` is omitted at input, current magnitudes of all elements are unity):

```
function AF = ArrayF(beta,d,theta,delta,N,I)
if nargin == 5;
I = ones(1,N);
```

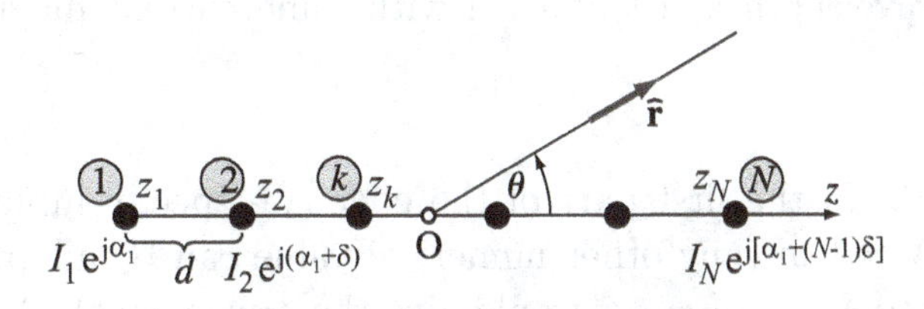

Figure 12.18 Linear array consisting of N point source elements with a uniform interelement spacing, uniform interelement phase difference, and generally nonuniform feed current magnitude along the array; for MATLAB Exercise 12.38.

```
end;
z = -(N-1)*d/2:d:(N-1)*d/2;
if (delta ~= 0)
alpha = -(N-1)*delta/2:delta:(N-1)*delta/2;
else
alpha = zeros(1:N);
end;
for k = 1:length(z);
if (k == 1)
AF(:,:)  = I(k)*exp(i*beta*z(k)*cos(theta)+ i*alpha(k));
else
AF(:,:)  = AF(:,:)  + I(k)*exp(i*beta*z(k)*cos(theta) + i*alpha(k));
end;
end;
```

MATLAB EXERCISE 12.39 **Broadside two-element array of point sources.** Consider a two-element array of point sources with feed currents of equal magnitudes and initial phases, and a half-wave interelement spacing. In MATLAB, obtain the normalized array factor of this array and present it graphically in pertinent polar diagrams. In specific, adopt a z-axis coinciding with the array axis (straight line passing through the element points), as in Fig.12.18, and plot a 3-D polar radiation pattern of the array, as well as its cuts in planes xy, yz, and xz, respectively. Show that the array pattern has nulls along the array axis, whereas the peak radiation is in directions normal to the array axis, so on the broad (long) side of the array; hence, this array is called a broadside antenna array. *(ME12_39.m on IR)* **H**

HINT:
Use function `ArrayF` (from the previous MATLAB exercise) with $N = 2$, $d = \lambda/2$, and $\delta = 0$ to compute the array factor, and then functions `polar3D` (from MATLAB Exercise 12.15) and `cutPattern` (from MATLAB Exercise 12.16) to generate polar diagrams.

MATLAB EXERCISE 12.40 **Endfire two-element array of point sources.** Repeat the previous MATLAB exercise but for a two-element array of point sources in counter-phase (elements are still spaced a half-wavelength apart and fed with same current magnitudes). *(ME12_40.m on IR)* **H**

TUTORIAL:
See Figs.12.19 and 12.20. In the first part of the code, we specify input data for the array; we adopt a wavelength of $\lambda = 1$ m (any other numerical value can be adopted as well), and set the interelement phase shift to $\delta = \pi$ ($\delta = -\pi$ would give the same result). A spherical spatial mesh is defined using arrays of `theta` and `phi` coordinates, with MATLAB function `meshgrid` generating two 2-D matrices, `theta1` and `phi1`, where rows of `theta1` are filled with the array `theta` and columns of `phi1` contain the array `phi`.

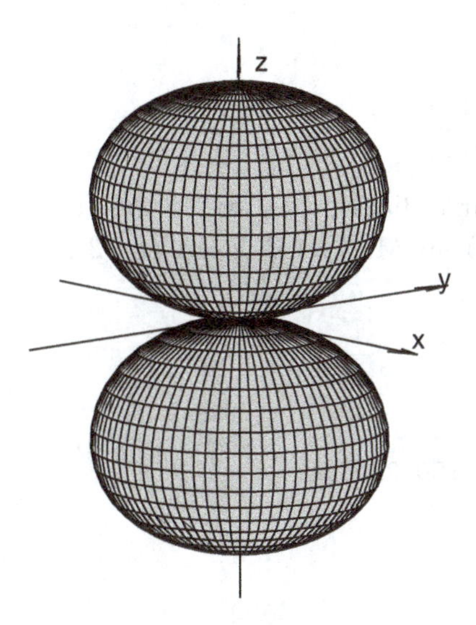

Figure 12.19 3-D polar plot of the normalized array factor of an endfire two-element antenna array – point sources with equal input powers, in counter-phase, and half-wave apart (note that z-axis is here presented as vertical, while it is horizontal in Fig.12.18); for MATLAB Exercise 12.40. *(color figure on CW)*

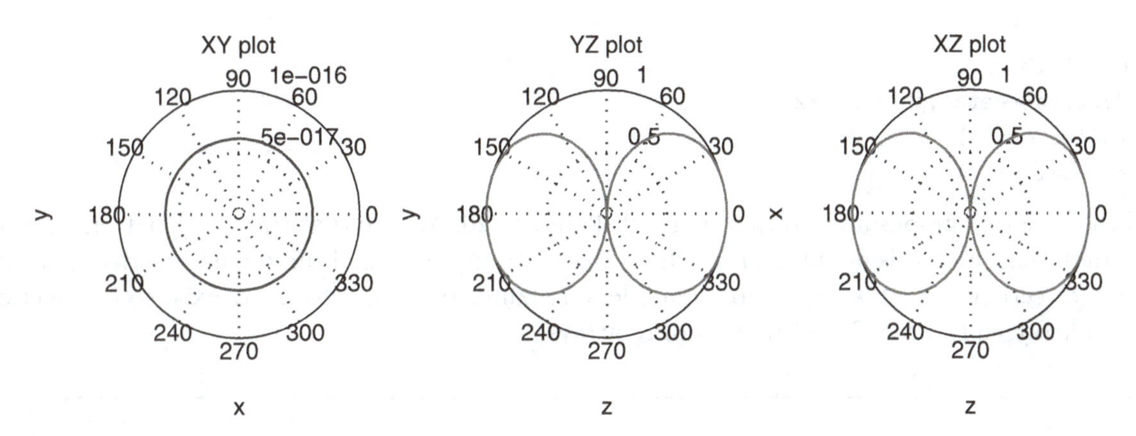

Figure 12.20 Cuts in three characteristic planes of the array pattern in Fig.12.19; for MATLAB Exercise 12.40. *(color figure on CW)*

```
lambda = 1;
N = 2;
delta = pi;
d = lambda/2;
beta = 2*pi/lambda;
theta = 0:0.025*pi:pi;
phi = 0:0.025*pi:2*pi;
[theta1,phi1] = meshgrid(theta,phi);
```

The array factor, $\underline{F}_{\mathrm{a}}(\theta)$ (`AF`), is computed by means of function `ArrayF` (from MATLAB Exercise 12.38), and is normalized for the maximum value of unity, according to Eq.(12.18), which gives $f_a(\theta)$ (`f`),

```
AF = ArrayF(beta,d,theta1,delta,N);
f = abs(AF)/max(max(abs(AF)));
```

Finally, we plot a 3-D polar radiation pattern of the array and its cuts in planes xy, yz, and xz using functions `polar3D` (from MATLAB Exercise 12.15) and `cutPattern` (from MATLAB Exercise 12.16),

```
figure(1);
polar3D(theta1,phi1,f);
fig2 = figure(2);
set(fig2,'Position',[500 500 750 220]);
subplot(1,3,1);
cutPattern(theta1,phi1,f,'xy');
title('XY plot');
axis equal;
subplot(1,3,2);
cutPattern(theta1,phi1,f,'yz');
title('YZ plot');
axis equal;
subplot(1,3,3);
cutPattern(theta1,phi1,f,'xz');
title('XZ plot');
axis equal;
```

and the resulting graphs are shown in Figs.12.19 and 12.20. We see that this array pattern exhibits zero radiation in broadside directions, while maximum in both axial directions (positive and negative z directions). Such arrays whose main lobe maxima are along the array axis, so in directions toward array ends, are referred to as endfire arrays.

MATLAB EXERCISE 12.41 **Full-wave interelement spacing and grating lobes.** Repeat MATLAB Exercise 12.39 but for a two-element array of point sources with a full-wave ($d = \lambda$) interelement spacing (elements are still fed with same current magnitudes and in phase). Show that the normalized array pattern has maxima (equal to unity) for $\theta = 0$, $90°$, and $180°$, and nulls for $\theta = 60°$ and $120°$, respectively. Note that this example is illustrative of the fact that multiple radiation lobes are formed in an array pattern for interelement spacings greater than $\lambda/2$. In addition, assuming that this array is principally meant as a broadside array, with a broadside 3-D main lobe, its endfire radiation lobes are considered as side (or minor) lobes. However, they are of the same intensity at their peaks as the main lobe. Such additional "main" lobes are called grating lobes, and in the majority of array applications they are undesirable. *(ME12_41.m on IR)*
H

MATLAB EXERCISE 12.42 Two-element array of point sources with cardioid pattern. Repeat MATLAB Exercise 12.39 but for a two-element array of point sources with a quarter-wave interelement spacing ($d = \lambda/4$) and in time-phase quadrature, where element 2 lags by 90° in phase with respect to element 1 (elements are still fed with same current magnitudes). *(ME12_42.m on IR)* **H**

HINT:
The resulting 2-D radiation pattern plots of the array are shown in Fig.12.21. We see that this is a predominantly endfire radiation pattern, with the maximum radiation only along one end of the array. In particular, the curve in the yz- and xz-planes in Fig.12.21 is the so-called cardioid pattern (it resembles the heart shape).

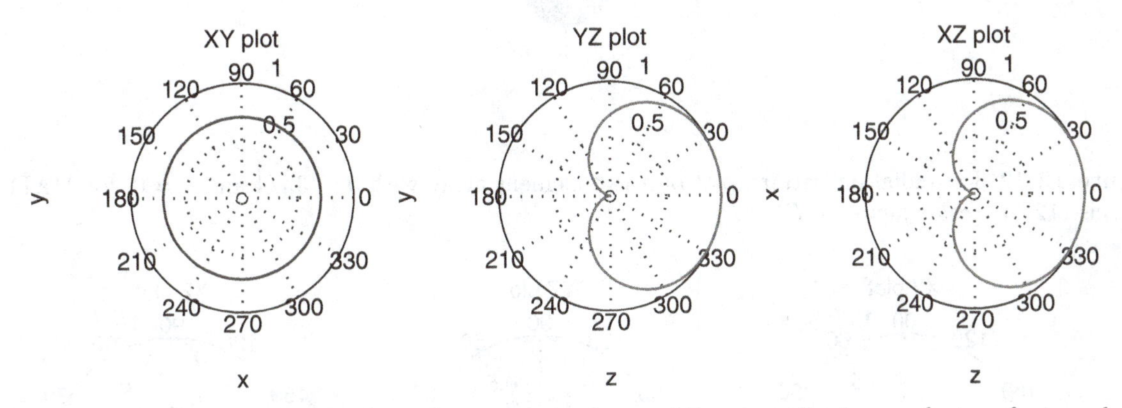

Figure 12.21 2-D polar plots in three characteristic planes of the normalized array factor of a two-element array of point sources with a quarter-wave interelement spacing and in time-phase quadrature; for MATLAB Exercise 12.42. *(color figure on CW)*

MATLAB EXERCISE 12.43 Full-wave interelement spacing and counter-phase excitation. In MATLAB, generate a 3-D radiation pattern plot and 2-D plots in planes xy, yz, and xz for a two-element array of point sources with $d = \lambda$ and $\delta = \pm 180°$. *(ME12_43.m on IR)* **H**

MATLAB EXERCISE 12.44 Three-quarter-wave separation between in-phase sources. Repeat the previous MATLAB exercise but for an array of two point sources with $d = 3\lambda/4$ and $\delta = 0$. *(ME12_44.m on IR)*

HINT:
The resulting plots are shown in Figs.12.22 and 12.23.

MATLAB EXERCISE 12.45 Visualization of the pattern multiplication theorem. Write a function `patternMultiplication()` that visualizes the pattern multiplication theorem for antenna arrays, Eq.(12.29). In specific, the function takes at input five matrices of equal dimensions: `theta`, `phi`, the element radiation function, `Felement`, the array factor, `Farray`, and the total radiation function, `Ftot`. It generates three figures, showing radiation plots in three charac-

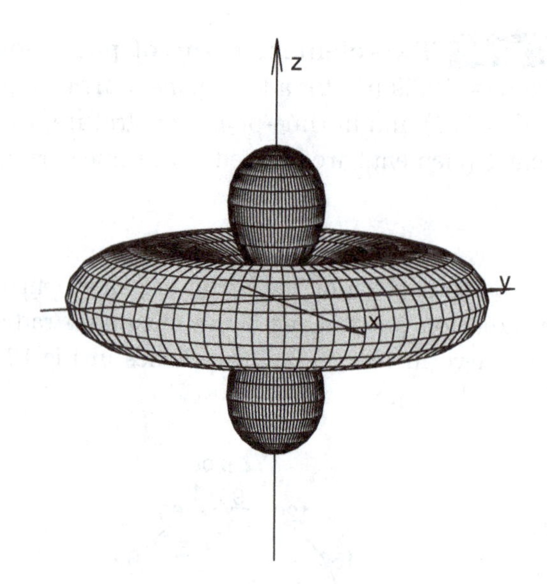

Figure 12.22 3-D radiation pattern plot of a two-element array with $d = 3\lambda/4$ and $\delta = 0$; for MATLAB Exercise 12.44. *(color figure on CW)*

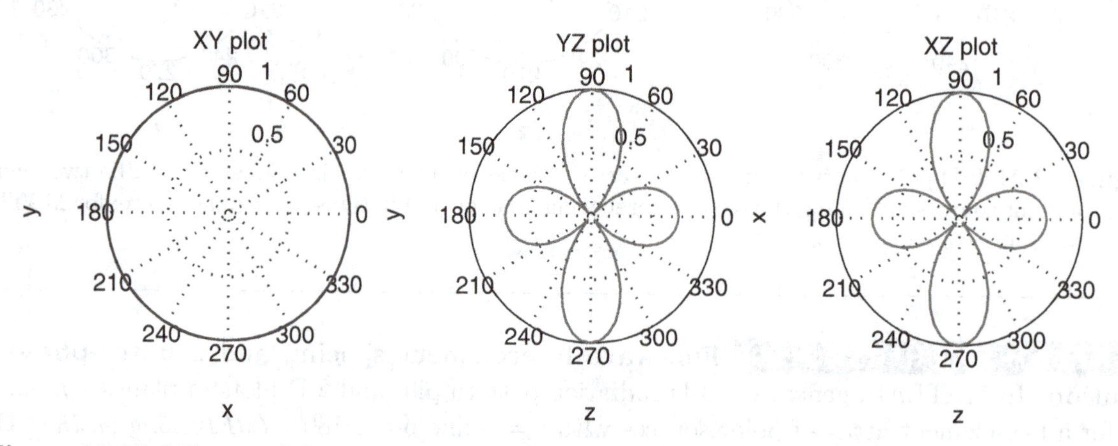

Figure 12.23 2-D cuts of the pattern in Fig.12.22; for MATLAB Exercise 12.44. *(color figure on CW)*

teristic planes, xy, xz, and yz, and each figure consists of three subplots, showing element, array, and total pattern cuts in the plane. Hence, the function shows a graphical form of Eq.(12.29): multiplying the element pattern and array factor we obtain the overall array pattern, and this is shown in three pattern cuts. *(patternMultiplication.m on IR)*

TUTORIAL:

We generate nine subplots, per directions above, each using function `cutPattern` (from MATLAB Exercise 12.16):

```
function patternMultiplication(phi,theta,Felement,Farray,Ftot)
fe = Felement/max(max(Felement));
fa = Farray/max(max(Farray));
ft = Ftot/max(max(Ftot));
```

```
figure(1);
% xy cut
subplot(1,3,1);
cutPattern(theta,phi,fe,'xy');
title('Element pattern');
subplot(1,3,2);
cutPattern(theta,phi,fa,'xy');
title('Array factor');
subplot(1,3,3);
cutPattern(theta,phi,ft,'xy');
title('Total pattern');
% xz cut
figure(2);
subplot(1,3,1);
cutPattern(theta,phi,fe,'xz');
title('Element pattern');
subplot(1,3,2);
cutPattern(theta,phi,fa,'xz');
title('Array factor');
subplot(1,3,3);
cutPattern(theta,phi,ft,'xz');
title('Total pattern');
% yz cut
figure(3);
subplot(1,3,1);
cutPattern(theta,phi,fe,'yz');
title('Element pattern');
subplot(1,3,2);
cutPattern(theta,phi,fa,'yz');
title('Array factor');
subplot(1,3,3);
cutPattern(theta,phi,ft,'yz');
title('Total pattern');
```

MATLAB EXERCISE 12.46 **Array of two collinear Hertzian dipoles – pattern multiplication.** An array of two collinear (coaxial) Hertzian dipole antennas, of length l, whose centers are spaced a half-wavelength apart, radiates in free space. The dipoles are fed with time-harmonic currents of equal complex intensities. In MATLAB, determine the total characteristic radiation function of the antenna array and plot element, array, and total normalized radiation pattern cuts in planes xy, xz, and yz, respectively. *(ME12_46.m on IR)* **H**

TUTORIAL:

We introduce a spherical coordinate system with the z-axis along the axes of dipoles and origin (O) at the array center (the array points are as in Fig.12.18, and the dipoles are z-directed).

Therefore, we can use function `ArrayF` (from MATLAB Exercise 12.38) to compute the array factor, and implement Eq.(12.17) to find the element characteristic radiation function and pattern for a Hertzian dipole. The total pattern is obtained multiplying the element pattern and array factor, Eq.(12.29). Matrices `theta1` and `phi1` describing the spherical mesh for plotting the patterns are filled in the same way as in MATLAB Exercise 12.40. Element, array, and total pattern plots are generated by function `patternMultiplication` (from the previous MATLAB exercise).

```
lambda = 1;
l = lambda/20;
N = 2;
delta = 0;
d = lambda/2;
beta = 2*pi/lambda;
theta = 0:0.025*pi:pi;
phi = 0:0.025*pi:2*pi;
[theta1,phi1] = meshgrid(theta,phi);
AF = ArrayF(beta,d,theta1,delta,N);
Felement = beta*l*sin(theta1)/2;
AFtot = AF.*Felement;
f = abs(AFtot)/max(max(abs(AFtot)));
patternMultiplication(phi1,theta1,Felement,AF,f);
figure(4);
polar3D(theta1,phi1,f);
```

MATLAB EXERCISE 12.47 **Array of two parallel dipoles – pattern multiplication.** Repeat the previous MATLAB exercise but for an array of two parallel Hertzian dipole antennas (dipole axes are perpendicular to the array axis, while the interelement spacing and phase shift are $d = \lambda/2$ and $\delta = 0$, respectively). *(ME12_47.m on IR)* **H**

HINT:

Since the dipoles are not along the array axis, if we keep the array axis coincident with the z-axis of a global spherical coordinate system in which the analysis is performed, then the dipoles will not be z-directed, and vice versa. So, let us adopt the system with dipole antennas parallel to the x-axis. Obviously, we cannot use now the expression in Eq.(12.17) for the element characteristic radiation function, and what we need rather is the one for an x-directed Hertzian dipole placed at the coordinate origin or reference point (O) of the array, which is given by $\underline{\mathbf{F}}_{\text{element}}(\theta,\phi) = \beta l(-\cos\theta\cos\phi\,\hat{\boldsymbol{\theta}} + \sin\phi\,\hat{\boldsymbol{\phi}})/2$. Implement this function and use functions `ArrayF` and `patternMultiplication` (in the same way as in the previous MATLAB exercise).

MATLAB EXERCISE 12.48 **Array of two collinear half-wave dipoles with full-wave spacing.** Repeat MATLAB Exercise 12.46 but for an array of two collinear half-wave dipole antennas with a full-wave separation between dipole centers (dipoles are fed in phase and with equal input powers). *(ME12_48.m on IR)* **H**

MATLAB EXERCISE 12.49 **Nonuniform array of three parallel half-wave dipoles.** Consider an array of three parallel half-wave wire dipole antennas lying in one plane, with a quarter-wave separation between adjacent array element points. In specific, the dipoles are z-directed and the element points are along the x-axis. The feed currents of adjacent dipoles are in time-phase quadrature (90° out of phase with respect to each other), with dipole 1 lagging and dipole 3 advancing in phase with respect to dipole 2. The current magnitudes are in the ratio $1:2:1$ along the array, so this is a nonuniform array – more precisely, a nonuniformly excited, equally spaced array. For this system, plot in MATLAB the element, array, and total pattern cuts in planes xy, xz, and yz. *(ME12_49.m on IR)*

HINT:
The array axis now coincides with the x-axis (and not the z-axis), and hence $\cos\theta$ in Eq.(12.28) needs to be replaced by $\sin\theta\cos\phi$. Since, also, $z_1 = -\lambda/4$, $z_2 = 0$, $z_3 = \lambda/4$, $\beta\lambda/4 = \pi/2$, $\alpha_1 = -\pi/2$, $\alpha_2 = 0$, $\alpha_3 = \pi/2$, $I_1 = 1$, $I_2 = 2$, and $I_3 = 1$, the array factor in Eq.(12.28) comes out to be

$$\underline{F}_{\mathrm{a}}(\theta,\phi) = \mathrm{e}^{-\mathrm{j}(\pi/2)(\sin\theta\cos\phi+1)} + 2 + \mathrm{e}^{\mathrm{j}(\pi/2)(\sin\theta\cos\phi+1)} \,. \tag{12.30}$$

Use function `HalfWaveDipoleF` (from MATLAB Exercise 12.20) for the element (half-wave dipole) characteristic radiation function and pattern, implement Eq.(12.30) to compute the AF, and multiply the patterns according to Eq.(12.29). The plots obtained by function `patternMultiplication` (MATLAB Exercise 12.45) are shown in Fig.12.24.

MATLAB EXERCISE 12.50 **Nonuniform three-element array of collinear dipoles.** Consider an array of three collinear (coaxial) half-wave wire dipole antennas, with centers spaced a half-wavelength apart (however, the dipoles are not quite touching). The feed currents of dipoles are all in phase, but their magnitudes are in the ratio $1:2:1$ along the array. Under these circumstances, compute and plot in MATLAB the element, array, and total radiation patterns of the system in three characteristic planes. *(ME12_50.m on IR)*

HINT:
As this is a nonuniform array that can readily be placed along the z-axis (element wire antennas are collinear), its array factor can be computed using function `ArrayF` – with the optional input `I` (array of magnitudes of feed currents) – see MATLAB Exercise 12.38.

MATLAB EXERCISE 12.51 **Universal pattern plot of a uniform linear array.** Linear arrays with a uniform interelement spacing (d), uniform feed current magnitude, and uniform interelement phase difference (δ) along the array are referred to as uniform linear antenna arrays, and the array in Fig.12.18 with $I_1 = I_2 = \ldots = I_N = 1$ is such an array. Expressing the phase of

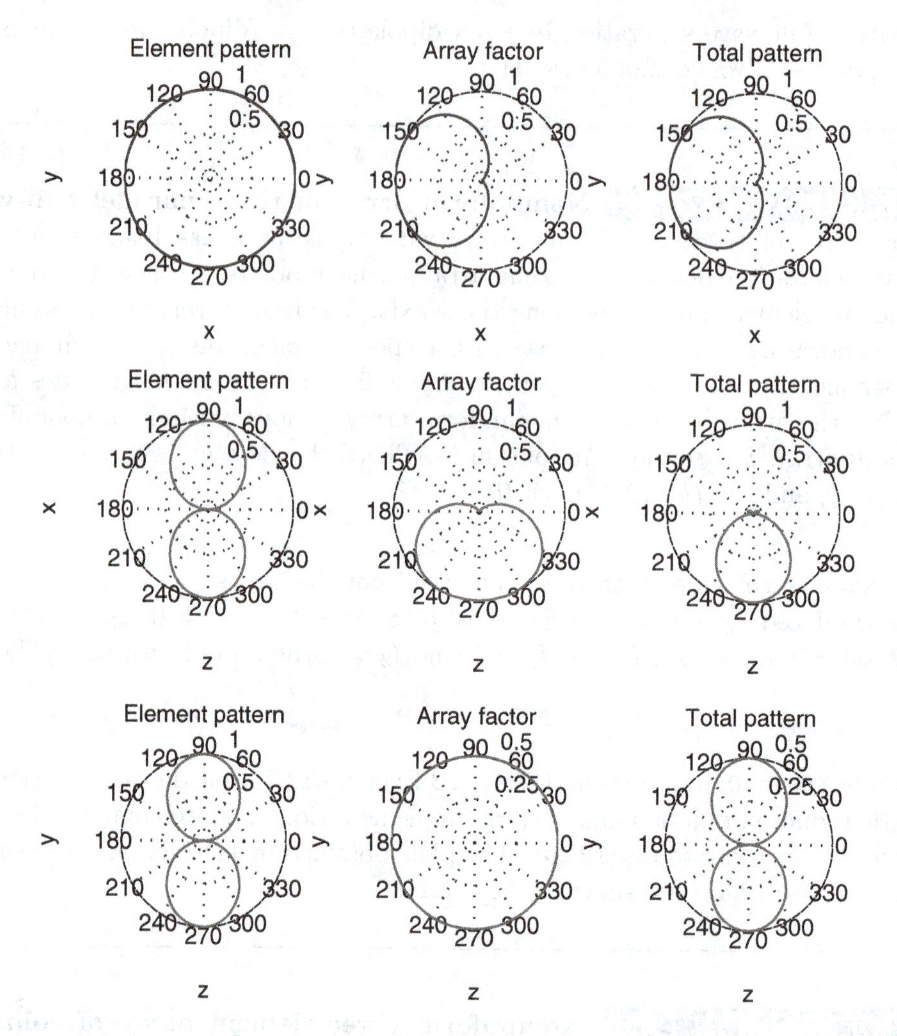

Figure 12.24 Element, array, and total pattern cuts in three characteristic planes of a nonuniform array (along the x-axis) of three parallel z-directed half-wave wire dipole antennas with a quarter-wave spacing and an excitation in time-phase quadrature between adjacent dipoles; for MATLAB Exercise 12.49. *(color figure on CW)*

the kth element in the array relative to the first one as $\alpha_k = (k-1)\delta$ $(k = 1, 2, \ldots, N)$, the array factor in Eq.(12.28) can be written in the following form:

$$\underline{F}_{\text{a}} = \sum_{k=1}^{N} \text{e}^{\text{j}(k-1)\psi}\,, \quad \text{where} \quad \psi = \beta d \cos\theta + \delta \quad (0 \le \theta \le 180^\circ) \tag{12.31}$$

(ψ can be interpreted as the phase difference between far fields due to adjacent point sources in the array). With the use of the formula for the sum of a geometric series, $1 + x + x^2 + \ldots + x^{N-1} = (1 - x^N)/(1 - x)$, with $x = \text{e}^{\text{j}\psi}$, and of Euler's identity, Eq.(6.26), the normalized array factor of a uniform linear antenna array comes out to be

$$f_{\text{a}} = \frac{|\underline{F}_{\text{a}}|}{|\underline{F}_{\text{a}}|_{\text{max}}} = \frac{|\sin(N\psi/2)|}{N|\sin(\psi/2)|} \quad \text{(uniform linear array)}\,. \tag{12.32}$$

Write a program in MATLAB that computes and visualizes the radiation pattern of an N-element uniform linear array. The input to the program, from the keyboard, consists of N, the electrical interelement spacing, $k = d/\lambda$, and δ (in degrees). The normalized factor f_a of the array is plotted both against the variable ψ defined in Eq.(12.31), for $0 \le \psi \le 2\pi$, and against the angle θ (2-D polar plot). Plot of $f_\mathrm{a}(\psi)$, based on Eq.(12.32), is called the universal pattern plot, since it is the same for all uniform linear arrays with N elements, regardless of k and δ. The function $f_\mathrm{a}(\theta)$, on the other side, can be computed either using function **ArrayF** (from MATLAB Exercise 12.38) or implementing Eqs.(12.32) and (12.31) – the result should be the same. *(ME12_51.m on IR)*

HINT:
The resulting graphs (for $N = 5$, $d = \lambda/2$, and $\delta = 0$) are shown in Fig.12.25.

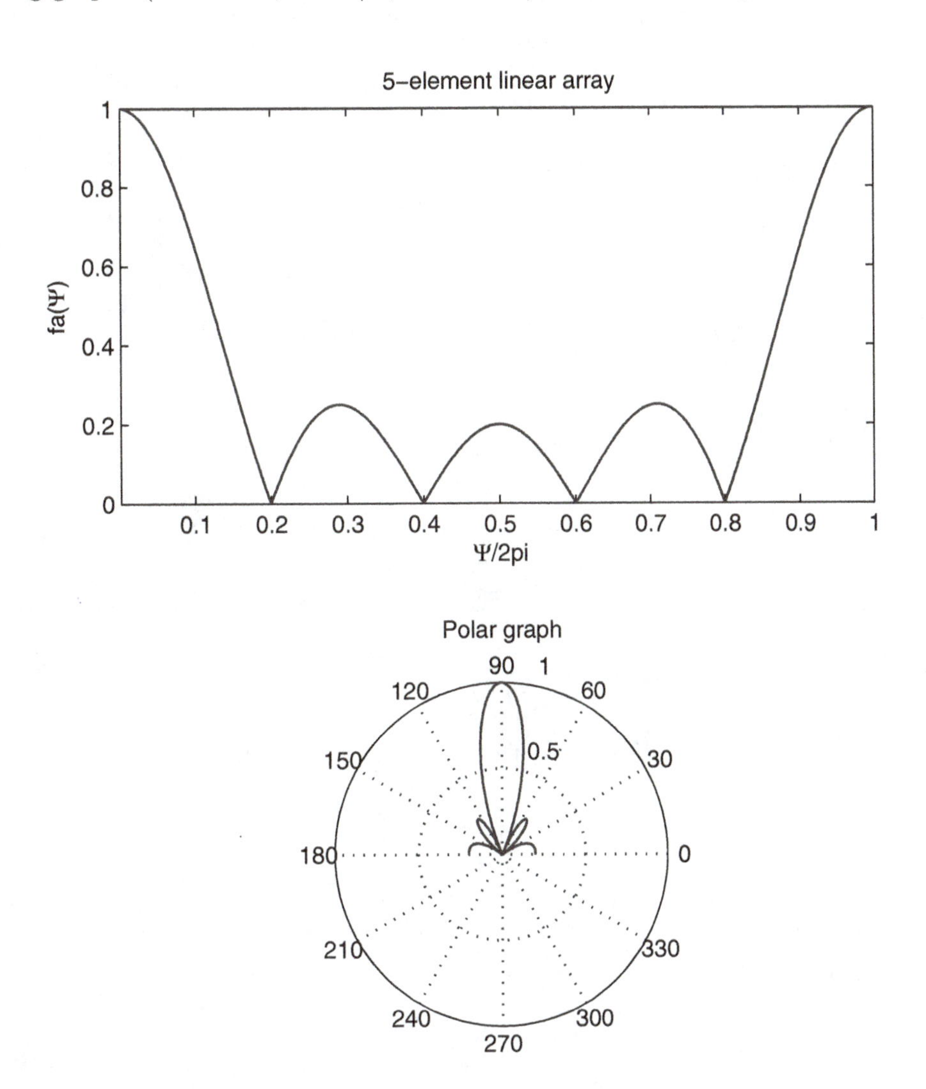

Figure 12.25 Normalized array factor of a uniform linear antenna array with $N = 5$ point source elements (Fig.12.18): universal pattern plot $[f_\mathrm{a}(\psi)]$, Eq.(12.32), and polar radiation plot $[f_\mathrm{a}(\theta)]$ for $d = \lambda/2$ and $\delta = 0$; for MATLAB Exercise 12.51. *(color figure on CW)*

APPENDIX 1: QUANTITIES, SYMBOLS, UNITS, AND CONSTANTS

Symbol	Quantity or Parameter	Alternative Notation*	SI Unit (and Value)
A	Scalar function (variable)		Appropriate unit
A	Scalar constant (single-valued), e.g., j, e		Appropriate unit
$\mathbf{A}$	Vector	$\vec{A}$	Appropriate unit
$\hat{\mathbf{A}}$	Unit vector of $\mathbf{A}$	$\mathbf{a}_A$, $\mathbf{i}_A$, $\mathbf{u}_A$	Appropriate unit
$\underline{A}$	Complex (phasor) quantity	$\tilde{A}$, A_s	Appropriate unit
$\underline{\mathbf{A}}$	Complex vector	$\tilde{\mathbf{A}}$, $\mathbf{A}_s$	Appropriate unit
$\mathbf{A}$	Magnetic vector potential		T · m (tesla-meter)
A_{dB}	Decibel attenuation (attenuation in dB)		dB (decibel)
$\mathbf{B}$	Magnetic flux density vector		T (tesla)
C	Capacitance		F (farad)
C'	Capacitance per unit length (p.u.l.)	C	F/m
c	Velocity of electromagnetic (EM) waves		m/s (s – second)
c_0	Velocity of EM waves in free space	c	299,792,458 m/s
$\mathbf{D}$	Electric flux density (displacement) vector		C/m^2 (C – coulomb)
D	Antenna directivity		Dimensionless
e	Charge of electron, magnitude		1.602×10^{-19} C
$\mathbf{E}$	Electric field intensity vector		V/m (volt per meter)
E_{cr}	Dielectric strength of a material		V/m
E_{cr0}	Dielectric strength of air		3 MV/m (M $\equiv 10^6$)
$\mathbf{E}_{\mathrm{ind}}$	Induced electric field intensity vector		V/m
$\mathcal{E}$, e	Electromotive force (emf) of a generator		V (volt)
e_{ind}	Induced emf	V_{emf}	V
$\mathbf{F}_{\mathrm{e}}$	Electric (Coulomb) force		N (newton)
$\mathbf{F}_{\mathrm{m}}$	Magnetic force		N
f	Frequency	ν	Hz (hertz)
f_{p}	Plasma frequency		Hz
f_{res}	Resonant frequency of an EM resonator		Hz
f_{c}	Cutoff frequency of a waveguide mode		Hz
$\underline{\mathbf{F}}$	Antenna characteristic radiation function		Dimensionless
f	Antenna normalized field pattern		Dimensionless
$\underline{F}_{\mathrm{a}}$	Antenna array factor (AF)		Dimensionless
G	Conductance		S
G'	Conductance per unit length	G	S/m
G	Antenna gain		Dimensionless
$\mathbf{H}$	Magnetic field intensity vector		A/m (amp per meter)
I, i	Current intensity (or current)		A (ampere or amp)
j or i	Imaginary unit ($\mathrm{j} = \sqrt{-1}$)		Dimensionless
$\mathbf{J}$	Current density vector (conduction current)		A/m^2
$\mathbf{J}_{\mathrm{d}}$	Displacement current density vector		A/m^2
$\mathbf{J}_{\mathrm{s}}$	Surface current density vector	$\mathbf{K}$	A/m
l	Length		m (meter)
L	Self-inductance		H (henry)
L_{21}	Mutual inductance	M	H

Alternative notations used by one or more other electromagnetics texts.

Symbol	Quantity or Parameter	Alternative Notation	SI Unit (and Value)
L'	Inductance per unit length	L	H/m
$\mathbf{m}$	Magnetic dipole moment		$\mathrm{A \cdot m^2}$
m	Mass		kg (kilogram)
n	Index of refraction		Dimensionless
$\mathbf{p}$	Electric dipole moment		$\mathrm{C \cdot m}$
P	Power (instantaneous)		W (watt)
P_{J}	Power of Joule's losses or ohmic losses		W
P_{ave}	Time-average power		W
$\boldsymbol{\mathcal{P}}$	Poynting vector	$\mathbf{S}$	$\mathrm{W/m^2}$
$\underline{\boldsymbol{\mathcal{P}}}$	Complex Poynting vector	$\underline{\mathbf{S}}$, $\tilde{\mathbf{S}}$	$\mathrm{W/m^2}$
$\boldsymbol{\mathcal{P}}_{\mathrm{ave}}$	Time-average Poynting vector	$\mathbf{S}_{\mathrm{ave}}$	$\mathrm{W/m^2}$
Q, q	Charge		C (coulomb)
Q'	Line charge density (charge p.u.l.)	ρ_{l}	C/m
Q	Quality factor of a resonator		Dimensionless
R	Source-to-field distance	R'	m
R	Resistance		Ω (ohm)
R_{rad}	Radiation resistance of an antenna		Ω
R_{ohmic}	Ohmic resistance of an antenna		Ω
R_{s}	Surface resistance of a good conductor		Ω/square (or Ω)
R'	Resistance per unit length	R	Ω/m
r	Normalized transmission-line resistance		Dimensionless
r	Radial distance cylindrical	ρ	m
r	Radial distance spherical	R	m
S	Surface area	s, A, a	$\mathrm{m^2}$
s	Standing wave ratio (SWR)		Dimensionless
t	Time		s (second)
T	Period of time-harmonic oscillation		s
T	One-way time delay period of a tr. line		s
T	Temperature		K (kelvin) or °C
v	Volume	V, τ	$\mathrm{m^3}$
V	Electric scalar potential (at a point)	Φ	V
V, v	Voltage (between two points)		V
V_{cr}	Breakdown voltage		V
$\underline{V}$	Complex rms (root-mean-square) voltage		V
$\underline{V}_{\mathrm{oc}}$	Open-circuit voltage of a receiving antenna		V
$\mathbf{v}$	Velocity (vector)	$\mathbf{u}$	m/s
v_{p}	Phase velocity		m/s
v_{g}	Group velocity		m/s
W	Work or energy	A (for work)	J (joule)
W_{e}	Electric energy		J
W_{m}	Magnetic energy		J
W_{em}	Electromagnetic energy		J
w_{e}	Electric energy density		$\mathrm{J/m^3}$
w_{m}	Magnetic energy density		$\mathrm{J/m^3}$
X	Reactance		Ω
x	Normalized transmission-line reactance		Dimensionless
$\underline{Y}$	Complex admittance		S
Y_0	Characteristic admittance of a tr. line	Y_{c}	S
$\underline{Z}$	Complex impedance		Ω

Symbol	Quantity or Parameter	Alternative Notation	SI Unit (and Value)
Z_0	Characteristic impedance of a tr. line	Z_c	Ω
Z_{TEM}	TEM wave impedance		Ω
Z_{TE}	TE wave impedance		Ω
Z_{TM}	TM wave impedance		Ω
$\underline{z}_n$	Normalized transmission-line impedance		Dimensionless
α	Attenuation coefficient		Np/m (neper per meter)
α_c	Attenuation coeff. due to conductor losses		Np/m
α_d	Attenuation coeff. due to dielectric losses		Np/m
β	Phase coefficient or wavenumber	k	rad/m
$\underline{\gamma}$	Complex propagation coefficient		m^{-1}
$\underline{\Gamma}$	Reflection coefficient	R, ρ	Dimensionless
δ	Skin depth		m
$\tan\delta_d$	Loss tangent		Dimensionless
ε	Permittivity of a dielectric material		F/m
ε_0	Permittivity of a vacuum (free space)		8.8542 pF/m ($p \equiv 10^{-12}$)
ε_r	Relative permittivity		Dimensionless
ε_{reff}	Effective relative permittivity of a tr. line		Dimensionless
η	Intrinsic impedance of a medium		Ω
η_0	Intrinsic impedance of free space		$\approx 120\pi\ \Omega \approx 377\ \Omega$
$\underline{\eta}$	Complex intrinsic impedance		Ω
$\underline{\eta}_w$	Wave impedance		Ω
η_{rad}	Radiation efficiency of an antenna		Dimensionless
λ	Wavelength		m
λ_0	Free-space wavelength		m
λ_z	Wavelength along a tr. line or waveguide		m
μ	Permeability of a magnetic material		H/m
μ_0	Permeability of a vacuum (free space)		$4\pi \cdot 10^{-7}$ H/m
μ_r	Relative permeability		Dimensionless
ρ	Volume charge density	ρ_v	C/m^3
ρ_s	Surface charge density	σ	C/m^2
ρ	Resistivity of a medium		$\Omega \cdot m$
σ	Conductivity of a medium		S/m
τ	Relaxation time or time constant		s
$\underline{\tau}$	Transmission coefficient	$\underline{T}$	Dimensionless
Φ	Magnetic flux	Ψ, λ	Wb (weber)
ψ	Phase angle of a reflection coefficient		rad (radian)
ω	Angular or radian frequency		rad/s

Powers of Ten as Multipliers of Fundamental Units

Multiple	Prefix	Symbol	Multiple	Prefix	Symbol
10^{18}	Exa	E	10^{-2}	centi	c
10^{15}	Peta	P	10^{-3}	milli	m
10^{12}	Tera	T	10^{-6}	micro	μ
10^{9}	Giga	G	10^{-9}	nano	n
10^{6}	Mega	M	10^{-12}	pico	p
10^{3}	kilo	k	10^{-15}	femto	f
10^{2}	hecto	h	10^{-18}	atto	a

APPENDIX 2: MATHEMATICAL FACTS AND IDENTITIES

A2.1 Trigonometric Identities

$$\sin(\alpha \pm \beta) = \sin\alpha\cos\beta \pm \cos\alpha\sin\beta\,, \quad \cos(\alpha \pm \beta) = \cos\alpha\cos\beta \mp \sin\alpha\sin\beta$$

$$2\sin\alpha\sin\beta = \cos(\alpha-\beta) - \cos(\alpha+\beta)\,, \quad 2\sin\alpha\cos\beta = \sin(\alpha+\beta) + \sin(\alpha-\beta)\,,$$

$$2\cos\alpha\cos\beta = \cos(\alpha+\beta) + \cos(\alpha-\beta)\,; \quad \sin\alpha \pm \sin\beta = 2\sin\frac{\alpha\pm\beta}{2}\cos\frac{\alpha\mp\beta}{2}\,,$$

$$\cos\alpha + \cos\beta = 2\cos\frac{\alpha+\beta}{2}\cos\frac{\alpha-\beta}{2}\,, \quad \cos\alpha - \cos\beta = -2\sin\frac{\alpha+\beta}{2}\sin\frac{\alpha-\beta}{2}$$

$$\sin^2\alpha = \frac{1-\cos 2\alpha}{2}\,, \quad \cos^2\alpha = \frac{1+\cos 2\alpha}{2}\,, \quad \sin^2\alpha + \cos^2\alpha = 1\,, \quad \sin 2\alpha = 2\sin\alpha\cos\alpha\,,$$

$$\cos 2\alpha = \cos^2\alpha - \sin^2\alpha\,, \quad \sin(-\alpha) = -\sin\alpha\,, \quad \cos(-\alpha) = \cos\alpha\,, \quad \sin(\alpha \pm 90^\circ) = \pm\cos\alpha\,,$$

$$\cos(\alpha \pm 90^\circ) = \mp\sin\alpha\,; \quad \tan\alpha = \frac{\sin\alpha}{\cos\alpha}\,, \quad \cot\alpha = \frac{1}{\tan\alpha}$$

$$\sin 0 = 0\,, \quad \sin 30^\circ = \frac{1}{2}\,, \quad \sin 45^\circ = \frac{\sqrt{2}}{2}\,, \quad \sin 60^\circ = \frac{\sqrt{3}}{2}\,, \quad \sin 90^\circ = 1\,, \quad \sin 180^\circ = 0\,,$$

$$\cos 0 = 1\,, \quad \cos 30^\circ = \frac{\sqrt{3}}{2}\,, \quad \cos 45^\circ = \frac{\sqrt{2}}{2}\,, \quad \cos 60^\circ = \frac{1}{2}\,, \quad \cos 90^\circ = 0\,, \quad \cos 180^\circ = -1$$

$c^2 = a^2 + b^2 - 2ab\cos\gamma$ (cosine formula, arbitrary triangle; angle γ is opposite to side c) ,

$c^2 = a^2 + b^2$ (Pythagorean theorem, right triangle; c is hypotenuse)

A2.2 Exponential, Logarithmic, and Hyperbolic Identities

$\mathrm{e}^x\mathrm{e}^y = \mathrm{e}^{x+y}\,, \quad (\mathrm{e}^x)^a = \mathrm{e}^{ax}\,, \quad \mathrm{e}^{\mathrm{j}x} = \cos x + \mathrm{j}\sin x$ (Euler's identity) ,

$\mathrm{e}^{\mathrm{j}x} + \mathrm{e}^{-\mathrm{j}x} = 2\cos x\,, \quad \mathrm{e}^{\mathrm{j}x} - \mathrm{e}^{-\mathrm{j}x} = 2\mathrm{j}\sin x\,, \quad \mathrm{e} = 2.71828\,, \quad \mathrm{j} = \sqrt{-1}$ (imaginary unit)

$\log x = \log_{10} x$ (common logarithm) , $\ln x = \log_{\mathrm{e}} x$ (natural logarithm) , $\ln \mathrm{e}^x = x$

$\log(xy) = \log x + \log y\,, \quad \log\frac{x}{y} = \log x - \log y\,, \quad \log x^a = a\log x$ (for logarithm of any base)

$\sinh x = \dfrac{\mathrm{e}^x - \mathrm{e}^{-x}}{2}\,, \quad \cosh x = \dfrac{\mathrm{e}^x + \mathrm{e}^{-x}}{2}$ (hyperbolic sine and cosine)

$$\sinh \mathrm{j}x = \mathrm{j}\sin x\,, \quad \cosh \mathrm{j}x = \cos x$$

$\tanh x = \dfrac{\sinh x}{\cosh x}\,, \quad \coth x = \dfrac{1}{\tanh x}$ (hyperbolic tangent and cotangent)

A2.3 Solution of Quadratic Equation

$$ax^2 + bx + c = 0 \quad \longrightarrow \quad x = \frac{-b \pm \sqrt{b^2 - 4ac}}{2a}$$

A2.4 Approximations for Small Quantities

$$\text{For } |x| \ll 1\,, \quad (1+x)^a \approx 1 + ax\,, \quad \sin x \approx x\,, \quad \cos x \approx 1 - \frac{x^2}{2}\,, \quad \mathrm{e}^x \approx 1 + x$$

A2.5 Derivatives

$$\frac{\mathrm{d}}{\mathrm{d}x} x^c = cx^{c-1}\,, \quad \frac{\mathrm{d}c}{\mathrm{d}x} = 0 \quad (c = \text{const})\,, \quad \frac{\mathrm{d}}{\mathrm{d}x}\mathrm{e}^x = \mathrm{e}^x\,, \quad \frac{\mathrm{d}}{\mathrm{d}x}\ln x = \frac{1}{x}\,,$$

$$\frac{\mathrm{d}}{\mathrm{d}x}\sin x = \cos x\,, \quad \frac{\mathrm{d}}{\mathrm{d}x}\cos x = -\sin x\,, \quad \frac{\mathrm{d}}{\mathrm{d}x}\tan x = \frac{1}{\cos^2 x}$$

$$\text{For } f = f(x) \text{ and } g = g(x)\,, \quad \frac{\mathrm{d}}{\mathrm{d}x}(cf) = c\frac{\mathrm{d}f}{\mathrm{d}x}\,, \quad \frac{\mathrm{d}}{\mathrm{d}x}(f+g) = \frac{\mathrm{d}f}{\mathrm{d}x} + \frac{\mathrm{d}g}{\mathrm{d}x}\,,$$

$$\frac{\mathrm{d}}{\mathrm{d}x}(fg) = \frac{\mathrm{d}f}{\mathrm{d}x}g + f\frac{\mathrm{d}g}{\mathrm{d}x}\,, \quad \frac{\mathrm{d}}{\mathrm{d}x}\left(\frac{f}{g}\right) = \frac{\frac{\mathrm{d}f}{\mathrm{d}x}g - f\frac{\mathrm{d}g}{\mathrm{d}x}}{g^2}\,,$$

$$\frac{\mathrm{d}}{\mathrm{d}x}f[g(x)] = \frac{\mathrm{d}f}{\mathrm{d}g}\frac{\mathrm{d}g}{\mathrm{d}x} \quad \text{(chain rule for taking derivatives)}$$

A2.6 Integrals

$$\int x^c\,\mathrm{d}x = \frac{x^{c+1}}{c+1} + C \quad (c \neq -1)\,, \quad \int \frac{\mathrm{d}x}{x} = \ln x + C\,, \quad \int \mathrm{e}^x\,\mathrm{d}x = \mathrm{e}^x + C\,,$$

$$\int \sin x\,\mathrm{d}x = -\cos x + C\,, \quad \int \cos x\,\mathrm{d}x = \sin x + C\,; \quad \int cf\,\mathrm{d}x = c\int f\,\mathrm{d}x\,,$$

$$\int (f+g)\,\mathrm{d}x = \int f\,\mathrm{d}x + \int g\,\mathrm{d}x\,, \quad \int f\,\mathrm{d}g = fg - \int g\,\mathrm{d}f \quad \text{(integration by parts)}$$

A2.7 Vector Algebraic Identities

For vectors **a** and **b**, and angle α between them, $\mathbf{a}\cdot\mathbf{b} = |\mathbf{a}||\mathbf{b}|\cos\alpha$ (dot product of vectors) ,

$\mathbf{a}\times\mathbf{b} = |\mathbf{a}||\mathbf{b}|\sin\alpha\,\hat{\mathbf{n}}$ (cross product of vectors; $\hat{\mathbf{n}}$ is the unit vector normal to the plane of **a** and **b**, directed according to the right-hand rule when **a** is rotated by the shortest route toward **b**)

$$\mathbf{a}\cdot\mathbf{a} = |\mathbf{a}|^2 = a^2\,, \quad \hat{\mathbf{a}} = \frac{\mathbf{a}}{a} \text{ (unit vector of } \mathbf{a};\ |\hat{\mathbf{a}}| = 1)\,, \quad \mathbf{a}\cdot\mathbf{b} = \mathbf{b}\cdot\mathbf{a}\,, \quad \mathbf{a}\times\mathbf{b} = -\mathbf{b}\times\mathbf{a}$$

$$(\mathbf{a}\times\mathbf{b})\cdot\mathbf{c} = (\mathbf{b}\times\mathbf{c})\cdot\mathbf{a} = (\mathbf{c}\times\mathbf{a})\cdot\mathbf{b} \quad \text{(scalar triple product)}$$

$$\mathbf{a} \times (\mathbf{b} \times \mathbf{c}) = \mathbf{b}(\mathbf{a} \cdot \mathbf{c}) - \mathbf{c}(\mathbf{a} \cdot \mathbf{b}) \quad \text{(vector triple product)}$$

For vectors in the Cartesian coordinate system, $\mathbf{a} = a_x\,\hat{\mathbf{x}} + a_y\,\hat{\mathbf{y}} + a_z\,\hat{\mathbf{z}}$ ($\hat{\mathbf{x}}$, $\hat{\mathbf{y}}$, and $\hat{\mathbf{z}}$ are coordinate unit vectors), $a = |\mathbf{a}| = \sqrt{a_x^2 + a_y^2 + a_z^2}$, $\mathbf{a}+\mathbf{b} = (a_x + b_x)\,\hat{\mathbf{x}} + (a_y + b_y)\,\hat{\mathbf{y}} + (a_z + b_z)\,\hat{\mathbf{z}}$,

$$\mathbf{a} \cdot \mathbf{b} = a_x b_x + a_y b_y + a_z b_z\,, \quad \mathbf{a} \times \mathbf{b} = (a_y b_z - a_z b_y)\,\hat{\mathbf{x}} + (a_z b_x - a_x b_z)\,\hat{\mathbf{y}} + (a_x b_y - a_y b_x)\,\hat{\mathbf{z}}$$

A2.8 Vector Calculus Identities

For a scalar function f (and g) and a vector function $\mathbf{a}$ (and $\mathbf{b}$), $\nabla f \equiv \operatorname{grad} f$ (gradient of f), $\nabla\cdot\mathbf{a} \equiv \operatorname{div}\mathbf{a}$ (divergence of $\mathbf{a}$), $\nabla\times\mathbf{a} \equiv \operatorname{curl}\mathbf{a}$ (curl of $\mathbf{a}$), $\nabla\cdot(\nabla f) \equiv \nabla^2 f$ (Laplacian of f),

$$\nabla \times (\nabla f) = 0\,, \quad \nabla \cdot (\nabla \times \mathbf{a}) = 0\,, \quad \nabla(fg) = (\nabla f)\,g + f\nabla g\,, \quad \nabla f(g) = \frac{\mathrm{d}f}{\mathrm{d}g}\nabla g \quad \text{(chain rule)},$$

$$\nabla\cdot(f\mathbf{a}) = (\nabla f)\cdot\mathbf{a} + f\nabla\cdot\mathbf{a}\,, \quad \nabla\times(f\mathbf{a}) = (\nabla f)\times\mathbf{a} + f\nabla\times\mathbf{a}\,, \quad \nabla\cdot(\mathbf{a}\times\mathbf{b}) = \mathbf{b}\cdot(\nabla\times\mathbf{a}) - \mathbf{a}\cdot(\nabla\times\mathbf{b})\,,$$

$$\nabla \times (\nabla \times \mathbf{a}) = \nabla(\nabla \cdot \mathbf{a}) - \nabla^2\mathbf{a} \quad (\nabla^2\mathbf{a} \text{ – Laplacian of } \mathbf{a})$$

$$\int_v \nabla \cdot \mathbf{a}\,\mathrm{d}v = \oint_S \mathbf{a} \cdot \mathrm{d}\mathbf{S} \quad \text{(divergence theorem; } S \text{ is the boundary surface of } v\text{)},$$

$$\int_S (\nabla \times \mathbf{a}) \cdot \mathrm{d}\mathbf{S} = \oint_C \mathbf{a} \cdot \mathrm{d}\mathbf{l} \quad \text{(Stokes' theorem; } C \text{ is the boundary contour of } S\text{)}$$

A2.9 Gradient, Divergence, Curl, and Laplacian in Orthogonal Coordinate Systems

Cartesian coordinate system $[f(x,y,z)\,, \quad \mathbf{a} = a_x(x,y,z)\,\hat{\mathbf{x}} + a_y(x,y,z)\,\hat{\mathbf{y}} + a_z(x,y,z)\,\hat{\mathbf{z}}]$,

$$\nabla = \frac{\partial}{\partial x}\,\hat{\mathbf{x}} + \frac{\partial}{\partial y}\,\hat{\mathbf{y}} + \frac{\partial}{\partial z}\,\hat{\mathbf{z}} \quad \text{(del operator)}, \quad \nabla f = \frac{\partial f}{\partial x}\,\hat{\mathbf{x}} + \frac{\partial f}{\partial y}\,\hat{\mathbf{y}} + \frac{\partial f}{\partial z}\,\hat{\mathbf{z}}\,,$$

$$\nabla \cdot \mathbf{a} = \frac{\partial a_x}{\partial x} + \frac{\partial a_y}{\partial y} + \frac{\partial a_z}{\partial z}\,, \quad \nabla \times \mathbf{a} = \left(\frac{\partial a_z}{\partial y} - \frac{\partial a_y}{\partial z}\right)\hat{\mathbf{x}} + \left(\frac{\partial a_x}{\partial z} - \frac{\partial a_z}{\partial x}\right)\hat{\mathbf{y}} + \left(\frac{\partial a_y}{\partial x} - \frac{\partial a_x}{\partial y}\right)\hat{\mathbf{z}}\,,$$

$$\nabla^2 f = \frac{\partial^2 f}{\partial x^2} + \frac{\partial^2 f}{\partial y^2} + \frac{\partial^2 f}{\partial z^2}\,, \quad \nabla^2\mathbf{a} = \nabla^2 a_x\,\hat{\mathbf{x}} + \nabla^2 a_y\,\hat{\mathbf{y}} + \nabla^2 a_z\,\hat{\mathbf{z}}$$

Cylindrical coordinate system $[f(r,\phi,z)\,, \quad \mathbf{a} = a_r(r,\phi,z)\,\hat{\mathbf{r}} + a_\phi(r,\phi,z)\,\hat{\boldsymbol{\phi}} + a_z(r,\phi,z)\,\hat{\mathbf{z}}]$,

$$\nabla f = \frac{\partial f}{\partial r}\,\hat{\mathbf{r}} + \frac{1}{r}\frac{\partial f}{\partial \phi}\,\hat{\boldsymbol{\phi}} + \frac{\partial f}{\partial z}\,\hat{\mathbf{z}}\,, \quad \nabla \cdot \mathbf{a} = \frac{1}{r}\frac{\partial}{\partial r}(r a_r) + \frac{1}{r}\frac{\partial a_\phi}{\partial \phi} + \frac{\partial a_z}{\partial z}\,,$$

$$\nabla \times \mathbf{a} = \left(\frac{1}{r}\frac{\partial a_z}{\partial \phi} - \frac{\partial a_\phi}{\partial z}\right)\hat{\mathbf{r}} + \left(\frac{\partial a_r}{\partial z} - \frac{\partial a_z}{\partial r}\right)\hat{\boldsymbol{\phi}} + \frac{1}{r}\left[\frac{\partial}{\partial r}(r a_\phi) - \frac{\partial a_r}{\partial \phi}\right]\hat{\mathbf{z}}\,,$$

$$\nabla^2 f = \frac{1}{r}\frac{\partial}{\partial r}\left(r\frac{\partial f}{\partial r}\right) + \frac{1}{r^2}\frac{\partial^2 f}{\partial \phi^2} + \frac{\partial^2 f}{\partial z^2}\,, \quad \nabla^2\mathbf{a} = \nabla(\nabla \cdot \mathbf{a}) - \nabla \times (\nabla \times \mathbf{a})$$

Spherical coordinate system $[f(r,\theta,\phi)\,, \quad \mathbf{a} = a_r(r,\theta,\phi)\,\hat{\mathbf{r}} + a_\theta(r,\theta,\phi)\,\hat{\boldsymbol{\theta}} + a_\phi(r,\theta,\phi)\,\hat{\boldsymbol{\phi}}]\,,$

$$\nabla f = \frac{\partial f}{\partial r}\,\hat{\mathbf{r}} + \frac{1}{r}\frac{\partial f}{\partial \theta}\,\hat{\boldsymbol{\theta}} + \frac{1}{r\sin\theta}\frac{\partial f}{\partial \phi}\,\hat{\boldsymbol{\phi}}\,, \quad \nabla\cdot\mathbf{a} = \frac{1}{r^2}\frac{\partial}{\partial r}(r^2 a_r) + \frac{1}{r\sin\theta}\frac{\partial}{\partial\theta}(\sin\theta\, a_\theta) + \frac{1}{r\sin\theta}\frac{\partial a_\phi}{\partial\phi}\,,$$

$$\nabla\times\mathbf{a} = \frac{1}{r\sin\theta}\left[\frac{\partial}{\partial\theta}(\sin\theta\, a_\phi) - \frac{\partial a_\theta}{\partial\phi}\right]\hat{\mathbf{r}} + \frac{1}{r}\left[\frac{1}{\sin\theta}\frac{\partial a_r}{\partial\phi} - \frac{\partial}{\partial r}(r a_\phi)\right]\hat{\boldsymbol{\theta}} + \frac{1}{r}\left[\frac{\partial}{\partial r}(r a_\theta) - \frac{\partial a_r}{\partial\theta}\right]\hat{\boldsymbol{\phi}}\,,$$

$$\nabla^2 f = \frac{1}{r^2}\frac{\partial}{\partial r}\left(r^2\frac{\partial f}{\partial r}\right) + \frac{1}{r^2\sin\theta}\frac{\partial}{\partial\theta}\left(\sin\theta\frac{\partial f}{\partial\theta}\right) + \frac{1}{r^2\sin^2\theta}\frac{\partial^2 f}{\partial\phi^2}\,, \quad \text{for } \nabla^2\mathbf{a}\text{, see the formula above}$$

A2.10 Vector Algebra and Calculus Index

APPENDIX 3: LIST OF MATLAB EXERCISES

ME 1.1 Vector magnitude. *TUTORIAL* *(vectorMag.m on IR)*[1]

ME 1.2 2-D vector plot. HINT *(vecPlot2D.m)*

ME 1.3 3-D vector plot. *TUTORIAL* *(vecPlot3D.m)*

ME 1.4 Electric force due to multiple charges. *TUTORIAL* *(ME1_4.m)*

ME 1.5 Four charges at tetrahedron vertices. HINT **H**[2] *(ME1_5.m)*

ME 1.6 Three point charges in Cartesian coordinate system. HINT **H** *(ME1_6.m)*

ME 1.7 Electric field due to multiple charges. *(ME1_7.m)*

ME 1.8 Three charges at rectangle vertices. HINT **H** *(ME1_8.m)*

ME 1.9 Charged ring and an equivalent point charge. HINT *(ME1_9.m)*

ME 1.10 Symbolic integration. *(integral.m)*

ME 1.11 Charged disk – symbolic and analytical solutions. *TUTORIAL* *(ME1_11.m)*

ME 1.12 Charged hemisphere – numerical integration. HINT **H** *(ME1_12.m)*

ME 1.13 Vector numerical integration and field visualization using `quiver`. *TUTORIAL* *(ME1_13.m)*

ME 1.14 Visualization of the electric field due to four point charges. HINT *(ME1_14.m)*

ME 1.15 Another field visualization using `quiver`. *(ME1_15.m)*

ME 1.16 Fields due to line charges of finite and infinite lengths. HINT *(ME1_16.m)*

ME 1.17 Dot product of two vectors. *(dotProduct.m)*

ME 1.18 Numerical integration of a line integral. *(LineIntegral.m)*

ME 1.19 Work in the field of a point charge. *TUTORIAL* **H** *(ME1_19.m)*

ME 1.20 Numerical proof that E-field is conservative – movie. *TUTORIAL* *(ME1_20.m)*

ME 1.21 Circulation of E-vector along a contour of complex shape. **H** *(ME1_21.m)*

ME 1.22 Electric potential due to multiple charges. HINT *(ME1_22.m)*

ME 1.23 Electric potential due to a charged ring. *(ME1_23.m)*

ME 1.24 Cartesian to cylindrical and spherical coordinate conversions. *(car2Cyl.m, car2Sph.m, cyl2Car.m, and sph2Car.m)*

ME 1.25 GUI for different coordinate conversions. HINT *[folder ME1_25(GUI) on IR]*

ME 1.26 Symbolic gradient in different coordinate systems. HINT *(gradCar.m, gradCyl.m, and gradSph.m)*

ME 1.27 Field from potential, in three coordinate systems. *(ME1_27.m)*

---

[1] *IR* = Instructor Resources (for the book); all m files are provided on IR. The material on IR is meant for instructors only.

[2] **H** = recommended to be done also "by hand," i.e., not using MATLAB.

ME 12.2 Poynting vector due to a Hertzian dipole. *(hertzianDipolePoynting.m)*

ME 12.3 Error in far electric field computation for a Hertzian dipole. *TUTORIAL* *(ME12_3.m)*

ME 12.4 Error in far magnetic field computation. HINT *(ME12_4.m)*

ME 12.5 Error in far-zone Poynting vector computation. *(ME12_5.m)*

ME 12.6 3-D E-vector visualization for a Hertzian dipole using `quiver3`. HINT *(ME12_6.m)*

ME 12.7 3-D H-vector visualization for a Hertzian dipole. *(ME12_7.m)*

ME 12.8 Symbolic radiation integral for an arbitrary straight wire antenna. *(radiationIntegral.m)*

ME 12.9 Symbolic magnetic potential of a Hertzian dipole. HINT **H** *(ME12_9.m)*

ME 12.10 Equivalent length of an arbitrary short wire antenna. *(equivLengthShortAntenna.m)*

ME 12.11 Equivalent length of a short dipole with a triangular current. **H** *(ME12_11.m)*

ME 12.12 Radiation and ohmic resistances of a Hertzian dipole. *TUTORIAL* *(ME12_12.m)*

ME 12.13 Radiation and ohmic resistances of a nonloaded short dipole. HINT **H** *(ME12_13.m)*

ME 12.14 Radiation efficiency of a short dipole with cosine current. HINT **H** *(ME12_14.m)*

ME 12.15 3-D polar radiation pattern plot, arbitrary radiation function. *TUTORIAL* *(polar3D.m)*

ME 12.16 Radiation pattern cuts in three characteristic planes. *TUTORIAL* *(cutPattern.m)*

ME 12.17 3-D and 2-D radiation patterns of a Hertzian dipole. HINT **H** *(ME12_17.m)*

ME 12.18 Radiation patterns of a traveling-wave antenna. *(ME12_18.m)*

ME 12.19 Movie demonstration that Hertzian dipole pattern cuts are circles. *TUTORIAL* *(ME12_19.m)*

ME 12.20 Characteristic radiation function of a half-wave dipole. *TUTORIAL* *(HalfWaveDipoleF.m)*

ME 12.21 Radiation patterns of a half-wave dipole. HINT **H** *(ME12_21.m)*

ME 12.22 Numerical integration by Simpson's rule. *TUTORIAL* *(Simpson.m)*

ME 12.23 Radiation resistance of a half-wave dipole using Simpson's rule. *(ME12_23.m)*

ME 12.24 Directivity of a half-wave dipole. *(ME12_24.m)*

ME 12.25 Radiation function and pattern plots of an arbitrary wire dipole. HINT *(ME12_25.m)*

ME 12.26 Dependence of radiation pattern on antenna length – movie. *TUTORIAL* *(patternVsLength-Movie.m)*

ME 12.27 Playing the pattern movie for different dipole lengths. *TUTORIAL* *(ME12_27.m)*

ME 12.28 Pattern movie for traveling-wave antennas of different lengths. *(ME12_28.m)*

ME 12.29 Radiation patterns of a quarter-wave monopole. **H** *(ME12_29.m)*

ME 12.30 Geometrical preprocessing for arbitrarily oriented T-R antennas. *TUTORIAL* *(antennaXYZ.m)*

ME 12.31 Visualization of a wireless system with two antennas. *TUTORIAL* *(antennaDraw.m)*

ME 12.32 Wireless link with nonaligned antennas – complete analysis in MATLAB. *TUTORIAL* **H** *(ME12_32.m)*

ME 12.33 Switching places of T-R antennas – complete analysis in MATLAB. **H** *(ME12_33.m)*

ME 12.34 Reception of a CP wave by a wire dipole – complete analysis in MATLAB. HINT **H** *(ME12_34.m)*

ME 12.35 Reception of an EP wave – complete analysis in MATLAB. **H** *(ME12_35.m)*

ME 12.36 Attenuation in a wireless link, Friis transmission formula. *TUTORIAL* *(attenuationFriis.m)*

ME 12.37 Wireless link with ideal antenna matching and orientation conditions. HINT *(ME12_37.m)*

ME 12.38 Array factor. *TUTORIAL* *(ArrayF.m)*

ME 12.39 Broadside two-element array of point sources. HINT **H** *(ME12_39.m)*

ME 12.40 Endfire two-element array of point sources. *TUTORIAL* **H** *(ME12_40.m)*

ME 12.41 Full-wave interelement spacing and grating lobes. **H** *(ME12_41.m)*

ME 12.42 Two-element array of point sources with cardioid pattern. HINT **H** *(ME12_42.m)*

ME 12.43 Full-wave interelement spacing and counter-phase excitation. **H** *(ME12_43.m)*

ME 12.44 Three-quarter-wave separation between in-phase sources. HINT *(ME12_44.m)*

ME 12.45 Visualization of the pattern multiplication theorem. *TUTORIAL* *(patternMultiplication.m)*

ME 12.46 Array of two collinear Hertzian dipoles – pattern multiplication. *TUTORIAL* **H** *(ME12_46.m)*

ME 12.47 Array of two parallel dipoles – pattern multiplication. HINT **H** *(ME12_47.m)*

ME 12.48 Array of two collinear half-wave dipoles with full-wave spacing. **H** *(ME12_48.m)*

ME 12.49 Nonuniform array of three parallel half-wave dipoles. HINT *(ME12_49.m)*

ME 12.50 Nonuniform three-element array of collinear dipoles. HINT *(ME12_50.m)*

ME 12.51 Universal pattern plot of a uniform linear array. HINT *(ME12_51.m)*

BIBLIOGRAPHY

[1] Notaroš, B. M., *Electromagnetics*, Prentice Hall, Upper Saddle River, NJ, 2010.

[2] Ulaby, F. T., E. Michielssen, and U. Ravaioli, *Fundamentals of Applied Electromagnetics*, Prentice Hall, Upper Saddle River, NJ, 2010, 6th edition.

[3] Sadiku, M. N. O., *Elements of Electromagnetics*, Oxford University Press, New York, 2009, 5th edition.

[4] Hayt, W. H. Jr., and J. A. Buck, *Engineering Electromagnetics*, McGraw-Hill, New York, 2011, 8th edition.

[5] Griffiths, D. J., *Introduction to Electrodynamics*, Addison Wesley, Upper Saddle River, NJ, 2012, 4th edition.

[6] Cheng, D. K., *Field and Wave Electromagnetics*, Addison-Wesley, Reading, MA, 1989, 2nd edition.

[7] Popović, Z., and B. D. Popović, *Introductory Electromagnetics*, Prentice Hall, Upper Saddle River, NJ, 2000.

[8] Rao, N. N., *Elements of Engineering Electromagnetics*, Prentice Hall, Upper Saddle River, NJ, 2004, 6th edition.

[9] Inan U. S., and A. S. Inan, *Engineering Electromagnetics*, Addison Wesley Longman, Menlo Park, CA, 1999.

[10] Wentworth, S. M., *Fundamentals of Electromagnetics with Engineering Applications*, John Wiley & Sons, New York, 2005.

[11] Paul, C. R., *Electromagnetics for Engineers with Applications*, John Wiley & Sons, New York, 2004.

[12] Demarest, K. R., *Engineering Electromagnetics*, Prentice Hall, Upper Saddle River, NJ, 1998.

[13] Kraus, J. D., and D. A. Fleisch, *Electromagnetics with Applications*, McGraw-Hill, New York, 1999, 5th edition.

[14] Iskander, M. F., *Electromagnetic Fields and Waves*, Waveland Press, Prospect Hills, IL, 2000.

[15] Lonngren, K. E., S. V. Savov, and R. J. Jost, *Fundamentals of Electromagnetics with MATLAB®*, SciTech Publishing, Raleigh, NC, 2007, 2nd edition.

[16] Johnk, C. T. A., *Engineering Electromagnetic Fields and Waves*, John Wiley & Sons, New York, 1988, 2nd edition.

[17] Ida, N., *Engineering Electromagnetics*, Springer, New York, 2004, 2nd edition.

[18] Stutzman, W. L., and G. A. Thiele, *Antenna Theory and Design*, John Wiley & Sons, New York, 1998, 2nd edition.

[19] Balanis, C. A., *Antenna Theory: Analysis and Design*, John Wiley & Sons, New York, 1997, 2nd edition.

[20] Kraus, J. D., and R. J. Marhefka, *Antennas for All Applications*, McGraw-Hill, New York, 2002, 3rd edition.

[21] Pozar, D. M., *Microwave Engineering*, John Wiley & Sons, New York, 2005, 3rd edition.

[22] Inan U. S., and A. S. Inan, *Electromagnetic Waves*, Prentice Hall, Upper Saddle River, NJ, 2000.

[23] Collin, R. E., *Field Theory of Guided Waves*, Wiley-IEEE Press, New York, 1990, 2nd edition.

[24] Ramo, S., J. R. Whinnery, and T. Van Duzer, *Fields and Waves in Communication Electronics*, John Wiley & Sons, New York, 1994, 3rd edition.

[25] Haus, H. A., and J. R. Melcher, *Electromagnetic Fields and Energy*, Prentice Hall, Upper Saddle River, NJ, 1989.

[26] Claycomb, J.R., *Applied Electromagnetics Using QuickField and MATLAB®*, Jones and Bartlett Publishers, Sudbury, MA, 2009.

[27] Balanis, C. A., *Advanced Engineering Electromagnetics*, John Wiley & Sons, New York, 2012, 2nd edition.

[28] Harrington, R. F., *Time-Harmonic Electromagnetic Fields*, Wiley-IEEE Press, New York, 2001 (classic reissue).

[29] Stratton, J. A., *Electromagnetic Theory*, Wiley-IEEE Press, New York, 2007 (classic reissue).

[30] Jin, J., *Theory and Computation of Electromagnetic Fields*, Wiley-IEEE Press, New York, 2010.

[31] Mahafza, B. R., and A. Z. Elsherbeni, *MATLAB® Simulations for Radar Systems Design*, Chapman & Hall/CRC, Boca Raton, FL, 2004.

[32] Sadiku, M. N. O., *Numerical Techniques in Electromagnetics with MATLAB®*, CRC Press, Boca Raton, FL, 2009, 3rd edition.

[33] Warnick, K. F., *Numerical Methods for Engineering: An Introduction Using MATLAB and Computational Electromagnetics Examples*, SciTech Publishing, Raleigh, NC, 2011.

[34] Peterson, A. F., S. L. Ray, and R. Mittra, *Computational Methods for Electromagnetics*, Wiley-IEEE Press, New York, 1997.

[35] McClellan, J. H., C. S. Burrus, A. V. Oppenheim, T. W. Parks, R. W. Schafer, and H. W. Schuessler, *Computer-Based Exercises for Signal Processing Using MATLAB® 5*, Prentice Hall, Upper Saddle River, NJ, 1997.

[36] Buck, J. R., M. M. Daniel, and A. C. Singer, *Computer Explorations in Signals and Systems Using MATLAB®*, Prentice Hall, Upper Saddle River, NJ, 2001, 2nd edition.

[37] Scharf, L. L. and Richard T. Behrens, *A First Course in Electrical and Computer Engineering with MATLAB® Programs and Experiments*, Addison-Wesley Publishing, Reading, MA, 1990.

[38] Ogata, K., *MATLAB® for Control Engineers*, Prentice Hall, Upper Saddle River, NJ, 2007.

[39] Davis T.A., *MATLAB® Primer*, Chapman and Hall/CRC, Boca Raton, FL, 2010, 8th edition.

[40] Higham, D. J., and N. J. Higham, *MATLAB Guide*, SIAM, Philadelphia, 2005, 2nd edition.

[41] Palm, W. J. III, *Introduction to MATLAB® for Engineers*, McGraw-Hill, New York, 2010, 3rd edition.

[42] The MathWorks, Inc., *MATLAB and Simulink Based Books*, mathworks.com/support/books/index.jsp (2012).

[43] The MathWorks, Inc., *Getting Started with MATLAB*, mathworks.com/help/matlab/getting-started-with-matlab.html (2012).

[44] The MathWorks, Inc., *MATLAB® Primer*, mathworks.com/help/pdf_doc/matlab/getstart.pdf (2012) ("Getting Started with MATLAB" pdf document available for download, or see [43]).

INDEX

D

E

F

G

H

I

J

K

L

M

N

O

P

Q

R

S

T

U

V

W